D1765866

Floodplain Processes

Floodplain Processes

Edited by

MALCOLM G. ANDERSON
University of Bristol, UK

DES E. WALLING
University of Exeter, UK

and

PAUL D. BATES
University of Bristol, UK

JOHN WILEY & SONS
Chichester · New York · Brisbane · Toronto · Singapore

Other Wiley Editorial Offices

John Wiley & Sons, Inc., 605 Third Avenue,
New York, NY 10158-0012, USA

Jacaranda Wiley Ltd, 33 Park Road, Milton,
Queensland 4064, Australia

John Wiley & Sons (Canada) Ltd, 22 Worcester Road,
Rexdale, Ontario M9W 1L1, Canada

John Wiley & Sons (Asia) Pte Ltd, 2 Clementi Loop #02-01,
Jin Xing Distripark, Singapore 129809

Library of Congress Cataloging-in-Publication Data

Floodplain processes / edited by Malcolm G. Anderson, Des E. Walling & Paul D. Bates.
        p.   cm.
    Includes bibliographical references and index.
    ISBN 0-471-96679-7
    1. Floodplains.   I. Anderson, M.G.   II. Walling, D.E.   III. Bates, P.D.
    GB561.F53 1996
    551.48'9—dc20   96-12987
                                                                    CIP

British Library Cataloguing in Publication Data

A catalogue record for this book is available from the British Library

ISBN 0-471-96679-7

Typeset in 10/12pt Times by Mathematical Composition Setters, Salisbury, Wiltshire
Printed and bound in Great Britain by Bookcraft (Bath) Ltd, Midsomer Norton, Somerset
This book is printed on acid-free paper responsibly manufactured from sustainable forestation,
for which at least two trees are planted for each one used for paper production.

Contents

List of Contributors

M.G. Anderson Department of Geography, University of Bristol, University Road, Bristol, BS8 1SS, UK

P.D. Bates Department of Geography, University of Bristol, University Road, Bristol, BS8 1SS, UK

K.J. Beven Institute of Environmental and Biological Sciences, University of Lancaster, Bailrigg, Lancaster, LA1 4YQ, UK

A. Brookes Environment Agency, Kings Meadow House, Kings Meadow Road, Reading, RG1 8DQ, UK

A.G. Brown Department of Geography, University of Exeter, Amory Building, Rennes Drive, Exeter, EX4 4RJ, UK

T.P. Burt Department of Geography, University of Durham, Durham, DH1 3LE, UK

Y. Chen Department of Civil and Environmental Engineering, University of Bradford, Bradford, West Yorkshire, BD7 1DP, UK

A.S. El-hames Department of Geography, University of Cambridge, Downing Place, Cambridge, CB2 3EN, UK

R.A. Falconer Department of Civil and Environmental Engineering, University of Bradford, Bradford, West Yorkshire, BD7 1DP, UK

J. Girel Institut de Geographie, Universite Joseph Fourier, BP 53, Grenoble Cedex, France 38041

R.J. Hardy Department of Geography, University of Bristol, University Road, Bristol, BS8 1SS, UK

T. Harris Department of Geography, University of Cambridge, Downing Place, Cambridge, CB2 3EN, UK

N.E. Haycock Quest Environmental, PO Box 45, Harpenden, Hertfordshire, UK

J.-M. Hervouet Fluvial Hydraulics Group, Laboratoire National d'Hydraulique, Electrictié de France, Direction des Etudes et Recherches, 6 Quai Watier, 78401 Chatou, Paris, France

Q. He Department of Geography, University of Exeter, Exeter, EX4 4RJ, UK

A.D. Howard Department of Environmental Sciences, University of Virginia, Charlottesville, VA 22903, USA

F.M.R. Hughes Department of Geography, University of Cambridge, Downing Place, Cambridge, CB2 3EN, UK

I.D. Jolly CSIRO Division of Water Resources, GPO Box 1666, Canberra ACT 2601, Australia

D.W. Knight School of Civil Engineering, University of Birmingham, Birmingham, B15 2TT, UK

M.G. Macklin School of Geography, University of Leeds, Woodhouse Lane, Leeds, LS2 9JT, UK

S.B. Marriott. Faculty of the Built Environment, University of the West of England, Cold Harbour Lane, Bristol, BS16 1QY, UK

A.P. Nicholas School of Geography, University of Leeds, Woodhouse Lane, Leeds, LS2 9JT, UK

G. Pautou Institut de Biologie Alpine, Universite Joseph Fourier, BP53, Grenoble Cedex 38041, France

J.-L. Peiry Institut de Geographie, Universite Joseph Fourier, BP53, Grenoble Cedex, France 38041

E.C. Penning-Rowsell Flood Hazard Research Centre, Middlesex University, Queens Way, Enfield, Middlesex, EN3 4SF, UK

G.E. Petts School of Geography, University of Birmingham, Edgbaston, Birmingham, B15 2TT, UK

D.A. Price Department of Geography, University of Bristol, University Road, Bristol, BS8 1SS, UK

K.S. Richards Department of Geography, University of Cambridge, Downing Place, Cambridge, CB2 3EN, UK

R. Romanowicz Institute of Environmental and Biological Sciences, University of Lancaster, Bailrigg, Lancaster, LA1 4YQ, UK

R.H.J. Sellin Department of Civil Engineering, University of Bristol, Bristol, BR8 1TR, UK

K. Shiono Department of Civil and Building Engineering, Loughborough University of Technology, Loughborough, Leicestershire, UK

C.N. Smith Department of Geography, University of Bristol, University Road, Bristol, BS8 1SS, UK

J. Tawn Institute of Environmental and Biological Sciences, University of Lancaster, Bailrigg, Lancaster, LA1 4YQ, UK

S.M. Tunstall Flood Hazard Research Centre, Middlesex University, Queens Way, Enfield, Middlesex, EN3 4SF, UK

L. Van Haren Fluvial Hydraulics Group, Laboratoire National d'Hydraulique, Electrictié de France, Direction des Etudes et Recherches, 6 Quai Watier, 78401 Chatou, Paris, France

D.E. Walling Department of Geography, University of Exeter, Exeter, EX4 4RJ, UK

B.B. Willetts Department of Engineering, University of Aberdeen, Fraser Noble Building, Kings College, Aberdeen, AB9 2UE, UK

B.A. Younis Department of Civil Engineering, City University, Northampton Square, London, EC1V 0HB, UK

1 The General Context for Floodplain Process Research

M.G. ANDERSON*, P.D. BATES* and D.E. WALLING†
*Department of Geography, University of Bristol, UK
†Department of Geography, University of Exeter, UK

1.1 ESTABLISHING RESEARCH INTEGRATION

Research is increasingly attempting to integrate those flow processes occurring within the river floodplain zone within a wider context. The context is complex – it relates to temporal questions of *floodplain topographic evolution*, *hydraulic modelling* of within-channel and out-of-bank flow processes, and consequential sediment *erosion and deposition and water quality* changes in what is recognized as a highly sensitive environment. This latter recognition has evolved both through scientific research (the perception that methodologies are becoming available for predictive purposes at suitable temporal and spatial scales) and heightened political awareness of the linkages between legislation and physical processes in floodplain environments. There is, therefore, the development of management approaches which *de facto* begin to imply a suitable knowledge base of process interaction with human interaction.

The floodplain landform assemblage thus has a potentially complex sedimentological background (Figure 1.1) with three-dimensional flow dynamics (Figure 1.2) that lead to high temporal and spatial variation in flow throughout the domain (Figure 1.3).

However, even in relatively low energy fluvial systems the topographic domain can be changed by relatively modest sized storm events. Figure 1.4 shows for a 1 in 150 year event the abandonment of the former meandering channel in favour of the flood-induced straighter channel. It follows that this dynamic change can impact within those time-scales typically used for both *management* and design. This then brings the floodplain dynamics full cycle to the effect such changes have on the sedimentological environment (Figure 1.1). Certain rivers have well-documented channel migration patterns of course, but it is of interest to additionally identify the dynamics of the sedimentation boundaries (Figure 1.5) – the spatial domain of the active floodplain zone can therefore change.

A further feature of the general context of research on floodplain flow processes is the manner in which such processes have been incorporated, if at all, within models of watershed hydrology. The frequent adherence to kinematic wave approximations for

Floodplain Processes. Edited by Malcolm G. Anderson, Des E. Walling and Paul D. Bates.
© 1996 John Wiley & Sons Ltd.

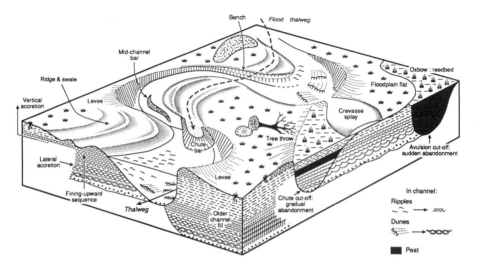

Figure 1.1   Floodplain landform assemblage (after Brown, Chapter 4, this volume)

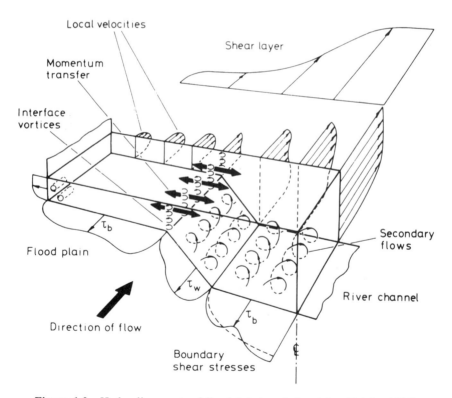

Figure 1.2   Hydraulic aspects of floodplain inundation (after Knight, 1989)

Figure 1.3   Flooding of the River Culm, Devon, UK

Figure 1.4   Straightening out of Hoaroak Water, Exmoor, UK, by the 150-year recurrence interval flood of 1952

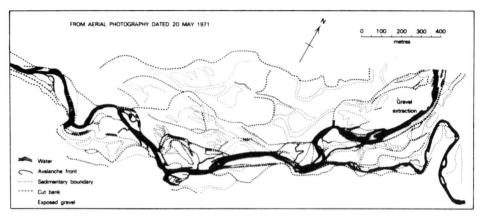

Figure 1.5   Migrating channel of the River Tywi, UK (after Lewin *et al.*, 1988, p. 445)

surface flow of all types, a model structure that typically makes no allowance for compound channel flow and a model grid scale that may well fail to capture key flood flow attributes have been contributory to the relative isolation of the potentially highly related aspects of physical hydrology and hydraulics. Within the longer-term geomorphological and sedimentological contexts we have described, it is important that progress is made in better representing floodplain flow processes.

In this text we seek to provide a structured approach to the four key areas of floodplain evolution, hydraulic modelling, sediment and water quality and management of the floodplain environment.

1.2   DATA CAPTURE ISSUES

> Theoretical development has outstripped empirical validation for reach length scales … In particular there is a rarity of reach-length studies of floodplain inundation and deposition at a fine enough resolution for validating models of overbank deposition. (Carling and Petts, 1992, p. xii)

Research integration needs within the full space–time context of floodplain development (long-term geomorphology) and detailed process operation (e.g. channel/floodplain momentum exchange) are exceptional in the demand for data to establish process models, parameterize and validate them. Following McLaughlin *et al.* (1993) we can identify some key attributes of this problem. Firstly, much of the required data is comparatively *inaccessible*. For example, for the subsurface, a limited number of hydraulic conductivity measurements are available, and monitoring of water quality and long-term sedimentation rates are restricted to a small number of sites. This poses a major question, more perhaps in the context of model parameterization than establishing a process model where other approaches can be involved (e.g. flume studies). *Heterogeneity* is a characteristic of the floodplain environment. Figure 1.1 illustrates the complex inheritance in the sedimentological environment. This variability can be structured or seemingly random; in either event because there is frequently only a weak relationship between the hydraulic and chemical properties controlling subsurface flow and transport rates, hydraulic and chemical heterogeneity are less well understood than sedimentological heterogeneity in the floodplain (McLaughlin *et al.*, 1993).

There is the need to *upscale* from the laboratory to the field floodplain environment. This approach has had considerable impact in establishing a number of testable hydraulic processes appropriate to field-scale rivers. Illustrative of this case are the experimental facilities now available for compound channel research. Figure 1.6 shows a summary of results from a 56-m long flume and highlights the spatial differences in water surface elevation, which can be interpreted in terms of energy and momentum exchange (see Ervine *et al.*, 1994, p. 459).

However, in contrast, isolated laboratory determinations of hydraulic conductivity and solute dispersivity do not adequately describe subsurface field conditions (see Sudicity, 1986; Le Blanc *et al.*, 1991; McLaughlin *et al.*, 1993). Developments in theoretical models are now, therefore, generating *new areas of data capture*. For example, recent developments in modelling stream–subsurface water exchange illustrate the need to acquire data at particular time-scales in order to field test theoretical developments. Bencala *et al.* (1993) review research into water and solute exchanges taking place between stream bed, stream banks and stream valley. Specifically, they define the hyporheic zone (Figure 1.7) within this stream–catchment continuum, as a zone in which subsurface–surface water exchange occurs because of uneven pressure distributions across upstream and downstream faces of small roughness elements in streams: just one area illustrating new data capture requirements.

Remote sensing is increasingly being utilized for floodplain environment research. Engman and Gurney (1991) reviewed and emphasized two major attributes of remote sensing that are relevant here. Foremost is the ability to measure spatial data as opposed

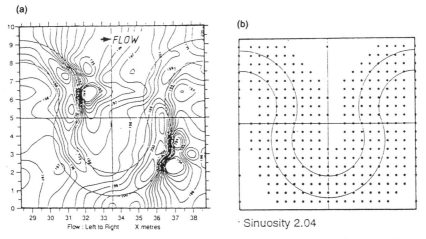

Figure 1.6 Spatial differences in water surface elevation in a 56-m flume study (a), and determined from measurement locations shown (b) (after Ervine *et al.*, 1994)

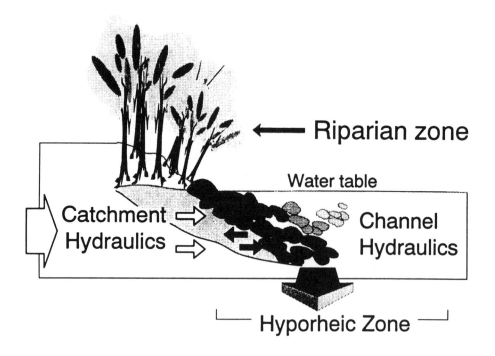

Stream - Catchment Continuum

Figure 1.7 Stream–catchment continuum and the hyporheic zone of water exchange (after Bencala *et al.*, 1993)

to point data from which most hydraulic concepts have been developed. Additionally, the ability to assemble long-term data sets is a further major advantage of remote sensing methods. Recent applications, albeit limited by satellite image frequency, have begun to focus on integrating floodplain flow modelling, and inundation extent as mapped onto the constituent land cover types.

Floodplain processes can thus be considered as a four-dimensional jigsaw (x–y–z through time) in which most of the pieces (data) are missing. "Validation", may thus not be possible in a manner that, classically, environmental scientists have presumed (Konikow and Bredehoeft, 1992). Moreover, as Bencala *et al.* (1993) conclude, "presently the limitations of models *and* directions for further research both focus on the need for obtaining appropriate field data".

1.3   THEORETICAL ISSUES

Whereas inbank flows may be treated as if they were predominantly one-dimensional flows in the streamwise direction, ... overbank flows must be treated differently as certain three-dimensional processes begin to be especially important, particularly the main channel/floodplain interaction. It is this interaction among others that makes the analysis of floodplain flows inherently difficult. (Knight and Shiono, Chapter 5, this volume p. 139)

Decisions of process inclusion, parameterization, dimensionality, discretization as well as others have to be made in the relatively data-sparse area of floodplain process modelling. Within the contemporary environment *computational fluid dynamics* (CFD) is playing an ever increasing role in this regard (see Chapters 5–7). Sophisticated finite element solutions are now capable of application to the river reach scale (Bates and Anderson, 1993; Anderson and Bates 1994; Hervouet *et al.*, 1994).

Behind such formulations are a number of approximations, and uncertainties. One approximation that is illustrative in concept of other constraints relates to the inability to conduct numerical simulations with mesh and time steps small enough to represent instantaneous motion. *Time averaging* removes this small scale-resolution need, but simultaneously introduces turbulent stresses within the flow that have to be approximated. Accordingly, flow turbulence models have received significant attention (Clifford *et al..*, 1993; Younis, Chapter 9) and for the reasons stated are a fundamental element in theoretical constructs or fluid flow processes.

As we examine, at a large scale, the *interaction between hillslopes and the floodplain*, further predictive difficulties are encountered. Firstly, current CFD modelling is restricted to floodplain and channel; presuming a zero flux boundary at the perimeter of the domain. Approaches are now being developed to relax this condition and couple hillslope and floodplain and thereby channel flows (Anderson and Bates, 1994; Bates *et al.*, Chapter 7). In this context Beven and Wood (1993) observe that for within-bank flows the floodplain will act to attenuate the hillslope hydrograph; for overbank flows the floodplain will lead to additional attenuation of the flood wave in the channel due to the effects of floodplain storage. There is a connection between this hillslope scale of input and the model structures that should be deployed in its representation. Wood (1995) utilizes the *representative elementary area (REA) concept*. This concept suggests that at some scale, the variance between hydrological responses of catchments of the same scale should reach a minimum. Wood *et al.* (1988) define the REA as "the critical scale at which implicit continuum assumptions can be used without explicit knowledge of the actual patterns of topographic, soil or rainfall fields". For a particular application (Figure 1.8) relating to runoff prediction modelling for a 11.7 km^2 catchment, the REA is of the order 1 km^2. This suggests that at scales larger than this responses can be modelled by a simpler macroscale model based on the statistical representation of heterogeneities in topography, soils, or rainfall, etc. The recent advances in scale issues for hydrological modelling (Kalma and Sivapalan, 1995) need careful consideration, therefore, in the large reach scale, as CFD techniques are extended and zero flux boundaries relaxed to incorporate hillslope inflows.

CFD approaches to modelling fluid flows presume non-deformable bottom boundary topography. There is exchange of both water and solutes across this boundary (the hyporheic zone – Figure 1.7) which we have already discussed, but additionally in terms of the boundary topography river channels both aggrade and degrade. Richards (1993) discusses the available evidence that is illustrative of wave-like patterns in sediment transport at the reach scale (and similar consequential patterns in sediment

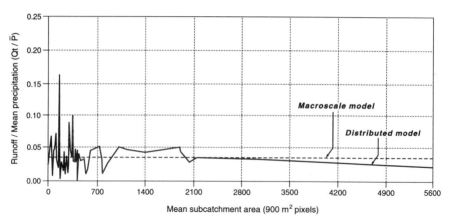

Figure 1.8 Comparison of storm runoff generated from a distributed model and a macroscale model illustrating the REA to be of the order of 1 km^2 (after Wood, 1995)

yield at the catchment and network scale). Pickup *et al.* (1983) show that a *sediment wave* undergoes a change of shape downstream by utilizing a sediment dispersion model – Richards (1993) notes the rarity of reliable data relating to sediment particle velocities. Perhaps of greatest significance in theoretical terms is the realization that "some sediment monitored as catchment yield will be rapidly routed recent sediment production, while some may be the 'tail' of a sediment wave initiated from unstable slopes in the Late Glacial period" (Richards, 1993). Thus bedload observations may not provide sufficient information on particle velocities, rest periods and residence times in storage that are required to calibrate models of sediment dispersion applicable to longer time-scales and larger reach lengths.

Consideration of such scour and fill processes leads to the broader question of channel migration prediction, requiring models of processes controlling bank erosion, flow patterns and bed topography. Howard (1992) developed a model of flow and bed topography in meander bends, incorporating bank erosion and sediment deposition in both stream channel and floodplain. This enables the simulation of the *topographic evolution* of floodplains. Howard in Chapter 2 advances this model further. In doing so he notes the difficulty of validating floodplain deposition models due to the lack of data – further support for the Konikow and Bredehoeft (1992) view.

The three themes of (1) integration of complex space-time processes, (2) data capture and (3) theoretical developments are increasingly central issues in the study of floodplain environments and are recognized as such by the contributors to this text. It is none the less appropriate, we believe, to structure the text in a thematic manner such that the research bases which drive these emergent considerations are treated in depth. Current research developments and areas of current limitations are thus structured in terms of floodplain evolution, hydraulic modelling, sediment and water quality, and ecology and management. Equally, however, these areas are seen as providing contributions to future research strategies that will emerge to specifically address process integration, model and process validation data needs, and theoretical developments consistent with both uncertainty and known process behaviour.

1.4 FLOODPLAIN EVOLUTION

Howard (1992) observes "meandering streams are one of the few geomorphological systems for which an abundant historical record exists of changes in channel pattern and associated floodplain erosion and deposition". We have noted above (Section 1.2) that despite the existence of such data, much remains that is relatively inaccessible in the context of many research investigations in this field. Models of meandering stream evolution through geomorphological time-scales are thus, relatively speaking, in their infancy. Howard (1992) outlines a preliminary model for such evolution, comprising a model of flow, bed topography and sediment transport, a relationship between near-bank velocity and depth (providing bank erosion and migration processes) with a third element being a process model of floodplain sedimentation. This model is further expanded in chapter 2. This chapter provides a central starting point for the text in that it fully portrays the context of the floodplain process problem – establishing the capability of integrating flow, sediment and channel evolution processes such that both short-term and long-term (1000 year) mechanisms are understood.

This understanding must in part be derived from developing models for deposition on floodplains that simulate alluvial architecture (Marriott, Chapter 3) and thereby facilitate a fuller understanding of complex assemblages (Figure 1.1) – floodplains being created both by river migration and flood events. The reconstruction of palaeoenvironments (Brown, Chapter 4) is of particular importance in being able to portray the changing impact of probable dominant processes over long periods (Holocene time-scales). In particular Brown raises the control of vegetation cover in the Holocene and cites such an environment and associated controls as an important context not only for current theoretical work (which requires clear specification of both vegetation condition and process relations) but also for land-use management policies. Thus Chapters 1–4 of the text raise the need for long-term integrative research (Section 1.1), point to the significance of data capture needs and difficulties (Section 1.2), and provide a fuller context for the theoretical developments (Section 1.3) that follow.

1.5 FLOODPLAIN FLOWS

Within the context outlined in Section 1.4, a fundamental understanding of the dynamics of both channel and floodplain flow is required. Of late, increased attention has been focused on integrating channel and floodplain flow into the continuum of hydraulic processes that exists in nature. It is this concept and advance that is the central theme of Chapters 5–10. Knight and Shiono (Chapter 5) review recent studies in this field and comment in particular upon dimensionality issues relating to flow representation (Section 5.4). Computational fluid dynamics (CFD) facilitates the solution of certain such questions and Hervouet and Van Haren (Chapter 6) exemplify such advances by examining two and three-dimensional computations of free surface flows. CFD methods present significant possibilities for advancing knowledge of floodplain processes. Both Knight and Shiono and Hervouet and Van Haren emphasize the urgent need for data to both benchmark and calibrate such advances. Bates *et al.* (Chapter 7) and Sellin and Willetts (Chapter 8) acknowledge this disjuncture in differing but complementary ways. Bates *et al.* introduce issues of "validating" sophisticated two-dimensional hydraulic models against available data in long (15 km) river reaches. Additionally, they seek to relax the hitherto rigid boundary conditions of such schemes. Hardware models are increasingly being used in a refined manner to both elicit specific process behaviour and to provide data sets for CFD models. Sellin and Willetts (Chapter 8) outline recent findings on flow mechanisms from a hardware flood channel facility and extend the analysis to field programmes in artificial compound channels.

As we move to better describe flow processes in terms of both spatial integration and temporal resolution by using CFD methods, we are forced to focus upon better representation of such aspects as roughness, turbulence (Younis, Chapter 9) and algorithm development. Younis sets out to demonstrate that utilization of turbulence models more sophisticated than those normally used (and not dependent upon the Boussinesq assumption) appears feasible and attractive in that they better represent flow processes without *ad hoc* modifications (see Section 9.1). This overall direction of research is placed into a broader context by Romanowicz *et al.* (Chapter 10) by emphasizing the problem of parameter calibration, and exploring the assertion that dominant in the desire to predict flood inundation is the representation of floodplain geometry and roughness specification. They argue that as the essential problem requires calibration, the generalised likelihood uncertainty estimation (GLUE) approach, among others, should be explored in the calibration procedure.

Floodplain flow research is thus seeking to develop the CFD methodology, assess implications for applications, provide a sharper focus for hardware models, and develop alternative models both at the specific (turbulence) and the broader parameterization (Bayesian) level. These *theoretical issues require resolution* if a truly satisfactory data acquisition framework is to be established. We might thereby augment the Bencala *et al.* (1993) quotation with which we ended Section 1.2 by adding the theoretical resolution needs identified here, since CFD and turbulence model developments have different conceptual elements from Bayesian emphases. The content of Chapters 5–10 thus sets out to identify key theoretical developments and establish a focus for a debate linking not only these developments, but associated data capture issues (Section 1.2).

1.6 FLOODPLAIN SEDIMENT TRANSPORT AND WATER QUALITY PROCESSES

There is a dual need to incorporate sediment and water quality considerations within flow modelling in compound channels. Firstly, there is the need to predict these fluxes and processes in time-scales that relate to currently perceived management horizons. Secondly, an understanding of floodplain sedimentation processes is essential for further developments in long-term floodplain evolution modelling. These two dimensions are reflected in chapters 11–14 of this text. Modelling hydrodynamic, sediment flux and water quality processes in complex tidal floodplains is considered by Falconer and Chen (chapter 11). Two and three-dimensional models are applied to field data with the continuing aim of improving bio-geochemical process representation.

We have already identified the complex theoretical issues involved in the sediment delivery process over long time-scales (Section 1.3). Walling *et al.* (Chapter 12) establish a clearer understanding of floodplains as suspended sediment sinks – providing for example, evidence of the need to account for the distinction between ultimate and effective grain size distributions in sediment modelling, as well as the role of the floodplain microtopography in controlling sedimentation. Just as in Section 1.5 we see the beginning of a debate on theoretical issues, so Walling *et al.* show the apparent inconsistency between recent floodplain sedimentation rates and existing alluvial evidence. Unlike the theoretical aspects, this inconsistency is not conceptual in origin, but rather the indicator of the potential magnitude of change occurring within the floodplain system. The isolation of cause here is important for future extrapolation of sedimentation. In particular, such issues receive particular attention when the sediment in question is associated with heavy metals. Thus many floodplains are major non-point sources of sediment-associated metal contaminants, where in the past they have been affected by mining or industrial activity. Macklin (Chapter 13) makes this point in stressing the need to understand processes of sediment-associated metal dispersal, and the resultant patterns of metal contamination storage in the floodplain. Such considerations are central to begin the process of involving metal contamination in environmental change scenarios.

It is necessary to establish within the hydrological framework, the process continuum from hillslopes to channel and floodplain taking advantage of the channel–floodplain process representation achieved by hydraulicians (see Chapters 5–7) in respect of flow in compound channels. Burt and Haycock (Chapter 14) examine the relevant processes of hydrological interaction between floodplain and hillslope and seek to couple this to issues of water quality in the riparian zone. The understanding of such buffer zones in the catchment context is of increased centrality to questions both of process controls and management of the floodplain.

Macklin and Burt and Haycock both seek to extend process investigations on the floodplain to issues of contamination and water quality that have clear connections with environmental management. This latter theme is the focus of Chapters 15–20.

1.7   FLOODPLAIN MANAGEMENT

Over the last two decades there has been a particular effort to establish close interaction between previously somewhat isolated disciplines. Such research concerns have focused in two principal areas – those relating to development and urbanization issues on floodplains and to hydroecology (see Harper and Ferguson, 1995). Penning-Rowsell and Tunstall (Chapter 15) discuss developments that are needed to achieve a better management of risk and resource in floodplain environments. They note the infancy of efforts geared specifically towards integrating land and water management targeted on sustainable futures and the incorporation of related decisions on the actual future use of floodplains. These questions of sustainability and the need to establish a scientific framework for management decisions are further explored by Petts (Chapter 16 – sustaining ecological integrity), Brookes (Chapter 17 – extending restoration issues to the floodplain) and Jolly (Chapter 18 – regulation impacts on arid and semi-arid floodplains). These four chapters (15–18) all stress the political imperative that is upon floodplain zones world-wide in so far as decisions for development and control increasingly focus in these zones. Equally clear however, is the urgent need to establish methodologies more closely including predictive aspects of geomorphology (Chapters 1–4), hydraulic processes (Chapters 5–10), hydrological, sediment and water quality (Chapters 11–14) with the question of developing a more sustainable future for floodplain zones. Richards *et al.* (Chapter 19) demonstrate the possibilities that increasingly present themselves in relation to developing linkages between process modelling and management of floodplain environments. They seek to prescribe river flow management options that are ecologically beneficial. In this example, numerical modelling of field water fluxes, complemented by laboratory determination of critical moisture conditions for riparian tree growth, leads to the prospect of developing management decisions based on sustainable futures.

1.8   CAPABILITIES AND DEBATES

This review, in introducing the structure and emphasis of this text has sought to present both current research capabilities and questions for debate and future research needs. Aspects of data capture and of management integration are at risk of being isolated by the respective demands for analytical "solutions" and political "solutions". And yet both of these "solutions" are, as we have seen, comparatively starved of the very data most often needed. This situation places at risk the methods, directions and outcomes of the science in the field of floodplain processes. Based upon the evidence provided in this text we must now purposefully utilize the available science outlined to develop new and more radical research programmes that more completely integrate analytical methodologies with management decision methodologies in floodplain process investigation. This text shows that all four areas of investigation have made significant recent progress. In also highlighting aspects of debate and omission, we hope that the following contributions provide a significant stimulus to future research.

REFERENCES

Anderson, M.G. and Bates, P.D. (1994) Initial testing of a two-dimensional finite element model for floodplain inundation. *Proc. Royal Soc. A*, **444**, 149–159.

Bates, P.D. and Anderson, M.G. (1993) A two dimensional finite element model for river flood inundation. *Proc. Royal Soc. A*, **440**, 481–491.

Bencala, K.E., Huff, J.H., Harvey, J.W., Jackman, A.P. and Tiska, F.J. (1993) Modelling within the stream catchment continuum. In: *Modelling Change in Environmental Systems*, Jakeman, A.J., Beck, M.B. and McAleer, M.J. (eds), John Wiley, Chichester, 163–187.

Beven, K. and Wood, E.F. (1993) Flow routing and the hydrological response of channel networks. In: *Channel Network Hydrology*, Beven, K. and Kirkby, M.J. (eds), John Wiley, Chichester, 99–128.

Carling, P. and Petts, G.E. (eds) (1992) *Floodplain Lowland Rivers*. John Wiley, Chichester, 302 pp.

Clifford, N.J., French, J.R. and Hardisty, J. (eds) (1993) *Turbulence – perspectives on Flow and Sediment Transport*. John Wiley, Chichester, 360 pp.

Engman, E.T. and Gurney, R.J. (1991) *Remote Sensing in Hydrology*. Chapman & Hall, London, 225 pp.

Ervine, D.A. and Ellis, J. (1987) Experimental and computational aspects of overbank floodplain flow. *Trans. Royal Society Edinburgh*, **78**, 315–325.

Ervine, D.A., Sellin, R.J. and Willetts, B.B. (1994) Large flow structures in meandering compound channels. In: *River Flood Hydraulics*, White, W.R. and Watts, J. (eds), John Wiley, Chichester, 459–469.

Harper, D.M. and Ferguson, A.J.D. (eds) (1995) *The Ecological Basis for River Management*. John Wiley, Chichester, 614 pp.

Hervouet, J.M., Hubert, J.L., Janin, J.M., Lepeintre, F. and Peltier, E. (1994) The computation of free surface water flows with TELEMAC. An example of evolution towards hydroinformatics. *J. International Assoc. for Research in Hydraulics*, **32**.

Howard, A. (1992) Modelling channel migration and floodplain sedimentation in meandering streams. In: *Lowland Floodplain Rivers: Geomorphological Perspectives*, Carling, P.A. and Petts, G.E. (eds), John Wiley, Chichester, 1–41.

Jakeman, A.J., Beck, M.B. and McAleer, M.J. (eds) (1993) *Modelling Change in Environmental Systems*. John Wiley, Chichester, 584 pp.

Kalma, J.D. and Sivapalan, M. (eds) (1995) *Scale Issues in Hydrological Modelling*. John Wiley, Chichester, 489 pp.

Knight, D.W. (1989) Hydraulics of flood channels. In: *Floods, Hydrological, Sedimentological and Geomorphological Implications*, Beven, K. and Carling, P. (eds), John Wiley, Chichester, 83–105.

Konikow, L.F. and Bredehoeft, J.P. (1992) Groundwater models cannot be validated. *Advances in Water Resources*, **15**, 75–83.

LeBlanc, D.R., Gorabedion, S.P., Hess, K.M., Gelher, L.W., Quadri, R.D., Stollenwork, K.G. and Wood, W.W. (1991) Large scale natural gradient tracer test in sand and gravel, Cape Cod, Massachusetts: 1. Experimental design and observed tracer movement. *Water Resources Research*, **27**, 895–910.

Lewin, J., Macklin, M.G. and Newson, M.D. (1988) Regime theory and environmental change – irreconcilable concepts. In: *International Conf. on River Regime*, White, W.R. (ed.), 431–445.

McLaughlin, D., Kinzelbach, W. and Ghassemi, F. (1993) Modelling subsurface flow and transport. In: *Modelling Change in Environmental Systems*, Jakeman, A.J., Beck, M.B. and McAleer, M.J. (eds), John Wiley, Chichester, 133–161.

Petts, G., Maddock, I. Bickerton, M. and Ferguson, A.J.D. (1995) Linking hydrology and ecology: the scientific basis for river management. In: *The Ecological Basis for River Management*, Harper, D.M. and Ferguson, A.J.D. (eds), John Wiley, Chichester, 1–16.

Pickup, G., Higgins, R.J. and Grant, I. (1983) Modelling sediment transport as a moving wave – the transfer and deposition of mining waste. *J. Hydrology*, **60**, 281–301.

Richards, K.S. (1993) Sediment delivery and the drainage network. In: *Channel Network Hydrology*, Beven, K. and Kirkby, M.J. (eds), John Wiley, Chichester, 221–254.

Sudicty, E.A. (1986) A natural gradient experiment on solute transport in a sand aquifer: spatial variability of hydraulic conductivity and its role in the dispersion process. *Water Resources Research*, **22**, 2069–2082.

Wood, E.F. (1995) Scaling behaviour of hydrological fluxes and variables: empirical studies using a hydrological model and remote sensing data. *Hydrological Processes*, **9**, 331–346.

Wood, E.F. Beven, K.J., Sivapalan, M. and Band, L. (1988) Effects of spatial variability and scale with implication to hydrologic modeling *J. Hydrology*, **102**, 29–47.

2 Modelling Channel Evolution and Floodplain Morphology

A.D. HOWARD
Department of Environmental Sciences, University of Virginia, Charlottesville, USA

2.1 INTRODUCTION

Floodplains result from the long-term cumulative action of the flow, erosion and depositional processes considered in this book. It is sometimes convenient in engineering or geomorphic practice to consider the floodplain (including the river channels) as a fixed geographic boundary over which flow properties and sediment transport are calculated to predict flow depths, scour potential, forces on structures, and rates and size distribution of sediment deposition. However, in many high energy fluvial regimes the rates of boundary modification from individual floods may be so great that the influence of recent and contemporaneous flooding in sculpting the floodplain cannot be ignored (e.g. Schumm and Lichty, 1963; Anderson and Calver, 1977, 1980; Graf, 1983; Kieffer, 1985; Nanson, 1986; Osterkamp and Costa, 1987; Baker, 1988; Miller and Parkinson, 1993). Even in lower energy fluvial systems, cumulative changes in channel pattern and floodplain morphology over time-scales of decades are often great enough to require their incorporation into engineering design and long-term sediment budgets. In addition, there are a range of issues such as groundwater flow in floodplains and petroleum reservoir characterization that require understanding of the internal stratigraphy of floodplains and hence of the history of floodplain development. The focus of this contribution is on recent developments and continuing research needs in the modelling of floodplain evolution over time-scales from years to centuries.

The long-term modelling approaches discussed here are based, directly or indirectly, upon the contemporary floodplain processes that are the topic of this book. But, of necessity, the long time-scales considered in this contribution require simplification of process laws, often to the point of using heuristic summaries of empirical observations. The major reason for the process simplification is the constraint that the long time intervals and large spatial scales place upon use of numerical models. Although this constraint is a receding horizon as computer capabilities improve, models based upon fundamental treatment of flow, erosion, transport, and deposition in channels and on floodplains extending over decades to thousands of years and over tens of square

Floodplain Processes. Edited by Malcolm G. Anderson, Des E. Walling and Paul D. Bates.
© 1996 John Wiley & Sons Ltd.

kilometres remain for a future generation. Even were such models practical, more modest model structure with simplified assumptions still might be preferred. The evolution of channel planform commonly depends sensitively on local boundary conditions and small differences in initial conditions. This is particularly true of flood chute development, meander cutoffs, creation and abandonment of braided-channel anastomoses, and channel avulsions, as well as crevasse-splay sedimentation. This chaotic behaviour suggests that even the most detailed model has limited predictive power when extrapolated over timespans of decades or longer. Our understanding of a number of constraints and processes occurring on floodplains is sufficiently uncertain that the predictive accuracy of any model is thereby limited; this includes transport of mixed bed sizes, weathering of sediment temporarily stored in floodplains, the interaction of vegetation with erosional and depositional processes, processes controlling bank erosion, and future history of discharge and sediment load. Finally, the use of simplified models allows wide exploration of the effects of variations in process assumptions as well as boundary and initial conditions, while promoting a humility about the accuracy and generality of conclusions.

Models of floodplain development can be constructed for a variety of purposes, such as prediction of future channel pattern changes and associated erosion and deposition for engineering purposes, or simulation of floodplain stratigraphy for groundwater characterization and/or for understanding of the stratigraphic record. The spatial and temporal scales, as well as the level of detail to which processes are considered, will depend upon the model purpose. The models considered here are primarily oriented towards understanding of the long-term, broad-scale evolution of floodplains and channels and hence many process details are very simplified. However, some discussion is devoted towards prospects for incorporating more explicit process representation.

2.1.1   A spectrum of channel planforms and floodplain morphology

Floodplains and their associated channel systems are remarkably varied. Channel patterns are commonly classified into meandering, braided, straight, and anastomosing. Each of these has a wide range of variation among natural channels, and all intergrade. Although there have been a wealth of studies of the influence upon planform of channel and valley gradient, flow regime, sediment supply (both amount and size range), erosional history (e.g. tectonic deformation and long-term aggradation or entrenchment), and local physiography, there are no universal criteria for predicting planform type or even a universal agreement on leading causative factors (see, for example, recent reviews by Ferguson, 1987; Gregory and Schumm, 1987; Bridge, 1993; Germanoski and Schumm, 1993; Knighton and Nanson, 1993; Smith and Perez-Arlucea, 1994). Floodplains are commonly classified by the associated channel planform type (e.g. Lewin, 1978), but physiographic constraints, particularly valley width, are also important. Nanson and Croke (1992) classify floodplains by a combination of sedimentary environment, formative processes, and stream power per unit channel width. Their main three divisions are: (a) high energy floodplains with non-cohesive sediments; (b) medium energy floodplains with non-cohesive sediments, and subdivided into those formed by (i) braided and (ii) meandering channels; and (c) low energy cohesive sediment floodplains of single or anastomosing channels. By further subdivision Nanson and Croke (1992) recognize a total of 15 floodplain types.

The most appropriate approaches to modelling floodplain temporal evolution vary with floodplain type. The selection of classes of floodplain models discussed here is based primarily on differences in model structure and technique, but these in turn are conditioned by channel planform, physiographic setting and the flow and sediment regime. Not all of the types of floodplains distinguished by Lewin (1978) and Nanson and Croke (1992) have been sufficiently well characterized to have been quantitatively modelled (e.g. the sandy floodplains subject to catastrophic channel widening during major floods discussed by Schumm and Lichty (1963) and Burkham (1972)). The broad categories of floodplains discussed here are: (1) strongly confined high energy; (2) braided channel; (3) meandering; and (4) avulsive and anastomosing.

The overall valley width is an important factor in all but the broadest of floodplains, such as the lower Mississippi River. A combination of five or more intergrading mechanisms may determine available valley width. In rapidly incising river systems the stream and the valley are essentially coincident, with valley width often no wider than can be scoured by the river in flood. Some channels are underfit, occupying valleys scoured by palaeofloods such as glacial meltwater (Dury, 1954, 1964a,b). Aggradation tends to widen valleys due to the layback of valley walls. Lateral channel migration, particularly in meandering channels, can undercut valley walls (Rich, 1914; Moore, 1926; Palmquist, 1975). Finally, if alluvial channels are restricted in their ability to downcut by a stable baselevel, valley walls will continue to erode, forming a widening pediment surface (as simulated by Howard, 1994, his Figure 19). This latter case is distinct from valley widening by lateral erosion, because the main channel complex does not necessarily occupy the entire valley width and fan-like pediments may extend from the valley edges to the centre of the valley. All of these processes may be influenced by tectonics (either active deformation or through rock structure) and lithology.

2.2   HIGH ENERGY CONFINED FLOODPLAINS

A wide spectrum of high energy confined floodplains exists, distinguished by floodplain width, flood frequency distribution, bed type, sediment characteristics and vegetation. The present discussion highlights three types selected from among this wide range, characterizing them slightly differently from Nanson and Croke (1992).

2.2.1   Chute canyon channels

Chute canyon channels are confined between narrow bedrock walls, usually within deep canyons, so that flood flows frequently extend from wall to wall. In some cases, such as portions of the granite narrows of the Colorado River in the Grand Canyon, even median or average flows fill the valley bottom, so that a floodplain is generally restricted to small fillets of deposited suspended sediment in slightly wider alcoves and an occasional gravel bar or small tributary fan exposed during low flow conditions. Slackwater sediments may be deposited along the lower course of tributary channels (e.g. O'Connor *et al.*, 1994). Channel beds may be bedrock, sand or gravel/boulder. Channel gradients range from very steep with many rapids in boulder or bedrock channels, to quite gentle in sand-bed sections such as portions of the Marble and Echo Canyon sections of the Colorado River (Howard and Dolan, 1981). Also within this category are somewhat wider canyons that expose some gravel, bedrock or sandy valley bottom during most flow conditions. Baker and Kochel (1988) call these bedrock fluvial systems, although similar features are found in narrow canyons with alluvial beds. Characteristic of such channels are discontinuous gravel-bar floodplains and finer slackwater deposits in alcoves and along tributaries (Baker, 1977; Baker and Kochel, 1988; Kochel and Baker, 1988). The alluvial deposits in such canyons are transient, being remoulded by floods large enough to occupy the valley width.

2.2.2 Tributary fan canyons

In deeply incised canyons, steep tributaries occasionally disgorge large quantities of mixed sediment sizes as debris flows into the main river channel. In the very narrow canyons discussed above, this debris flow material may be rapidly eroded by main-stem flood flows, or it may form largely submerged bars and associated rapids (Dolan *et al.*, 1978; Howard and Dolan, 1981). In somewhat wider canyon sections, however, this debris flow material is deposited as a fan that laterally constricts the main-stem flow, generally creating rapids. The coarsest boulders in such debris flows are generally much coarser than the gravels transported through the main stem, so that main-stem flood flows winnow the finer sediment and create rapids with gradients at the threshold of motion of the boulders (Graf, 1979; Howard and Dolan, 1981; Kieffer, 1985). The steepness and length of the rapids thus depend upon the size and quantity of the supplied boulders, the highest post-deposition main-stem floods, and subsequent weathering, sorting, and abrasion of the fan deposits. Such systems of rapids are therefore temporally varying as older deposits are gradually eroded and local flooding creates new fans or replenishes existing ones (Howard and Dolan, 1981; Kieffer, 1985). The fans and associated rapids in turn provide a framework for suspended load deposition. Thin, discontinuous fillets of sandy to clayey sediments may accumulate on the valley walls along the pools between rapids, but most fine sediments accumulate in association with low-velocity zones associated with the tributary fans. Recirculation zones develop against the fans both upstream and downstream from the rapids (Figure 2.1), encouraging deposition of a veneer of suspended load deposits (Howard and Dolan, 1981; Rubin *et al.*, 1990; Schmidt, 1990; Bauer and Schmidt, 1993; Schmidt and Rubin, 1995). The fine sediments are generally deposited during flood flows, resulting in low benches exposed during low flows. Very high floods capable of moving the fan boulders will cause complete reworking and possibly partial removal of the suspended load benches. The extent of suspended load deposits depends upon the balance between addition of suspended sediments into the canyon (generally by summer and autumn flooding of desert tributaries in the case of the Colorado River) and redistribution and removal by large floods (snowmelt spring floods on the Colorado River). In the case of the Colorado River, disruption of sediment input by Glen Canyon Dam has resulted in partial erosion of the benches (Howard and Dolan, 1981; Schmidt and Graf, 1990).

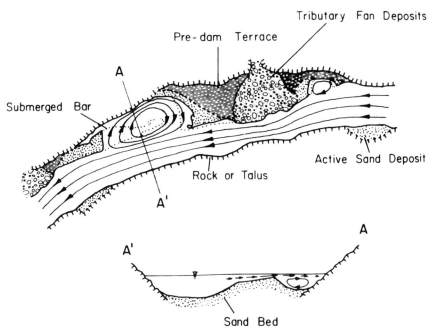

Figure 2.1   A tributary fan complex of the Colorado River in the Grand Canyon. Tributary fan deposits originate from debris flows from tributary side canyons. Both pre-dam terraces and active sand deposits are suspended load deposits. Sand deposits on the upstream side of the eddy are called separation bars, and the partially submerged bar on the downstream end is a reattachment bar (Schmidt, 1990). Illustration from Howard and Dolan (1981)

2.2.3   Vertical accretion floodplains

In somewhat wider and less steep confined valleys a pattern of vertical floodplain accretion during modest floods may be interrupted by widespread floodplain stripping during very large floods (Kochel *et al.*, 1982; Nanson, 1986; Kochel, 1988). Because of the time dependent floodplain height, Nanson (1986) terms these disequilibrium floodplains. Floodplain sediments coarsen upwards from gravels to sandy loam, with pronounced natural levees. Vertical accretion rates diminish as floodplain height increases, in accordance with observations in many floodplain environments (Wolman and Leopold, 1957; Everitt, 1968; Nanson, 1980). Stripping during large floods occurs most extensively near the main channel, and concomitant deposition may occur on more remote portions of the floodplain (Nanson, 1986).

Large floods also play a major role in somewhat wider valleys in hilly or mountainous terrain occupied by gravel or bedrock channels. The role of floods in such valleys has been described by Hack and Goodlett (1960), Williams and Guy (1973), Ritter and Blakely (1986), Ritter (1988), Miller (1990b), Harper (1991), and, most comprehensively, by Miller and Parkinson (1993). Channels in such valleys are commonly meandering where valley width permits, and often weakly braided. The meandering apparently occurs slowly enough that vertical accretion and stripping are the dominant floodplain-forming processes. Tributary fans occur locally in these valleys (Miller and Parkinson, 1993), but do not control depositional and erosional processes as strongly as in tributary fan canyons. As with the narrower vertical accretion floodplains described above, floods with recurrence intervals less than about 50 years are primarily depositional, and floodplains are composed of a gravel base with generally less than 2 m of suspended load accretionary deposits. Major floods strip vegetation from and erode channel banks. Overbank flow locally scours the floodplains, often along pre-existing flood chutes and floodplain depressions, exposing the underlying gravel/boulder framework. These chutes appear to be semi-permanent features in which modest floods deposit sediment, but they are reoccupied and rescoured by large floods (Harper, 1991; Miller and Parkinson, 1993). Localized scour is also triggered by vortices generated by floodplain topography, trees and man-made structures. The most intense scour and most common location for chutes occurs on floodplain enclosed within meander bends (Figure 2.2). Scour also becomes concentrated at locations where flow leaves or returns to the main channel. In narrowly confined sections or across sharp bends vegetation destruction and floodplain stripping may be general enough to approach the catastrophic erosion described by Nanson (1986). Gravel scoured from the main channel or from stripped areas of the floodplain may be spread as a thin layer on essentially uneroded portions of the floodplain (Ritter, 1975, 1988; Miller and Parkinson, 1993). Modest amounts of suspended load deposition may occur along the margins of the floodplain where flow is slow. This pattern is somewhat modified in locations where extensive, concomitant debris

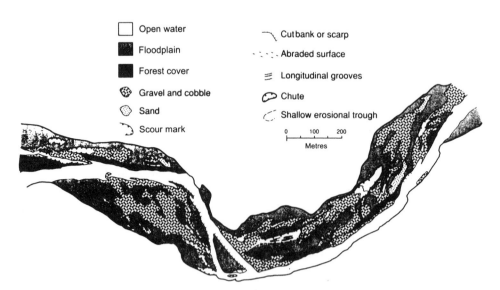

Figure 2.2 Patterns of flood scour and deposition on the North Fork of the South Branch of the Potomac River, West Virginia, resulting from 1985 flooding. Flow is from left to right. Gravel and cobble areas are partially exposed basal floodplain deposits and partially gravels transported onto the floodplain. Most of the forest cover originally bordering the channel has been stripped, and flood chutes have been excavated locally. Illustration from Miller and Parkinson (1993)

avalanches occur on steep mountain slopes due to high rainfall intensities (Hack and Goodlett, 1960; Williams and Guy, 1973; Miller, 1990b; Jacobson et al., 1993), in that more extensive valley bottom sedimentation occurs, often accompanied by construction of levees and dams of trees stripped from debris avalanche sites and the floodplain. Miller (1990a) distinguishes a continuum of flood events ranging from very localized flooding due to high-intensity thunderstorms that have catastrophic effects only on headwater channels (e.g. Hack and Goodlett, 1960), through hurricane-induced rainfall extending over somewhat larger areas resulting in rainfall exceeding 300 mm in a few hours (e.g. Williams and Guy, 1973), to more regional rainfalls exceeding 100 mm over a period of a few days that primarily affect larger rivers (Wolman and Eiler, 1958; Costa, 1974; Gupta and Fox, 1974; Miller and Parkinson, 1993). Costa and O'Connor (1995) suggest that flood duration may be as important as flood magnitude in determining the degree of floodplain modification.

2.2.4 Prospects for modelling

Most of the model components necessary to quantitatively simulate geomorphic evolution of high energy floodplains are in a reasonable state of development, as manifested by the chapters of the present book. Because most of the geomorphic modification of high energy floodplains occurs during overbank floods, flow models capable of treating both within-channel and overbank flow are necessary (e.g. Knight and Demetriou, 1983; Yen and Yen, 1984; Ervine and Ellis, 1987; Knight, 1989; Gee et al., 1990; Bates et al., 1992; Miller, 1995; Knight and Shiono, Chapter 5, this volume; Sellin and Willetts, Chapter 8, this volume; Younis, Chapter 9, this volume). Routing of suspended sediment and its deposition on the floodplain can be treated by a variety of approaches ranging from the heuristic rules (Howard, 1992, and discussion below) to mechanistic treatment of advection and deposition from suspension (e.g. Carey, 1969; Pizzuto, 1987; Marriott, 1992, Chapter 3, this volume; Nicholas and Walling, 1995). However, the most challenging component of geomorphic models will probably be treatment of floodplain scour. The resistance offered by cohesive sediments and floodplain vegetation induce a threshold to scour that is difficult to quantify. Furthermore, scour is strongly affected by subtle variations in floodplain topography and vortices generated by flow obstacles (Ritter, 1988; Miller and Parkinson, 1993; Miller, 1995).

The greatest efforts in modelling of high energy floodplains have been directed towards understanding the balance of suspended load deposition and erosion in tributary fan canyons, particularly the Colorado River in the Grand Canyon. The mechanics of flow and suspended load transport and deposition has been investigated in field studies (McDonald et al., 1994) flume experiments (Schmidt et al., 1993) and theoretical studies (Nelson et al., 1994; Smith, 1994). Miller (1994, 1995) has used depth-averaged flow modelling to investigate shear-stress distribution in variable-width valleys and in canyon reaches with tributary fans, showing that maximum stresses are much greater than in uniform reaches. However, modelling is still far short of full simulation of suspended load benches and recirculation bar development, or of the emplacement and subsequent modification and erosion of tributary fan deposits.

2.3   MODELLING OF BRAIDED STREAM DEVELOPMENT

Straight or irregular single-thread channel patterns are the exception on wide floodplains, where braiding, meandering and/or anastomosing patterns are most common. These more complicated channel patterns require both process(es) leading to the instability of straight, single-thread channels as well as other constraints or processes which prevent infinite elaboration of the pattern. Processes which may promote instability include (1) the non-linearity of sediment transport in response to applied stress, including the threshold of motion; (2) Interaction of grain sizes during transport; (3) secondary and recirculating fluid flows; (4) topographic effects upon fluid motion, including direction and magnitude of flow; (5) topographic effects upon sediment transport, including the threshold of motion, transport rate, and transport direction; (6) lags between change of bed shear and sediment transport rate; (7) effects of surface waves on flow and sediment transport, including out-of-phase relationships between the water surface and bed topography (Froude number effects); (8) discharge fluctuation; and (9) valley aggradation. Many of these instability mechanisms are outlined in Nelson (1990) and McLean (1990). Processes and constraints which limit the degree of planform elaboration include: (1) a finite valley width; (2) cutoffs; (3) finite quantity and depth of flow; and (4) topographic effects upon fluid motion and sediment transport.

Stream braiding results from two processes: (1) channel avulsion, that is, the partial or full diversion of flow across the channel bank (analogous to chute or neck cutoffs in meanders), and (2) growth of and emergence of within-channel bars and their subsequent dissection into two or more subchannels (Brice, 1964; Ashmore, 1991; Bridge, 1993; Bristow and Best, 1993; Ferguson, 1993; Leddy et al., 1993). Many types of avulsion and bar dissection have been distinguished based upon channel geometry, origin by blocking of flow by sediment deposition versus headward erosion along the diversion path, degree of channel curvature and channel asymmetry, etc. So long as the emergent bars remain unvegetated, braided channel patterns are very changeable, with frequent addition and abandonment of anabranches (see Figure 2.4). The role of bars in creation of braiding has been documented in many field studies (e.g. Williams and Rust 1969; Cant and Walker, 1978; Ashmore, 1982; Church and Jones, 1982) and flume experiments (e.g. Ashmore, 1982, 1993; Fujita, 1989; Hoey and Sutherland, 1991; Germanoski and Schumm, 1993). Leopold and Wolman (1957) cite the development of braiding in gravel streams as resulting from growth of medial bars due to local accumulations of coarse grains that the stream is incompetent to transport. However, braiding also occurs in uniform sand channels at transport levels well above threshold, so that incompetence cannot be a universal explanation.

Most theoretical studies view braids as developing from bar-form instabilities resulting from bed topography effects on flow and sediment transport. Early theoretical approaches utilized linear perturbation analysis (e.g. Engelund and Skovgaard, 1973; Parker, 1976; Callander, 1978; Fredsoe, 1978; Blondeaux and Seminara, 1985; Struiksma and Crosato, 1989), and more recent approaches have broadened the analysis to finite amplitude bars, either mathematically or through mathematical simulation (Fukuoka, 1989; Nelson and Smith, 1989a,b; Seminara and Tubino, 1989; Nelson, 1990). The initial instability is due to topographic steering of flow and sediment transport (Nelson, 1990), which is greater the shorter the bar wavelength. However, due to the lag between downstream changes in bed elevation and resulting boundary shear stress, an optimal bar growth rate occurs at a finite wavelength (Nelson, 1990). Finite-amplitude models include non-linear effects of flow and sediment transport as well as secondary flows (Nelson, 1990). These studies suggest that the degree of bar development is a strong function of the channel width W to depth H ratio (W/H) and a weaker function of channel gradient and flow stage $(\tau/\tau_c$ where τ is bed shear stress and τ_c is the critical shear stress for sediment movement). For very narrow channels $(W/H < 10)$ systematic bars cannot form, although curvature-induced (point) bars can occur. For larger W/H values alternate bars form, and for very wide channels, multiple bars occur, including both alternate bars at channel banks and linguoid bars in the central portions of the channel (Figure 2.3). The intensity of braiding has been described by a number of indices, including Brice's (1964) braiding index (twice the total length of bars divided by reach length), a total sinuosity defined as total channel length divided by reach length (Richards, 1982; Friend and Sinha, 1993; Robertson-Rintoul and Richards, 1993), and the average number of anabranches minus one (Howard *et al.*, 1970). The stability analyses suggest the bar mode, m, as a more fundamental quantity, which is related to the average number of scour holes and

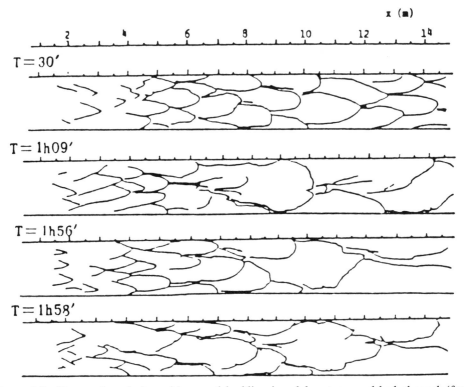

Figure 2.3 Temporal evolution of bars and braiding in a laboratory sand-bed channel (from Fujita, 1989). The linguoid bars in the initial stages are submerged, but portions of the bars in the final channel have emerged, forming a braided channel

channel thalwegs per cross-section, n (Fujita, 1989):

$$m = 2n - 1 \qquad (2.1)$$

This index offers the advantage that it takes into account both distinct channels as well as submerged bars and it should not be strongly dependent upon flow stage, but it has the disadvantage that in-channel measurement is required (Bridge, 1993; Ferguson, 1993).

Field observations and flume experiments indicate that channels with alternate bars do not necessarily develop into braided channels. In such cases the finite channel width and non-linear effects of flow depth and bed topography stabilize the height of bars before they become emergent. However, for higher bar modes flume experiments suggest that, starting from a plane bed, braiding develops from a fairly regular pattern of generally linguoid bars. These bars tend to lengthen and increase in amplitude with decrease in m with time, as well as becoming more irregular through the development of smaller superimposed bars, until the largest bars may become exposed and relatively fixed in location, creating a braided pattern (Figure 2.3). Fujita (1989) relates the submerged bar mode in flume studies to channel width, depth and sediment grain size, d:

$$2.2 \ m^{2/3} < (W/d)^{2/3}(d/H) < 6.7 \ m^{2/3} \qquad (2.2)$$

where the inequalities indicate that there is an overlapping range of widths and depths for each bar mode. Braided channels tend to have slightly higher values of width to depth ratio for a given bar mode than the relationship above (Fujita, 1989).

The above analysis suggests that braiding is related to a greater freedom of adjustment in wide channels in which the negative feedbacks limiting bar growth are insufficient to prevent bar emergence and the chaos of braiding. A fundamental unit in development of braids is the combination of a scour hole (often located at braid confluences; Ashmore, 1993) and a downstream depositional bar (Ashmore, 1982, 1991, 1993; Bridge, 1993; Ferguson, 1993), often called a pool-bar complex or a chute and lobe. The following enumeration of braiding behaviour is largely summarized from Ashmore (1991, 1993), Bridge (1993), Bristow and Best (1993), Ferguson (1993), Germanoski and Schumm (1993), and Leddy et al. (1993). Flow is usually convergent in the pools and divergent on the bars. The bars can become emergent even at constant discharge due to channel widening and flowstage lowering at the bar or migration and stalling of dunes or bedload sheets on the bar crest. Multiple channels can develop directly due to mid-channel bar emergence or by channelling and headward erosion by water flowing over the top of the bar. Incision may be aided by lowering river stage due to decrease in discharge. Individual braid channels can be curved, leading to asymmetric bar development (including point bars) and lateral channel migration. Pulses of bedload and temporal variation of water and sediment influx at junctions can cause aggradation and avulsions, blocking of channels, and migration and reorientation of confluences. Overall aggradation leads to a greater degree of braiding and more frequent channel shifting.

Braided channels are traditionally viewed as being distinguished from meandering channels by a threshold value of channel gradient and/or discharge (Leopold and Wolman, 1957; and numerous subsequent studies reviewed by Ferguson, 1987, 1993, and Bridge, 1993). Additional criteria are bank strength, sediment size and quantity, stream competence, and width–depth ratio (e.g. Equation 2.2). Ferguson (1987) points out that most of these criteria are strongly interrelated by channel hydraulics. Although most studies have emphasized a threshold between meandering and braiding, the indices of braiding discussed above suggest a continuum from single-channel to strongly braided, and a few studies (e.g. Howard *et al.*, 1970; Moseley, 1981; Richards, 1982) have related degree of braiding to hydraulic parameters.

Where valleys occupied by braided rivers are wider than the active channel complex, much of the floodplain is often essentially unmodified channels left behind after the active braid-train migrates to a new location (Figure 2.4). Portions of the floodplain which have been inactive for longer periods may become vegetated with thin caps of accreted fines from overbank deposition (Reinfelds and Nanson, 1993). Creation, abandonment and shifting of anabranches is a continuing process in braided rivers with negligible braid-bar vegetation. Islands are generally about the same dimensions as the anabranches, although the extent of bars/islands diminishes as they become submerged at higher stages (e.g. Mosley, 1982). In such braided systems small islands generally can be considered to have originated as submerged bars that have become emergent either by falling stage or by erosion and water surface lowering by the bordering anastomoses. Larger islands are generally just inactive portions of the braid-train. However, when climate, flood regime and sedimentary characteristics permit vegetational colonization of braid bars, these bars acquire a resistance to scour, and often increase in size by both lateral and vertical accretion. As a result, active channel width becomes restricted, with fewer, deeper and larger anastomoses. Furthermore, the frequency of channel shifting diminishes, often occurring only during large floods. Creation of new channels is primarily by creation of chutes and avulsions across existing vegetated islands.

The development of a simulation model of braiding and associated sedimentation and floodplain construction that incorporates a detailed description of flow and sediment transport would be a formidable undertaking because of the multiple channels, constantly changing boundary conditions (including creation and abandonment of channels and some meandering in the larger anabranches), and the sensitivity of the braiding pattern to discharge variation. As a result, extant modelling approaches have been very heuristic, with most having no explicit treatment of flow and sediment transport, and one (the Murray and Paola, 1994 model discussed below) having a very simplified hydraulic formulation. All of these models are targeted towards braided channels lacking appreciable vegetated islands.

Early attempts at modelling braided channel complexes utilized a downstream random walk with stochastic channel branching and recombination (Howard *et al.*, 1970; Krumbein and Orme, 1972; Rachocki, 1981). The channel pattern is generated by marching downstream with existing channels extending downstream with an assigned probability of a lateral component of motion and of branching. The lateral component of motion may cause neighbouring channels to intersect and combine. The combination of branching and recombining generates a braided pattern. In order to create a braided planform with a statistically constant frequency of channels, there must either be a lateral constraint on floodplain width or a central tendency for lateral movements. Such stochastically generated braided planforms have statistical characteristics of braid

Figure 2.4 A braided tributary of the Toklat River, Denali National Park, Alaska. Flow from top to bottom

geometry that are similar to natural braided channel complexes (Howard *et al.*, 1970; Krumbein and Orme, 1972). A similar approach can be used to simulate alluvial fans (Rachocki, 1981) and deltaic distributaries (Smart and Moruzzi, 1972). This modelling, although static, indicates that branching and recombination of braided channels can be viewed as a stochastic, or possibly chaotic, process.

Webb (1994, 1995) has extended the random-walk model by accounting for the discharge and cross-sectional shape within each channel. Furthermore, the stratigraphy of aggradational braided channel floodplains is simulated by accounting for the deposition from each channel. Successive deposits are simulated by via generations of random-walk channel systems, with pattern changes between each generation being constrained by the previous pattern. Braided patterns simulated by Webb (1994, 1995) have morphometric characteristics that are closer to those of natural braided channels than the earlier models of Howard *et al.* (1970) and Krumbein and Orme (1972).

The major limitation of the random-walk models is that they do not account directly for the mechanics of flow and sediment transport that cause braiding, whereas the major advantage is their computational efficiency. Such models may prove to be useful for simulation of floodplain stratigraphy and topography if the rules of generation are empirically adjusted to replicate natural channel kinematics. The Webb (1994, 1995) approach seems promising in this respect, although the validation and calibration are limited by the sparsity of observations. The random-walk models offer little insight into the flow and transport mechanics that create braiding.

A novel approach to simulating braiding has been developed by Murray and Paola (1994). The model simulates flow and sediment transport over a mesh of nodes. Flow and sediment from each node is routed downstream, being divided among the three nodes lying directly downstream and to the left and right. The flow Q_i from a given node is apportioned between the downstream nodes in proportion to the gradient S_i to the downstream node:

$$Q_i = Q_u S_i^n / \sum S_i^n \qquad (2.3)$$

where Q_u is the total flow reaching the node from upstream and the exponent n is less than or equal to unity (flow is routed only to those nodes with positive S_i and the summation occurs only for positive S_i). An important additional feature is that if all S_i are negative (the local node is a depression), flow out of the node is apportioned among the downstream nodes, presumably accounting for flow momentum. Sediment discharge Q_{si} to downstream nodes is assumed to be a power function of gradient and discharge:

$$Q_{si} = K[Q_i(S_i + C)]^m \qquad (2.4)$$

where K is a constant, C is an additive constant on the order of the average gradient, and the exponent m is typically 2 to 3. The C term allows for a certain amount of upslope sediment transport. Certain other transport relationships also result in braided patterns. The realism of the braiding can be improved if a small amount of lateral transport is also permitted. Figure 2.5 (Plate I) shows bed topography and flow discharge produced by a simulation utilizing the Murray and Paola (1994) algorithm.

Braiding in the Murray and Paola (1994) model results from scour where flow converges and downstream deposition as a broad bar around which the flow ultimately

becomes diverted when the bar become sufficiently high. These high bars can become emergent when flow becomes diverted through adjacent scour holes. In this respect the model captures the evolutionary sequence observed in flume studies, including the reduction in bar mode through time and a continuing creation and abandonment of braid channels (cf. Figures 2.5a and 2.5b). The model does not account for curvature-related effects on flow and transport, including lateral migration and point-bar development. The channels simulated by the Murray and Paola (1994) model have a visual similarity to natural and laboratory braiding, but it remains to be seen how successfully the model can replicate the morphometry of braided streams, channel changes through time, and braid deposit sedimentology. None the less, the richness of pattern of the simulated braiding suggests that much of the diversity of braid geometry and of kinematic patterns of initiation, growth and decay of braid channels results from a single basic instability in flow and transport.

Despite the mathematical simplicity of the Murray and Paola (1994) model, restrictions on the allowable iteration time-step due to requirements for numerical stability severely limit the spatial size and temporal duration of simulations. An even simpler approach might be possible using the techniques recently developed for simulating aeolian dune fields by Werner (1995). In this approach sediment movement occurs as small unit sheets and the interaction of transport with topography and the fluid flow is highly idealized using heuristic rules.

Bridge (1993) shows how models of braid kinematics can be combined with sediment transport relationships and associated bedforms to predict braided river sedimentology.

2.4   MODELLING OF MEANDERING STREAM FLOODPLAIN DEVELOPMENT

Mackin (1937), Fisk (1944, 1947) and Wolman and Leopold (1957) specified the archetype for a meandering channel floodplain (type B3 of Nanson and Croke, 1992) formed by lateral channel migration with floodplain sediments dominated by sand-to-gravel point-bar deposits and finer sand, silts and clays deposited by overbank flooding. Subsidiary facies include sandy splays from levee crevasses, fine-grained fills in abandoned oxbow lakes, and organic-rich backswamp deposits.

The leading processes in the development of meandering channel floodplains are the lateral migration of meander loops and attendant cutoffs (Figure 2.6). The primary causative factor for meander development is the secondary circulation created by channel planform curvature, which leads to asymmetry of flow and bed topography in bends. Flow momentum causes greater shear stresses on the outer bank in a long bend, leading to enhanced erosion and concomitant point-bar deposition on the inside bank. This curvature-induced flow asymmetry is a sufficient explanation for development of

---

Figure 2.6 (*opposite*)   An unknown tributary of the Yukon River, Alaska. Note recent cutoff and oxbow lake at lower left and impending neck cutoff in lower centre of picture. Flow is from top to bottom

meanders (Ikeda *et al.*, 1981; Howard, 1984), although flow patterns associated with alternate bars may aid in initial stages of bank erosion (Lewin, 1976, 1978; Nelson, 1990). Bank erosion may also be affected by channel depth, which is also affected by secondary circulation. As will be discussed more fully below, cutoffs (and sometimes limited valley width) prevent infinite growth of meander bends.

The earliest models of meandering planforms were static simulations utilizing constrained random walks (Ferguson 1976, 1977). The Ferguson disturbed periodic model demonstrated that meanders can be considered to be generated by a periodic migration with superimposed irregularity. However, such static simulations cannot be used to simulate planform evolution or the topography and stratigraphy of the floodplain.

Prediction of channel migration requires models of the processes controlling bank erosion as well as flow patterns and bed topography in the meandering channel. Important observations of migration rates in natural channels were obtained by Hickin and Nanson (1975, 1984), Nanson and Hickin (1983, 1986), Hooke (1987), and Biedenharn *et al.* (1989), who showed that migration rates increase with bend curvature, reaching a maximum value at R/W (R is mean bend radius and W is channel width) of about 2–3, and decreasing rapidly for tighter curves (even becoming negative in the sharpest curves). However, Howard (1984), Howard and Knutson (1984), Parker (1984) and Furbish (1988) show that lateral migration rates must be an integral function of both local and upstream curvatures. Howard and Knutson (1984) and Ferguson (1984) introduced heuristic models relating local migration rates to weighted integrals of local and upstream curvatures. In addition, Odgaard (1987), Pizzuto and Meckelnberg (1989) and Hasegawa (1989a,b) found that spatial variations in migration rates in natural channels are linearly related to the magnitude of near-bank velocity.

These sets of observations relating relating migration rates to channel curvature and to near-bank velocities were shown to be consistent as models of flow in meandering channels were developed. In a seminal paper Ikeda *et al.* (1981) provided an analysis that showed that the cross-sectional velocity distribution and cross-sectional channel slope can be expressed as a function of local and upstream planform curvatures. Howard (1984), Howard and Knutson (1984), Parker (1984) and Furbish (1988) demonstrated that the Ikeda *et al.* (1981) model coupled with the assumption that bank erosion is proportional to near-bank velocity produces realistic simulated meanders with a relationship between rate of meander loop migration and loop curvature that is similar to that observed by Hickin and Nanson (1975).

2.4.1 Modelling meander migration and floodplain deposition

A simulation model of flow and bed topography in meander bends was coupled by Howard (1992) with simple assumptions about the rate of bank erosion and sediment deposition rates in the stream channel and on the floodplain to simulate the topographic evolution of floodplains. This model is briefly summarized below, together with extensions of the original model to permit the possibility of chute cutoffs and channel aggradation, as well as spatially variable bank erosional resistance due to valley walls or oxbow lake clay plugs.

Within-channel flow and bed topography

Several models have been developed to predict bed topography and flow structures in channels with arbitrary planform. The models of Ikeda *et al.* (1981), Johannesson and Parker (1989), Odgaard (1989a,b) and Parker and Johannesson (1989) are linearized one-dimensional (along-stream) models that treat downstream variations in velocity and depth structure explicitly but make simplifying assumptions about cross-sectional variations. A model of this type (the Johannesson and Parker, 1989, abbreviated J&P) was utilized by Howard (1992) for reasons of computational efficiency, although more general models exist that explicitly treat cross-sectional variations in depth and velocity (e.g. Nelson and Smith, 1989a,b). The J&P model treats only time-averaged, curvature-forced bed variations and thus does not account for migrating dunes and alternate bars. Local depth (h) and downstream vertically averaged velocity (u) are resolved into section means (H and U) and a dimensionless perturbation (h_1 and u_1):

$$u = U(1 + u_1) \tag{2.5}$$

$$h = H(1 + h_1) \tag{2.6}$$

Near-bank values of the perturbations are indicated by h_{1b} and u_{1b}. At the channel centreline u_1 and h_1 are assumed to be zero. The model also makes the simplifying assumptions that H, U and channel width, W, are constant downstream, that depth and depth-averaged velocity vary linearly across the channel, and that there are negligible sidewall effects on near-bank flows.

Bank erosion rate law

Howard (1992) notes that four processes may limit the rate at which channel banks erode to produce channel migration: (1) the rate of deposition of the point bar; (2) the ability of the stream to remove the bedload component of the sediment eroded from banks; (3) the rate of detachment of *in situ* or mass-wasted bank deposits; and (4) the rate of weathering of bank sediment. A similar assessment was made by Ikeda (1989). Because banks in meandering channels are generally cohesive, bank erosion is slow enough that (1) is seldom limiting. Constraint (2) would be appropriate for readily disaggregated banks composed dominantly of bedload-sized sediment. Even though stream banks are commonly initially undermined by erosion of bed sediment exposed at the foot of cut banks (Laury, 1971; Thorne and Lewin, 1979; Thorne and Tovey, 1981; Thorne, 1982; Pizzuto, 1984; Ullrich *et al.*, 1986; Osman and Thorne, 1988; Thorne and Osman, 1988), the cohesive upper bank sediment must be subsequently detached before further bank erosion can occur, governed by constraint (3). If banks are indurated then erosion may be limited by subaerial weathering, so that erosion rates may not be directly related to flow characteristics (constraint 4). All four constraints may vary in importance among streams, from place to place along a given stream, and through time at a given location.

Howard (1992) suggests a general relationship relating lateral bank erosion rate, $\partial n/\partial t$, to the near-bank velocity and depth perturbations:

$$\partial n/\partial t = E(au_{1b} + bh_{1b}) \tag{2.7}$$

where a and b are coefficients and E is intrinsic bank erodibility. Most studies have suggested that the velocity perturbation is the important factor, so that $a = 1$ and $b = 0$ (Ikeda *et al.*, 1981; Beck, 1984; Beck *et al.*, 1984; Howard and Knutson, 1984; Parker, 1984; Parker and Andrews, 1986; Odgaard, 1987; Hasegawa, 1989a,b; Pizzuto and Mechelnberg, 1989). Hickin and Nanson (1984) and Nanson and Hickin (1986) relate bank erodibility to median grain size at the base of cut banks in the deeper scour holes. This suggests that bank erosion should be a positive function of depth (positive b in Equation 2.7); a similar assumption was made by Odgaard (1989a,b). On the other hand, Hasegawa (1989a,b) shows that if constraint (2) is governing, then b in Equation 2.7 should be negative; this occurs because the quantity of eroded bank sediment is proportional to channel depth.

If erosion rates are related to h_{1b} rather than u_{1b}, then meanders would tend to grow in amplitude with little downstream translation, because the depth perturbation is nearly in phase with, or may even lead the curvature for flow and sediment characteristics that are characteristic of natural channels (Howard, 1992). By contrast, u_{1b} lags changes in curvature significantly, so that maximum outer-bank velocities are downstream of the meander axis of symmetry, resulting in downstream migration. The simulations conducted below assume that $a = 1$ and $b = 0$.

Floodplain deposition

The third major component of the Howard (1992) model is an assumed rate law for floodplain sedimentation. Several studies have shown that the rate of overbank deposition on floodplains diminishes with floodplain age and floodplain elevation, because the higher locations are less frequently and less deeply flooded (Wolman and Leopold, 1957; Everitt, 1968; Nanson, 1980; Hupp and Bazemore, 1993). Further-more, floodplain deposition rates diminish with distance from the channel supplying water and sediment to the floodplain (Kesel *et al.*, 1974; Pizzuto, 1987; Walling *et al.*, 1992, Chapter 12, this volume; Mertes, 1994). Similar patterns have been observed in tidal marshes (Allen, 1990, 1992; French, 1993; French and Spencer, 1993; Falconer and Chen, Chapter 11, this volume). These observations were incorporated into a heuristic relationship for long-term overbank sedimentation rate, ϕ, by Howard (1992):

$$\phi = (E_{max} - E_{act})[v + \mu \exp(-D/\lambda)] \tag{2.8}$$

where E_{max} is a maximum floodplain height, E_{act} is the local floodplain elevation, v is a position-independent deposition rate of fine sediment, μ is the deposition rate of coarser sediment by overbank diffusion, λ is a characteristic diffusion/advection length scale, and D is the distance to the nearest channel (both measured in channel-width equivalent units). The upper limit to deposition (E_{max}) in the Howard (1992) model is a rather arbitrary limit to maximum floodplain height. For the present contribution the deposition rate was given a more general formulation:

$$\phi = [v + \mu \exp(-D/\lambda)]\exp[-\gamma(E_{act} - E_{bed})^w], \tag{2.9}$$

where v, μ, λ, D and E_{act} are as above, E_{bed} is the mean bed elevation, γ is the elevation decay rate, and w is an exponent, set in these simulations to 2.0. If $E_{act} < E_{bed}$ then the second exponential term is set to unity. Thus there is no absolute upper elevation for sediment deposition.

Sedimentation occurs generally over the floodplain using Equation 2.9, but locations occupied by the channel have elevations determined by the flow and bed elevation model. When a channel migrates through and beyond a given floodplain cell, the initial elevation on which floodplain sedimentation occurs is the near-bank bed elevation of the channel predicted by the flow model on the side opposite to the direction of migration. This initial elevation is set to

$$E_{act} = E_{bed} + \eta_b \qquad (2.10)$$

where η_b is the relative bed elevation predicted by the flow model at the bank opposite to the direction of migration.

In the simulations reported below the deposition rate ϕ is assumed to be a long-term average resulting from the natural spectrum of flood flows; in other words, the effects of individual floods are not explicitly modelled, and all floods are assumed to result in net deposition.

In summary, the depositional model incorporates both a crude model of point-bar sedimentation expressed as a variable advancing bank initial elevation (Equation 2.10) and an overbank (vertical accretion) component (Equation 2.9).

2.4.2 Simulation results

This section summarizes the results of the Howard (1992) model, and subsequent sections explore modifications to that model. When the initial stream is straight except for small, random normal perturbations (Figure 2.7a), the early meandering pattern is very regular and nearly sinusoidal, but rapidly develops a classic very sinuous, upstream-skewed shape prior to cutoff. This shape is characteristic of the solutions to the governing equations (Parker, 1984; Parker and Andrews, 1986) and common in natural streams (Carson and Lapointe, 1983). Local differences in loop growth rate are due to the random perturbations of the initial input stream. After neck cutoffs begin (when the centreline of two channel segments approach within 1.5 channel widths), however, the stream pattern becomes much more varied in the form of the meanders, due to the disturbances that propagate throughout the meander pattern as a result of cutoffs. At these advanced stages, the pattern becomes much more similar to natural meandering streams with their commonly complex loop shapes (Figures 2.7b and 2.7c). As a result of chance occurrence of two or more cutoffs on the same side of the valley, the overall meander belt can develop a wandering path, as noted by Howard and Knutson (1984). The sharp bends that result from cutoffs are very rapidly converted to more gentle bends, commonly by reverse migration caused by maximum flow velocities occurring on the inside of very sharp bends. The development of varied meander forms from an initially regular pattern indicates that the combination of meander growth, the occurrence of cutoffs, and the influence of complicated initial and boundary conditions (including variations in bank resistance which are here held constant) implies a "sensitivity to initial conditions". That is, small differences in initial geometry or boundary conditions between two otherwise identical streams will cause different meander patterns. Also, predictability of future meander patterns decreases with time.

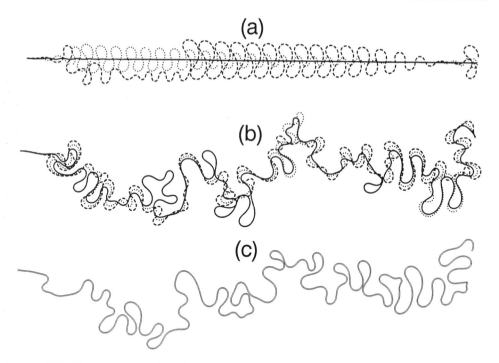

Figure 2.7 Simulated meander planform development. (a) Channel centrelines after 250 (solid), 500 (dotted) and 750 (dashed) iterations; (b) Channel centrelines at 4400, 4600 and 4800 iterations; (c) final channel at 5000 iterations

Floodplain age and elevation for the deposition model using Equations 2.9 and 2.10 are illustrated in Figure 2.8 (Plate II), which portrays the evolution of the central region of the stream shown in Figure 2.7. White areas have not been occupied by the meandering channel during the simulation. Similarly, Figure 2.9 (Plate III) shows meander and floodplain evolution for confined meanders developed between non-erodible valley walls. Confined meanders develop a characteristic asymmetric pattern with gentle bends terminating abruptly in sharp bends at valley walls, similar to natural cases (Lewin, 1976; Lewin and Brindle, 1977; Allen, 1982; Howard and Knutson, 1984).

These simulations exhibit many of the features of natural meandering streams, including overbank deposits gradually increasing in elevation away from the channel, rapid isolation of abandoned channels by filling near the main channel (modelled here as resulting from sediment diffusion from the main channel, but in natural channels advectional transport through the abandoned channel would also occur), and slower infilling of oxbow lakes primarily by deposition from suspension. Channel migration rates vary considerably from bend to bend, and the slope of the floodplain surface in the interior of bends is generally steeper the slower the migration. For high rates of overbank deposition (Figures 2.8b and 2.9b) most of the floodplain rapidly reaches a nearly uniform elevation and cutoffs are rapidly isolated into oxbows. The simulations with a low rate of overbank sedimentation (Figure 2.8c) exhibit a depositional feature that is a consequence of out-of-phase relationships between near-bank velocity and bed

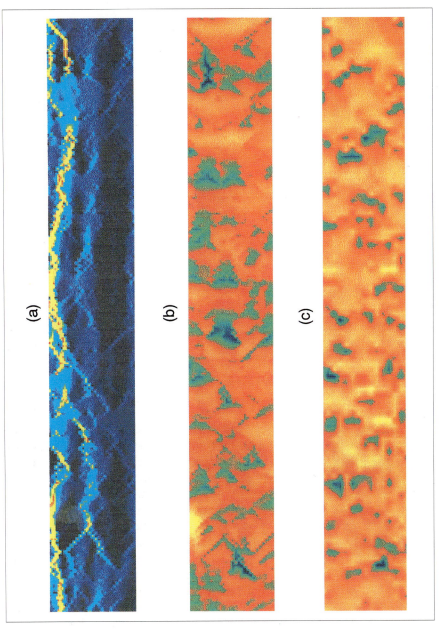

(a)

(b)

(c)

Plate I Figure 2.5 Simulation model of braided channel development. (a) Discharge over bed topography shown in (b); (b) final bed topography; (c) bed topography at an early stage of development. Areas of high topography and high discharge are yellow, low values are blue and green. Simulation is based upon the model of Murray and Paola (1994)

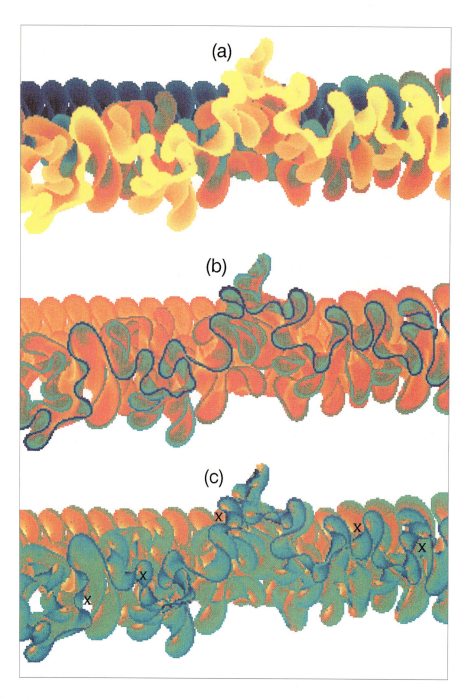

Plate II Figure 2.8 Map view of floodplain age (a) and floodplain elevation under conditions of high (b) and low (c) rates of overbank deposition. There are no lateral constraints on channel migration and no chute cutoffs in this simulation. Yellow areas are young in (a) and high areas are red and yellow in (b) and (c). All figures have been given a non-linear contast stretch to provide greater detail in young or high areas. This figure covers the central portions of the simulation shown in Figure 2.7

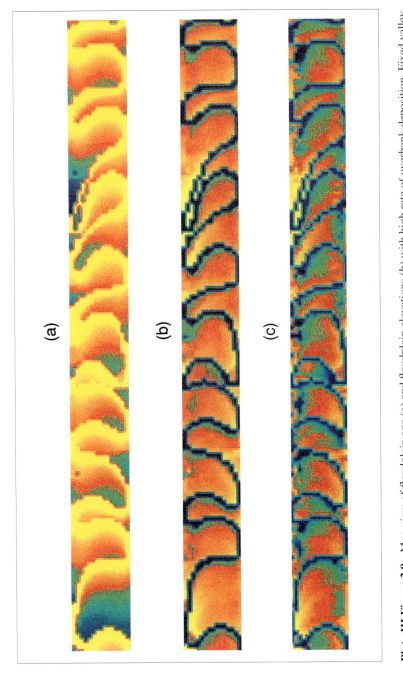

Plate III Figure 2.9 Map view of floodplain age (a) and floodplain elevations (b) with high rate of overbank deposition. Fixed valley walls restrict meander enlargement

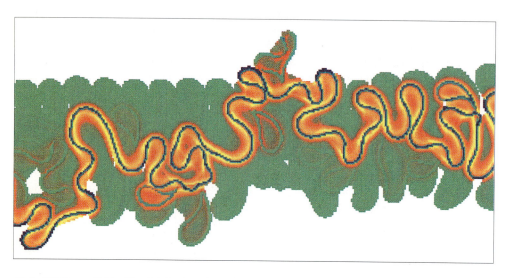

Plate IV Figure 2.14 Floodplain elevation (high areas yellow) in a simulation similar to that shown in Figure 2.8b but with superimposed channel aggradation. Picture is contrast stretched to enhance portrayal of high areas

elevation perturbations. Where curvature changes abruptly downstream, the depth adjusts quite rapidly on the new outer bank, and generally overshoots its value for constant curvature, but velocity responds more slowly and migration is directed towards the inside of the bend. This means that, for short sections at the beginning of a sharp bend, deposition on the newly created floodplain on the outside bank must start from very low relative elevations (a scour hole). This zone of very low initial elevations is only about 2.5 width equivalents in length, followed by the more normal pattern of migration towards the outside bank. These short zones with lower-than-average initial floodplain elevations are located in consistent positions relative to bends as the channel migrates, in places leaving behind depressions, or sloughs, in the floodplain deposits. These sloughs are most commonly located in the axial position of sharp meander bends, and are best developed on the downstream end of the point bar near the curvature inflection leading to the next bend. Several of these depressions are labelled with an "X" in Figure 2.8c. These depressions are best developed in short abrupt bends near the site of recent neck cutoffs (Figure 2.8c) or where meander migration is confined. Howard (1992) points out that natural and experimental channels abound in similar features which may result from a mechanism similar to that incorporated in the model (e.g. the experimental runs of Friedkin (1945) and Wolman and Brush (1961)). Fine sediments accumulating in such sloughs and floodplain lows have been called concave bank benches or counterpoint bars (Cary, 1969; Woodyer, 1975; Hickin, 1979; Nanson and Page, 1983). Lewin (1978) attributes floodplain sloughs extending upstream from the outside bank of confined and unconfined sharp bends to formation as residual depressions from migrating deep scour pools – essentially the same mechanism as occurs in the simulations. These sloughs are best developed when vertical accretion deposition rates are modest (Nanson and Page, 1983). If this is not the case (large values of v and/or μ in Equation 2.9), then bank and floodplain deposition is more uniform (Figures 2.8b and 2.9b).

The simulations suggest that bar and low floodplain features of wide meandering streams cannot be solely understood in terms of adjustments of bar form to contemporary flow and sediment transport, but are in addition intimately related to the kinematics of bank migration. The topography of these features should affect location of chute cutoffs, as is explored below.

Cutoffs

Loop cutoffs in the Howard (1992) model were limited to *neck cutoffs*, occurring when the growth of two loops cause mutual intersection (the criterion in these simulations was centreline approach to within 1.5 channel widths). The resulting channel pattern is very sinuous ($\mu > 3.5$, as shown in Figures 2.7 and 2.8); such highly sinuous meandering is rare in natural channels. The model has been revised to incorporate probabilistic *chute cutoffs*, which are longer flow diversions which generally result from formation and enlargement of a depression across a point-bar complex. Because the chutes have steeper overall gradient than the existing meander bend, the chute can often gradually increase in size during successive floods until it eventually accommodates almost all of the flow and the former channel becomes choked with deposited sediment. The cutoff model does not model the flood flow across the point bar which creates the chute, but rather predicts the influence of planimetric and relief properties that determine where and when chute cutoffs might occur. Five such properties have been incorporated into the model: (1) the ratio R_c of gradients across the potential chute path to existing channel gradient; (2) the distance D_c across the potential cutoff; (3) the elevation E of the floodplain across which the chute develops; (4) the planimetric angle ψ between the existing channel direction and the path of the chute; and (5) the relative magnitude of the near-bank velocity at the site of the potential chute. The probability P of chute development at a given time and across a given chute path is expressed as:

$$P = K_c R_c e^{(-K_d D_c - K_e E + K_a \cos \psi + K_v u_{1b})} \tag{2.11}$$

where

$$R_c = (D_p - D_c)/D_p = (\mu_c - 1)/\mu_c \tag{2.12}$$

The gradient ratio R_c depends upon the relative distances (and hence also overall gradients) across the existing flowpath D_p and across the potential chute D_c, and it can also be expressed in terms of the cutoff sinuosity $\mu_c = D_p/D_c$. In evaluating Equation 2.12 the downstream channel gradient is assumed to be uniform. The terms in the exponential reflect the presumed smaller probability of cutoff across longer chute lengths and floodplains with higher elevation (E can be specified either as the maximum or average elevation across the chute, measured relative to the average bed elevation). Additionally, the probability of cutoff is assumed to decrease as the angle between the existing flowpath and the cutoff path increases. Finally, the probability of cutoff is assumed to increase if the near-bank velocity perturbation u_{1b} at the site of potential capture is large and positive. The coefficients K_c, K_d, K_e, K_a and K_v are assumed to be temporally and areally invariant. The sum of the exponent terms in Equation 2.11 is restricted to values ≤ 0.

Although there have been a number of studies of the relative frequencies of neck and chute cutoff in natural meandering channels and of their temporal development, no systematic observations of cutoffs in natural meandering channels have been made which would permit evaluation of the relative importance of the coefficients in Equation 2.11 or of the adequacy of the assumed functional form. Lewis and Lewin (1983) provide most comprehensive observations. Over half of the observed cutoffs in the Welsh rivers observed occurred as chutes. Cutoff locations vary from chord sites (across the point bar), *axial* sites (middle of loop), and *tangential* sites (loop-to-loop, connecting outer bends). Most cutoffs occurred across tight bends at positions from axial to tangential.

For the present contribution a number of simulations have been conducted to explore the characteristics of cutoffs and channel form resulting from the application of Equation 2.11. At regular intervals during the simulation (every 50 iterations in the present case) each point along the stream is examined in turn for the possibility of chute cutoff, progressing downstream. At each point all downstream locations up to a practical limit of 50 width equivalents are evaluated with respect to the probability P of chute development. For that point the potential cutoff path with the greatest probability P of chute development is compared with a randomly generated number from a uniform distribution. If that random number is $\leq P$, then the chute cutoff occurs. Locations downstream from the cutoff are also examined in the same manner for additional cutoff. The probability P is assumed to be unity if D_c is less than 1.5 channel widths (a neck cutoff). In the model the chute cutoff is assumed to be instantaneous relative to the simulation time-scale (although natural chutes require an appreciable time to fully divert the main-stem flow).

A number of simulations were made with a range of values for the coefficients in Equation 2.11, generally by varying one coefficient while holding the others constant at zero, except that K_c was held at unity for runs with varying K_d, K_e, K_a or K_v. In addition, for runs with varying K_a or K_v the value of K_d was set at a value producing an intermediate degree of sinuosity of $\mu \approx 2$. Some of the simulated planforms are illustrated in Figure 2.10. The planforms are plotted at times just before the occurrence of cutoffs is allowed (every 50 iterations); this was done so that the very sharp bends that occur immediately after cutoffs would be smoothed by some channel migration. As a result, the plotted figures slightly exaggerate the average sinuosity during the simulation. By varying the coefficients any degree of overall sinuosity from near unity

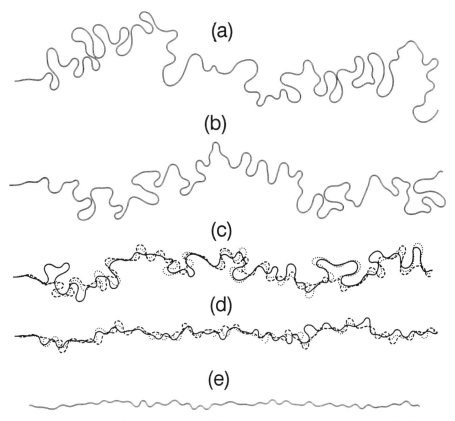

Figure 2.10   Final channel planforms after 5000 iterations created by model with chute cutoffs controlled by Equation (2.11). For these simulations $K_c = 1$, $K_e = 0$, $K_a = 0$, $K_v = 0$ and K_d takes the following values: (a) ∞ (no chute cutoffs); (b) 0.3; (c) 0.2; (d) 0.1; (e) 0.03. Planforms at successive 200 iteration intervals are shown in (c) and (d). Statistical characteristics of cutoffs are shown in Table 2.1

to greater than three can be obtained, and in general the planforms for simulations with restricted sinuosity resemble natural examples with the same degree of sinuosity. As the probability of chute cutoffs is increased, not only does the sinuosity diminish, but the width of the meander belt diminishes.

Cutoff characteristics As K_c increases and K_d or K_e decreases, chute development becomes more probable and the planform becomes less sinuous (Table 2.1). However, variations in K_a or K_v have little effect on total sinuosity. Decreases in sinuosity are accompanied by decreases in the averages of cutoff path length D_p, cutoff sinuosity μ_c, and cutoff angle ψ, whereas the frequency of cutoffs and the average chute length D_c increase. Variations in K_c, K_d, K_e and K_a have little effect on the average value of the velocity perturbation u_{1b} at the cutoff sites; also, as noted above, variations in K_v have

Table 2.1 Summary statistics of chute cutoffs

Input parameters[*]					Summary statistics[†]							
K_c	K_d	K_e	K_a	K_v	μ	μ_c	f	D_c	D_p	E	ψ	u_{1b}
1.0	∞	–	–	–	3.56‡	–	–	–	–	–	–	–
1.0	0.43	–	–	–	3.17	18.89	0.024	3.69	57.5	13.4	84.9	0.02
1.0	0.30	–	–	–	2.88	8.64	0.050	5.31	36.2	13.9	77.3	-0.03
1.0	0.20	–	–	–	1.98	3.73	0.089	7.28	24.6	13.6	67.0	-0.01
1.0	0.10	–	–	–	1.39	1.50	0.171	9.17	13.9	13.7	48.3	0.00
1.0	0.09	–	–	–	1.22	1.43	0.181	9.55	13.6	13.8	43.0	-0.01
1.0	0.03	–	–	–	1.10	1.16	0.179	13.53	15.6	14.2	23.8	-0.02
1.0	0.06	0.15	–	–	2.14	5.27	0.069	7.8	28.3	6.9	66.1	-0.02
1.0	0.06	0.10	–	–	1.73	2.33	0.107	8.78	19.0	10.6	62.6	0.00
1.0	0.06	0.07	–	–	1.58	1.79	0.137	9.44	16.7	11.5	55.8	0.01
1.0	0.06	0.05	–	–	1.42	1.61	0.152	9.40	15.1	12.4	51.4	-0.01
1.0	0.06	0.035	–	–	1.29	1.49	0.157	10.04	14.8	13.6	45.7	0.03
1.0	0.20	–	0.5	–	1.94	3.11	0.104	7.16	20.7	13.5	53.2	0.00
1.0	0.20	–	1.0	–	2.19	2.51	0.126	7.26	17.7	13.0	46.4	0.00
1.0	0.20	–	2.0	–	1.69	1.71	0.193	7.42	13.2	12.4	38.2	0.00
1.0	0.20	–	–	4.0	2.41	2.86	0.113	8.03	22.1	13.7	60.4	0.30
1.0	0.20	–	–	2.0	2.40	3.09	0.102	7.42	21.1	14.1	65.4	0.17
1.0	0.20	–	–	1.0	2.05	3.20	0.091	7.34	22.1	13.5	68.5	0.11
1.0	0.20	–	–	-1.0	2.04	3.34	0.097	7.25	22.0	13.6	67.8	-0.09
1.0	0.20	–	–	-2.0	2.08	3.32	0.086	7.41	22.5	13.5	65.5	-0.18
1.0	0.20	–	–	-4.0	2.30	2.44	0.117	7.89	17.2	14.0	58.6	-0.31
0.8	0.03	–	–	–	1.11	1.20	0.185	12.67	15.1	14.5	23.4	0.00
0.5	0.03	–	–	–	1.14	1.27	0.163	13.07	16.4	15.2	29.3	0.01
0.25	0.03	–	–	–	1.19	1.42	0.133	12.76	17.9	15.4	37.2	0.00
0.1	0.03	–	–	–	1.49	1.81	0.102	12.31	21.4	16.4	48.3	0.03
0.05	0.03	–	–	–	1.41	2.31	0.081	12.41	26.3	16.6	55.8	-0.01
0.025	0.03	–	–	–	1.69	2.92	0.061	12.47	32.0	16.9	58.3	-0.01
0.01	0.03	–	–	–	2.41	4.16	0.038	12.12	39.9	17.0	67.8	0.03

[*] Input parameters defined in Equations 2.11 and 2.12.
[†] Average values of cutoff properties. Properties defined in Equations 2.11 and 2.12 except for f, which is the frequency of cutoffs (number per iteration).
‡ Run with only neck cutoffs.

little effect on sinuosity, although it would be expected that high K_v would favour axial to tangential cutoffs, because u_{1b} is likely to be positive on the outer bank. Similarly, a high weighting on K_a would tend to favour tangential cutoffs, because ψ is small. Simulations with cutoffs due to high K_c resulted in smaller cutoff frequency, larger D_c, larger D_p and greater E than simulations of similar sinuosity resulting from high values of K_c, K_d, K_e or K_a. This is because the lack of restriction on distance or elevation across the cutoff for the K_c runs leads to longer cutoffs and K_c has to be set to a low value (low frequency of cutoffs) in order to maintain sinuosity appreciably greater than unity. Increases in sinuosity with variations in K_c and K_d are accompanied by increases in the average and maximum elevations across the cutoffs, whereas, as would be expected, average and maximum elevations decrease with increasing sinuosity as K_e is increased.

Planimetric and elevation characteristics of floodplains with chute cutoffs An obvious characteristic of simulations with frequent chute cutoffs is the restricted width of the meander belt (Figure 2.10). Even though there are no lateral restrictions imposed upon the simulated meandering (other than the fixed point at the upstream end of the simulated channel), when chute cutoffs are frequent the meander belt increases in width very slowly, and the average elevation of the floodplain is higher for a given average distance from the active channel. This is because each site within the meander belt stays closer to the active channel on the average.

Both the amplitude and wavelength (measured *following* the channel centreline) are smaller for meander planforms dominated by chute cutoffs as compared to those solely shortened by neck cutoffs for equivalent flow, bed topography and bank erosion parameterization. This results because meanders grow primarily in amplitude rather than along-valley wavelength (Howard and Knutson, 1984; Parker, 1984; Howard, 1992), and chute cutoffs primarily limit amplitude growth. Furthermore, the disturbances in the planform caused by cutoffs encourages growth of new meander loops. In addition, the simulations assume that the overall valley gradient is fixed over the time-scales necessary for full development of the meander pattern. As a result, the channel gradient is inversely related to sinuosity. Thus high sinuosity planforms (no chute cutoffs) have low velocity and relatively large channel depth; the preferred wavelength of meandering increases with increase in channel depth (Howard and Knutson, 1984).

Spatially variable bank erodibility

The simulations reported above assume that the erodibility of bank materials is spatially uniform. However, the pattern of meandering may be affected by at least two types of non-uniformity. The first of these occurs where meandering locally impinges on resistant materials such as bedrock or terraces bordering the active floodplain. This is termed confined meandering (or restricted meandering by Ikeda, 1989). Lateral migration may also be locally inhibited by exposure of resistant floodplain sediments, most commonly the cohesive sediments (clay plugs) deposited in oxbow lakes by overbank deposition (Fisk, 1947; Ikeda, 1989; Thorne, 1992), but sometimes older bankswamp deposits (Ikeda, 1989; Thorne, 1992). Ikeda (1989) calls this case confined free meandering. Simulations incorporating spatially variable bank resistance are reported below.

Confined meanders   Narrow meander belts confined within valley walls or high terraces occur commonly. Meanders in such situations commonly develop an asymmetric pattern, with gentle bends deflecting abruptly at valley walls (Lewin, 1976; Lewin and Brindle, 1977; Allen, 1982). The effects upon meander pattern of a narrow floodplain was simulated by Howard and Knutson (1984) and Howard (1992) by the simple expedient of disallowing erosion beyond a fixed belt surrounding the centre of the floodplain (Figure 2.9), which is a reasonable model for some underfit valleys. The effect of resistant sidewalls is investigated using a more general approach in this contribution (a similar approach is used by (Sun *et al.*, 1996). The spatial variation in bank erodibility is explicitly modelled by overlaying a matrix of square cells on the floodplain and surrounding valley walls (or terraces), with each cell having dimensions equivalent to one channel width. Thus a floodplain of arbitrary width and pattern could be set as initial conditions. In the simulations reported here, meandering and floodplain development starts from a nearly straight initial channel with a floodplain width just equalling the channel width. Valley wall erodibility is assigned an areally uniform value of 0.2 times the erodibility of floodplain deposits. If the node representing a given location along the stream is located on a cell marked as a valley wall and if the bank erosion is directed towards the valley wall, then the rate of erosion is governed by the valley wall erodibility. However, if the channel node is located within a floodplain cell or is on a valley wall cell but erosion is directed away from the valley wall, the rate of erosion is proportional to the floodplain bank erodibility. The direction of erosion relative to the valley wall is determined in the following manner. Assume the direction of bank erosion is directed towards the east and north (so that the stream is flowing southeast or northwest). East and north are in the positive I and J directions respectively. If the local cell (I, J) is a valley wall cell and if any of the bordering cells at $(I, J + 1)$, $(I + 1, J + 1)$ or $(I + 1, J)$ is a valley wall cell, then the rate of erosion is governed by the valley wall erodibility, otherwise it is proportional to the floodplain sediment erodibility. Two variations of these rules were utilized; in the first, "strict" rules the bordering valley wall cells must not be occupied by the stream, whereas in the "liberal" rules the bordering cells may be occupied by the stream. Only slight differences in pattern resulted from the two sets of rules. Only the simulations for the strict rules are shown.

Figure 2.11 shows an early and later stage in the enlargement of the meander belt and retreat of the valley walls. In the earliest stages of valley wall enlargement (right side of Figure 2.11a) the channel pattern shows the asymmetric pattern discussed above. As valley walls retreat further, meander loops tend to become locked into reentrants eroded into the valley wall. This pattern is often seen in entrenched meandering. Upon further meander bend enlargement the stream impinges only locally against the valley walls (Figure 2.11b). Where this occurs the meander enlargement commonly becomes constrained against projections in the valley wall, resulting in highly distorted meander loops whose axes often point upstream. In some cases these distortions of the meander pattern induce neck cutoffs earlier than if the meandering were unconstrained, but in some locations highly sinuous loops form due to intervening valley walls. In the above

(a)

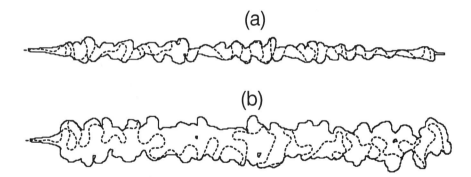

(b)

Figure 2.11   Successive stages (a and b) of valley widening and planform development with valley walls five times less erodible than active floodplain. Chute cutoffs are not permitted

simulations chute cutoffs are prohibited; when such cutoffs are permitted, the rate of floodplain enlargement is slower, the highly sinuous distorted loops are rare, and the valley width is more uniform with few deep reentrants (Figure 2.12).

The simulations do not show any tendency for development of "underfit" valleys, that is, a long wavelength valley meandering (generally with valley walls of fairly uniform width) with shorter wavelength channel meandering. Such a pattern is fairly common in nature and is attributed to valley enlargement during a past epoch of very high discharges (e.g. from glacial meltwater) followed by more moderate discharges related to the present meandering (Dury, 1954, 1964a,b). Thus the simulations provide circumstantial support for this interpretation.

Cohesive oxbow lake plugs   Other simulations were conducted in which it was assumed that the cohesive sediments deposited in oxbow lakes are less erodible by a factor of five than normal point-bar and overbank sediments. The rules governing determination of local bank erosion rates are the same as for the valley wall simulations. Immediately after a meander loop is cutoff all of the cells along the cutoff path are assigned an erodibility of 0.2 except for portions of the loop closer to the existing channel of three channel-width equivalents. The portions of the cutoff loop close to the existing channel are assumed to be infilled with relatively coarse suspended sediment

(a)

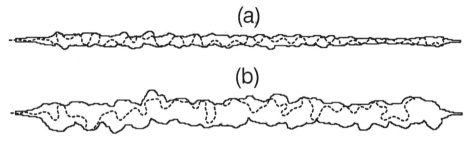

(b)

Figure 2.12   As in Figure 2.11 but with chute cutoffs permitted

diffusing and advecting from the main channel. Sun *et al.* (1996) utilize a similar approach to investigate the effects of clay plugs.

Figure 2.13a shows the channel pattern and location of resistant clay plug deposits for an advanced stage of meander belt development, under the conditions where only neck cutoffs are permitted. As with the valley wall simulations, impingement on clay plugs creates locally tortuous and distorted meander loops. The pattern of clay plugs over the floodplain becomes very complicated, with some locations evidencing a very heavy areal density of plugs. In some cases, a growing loop penetrates into the interior of a previously abandoned loop, whereupon the loop grows rapidly, more or less duplicating the earlier pattern of loop enlargement. The rate of enlargement then diminishes significantly when the channel reaches the plug on the far side of the filled oxbow.

Although there is some tendency for the channel to become constrained by the development of the resistant clay plugs (Ikeda, 1989), the simulations suggest that the channel can break through the plug belt on occasions, and in some locations the meander belt may migrate laterally, if plugs have formed preferentially on one side of the valley.

The pattern of lateral groundwater flow on the floodplain would be strongly affected by the areal distribution of plugs. As can be seen from the map, many floodplain regions are surrounded by one or more plugs; these would be expected to have relatively high water tables. On the other hand, in some locations a path extends from the active channel well into the floodplain without encountering plugs; such areas should be well drained. In addition, there is a zone surrounding the active channel with relatively few or only short, disconnected plugs that would be well drained.

In the above simulations chute cutoffs were not permitted. When chute cutoffs are allowed (Figure 2.13b), the overall width of the meander belt diminishes and the average size of oxbow plugs is smaller, as discussed above. In addition, the lower sinuosity of the channel system means that the axes of meanders loops and the resultant oxbow plugs are, on the average, nearly perpendicular to the valley axis. This contrasts with the greater variability of oxbow axes when only neck cutoffs occur (Figure 2.13a). Because of the narrower floodplain and more frequent cutoffs, oxbow plugs have a

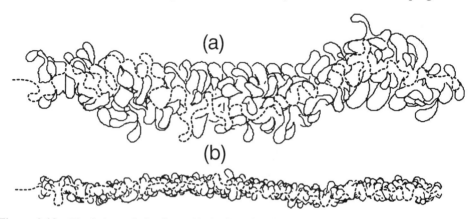

Figure 2.13 Final channel planform (dashed) and oxbow plug deposits (solid) in a simulation in which oxbow clay plugs are assumed to be five times less erodible than point-bar and overbank deposits: (a) after 5000 iterations with only neck cutoffs; (b) after 5000 iterations with chute cutoffs permitted

greater areal density when chute cutoffs are frequent. This does not necessarily mean that the prevalence of plugs would inhibit lateral migration strongly when chutes cutoffs are frequent, because in natural channels chute cutoffs are more common in stream systems with a relatively low supply of cohesive overbank sediments. Thus plugs are probably more erodible than where neck cutoffs predominate.

Natural levees and the effects of aggradation

Simulations of floodplain development by stream meandering under conditions in which the channel bed elevation remains constant fail to produce pronounced natural levees using the deposition model of Equation 2.9, even though the deposition rate is much higher near the active channel than at more distant locations. Slightly elevated banks do occur at crossings (inflection points), where the channel migration rate is very low (Figure 2.8b), but along bends levee development is discouraged because sediment deposited on the outside of bends is eroded by lateral migration soon after deposition, and deposition on the inside, or point-bar locations, occurs at low elevations because of the floodplain youth.

In a number of simulations the mean channel bed elevation was assumed to rise at a constant rate relative to the floodplain elevation. In these simulations pronounced levees were created (Figure 2.14 (Plate IV)). This suggests that channels with high, extensive natural levees (such as the lower Mississippi River) primarily occur where the channel bed is increasing relative to the floodplain level due to aggradation or floodplain lowering (due, perhaps, to sediment consolidation). The other main circumstance producing high natural levees occurs during disequilibrium floodplain sedimentation following catastrophic stripping of floodplain deposits (Nanson, 1986, and previous discussion).

Temporal evolution of the meander belt

The meander simulations typically start from a nearly straight initial stream. As the simulation progresses the stream migrates through an increasingly wide section of floodplain. Initially the growth of the meander belt width (defined as the maximum width of floodplain, measured normal to the valley axis, occupied by the meandering stream during the elapsed time) is slow because of the low sinuosity. As average bend size increases, the average rate of migration and the meander belt width increase rapidly. Cutoffs, however, begin to limit the rate of migration and the rate of meander belt growth slows. Figure 2.15 depicts the overall average width of the meander belt as a function of simulation time (arbitrary units). Meander belt width, W_b, during the nominal run, with no chute cutoffs, is well fit by a logarithmic growth function:

$$W_b = W + K_b \ln(t/t_0), \qquad \text{for } t \geq t_0$$
$$W_b = W, \qquad \text{for } t < t_0 \tag{2.13}$$

where W is the channel width, t is simulation time, t_0 is a "lag time" for initiation of well-developed meandering from a nearly straight initial channel, and K_b is the growth rate constant. The inclusion of chute cutoffs decreases the rate constant, but not the overall logarithmic pattern of growth. Resistant valley walls delay the onset (t_0) and diminish the growth rate (K_b). Resistant oxbow plugs do not form until appreciable

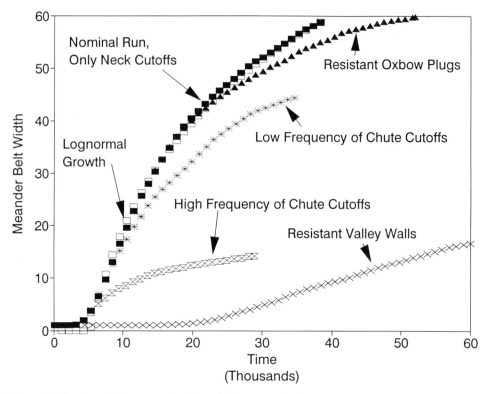

Figure 2.15 Cumulative meander belt width versus simulation time (arbitrary units). Meander belt width is total floodplain width (measured perpendicular to valley axis) occupied by the channel since the start of the simulation. Filled boxes for simulated meandering with uniform bank erodibility and no lateral constraints. Open boxes are a logarithmic growth curve (Equation 2.13) fit to the simulation results. Other curves show effects on growth rate of chute cutoffs, resistant clay plugs associated with oxbow lakes, and resistant valley walls

neck cutoffs occur (after some time $t_n > t_0$), and the growth constant (K_b) decreases for $t_n > t_0$. Ikeda (1989) suggested that meander belts in some rivers may be confined by cohesive fine-grained overbank sediments. The simulations suggest that, even in the absence of lateral constraints or of other controls such as tectonic tilting (Leeder and Alexander, 1987: Alexander *et al.*, 1994), the meander belt width is slow to increase beyond about two to three times the size of the largest meander loops.

The cumulative growth rate of the meander belt is of interest for several reasons: (1) the use of meander belt width to date the valley age; (2) inferring the geometry of sandstone aquifers and petroleum reservoirs; and (3) inferring the storage time of pollutants in fluvial deposits.

Comparison of simulated and natural meander morphometry

Because of the numerous simplifying assumptions made during the development of the meander simulation model, it is important to validate the model. In particular, the prediction of the pattern of meander migration and attendant cutoffs depends both on assumptions regarding in-channel flow and bed morphology as well as assumptions regarding bank erosion processes. The most direct validation compares predicted patterns of channel migration with those observed; a number of studies have indicated that the combination of the Johannesson and Parker (1989) flow model with the assumption that bank erosion is proportional to the velocity perturbation (Equation 2.7) gives reasonable predictions of short-term migration patterns (Howard and Knutson, 1984; Hasegawa, 1989a,b; Pizzuto and Meckelnberg, 1989; Shane Cherry, personal communication, 1995). However, these techniques are limited by the duration of the historical record of channel migration.

Howard and Hemberger (1991) compared the planform of simulated and natural channel patterns using a suite of morphometric variables. Two linear discriminant functions composed of weighted combinations of these variables are quite successful in separating natural channel planforms (indicated by open circles in Figure 2.16) from those simulated by the Howard (1992) and Howard and Knutson (1984) model (open inverted triangles in Figure 2.16) and Ferguson's (1976, 1977) disturbed periodic model (which simulates meander planforms, but not migration, using an autoregressive approach; open diamonds in Figure 2.16). Discriminant function 1 primarily measures sinuosity and degree of upstream bend asymmetry, whereas function 2 is an index of the irregularity of the channel pattern (more irregular for positive values). The addition of chute cutoffs and spatially variable bank resistance to the model creates meandering patterns that approach closer to those of natural channels. Simulation runs with varying degrees of chute cutoff frequency (filled triangles in Figure 2.16) approach natural channels in having smaller sinuosity and less marked upstream bend asymmetry, but the patterns are still more regular than those of natural channels. Spatially variable bank resistance, due either to valley walls or oxbow lakes (filled boxes in Figure 2.16), both decreases sinuosity and increases pattern irregularity, particularly when combined with chute cutoffs. In some cases simulation runs with spatially variable bank resistance and chute cutoffs are classified as natural channels (Figure 2.16).

Not all meandering channels migrate

The meandering model presented above assumes that lateral migration is continuously active and that point-bar accretion is an important component of floodplain construction. Some highly sinuous channels, however, appear to have little to no active bank erosion. Examples include streams on the intermontane "parks" of the Rocky Mountains (Figure 2.17), channels on the lower coastal plain of the southeastern United States, low-gradient channels in England (Ferguson, 1981, 1987), and possibly meandering tidal channels (Ikeda, 1989). Channel pattern changes over scales of decades are slight to unnoticeable, the floodplains exhibit few scars from past cutoffs, and subaerial point bars are poorly developed or absent (inside banks generally are steep and well vegetated). These floodplains and channels share the following characteristics: (1) low supply rate of bedload; (2) a floodplain that is low and wide compared to channel width; (3) cohesive, strongly vegetated banks and floodplain surface; (4) low valley gradient; and, possibly, (5) a low frequency of large floods. These characteristics suggest that such channels develop from an initially nearly

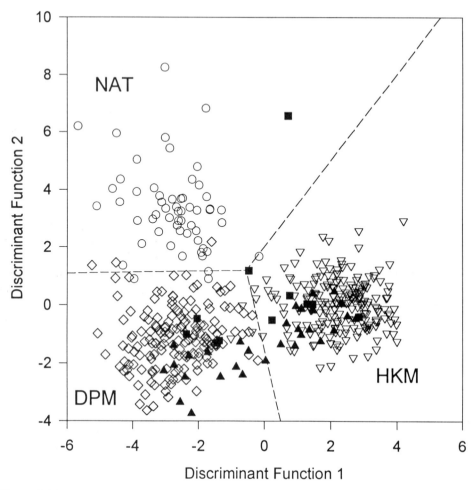

Figure 2.16 Classification of natural and simulated streams using two multivariate discriminant functions. Open circles are natural stream channels (domain classified as natural channels labelled NAT); inverted open triangles are simulated by the model of Howard and Knutson (1984) and Howard (1992) (domain labelled HKM); open diamonds are disturbed periodic simulations (Ferguson, 1976, 1977) (domain labelled DPM). Filled triangles are simulated channels with chute cutoffs, and filled boxes are simulations with resistant valley walls or resistant oxbow lakes

straight channel in which bankfull flows are barely competent to induce systematic bank erosion, but as sinuosity increases the decrease in flow velocity reduces the bank erosion rate to nearly zero before sinuosity reaches a value sufficient to induce cutoffs. During flood flows the low banks and wide floodplain limit within-channel flows. In addition, the downstream flow over the floodplain may interfere with flow in the channel, particularly since the overbank flow is largely perpendicular to the meander arms. Jarrett and Costa (1986) note that a dam break flood passing over the floodplain shown in Figure 2.17 had little effect upon channel pattern and bank erosion. The

Figure 2.17 Fall Creek, Estes Park, Colorado. Note the highly sinuous channel pattern with few oxbow lakes. Flow from bottom left to upper centre

slowness of lateral shifting may either be due to high bank resistance (constraint 3, above) or to a low supply of sediment to construct point bars (constraint 1), or both. An examination of the channel in Figure 2.17 reveals that banks occasionally fail by undermining and toppling, but the collapsed bank remains in the channel bed, anchored by the still-living grass, and apparently surviving with minimal break-up and erosion for several years (Deborah Anthony, personal communication). Low point bars form in these channels during high flows, but are eroded as flows diminish (Anthony, 1991). Lateral shifting in the channel system may have increased in recent years due to increased bed sediment supply originating from a debris fan constructed by the 1982 dam Break flood (Deborah Anthony, personal communication), suggesting that the ability to construct point bars may play a role in limiting bank erosion rates. Floodplains constructed by such relatively fixed meandering channels presumably occur primarily by vertical accretion, at least during their final stages.

Proposed model enhancements

Howard (1992) discussed a number of shortcomings of the meandering floodplain evolutionary model and strategies for overcoming these, as well as possible model enhancements. Only a synopsis is considered here.

Howard (1992) and the present contribution utilize the linearized J&P (Johannesson and Parker, 1989) flow and bed topography model. This model is utilized primarily because of its computational efficiency, which is of primary concern when flow and bed topography must be recalculated thousands of times per simulation. However, the J&P model considers only time-averaged, curvature-forced bedforms. In wider meandering channels alternate bars may migrate through the channel (Kinoshlta, 1961; Fukuoka, 1989; Ikeda, 1989; Tubino and Seminara, 1990; Whiting and Dietrich, 1993a,b,c) and these bars may systematically affect bank erosion (Whiting and Dietrich, 1993c). For certain combinations of width/depth ratio and flow parameters, alternate bars become stationary, and, if their natural wavelength is the same as the meander wavelength, a "resonance" occurs under which conditions the linearized models, such as J&P, predict very large amplitude bars (Blondeaux and Seminara, 1985; Columbini et al., 1987; Parker and Johannesson, 1989; Seminara and Tubino, 1989; Tubino and Seminara, 1990; Columbini et al., 1992). The J&P model fails to converge to a stable solution under such conditions, although this might be correctable In a future version (G. Parker, personal communication). Furthermore, multiple bars may occur in large-amplitude meander bends (Whiting and Dietrich, 1993a,b), and the migration of alternate bars can be suppressed in sinuous meanders (Kinoshita, 1961; Fukuoka, 1989; Tubino and Seminara, 1990, Whiting and Dietrich, 1993c), both of which might affect near-bank flow and depth (and hence bank erosion) in systematic ways. The effects of alternate bars could be addressed by use of a flow and bed topography model that explicitly treats cross-stream topography and which has time-steps sufficiently short to resolve alternate bar migration (e.g. Nelson and Smith, 1989a,b; Shimizu and Itakura, 1989), but computational costs may prohibit incorporation into long-term evolutionary models. An alternative approach is to use the more explicit models to develop heuristic "corrections" to linearized models (such as J&P, 1989, and Odgaard, 1989a,b).

Bank erosion rates are assumed to respond to a dominant discharge. Discharge variation affects not only the magnitude of the velocity and depth perturbations, but also their distribution around the bend (lags in response to curvature changes become greater at higher discharges), as well as influencing the development and migration of bars. A frequency distribution of flow discharges could be incorporated into the model.

Equations 2.9 and 2.10 are based upon a simple diffusional conceptualization of floodplain deposition as embodied in the approach of Carey (1969) and Pizzuto (1987). However, as Pizzuto (1987) points out, advectional flow transport can lead to patterns of deposition rates and grain sizes not describable by the above parameters. This is particularly important for flows in chutes and sloughs, where both suspended load advection and bedload transport may occur. Techniques for modelling of overbank flows have been developed (Knight and Demetriou, 1983; Yen and Yen, 1984; Ervine and Ellis, 1987; Knight, 1989; Gee et al., 1990; Miller, 1994; Bates et al., 1992; Knight and Shiono, Chapter 5, this volume; Sellin and Willetts, Chapter 8, this volume; Younis, Chapter 9, this volume), and these can be coupled with models for suspended load deposition on floodplains (Carey, 1969; Pizzuto, 1987; Marriott, 1992, Chapter 3, this volume) to predict the spatial pattern and grain size distribution of deposited sediment (Nicholas and Walling, 1995). A frequency distribution of flood flows could be assumed, with deposition rates being based on a steady-state solution of floodplain flow for the flood peak.

An obvious extension would be to include grain size and stratigraphic information by modelling deposit thickness, bedform and grain size. Grain size and bedforms of point-bar deposits can be related to within-channel sorting and flow parameters much in the manner adopted by Allen (1970), Bridge (1975, 1978, 1984a) and Bridge and Jarvis (1977, 1982).

2.5   ANASTOMOSING AND AVULSIVE FLOODPLAINS

Very-low-gradient channel systems are commonly straight to weakly meandering with well-separated multiple channels, or anastomoses (Knighton and Nanson, 1993). Low gradients are associated with low bedload conveyance, tectonic backtilting (Burnett and Schumm, 1983; Gregory and Schumm, 1987), blocked valleys (Smith and Smith, 1980), and, usually, an aggradational regime (Smith *et al.*, 1989; Knighton and Nanson, 1993). Channel meandering is weak due to low stream power coupled with resistant banks (cohesive and/or strongly vegetated) (Harwood and Brown, 1993; Knighton and Nanson, 1993; Stanistreet *et al.*, 1993). Due to the aggradational regime, natural levees are usually well developed. Anastomoses generally develop due to avulsion at breaches in the natural levees. Also due to the aggradational regime (or, possibly, floodplain sinking), non-active portions of the floodplain are commonly swampy or occupied by shallow lakes. Splay deposits associated with multiple small distributaries are a common occurrence (Smith and Putnam, 1980; Smith and Perez-Arlucea, 1994).

Avulsions are not limited to the classic low-gradient anastomosed floodplains. Richards *et al.* (1993) note that avulsions may even occur in braided channel systems.

Geomorphic modelling of anastomosed or avulsive floodplains apparently has not been attempted, except for large-scale stratigraphic models of sedimentary facies (Allen, 1978; Bridge and Leeder, 1979). Some of the modelling techniques utilized for meandering channels, described above, could be adapted to anastomosed floodplains. Although a variety of processes and constraints may be involved in avulsions (see, for example, Richards *et al.*, 1993), most are strongly influenced by floodplain topography. Thus the development of avulsions could be modelled stochastically, similar to the approach used for chute cutoffs, with the probability of a diversion being a function of levee height and relative channel and floodplain elevations (e.g. Mohrig *et al.*, 1994).

2.6   LARGE SPATIAL DOMAIN MODELS

In the previous discussion the focus has been on models that attempt to simulate the history of individual channels and details of floodplain topography, and in some cases, floodplain stratigraphy. If the concern is for larger spatial and temporal domains, such as valley aggradation and entrenchment, or the overall development of deltas, alluvial fans or pediments, then faithful reproduction of channel morphology, channel pattern changes and detailed stratigraphy may be abandoned in order to efficiently simulate regional morphological and stratigraphical relationships. Models of sedimentary basin development (e.g. Flemings and Jordan, 1989; Paola, 1989; Jordan and Flemings, 1990; Paola *et al.*, 1992; Slingerland *et al.*, 1994) have long used simplified approaches generally ignoring effects of individual channels and their deposits, or, in some cases, representing deposits of individual channels only in their bulk as channel complexes (e.g. Leeder, 1978; Bridge and Leeder, 1979). The present discussion will concern

Figure 2.18 Simulation of drainage basin erosion and concomitant deltaic deposition using the simulation model of Howard (1994). Sea level is at an elevation of 0.0

spatially explicit modelling of fluvial depositional environments in which the topography of the deposit as well as, possibly, the stratigraphy are of interest.

In depositional environments, particularly deltas and alluvial fans, channel aggradation, possibly coupled with land surface lowering by sediment consolidation, results in frequent avulsions. In addition, the channel system commonly has multiple active channels as anabranches or distributaries. Rachocki (1981) simulated braided channels on alluvial fans with a random-walk model. However, this model was not coupled with deposition. Price (1974) simulated alluvial fan deposition with a probabilistic model for uplift, precipitation and fan deposition. In this model flow is routed down the fan, following a single course during each event, with the relative probabilities among possible directions at each node representing the fan surface being proportional to the slope gradients to adjacent nodes. Howard (1994) abstracts the depositional process further in a coupled model of basin erosion and sedimentation that simulates development of alluvial fans and alluvial pediments. Figure 2.18 shows an example of such a coupled model with delta development at a presumed shoreline. This model does not account for suspended load deposition as bottomset deposits, so that it is most applicable to the bedload-dominated Gilbert type of delta (Gilbert, 1890). The models of Howard (1994) utilize an approach to sediment routing and deposition that is computationally three orders of magnitude more efficient than finite difference modelling. In these models of alluvial fans, pediments and deltas, channels during each iteration follow the steepest path downstream across the alluvial cover. The deposition during each iteration is not intended to represent an actual flow and sedimentation event, but rather, the cumulative effect of multiple events. In aggradational settings the sedimentation occurring during each iteration tends to raise the current channel relative to older sections of the fan or delta, so that during the next iteration the steepest path follows a different course. In the Howard (1994) model the channels shift so frequently that the resulting landform is very symmetrical and often strongly affected by the restrictions of flow from any cell to one of eight directions (Figure 2.16). A more realistic simulation and presumably also a more realistic morphology (including, perhaps, birdfoot deltas) would occur if channel shifting was made more difficult, corresponding to the development of natural levees that constrain avulsions until the channel bed is well aggraded. Rules for such avulsions were discussed above. As with the floodplain meandering model described above, only the net quantity of sediment deposition is accounted for, and no attempt has been made to model sediment size or stratigraphy, although such features could be added.

2.7 DISCUSSION

Modelling of the geomorphic and stratigraphic development of floodplains is presently more of a promise than a reality, but most of the quantitative tools are at hand and computers are sufficiently capable that model development should mature rapidly. These concluding remarks address issues of model development and validation.

In the early days of the quantitative revolution in geomorphology, characterization of landform morphology and geomorphic processes revealed numerous intriguing relationships and patterns that exceeded our ability to explain. The rapid development of theory and computational resources during the last two decades has often reversed this situation, such that validation of models is handicapped by the lack of appropriate empirical data. Sometimes this is unavoidable because of large spatial or temporal scales, as with models of drainage basin evolution. However, in the case of modelling of floodplain evolution, changes occur rapidly enough that short-term observation coupled with longer-term evidence from surveys, aerial photographs and floodplain stratigraphy can be utilized for model calibration and validation. Even so, there are few systematic databases on floodplain evolution that can be directly utilized in model validation.

Howard (1992) discussed a variety of validation approaches within the context of meandering floodplain evolution. Although many studies of stream meandering and braiding have collected information from individual short reaches, validation of large-scale models requires "reach-length" studies, extending downstream through multiple braids or meander bends. One such approach utilizes static comparison of simulated floodplain features with natural analogues, generally using multivariate statistical methods. For example, Howard and Hemberger (1991) showed that meander planforms simulated by the Howard (1992) model and the Ferguson (1976, 1977) disturbed periodic model are statistically distinguishable from natural meanders. Similar multivariate morphometric measurements have been applied to braided channels by Howard *et al.* (1970) and Webb (1994, 1995), and these could be used for validation of the braided channel model of Murray and Paola (1994).

A more powerful validation technique compares the temporal changes, or kinematics, predicted by a model with observed natural changes, thereby offering important clues as to specific sources of model deficiencies. Unfortunately, although a rich database of historical meander change exists, analyses of meander kinematics to date have been primarily qualitative or have yielded only summary statistics, such as average bank erosion rates (e.g. Brice, 1974a,b; Hooke, 1977, 1979, 1980, 1984; Dort, 1978). Studies of systematic variation in erosion rates with channel curvature and sediment properties utilized isolated bends (Hickin and Nanson, 1975, 1984; Nanson and Hickin, 1983, 1986) and have not fully accounted for upstream control of local flow and bed characteristics implied in theoretical models (such as the J&P model used here) (Parker, 1984; Furbish, 1988). Howard and Knutson (1984) have used an earlier version of the flow model to simulate several decades of channel shifting on the White River of Indiana, with generally encouraging results. Short-term predictions of channel shifting have also been made by Parker (1982), Beck (1984), and Beck *et al.* (1984). The flow and transport model has been compared to flume studies of meandering channels (Johannesson and Parker, 1989). Furbish (1988), Hasegawa (1989a,b), Pizzuto and Meckelnberg (1989), and Shane Cherry (personal communication, 1995) have compared observed bank erosion for individual bends or short reaches with model predictions. Although these comparisons are generally encouraging, they are too few and too rudimentary to comprise a thorough testing of the flow and erosion model. What is needed is systematic analysis of meander kinematics on a number of long reaches of natural and simulated channels and a comparison with model predictions. Howard (1992) outlines several direct and indirect techniques of reach-length comparison between historical records of change in natural meandering channels with simulated changes. Similar techniques could be utilized for braided channels, although highly braided channels change so rapidly that year-to-year aerial photographic comparisons may not be useful.

By extending the types of observation made by Lewis and Lewin (1983), spatio-temporal information on neck and chute cutoffs and avulsions, together with information on floodplain topography and channel geometry could be used to quantify models of channel pattern changes in meandering and braided streams.

The validation of floodplain deposition models will prove troublesome because relevant data are difficult to obtain. Most easily assembled are statistics relating floodplain age, elevation and distance from stream channels (e.g. Everitt, 1968; Kesel *et al.*, 1974; Hickin and Nanson, 1975; Nanson, 1980). Floodplain sedimentology and stratigraphy is more difficult to characterize. Cores and trenches are the obvious but tedious method, supplemented with archaeological and age-dating techniques to provide rate information (e.g. Brakenridge, 1984, 1985; Brown 1987, 1990; Walling and Bradley, 1989). Judicious use of sections exposed in cutbanks can also be useful. Ancient deposits in the sedimentary record serve as comparisons. Relatively short duration studies of rates and spatial patterns of deposition and erosion are quite practical for reach-length studies, using surveying, coring, and use of markers such as sand or gypsum (e.g. Hupp and Bazemore, 1993) or isotopic markers (e.g. Walling *et al.*, 1992).

Geomorphologists developing theoretical models often find that appropriate empirical data are lacking. Sometimes this occurs because relevant data are difficult to obtain, but often geomorphologists have not appreciated the relevance of particular types of data or that data that have been collected are in the wrong form for model validation purposes. Hopefully, as it has in other branches of geomorphology, development of theoretical models will spur new field or experimental studies to acquire appropriate data sets on floodplain evolution.

REFERENCES

Alexander, J., Bridge, J.S., Leeder, M.R., Collier, R.E. and Gawthorpe, R.L. (1994) Holocene meander-belt evolution of an active extensional basin, southwestern Montana. *Journal of Sedimentary Research*, **B64**, 542–559.

Allen, J.R.L. (1970) A quantitative model of grain size and sedimentary structures in lateral deposits. *Geological Journal*, **7**, 129–146.

Allen, J.R.L. (1978) Studies in fluviatile sedimentation: an exploratory quantitative model for the architecture of avulsion-controlled alluvial suites. *Sedimentary Geology*, **21**, 129–147.

Allen, J.R.L. (1982) Free meandering channels and lateral deposits. In: *Sedimentary Structures: Their Character and Physical Basis*, Vol. 2, New York, Elsevier, 53–100.

Allen, J.R.L. (1990) Salt-marsh growth and stratification: a numerical model with special reference to Severn Estuary, southwest Britain. *Marine Geology*, **95**, 77–96.

Allen, J.R.L. (1992) Large-scale textural patterns and sedimentary processes on tidal salt marshes in the Severn Estuary, southwest Britain. *Sedimentary Geology*, **81**, 299–318.

Anderson, M.G. and Calver, A. (1977) On the persistence of landscape features formed by a large flood. *Institute of British Geographers Transactions*, N.S., **2**, 243–254.

Anderson, M.G. and Calver, A. (1980) Channel plan changes following large floods. In: *Timescales in Geomorphology*, R.A. Cullingford, D.A. Davidson and J. Lewin (eds), Chichester, John Wiley & Sons, 43–52.

Anthony, D.J. (1991) Stage-dependent cross-section adjustments in a meandering reach of Fall River, Colorado. *Geomorphology*, **4**, 187–203.

Ashmore, P.E. (1982) Laboratory modelling of gravel braided stream morphology. *Earth Surface Processes*, **7**, 201–225.

Ashmore, P.E. (1991) How do gravel-bed rivers braid?. *Canadian Journal of Earth Science*, **28**, 326–341.

Ashmore, P.E. (1993) Anabranch confluence kinetics and sedimentation processes in gravel-braided streams. In: *Braided Rivers*, J.L. Best and C.S. Bristow (eds), Geological Society Special Publication 75, 129–146.

Baker, V.R. (1977) Stream channel response to floods with examples from central Texas. *Geological Society of America Bulletin*, **88**, 1057–1071.

Baker, V.R. (1988) Flood erosion. In: *Flood Geomorphology*, V.R. Baker, R.C. Kochel and P.C. Patton (eds), New York, John Wiley & Sons, 81–95.

Baker, V.R. and Kochel, R.C. (1988) Flood sedimentation in bedrock fluvial systems. In: *Flood Geomorphology*, V.R. Baker, R.C. Kochel and P.C. Patton (eds), New York, John Wiley & Sons, 123–138.

Bates, P.D., Anderson, M.G., Baird, L., Walling, D.E. and Simm, D. (1992) Modelling floodplain flows using a two-dimensional finite element model. *Earth Surface Processes and Landforms*, **17**, 575–588.

Bauer, B.O. and Schmidt, J.C. (1993) Waves and sandbar erosion in the Grand Canyon: applying coastal theory to a fluvial system. *Annals of the Association of American Geographers*, **83**, 475–497.

Beck, S. (1984) Mathematical modeling of meander interaction. In: *River Meandering*, C.M. Elliott (ed.), New York, American Society of Civil Engineers, 932–941.

Beck, S., Mefli, D.A. and Yalamanchili, K. (1984) Lateral migration of the Genessee River, New York. In: *River Meandering*, C.M. Elliott (ed.), New York, American Society of Civil Engineers, 510–517.

Biedenharn, D.S., Combs, P.G., Hill, G.J., Pinkard, C.F. and Pinkstone, C.B. (1989) Relationship between channel migration and radius of curvature in the Red River. In: *Sediment Transport Modeling*, S.S.Y. Wang (ed.), New York, American Society of Civil Engineers, 536–541.

Blondeaux, P. and Seminara, G. (1985) A unified bar-bend theory of river meanders. *Journal of Fluid Mechanics*, **157**, 449–470.

Brakenridge, G.R. (1984) Alluvial stratigraphy and radiocarbon dating along the Duck River, Tennessee: implications regarding flood-plain origin. *Geological Society of America Bulletin*, **95**, 9–25.

Brakenridge, G.R. (1985) Rate estimates for lateral bedrock erosion based on radiocarbon ages, Duck River, Tennessee. *Geology*, **13**, 111–114.

Brice, J.C. (1964) Channel patterns and terraces of the Loup Rivers in Nebraska. *US Geological Survey Professional Paper* 422-D, 41 pp.

Brice, J.C. (1974a) Meandering pattern of the White River in Indiana – an analysis. In: *Fluvial Geomorphology*, M. Morisawa (ed.), Binghamton, State University of New York, 178–200.

Brice, J.C. (1974b) Evolution of meander loops. *Geological Society of America Bulletin*, **85**, 581–586.

Bridge, J.S. (1975) Computer simulation of sedimentation in meandering streams. *Sedimentology*, **22**, 3–43.

Bridge, J.S. (1978) Palaeohydraulic interpretation using mathematical models of contemporary flow and sedimentation in meandering channels. In: *Fluvial Sedimentology*, A.D. Miall (ed.), Calgary, Canada, Canadian Society of Petroleum Geologists, 723–742.

Bridge, J.S. (1984a) Flow and sedimentary processes in river bends: comparison of field observations and theory. In: *River Meandering*, C.M. Elliott (ed.), New York, American Society of Civil Engineers, 857–872.

Bridge, J.S. (1993) The interaction between channel geometry, water flow, sediment transport and deposition in braided rivers. In: *Braided Rivers*, J.L. Best and C.S. Bristow (eds), Geological Society Special Publication 75, 13–71.

Bridge, J.S. and Jarvis, J. (1977) Velocity profiles and bed shear stress over various bed configurations in a river bend. *Earth Surface Processes*, **2**, 281–294.

Bridge, J.S. and Jarvis, J. (1982) The dynamics of a river bend: a study of flow and sedimentary processes. *Sedimentology*, **29**, 499–541.

Bridge, J.S. and Leeder, M.R. (1979) A simulation model of alluvial stratigraphy. *Sedimentology*, **26**, 617–644.

Bristow, C.S. and Best, J.L. (1993) Braided rivers: perspectives and problems. In: *Braided Rivers*, J.L. Best and C.S. Bristow (eds), Geological Society Special Publication 75, 1–11.

Brown, A.G. (1987) Holocene floodplain sedimentation and channel response of the lower River Severn, United Kingdom. *Zeitschrift für Geomorphologie*, **31**, 293–310.

Brown, A.G. (1990) Holocene floodplain diachronism and inherited downstream variations in fluvial processes: a study of the river Perry, Shropshire, England. *Journal of Quaternary Science*, **5**, 39–51.

Burkham, D.E. (1972) Channel changes of the Gila River in Safford Valley, Arizona 1846–1970. *US Geological Survey Professional Paper* 655-G, 24 pp.

Burnett, A.W. and Schumm, S.A. (1983) Alluvial river response to neotectonic deformation in Louisiana and Mississippi. *Science*, **222**, 49–50.

Callander, R.A. (1978) River meandering. *Annual Review of Fluid Mechanics*, **10**, 129–158.

Cant, D.J. and Walker, R.G. (1978) Fluvial processes and facies sequences in the sandy braided South Saskatchewan River, Canada. *Sedimentology*, **25**, 625–648.

Carey, W.C. (1969) Formation of flood plain lands. *Journal of the Hydraulics Division, Proceedings of the American Society of Civil Engineers*, **95**, 981–994.

Carson, M.A. and Lapointe, M.F. (1983) The inherent asymmetry of river meanders. *Journal of Geology*, **91**, 41–55.

Church, M. and Jones, D. (1982) Channel bars in gravel bed rivers. In: *Gravel Bed Rivers*, R.D. Hey, H.C. Bathhurst and C.R. Thorne (eds), Chichester, John Wiley & Sons, 291–338.

Colombini, M., Seminara, G. and Tubino, M. (1987) Finite-amplitude alternate bars. *Journal of Fluid Mechanics*, **181**, 213–232.

Columbini, M., Tubino, M. and Whiting, P.J. (1992) Topographic expression of bars in meandering channels. In: *Dynamics of Gravel Bed Rivers*, P. Billi, R.D. Hey, C.R. Thorne and P. Tacconi (eds), New York, John Wiley & Sons, 457–474.

Costa, J.E. (1974) Response and recovery of a Piedmont watershed from tropical storm Agnes, June 1972. *Water Resources Research*, **10**, 106–112.

Costa, J.E. and O'Connor, J.E. (1995) Geomorphically effective floods. In: *Natural and Anthropogenic Influences in Fluvial Geomorphology*, J.E. Costa *et al.* (eds), American Geophysical Union, Geophysical Monograph 89, 45–56.

Dolan, R., Howard, A. and Trimble, D. (1978) Structural control of the rapids and pools of the Colorado River in the Grand Canyon. *Science*, **202**, 629–631.

Dort, W., Jr (1978) *Channel Migration Investigation: Historic Channel Change Maps*. Kansas City District, US Army Corps of Engineers, 50 pp.

Dury, G.H. (1954) Contribution to a general theory of meandering valleys. *American Journal of Science*, **252**, 193–224.

Dury, G.H. (1964a) Principles of underfit streams. *US Geological Survey Professional Paper* 452-A, 67 pp.

Dury, G.H. (1964b) Theoretical implications of underfit streams. *US Geological Survey Professional Paper* 452-C, 43 pp.

Engelund, F. and Skovgaard, O. (1973) On the origin of meandering and braiding in alluvial streams. *Journal of Fluid Mechanics*, **57**, 289–302.

Ervine, D.A. and Ellis, J. (1987) Experimental and computational aspects of overbank floodplain flow. *Transactions of the Royal Society of Edinburgh: Earth Sciences*, **78**, 315–325.

Everitt, B.L. (1968) Use of the cottonwood in an investigation of the recent history of a flood plain. *American Journal of Science*, **266**, 417–439.

Ferguson, R.I. (1976) Disturbed periodic model for river meanders. *Earth Surface Processes*, **1**, 337–347.

Ferguson, R.I. (1977) Meander migration: equilibrium and change. In: *River Channel Changes*, K.J. Gregory (ed.), Chichester, John Wiley & Sons, 235–263.

Ferguson, R.I. (1981) Channel form and channel changes. In: *British Rivers*, J. Lewin (ed.), London, Allen & Unwin, 90–125.

Ferguson, R.I. (1984) The threshold between meandering and braiding. In: *Proceedings of the 1st International Conference on Hydraulic Design*, K.V.H. Smith (ed.), Berlin, Springer Verlag, 6.15–6.29.

Ferguson, R. (1987) Hydraulic and sedimentary controls on channel pattern. In: *River Channels: Environment and Process*, K. Richards (ed.), Oxford, Basil Blackwell, 129–158.

Ferguson, R. (1993) Understanding braiding processes in gravel-bed rivers: progress and unsolved problems. In: *Braided Rivers*, J.L. Best and C.S. Bristow (eds), Geological Society Special Publication 75, 73–87.

Fisk, H.N. (1944) *Geological Investigations of the Alluvial Valley of the Lower Mississippi River*. Vicksburg, Mississippi River Commission, 78 pp.

Fisk, H.N. (1947) *Fine-Grained Alluvial Deposits and their effect on Mississippi River Activity*. Vicksburg, Waterways Experiment Station, US Army Corps of Engineers, 82 pp.

Flemings, P.B. and Jordan, T.E. (1989) A synthetic stratigraphic model of foreland basin development. *Journal of Geophysical Research*, **94**, 3851–3866.

Fredsoe, J. (1978) Meandering and braiding of rivers. *Journal of Fluid Mechanics*, **84**, 609–624.

French, J.R. (1993) Numerical simulation of vertical marsh growth and adjustment to accelerated sea-level rise, North Norfolk, U.K. *Earth Surface Processes and Landforms*, **18**, 63–81.

French, J.R. and Spencer, T. (1993) Dynamics and sedimentation in a tide-dominated backbarrier salt marsh, Norfolk, U.K. *Marine Geology*, **110**, 315–331.

Friedkin, J.F. (1945) *A Laboratory Study of the Meandering of Alluvial Rivers*. Vicksburg, Mississippi River Commission, Waterways Experiment Station, US Army Corps of Engineers, 40 pp.

Friend, P.F. and Sinha, R. (1993) Braiding and meandering parameters. In: *Braided Rivers*, J.L. Best and C.S. Bristow (eds), Geological Society Special Publication, 75, 105–111.

Fujuta, Y. (1989) Bar and channel formation in braided streams. In: *River Meandering*, S. Ikeda and G. Parker (eds), Washington, DC, Water Resources Monograph 12, American Geophysical Union, 417–462.

Fukuoka, S. (1989) Finite amplitude development of alternate bars. In: *River Meandering*, S. Ikeda and G. Parker (eds), Washington, DC, Water Resources Monograph 12, American Geophysical Union, 237–266.

Furbish, D.J. (1988) River-bed curvature and migration: how are they related? *Geology*, **16**, 752–755.

Gee, D.M., Anderson, M.G. and Baird, L. (1990) Large-scale floodplain modeling. *Earth Surface Processes and Landforms*, **15**, 513–523.

Germanoski, D. and Schumm, S.A. (1993) Changes in braided river morphology resulting from aggradation and degradation. *Journal of Geology*, **101**, 451–466.

Gilbert, G.K. (1890) *Lake Bonneville*. US Geological Survey Monograph 2, 438 pp.

Graf, W.L. (1979) Rapids in canyon rivers. *Journal of Geology*, **87**, 533–551.

Graf, W.L. (1983) Flood-related channel change in an arid-region river. *Earth Surface Processes and Landforms*, **8**, 125–139.

Gregory, D.I. and Schumm, S.A. (1987) The effect of active tectonics on alluvial river morphology. In: *River Channels: Environment and Process*, K. Richards (ed.), Oxford, Basil Blackwell, 41–68.

Gupta, A. and Fox, H. (1974) Effects of high-magnitude floods on channel form. A case study in the Maryland Piedmont. *Water Resources Research*, **10**, 499–509.

Hack, J.T. and Goodlett, J.C. (1960) Geomorphology and forest ecology of a mountain region in the central Appalachians. *US Geological Survey Professional Paper* 347, 66 pp.

Harper, J.M. (1991) Geomorphic floodplain modification due to an extreme flood, South Branch Potomac River basin, West Virginia. Unpublished MS thesis, Charlottesville, University of Virginia, 254 pp.

Harwood, K. and Brown, A.G. (1993) Fluvial processes in a forested anastomosing river: flood partitioning and changing flow patterns. *Earth Surface Processes and Landforms*, **18**, 741–748.

Hasegawa, K. (1989a) Studies on qualitative and quantitative prediction of meander channel shift. In: *River Meandering*, S. Ikeda and G. Parker (eds), Washington, DC, Water Resources Monograph 12, American Geophysical Union, 215–236.

Hasegawa, K. (1989b) Universal bank erosion coefficient for meandering rivers. *Journal of Hydraulic Engineering*, **115**, 744–765.

Hickin, E.J. (1979) Concave-bank benches in the Squamish River, British Columbia, Canada. *Canadian Journal of Earth Science*, **16**, 200–203.

Hickin, E.J. and Nanson, G.C. (1975) Character of channel migration on the Beatton River, northwest British Columbia, Canada. *Geological Society of America Bulletin*, **86**, 487–494.

Hickin, E.J. and Nanson, G.C. (1984) Lateral migration rates of river bends. *Journal of Hydraulic Engineering*, **110**, 1557–1567.

Hoey, T.B. and Sutherland, A.J. (1991) Channel morphology and bedload pulses in braided rivers: a laboratory study. *Earth Surface Processes and Landforms*, **16**, 447–462.

Hooke, J.M. (1977) The distribution and nature of changes in river channel patterns: the example of Devon. In: *River Channel Changes*, K.J. Gregory (ed.), Chichester, John Wiley & Sons, 265–280.

Hooke, J.M. (1979) An analysis of the processes of river bank erosion. *Journal of Hydrology*, **42**, 39–62.

Hooke, J.M. (1980) Magnitude and distribution of rates of river bank erosion. *Earth Surface Processes*, **5**, 143–157.

Hooke, J.M. (1984) Meander behavior in relation to slope characteristics. In: *River Meandering*, C.M. Elliott (ed.), New York, American Society of Civil Engineers, 67–76.

Hooke, J.M. (1987) Changes in meander morphology. In: *International Geomorphology, 1986*, V. Gardiner (ed.), Chichester, John Wiley & Sons, Part I, 591–609.

Howard, A.D. (1984) Simulation model of meandering. In: *River Meandering*, C.M. Elliott (ed.), New York, American Society of Civil Engineers, 952–963.

Howard, A.D. (1992) Modelling channel migration and floodplain development in meandering streams. In: *Lowland Floodplain Rivers*, P.A. Carling and G.E. Petts (eds), Chichester, John Wiley & Sons, 1–42.

Howard, A.D. (1994) A detachment-limited model of drainage basin development. *Water Resources Research*, **30**, 2261–2285.

Howard, A.D. and Dolan, R. (1981) Geomorphology of the Colorado River in the Grand Canyon. *Journal of Geology*, **89**, 269–298.

Howard, A.D. and Hemberger, A.T. (1991) Multivariate characterization of meandering. *Geomorphology*, **4**, 161–186.

Howard, A.D. and Knutson, T.R. (1984) Sufficient conditions for river meandering: a simulation approach. *Water Resources Research*, **20**, 1659–1667.

Howard, A.D., Keetch, M.E. and Vincent, L. (1970) Topological and geometrical properties of braided streams. *Water Resources Research*, **6**, 1674–1688.

Hupp, C.R. and Bazemore, D.E. (1993) Temporal and spatial patterns of wetland sedimentation, West Tennessee. *Journal of Hydrology*, **141**, 179–196.

Ikeda, H. (1989) Sedimentary controls on channel migration and origin of point bars in sand-bedded meandering rivers. In: *River Meandering*, S. Ikeda and G. Parker (eds), Washington, DC, Water Resources Monograph 12, American Geophysical Union, 51–68.

Ikeda, S., Parker, G. and Sawai, K. (1981) Bend theory of river meanders, 1, linear development. *Journal of Fluid Mechanics*, **112**, 363–377.

Jacobson, R.B., McGeehin, J.P., Cron, E.D., Carr, C.E., Harper, J.M. and Howard, A.D. (1993) Landslides triggered by the storm of November 3–5, 1985, Wills Mountain Anticline, West Virginia and Virginia. *US Geological Survey Professional Paper* 1981-C, 33 pp.

Jarrett, R.D. and Costa, J.E. (1986) Hydrology, geomorphology and dam-break modeling of the July 15, 1982 Lawn Lake Dam and Cascade Lake Dam failures, Larimer County, Colorado. *US Geological Survey Professional Paper* 1369, 78 pp.

Johannesson, J. and Parker, G. (1989) Linear theory of river meanders. In: *River Meandering*, S.

Ikeda and G. Parker (eds), Washington DC, Water Resources Monograph 12, American Geophysical Union, 181–214.

Jordan, T.E. and Flemings, P.B. (1990) From geodynamic models to basin fill – a stratigraphic perspective. In: *Quantitative Dynamic Stratigraphy*, T.A. Cross (ed.), Englewood Cliffs, New Jersey, Prentice-Hall, 149–163.

Kesel, R.H., Dunne, K.C., McDonald, R.C. Allison, K.R. and Spicer, B.E. (1974) Lateral erosion and overbank deposition on the Mississippi River in Louisiana caused by 1973 flooding. *Geology*, **2**, 461–464.

Kieffer, S.W. (1985) The 1983 hydraulic jump in Crystal Rapid: implications for river-running and geomorphic evolution in the Grand Canyon. *Journal of Geology*, **93**, 385–406.

Kinoshita, R. (1961) *Investigation of Channel Deformation in Ishikari River*. Report of Bureau of Resources, Department of Science and Technology, Japan (in Japanese), 174 pp.

Knight, D.W. (1989) Hydraulics of flood channels. In: *Floods: Hydrological, Sedimentological and Geomorphological Implications*, K. Beven and P. Carling (eds), Chichester, John Wiley & Sons, 83–105.

Knight, D.W. and Demetriou, J.D. (1983) Flood plain and main channel flow interaction. *Journal of the Hydraulics Division, Proceedings of the American Society of Civil Engineers*, **109**, 1073–1082.

Knighton, A.D. and Nanson, G.C. (1993) Anastomosis and the continuum of channel patterns. *Earth Surface Processes and Landforms*, **18**, 613–625.

Kochel, R.C. (1988) Geomorphic impact of large floods: review and new perspectives on magnitude and frequency. In: *Flood Geomorphology*, V.R. Baker, R.C. Kochel and P.C. Patton (eds), New York, John Elley & Sons, 169–187.

Kochel, R.C. and Baker, V.R. (1988) Paleoflood analysis using slackwater deposits. In: *Flood Geomorphology*, V.R. Baker, R.C. Kochel and P.C. Patton (eds), New York, John Wiley & Sons, 357–376.

Kochel, R.C., Baker, V.R. and Patton, P.C. (1982) Paleo-hydrology of southwestern Texas. *Water Resources Research*, **18**, 1165–1183.

Krumbein, W.C. and Orme, A.R. (1972) Field mapping and computer simulation of braided stream networks. *Geological Society of America Bulletin*, **83**, 3369–3380.

Laury, R.L. (1971) Stream bank failure and rotational slumping: preservation and significance in the geologic record. *Geological Society of America Bulletin*, **82**, 1251–1266.

Leddy, J.O., Ashworth, P.J. and Best, J.L. (1993) Mechanisms of anabranch avulsion within gravel-bed braided rivers: observations from a scaled physical model. In: *Braided Rivers*, J.L. Best and C.S. Bristow (eds), Geological Society Special Publication 75, 119–127.

Leeder, M.R. (1978) A quantitative stratigraphic model for alluvium, with special reference to channel deposit density and interconnectedness. In: *Fluvial Sedimentology*, A.D. Miall (ed.), Calgary, Canadian Society of Petroleum Geologists, 587–596.

Leeder, M.R. and Alexander, J. (1987) The origin and tectonic significance of asymmetrical meander-belts. *Sedimentology*, **34**, 217–226.

Leopold, L.B. and Wolman, M.G. (1957) River channel patterns – braided, meandering, and straight. *US Geological Survey Professional Paper* 282-B.

Lewin, J. (1976) Initiation of bed forms and meanders in coarse-grained sediment. *Geological Society of America Bulletin*, **87**, 281–285.

Lewin, J. (1978) Meander development and floodplain sedimentation: a case study from mid-Wales. *Geological Journal*, **13**, 25–36.

Lewin, J. and Brindle, B.J. (1977) Confined meanders. In: *River Channel Changes*, K.J. Gregory (ed.), Chichester, John Wiley & Sons, 221–233.

Lewis, G.W. and Lewin, J. (1983) Alluvial cutoffs in Wales and the Borderlands. In: *Modern and Ancient Fluvial Systems*, J.D. Collinson and J. Lewin (eds), Oxford, Blackwell Scientific Publications, 145–154.

Mackin, J.H. (1937) Erosional history of the Bighorn Basin, Wyoming. *Geological Society of America Bulletin*, **48**, 813–894.

Marriott, S. (1992) Textural analysis and modelling of a flood deposit: River Severn, U.K. *Earth Surface Processes and Landforms*, **17**, 687–697.

McDonald, R.R., Nelson, J.M., Andrews, E.D., Campbell, D.H. and Rubin, D.M. (1994) Field study of the flow in lateral separation eddies in the Colorado River [abstract]. *EOS, Transactions, American Geophysical Union*, **75**(44), 268.

McLean, S.R. and Smith, J.D. (1986) A model of flow over two-dimensional bedforms. *Journal of Hydraulic Engineering*, **112**, 300–317.

McLean, S.R. (1990) The stability of ripples and dunes. *Earth Science Reviews*, **29**, 131–144.

Mertes, L.A K. (1994) Rates of flood-plain sedimentation on the central Amazon River. *Geology*, **22**, 171 –174.

Miller, A.J. (1990a) Flood frequency and geomorphic effectiveness in the central Appalachians. *Earth Surface Processes and Landforms*, **15**, 119–134.

Miller, A.J. (1990b) Fluvial response to debris associated with mass wasting during extreme floods. *Geology*, **18**, 599–602.

Miller, A.J. (1994) Debris-fan constrictions and flood hydraulics in river canyons: some implications from two-dimensional flow modelling. *Earth Surface Processes and Landforms*, **19**, 681–697.

Miller, A.J. (1995) Valley morphology and boundary conditions influencing spatial patterns of flood flow. In: *Natural and Anthropogenic Influences in Fluvial Geomorphology*, J.E. Costa et al. (ed.), American Geophysical Union, Geophysical Monograph 89, 57–81.

Miller, A.J. and Parkinson, D.J. (1993) Flood hydrology and geomorphic effects on river channels and flood plains: the flood of November 4–5, 1985 in the South Branch Potomac River Basin of West Virginia. *US Geological Survey Professional Paper* 1981-E, 96 pp.

Mohrig, D., Paola, C. and Heller, P.L. (1994) Constraints on river avulsion from a well exposed fluvial system [Abstract]. *EOS, Transactions, American Geophysical Union*, **75**(44), 303.

Moore, R.C. (1926) Origin of inclosed meanders on streams of the Colorado Plateau. *Journal of Geology*, **34**, 29–57.

Mosley, J.P. (1981) Semi-determinate hydraulic geometry of river channels, South Island, New Zealand. *Earth Surface Processes and Landforms*, **6**, 127–157.

Mosley, J.P. (1982) Analysis of effect of changing discharge on channel morphology and instream uses in a braided river, Ohau River, New Zealand. *Water Resources Research*, **18**, 800–812.

Murray, A.B. and Paola, C. (1994) A cellular model of braided rivers. *Nature*, **371**(6492), 54–57.

Nanson, G.C. (1980) Point bar and floodplain formation of the meandering Beatton River, northeastern British Columbia, Canada. *Sedimentology*, **27**, 3–29.

Nanson, G.C. (1986) Episodes of vertical accretion and catastrophic stripping: a model of disequilibrium flood plain development. *Geological Society of America Bulletin*, **97**, 1467–1475.

Nanson, G.C. and Croke, J.C. (1992) A genetic classification of floodplains. *Geomorphology*, **4**, 459–486.

Nanson, G.C. and Hickin, E.J. (1983) Channel migration and incision on the Beatton River. *Journal of Hydraulic Engineering*, **109**, 327–337.

Nanson, G.C. and Hickin, E.J. (1986) A statistical analysis of bank erosion and channel migration in western Canada. *Geological Society of America Bulletin*, **97**, 497–504.

Nanson, G. and Page, K. (1983) Lateral accretion of fine-grained concave benches on meandering rivers. In: *Modern and Ancient Fluvial Systems*, J.D. Collinson and J. Lewin (eds) Oxford, Blackwell Scientific Publications, 133–143.

Nelson, J.M. (1990) The initial stability and finite-amplitude stability of alternate flows in straight channels. *Earth Science Reviews*, **29**, 97–115.

Nelson, J.M. and Smith, J.D. (1989a) Flow in meandering channels with natural topography. In: *River Meandering*, S. Ikeda and G. Parker (eds), Washington, DC, Water Resources Monograph 12, American Geophysical Union, 69–102.

Nelson, J.M. and Smith, J.D. (1989b) Evolution and stability of erodible channel beds. In: *River Meandering*, S. Ikeda and G. Parker (eds), Washington, DC, Water Resources Monograph 12, American Geophysical Union, 321–378.

Nelson, J.M., McDonald, R.R. and Rubin, D.M. (1994) Computational prediction of flow and sediment transport patterns in lateral separation eddies [abstract]. *EOS, Transactions, American Geophysical Union*, **75**(44), 268.

Nicholas, A.P. and Walling, D.E. (1995) Modelling contemporary overbank sedimentation on floodplains: some preliminary results. In: *River Geomorphology*, E.J. Hickin (ed.), Chichester, John Wiley & Sons, 131–153.

O'Connor, J.E., Ely, L.L., Wohl, E.E., Stevens, L.E., Melis, T.S., Kale, V.S. and Baker, V.R. (1994) A 4500-year record of large floods on the Colorado River in the Grand Canyon, Arizona. *Journal of Geology*, **102**, 1–9.

Odgaard, A.J. (1987) Streambank erosion along two rivers in Iowa. *Water Resources Research*, **23**, 1225–1236.

Odgaard, A.J. (1989a) River-meander model. I: Development. *Journal of Hydraulic Engineering*, **115**, 1433–1450.

Odgaard, A.J. (1989b) River-meander model. II: Applications. *Journal of Hydraulic Engineering*, **115**, 1451–1464.

Osman, A.M. and Thorne, C.R. (1988) Riverbank stability analysis. I: Development. *Journal of Hydraulic Engineering*, **114**, 134–150.

Osterkamp, W.R. and Costa, J.E. (1987) Changes accompanying an extra-ordinary flood on a sand-bed stream. In: *Catastrophic Flooding*, L. Mayer and D. Nash (eds), Boston, Allen & Unwin, 201–224.

Palmquist, R.C. (1975) Preferred position model and subsurface asymmetry of valleys. *Geological Society of America Bulletin*, **86**, 1392–1398.

Paola, C. (1989) A simple basin-filling model for coarse-grained alluvial systems. In: *Quantitative Dynamic Stratigraphy*, T.A. Cross (ed.), Englewood Cliffs, New Jersey, Prentice-Hall, 363–374.

Paola, C., Heller, P.L. and Angevine, C.L. (1992) The large-scale dynamics of grain-size variation in alluvial basins, 1, Theory. *Basin Research*, **4**, 73–90.

Parker, G. (1976) On the cause and characteristic scales of meandering and braiding in rivers. *Journal of Fluid Mechanics*, **76**, 457–480.

Parker, G. (1982) *Stability of the Channel of the Minnesota River Near State Bridge No. 93, Minnesota*. Project Report No. 205, St. Anthony Falls Hydraulic Laboratory, University of Minnesota, 33 pp.

Parker, G. (1984) Theory of meander bend deformation. In: *River Meandering*, C.M. Elliott (ed.), New York, American Society of Civil Engineers, 722–732.

Parker, G. and Andrews, E.D. (1986) On the time development of meander bends. *Journal of Fluid Mechanics*, **162**, 139–156.

Parker, G. and Johannesson, H. (1989) Observations on several recent theories of resonance and overdeepening in meandering channels. In: *River Meandering*, S. Ikeda and G. Parker (eds), Washington, DC, Water Resources Monograph 12, American Geophysical Union, 379–416.

Pizzuto, J.E. (1984) Bank erodibility of sand-bed streams. *Earth Surface Processes and Landforms*, **9**, 113–124.

Pizzuto, J.E. (1987) Sediment diffusion during overbank flows. *Sedimentology*, **34**, 301–317.

Pizzuto, J.E. and Meckeinburg, T.S. (1989) Evaluation of a linear bank erosion equation. *Water Resources Research*, **25**, 1005–1013.

Price, W.E., Jr (1974) Simulation of alluvial fan deposition by a random walk model. *Water Resources Research*, **10**, 263–274.

Rachocki, A. (1981) *Alluvial Fans: An Attempt at an Empirical Approach*. New York, John Wiley & Sons, 157 pp.

Reinfelds, I. and Nanson, G. (1993) Formation of braided-river floodplains, Waimakariri River, New Zealand. *Sedimentology*, **40**, 1113–1127.

Rich, J.L (1914) Certain types of stream valleys and their meandering. *Journal of Geology*, **22**, 469–497.

Richards, K.S. (1982) *Rivers: Form and Process in Alluvial Channels*. London, Methuen.

Richards, K., Chandra, S. and Friend, P. (1993) Avulsive channel systems: characteristics and examples. In: *Braided Rivers*, J.L. Best and C.S. Bristow (eds), Geological Society Special Publication 75, 195–203.

Ritter, D.F. (1975) Stratigraphic implication of coarse-grained gravel deposited as overbank sediment, southern Illinois. *Journal of Geology*, **83**, 645–650.

Ritter, D.F. (1988) Floodplain erosion and deposition during the December 1982 floods in southeast Missouri. In: *Flood Geomorphology*, V.R. Baker *et al.* (eds), New York, John Wiley & Sons, 243–259.

Ritter, D.F. and Blakely, D.S. (1986) Localized catastrophic disruption of the Gasconade River flood plain during the December 1982 flood, southeast Missouri. *Geology*, **14**, 472–476.

Robertson-Rintoul, M.S.E. and Richards, K.S. (1993) Braided-channel pattern and palaeohydrology using an index of total sinuosity. In: *Braided Rivers*, J.L. Best and C.S. Bristow (eds), Geological Society Special Publication 75, 113–118.

Rubin, D.M., Schmidt, J.C. and Moore, J.N. (1990) Origin, structure, and evolution of a reattachment bar, Colorado River, Grand Canyon, Arizona. *Journal of Sedimentary Petrology*, **60**, 982–991.

Schmidt, J.C. (1990) Recirculating flow and sedimentation in the Colorado River in Grand Canyon, Arizona. *Journal of Geology*, **98**, 709–724.

Schmidt, J.C. and Graf, J.B. (1990) Aggradation and degradation of alluvial sand deposits, 1965 to 1986, Colorado River, Grand Canyon National Park, Arizona. *US Geological Survey Professional Paper* 1493, 74 pp.

Schmidt, J.C. and Rubin, D.M. (1995) Regulated streamflow, fine-grained deposits, and effective discharge in canyons with abundant debris fans. In: *Natural and Anthropogenic Influences in Fluvial Geomorphology*, J.E. Costa *et al.* (eds), American Geophysical Union, Geophysical Monograph, 89, 177–195.

Schmidt, J.C., Rubin, D.M. and Ikeda, H. (1993) Flume simulation of recirculating flow and sedimentation. *Water Resources Research*, **29**, 2925–2939.

Schumm, S.A. and Lichty, R.W. (1963) Channel widening and flood-plain constriction along Cimarron River in southwestern Kansas. *US Geological Survey Professional Paper* 352-D, 71–88.

Seminara, G. and Tubino, M. (1989) Alternate bars and meandering: free, forced and mixed interactions. In: *River Meandering*, S. Ikeda and G. Parker (eds), Washington, DC, Water Resources Monograph 12, American Geophysical Union, 267–320.

Shimizu, Y. and Itakura, T. (1989) Calculation of bed variation in alluvial channels. *Journal of Hydraulic Engineering*, **115**, 367–384.

Slingerland, R., Harbaugh, J. and Furlong, K. (1994) *Simulating Classic Sedimentary Basins*. Englewood Cliffs, New Jersey, Prentice-Hall, 220 pp.

Smart, J.S. and Moruzzi, V.L. (1972) Quantitative properties of delta channel networks. *Zeitschrift für Geomorphologie*, **16**, 268–282.

Smith, D.G. and Putnam, P.E. (1980) Anastomosed river deposits: modern and ancient examples in Alberta, Canada. *Canadian Journal of Earth Sciences*, **17**, 1396–1406.

Smith, D.G. and Smith, N.D. (1980) Sedimentation in anastomosed river systems: examples from alluvial valleys near Banff, Alberta. *Journal of Sedimentary Petrology*, **50**, 157–164.

Smith, N.D., Cross, T.A., Dufficy, J.P. and Clough, S.R. (1985) Anatomy of an avulsion. *Sedimentology*, **36**, 1–23.

Smith, J.D. (1994) Exchange of sand between the bed and margins of rivers with perimeters composed predominantly of gravel and bedrock [abstract]. *EOS, Transactions, American Geophysical Union*, **75**(44), 269.

Smith, N.D. and Perez-Arlucea, M. (1994) Fine-grained splay deposition in the avulsion belt of the lower Saskatchewan River, Canada. *Journal of Sedimentary Research*, **B64**, 159–168.

Stanistreet, I.G., Cairncross, B. and McCarthy, T.S. (1993) Low sinuosity and meandering bedload rivers of the Okavango Fan: channel confinement by vegetated levees without fine sediment. *Sedimentary Geology*, **85**, 135–156.

Struiksma, N. and Crosato, A. (1989) Analysis of 2-D bed topography model for rivers. In *River Meandering*, S. Ikeda and G. Parker (eds), Washington, D.C., Water Resources Monograph 12, American Geophysical Union, 153–180.

Sun, T., Meakin, P., Jossang, T. and Schwarz, K. (1996) A simulation model for meandering rivers. *Water Resources Research* (in press).

Thorne, C.R. (1982) Processes and mechanisms of river bank erosion. In: *Gravel-Bed Rivers*, R.D. Hey, J.C. Bathurst and C.R. Thorne (eds), Chichester, John Wiley & Sons, 227–272.

Thorne, C.R. (1992) Bend scour and bank erosion on the meandering Red River, Louisiana. In: *Lowland Floodplain Rivers*, P.A. Carling and G.E. Petts (eds), Chichester, John Wiley & Sons, 95–115.

Thorne, C.R. and Lewin, J. (1979) Bank processes, bed material movement, and planform development in a meandering river. In: *Adjustments of the Fluvial System*, D.D. Rhodes and G.P. Williams (eds), Dubuque, Iowa, Kendall/Hunt, 117–137.

Thorne, C.R. and Osman, A.M. (1988) Riverbank stability analysis. II: Applications. *Journal of Hydraulic Engineering*, **114**, 151–172.

Thorne, C.R. and Tovey, N.K. (1981) Stability of composite river banks. *Earth Surface Processes and Landforms*, **6**, 469–484.

Tubino, M. and Seminara, G. (1990) Free-forced interactions in developing meanders and suppression of free bars. *Journal of Fluid Mechanics*, **214**, 131–159.

Ullrich, C.R., Hagerty, D.J. and Holmberg, R.W. (1986) Surficial failures of alluvial stream banks. *Canadian Geotechnical Journal*, **23**, 304–316.

Walling, D.E. and Bradley, S.B. (1989) Rates and patterns of contemporary floodplain sedimentation: A case study of the river Culm, Devon, UK. *Geojournal*, **19**(1), 53–62.

Walling, D.E., Quine, T.A. and He, Q. (1992) Investigating contemporary rates of floodplain sedimentation. In: *Lowland Floodplain Rivers*, P.A. Carling and G.E. Petts (eds), Chichester, John Wiley & Sons, 165–184.

Webb, E.K. (1994) Simulating the three-dimensional distribution of sediment units in braided-stream deposits. *Journal of Sedimentary Research*, **B64**, 219–231.

Webb, E.K. (1995) Simulation of braided-channel topology and topography. *Water Resources Research*, **31**, 2603–2611.

Werner, B.T. (1995) Eolian dunes: computer simulations and attractor interpretation. *Geology*, **23**, 1107–1110.

Whiting, P.J. and Dietrich, W.E. (1993a) Experimental studies of bed topography and flow patterns in large-amplitude meanders, 1, Observations. *Water Resources Research*, **29**, 3605–3614.

Whiting, P.J. and Dietrich, W.E. (1993b) Experimental studies of bed topography and flow patterns in large-amplitude meanders, 1, Mechanisms. *Water Resources Research*, **29**, 3615–3624.

Whiting, P.J. and Dietrich, W.E. (1993c) Experimental constraints on bar migration through bends: implications for meander wavelength selection. *Water Resources Research*, **29**, 1091–1102.

Williams, G.P. and Guy, H.P. (1973) Erosional and depositional aspects of Hurricane Camille in Virginia, 1969. *US Geological Survey Professional Paper* 804, 80 pp.

Williams, P.E. and Rust, B.R. (1969) The sedimentology of a braided river. *Journal of Sedimentary Petrology*, **39**, 649–679.

Wolman, M.G. and Brush, L.M. Jr (1961) Factors controlling the size and shape of stream channels in coarse noncohesive sands. *US Geological Survey Professional Paper* 282-G, 183–210.

Wolman, M.G. and Eiler, J.P. (1958) Reconnaissance study of erosion and deposition produced by the flood of August 1955 in Connecticut. *American Geophysical Union Transactions*, **39**, 1–14.

Wolman, M.G. and Leopold, L.B. (1957) River flood plains: some observations on their formation. *US Geological Survey Professional Paper* 282-C, 87–109.

Woodyer, K.D. (1975) Concave-bank benches on Barwon River, N.S.W. *Australian Geographer*, **13**, 36–40.

Yen, B.C. and Yen, C.-L. (1984) Flood flow over meandering channels. In: *River Meandering*, C.M. Elliott (ed.), New York, American Society of Civil Engineers, 554–561.

3 Analysis and Modelling of Overbank Deposits

S.B. MARRIOTT

Faculty of the Built Environment, University of the West of England, UK

3.1 INTRODUCTION

Despite the fact that rivers flood regularly (on average every one to two years according to Wolman and Leopold, 1957) sometimes leaving thick deposits of sediment on agricultural and densely populated areas, comparatively little is known about the processes effecting the transport of sediment by flood water. Investigation into the nature of deposition on floodplains has generally involved a theoretical approach, with tests of the theory carried out in flume experiments rather than field studies, due largely to the unpredictability of floods and the difficulty in collecting samples and making observations during such events. Within the past few years, however, there has been an upsurge of interest in rates and processes of flood flow and sediment deposition on floodplains, as indicated by the increase in related publications (e.g. Baker *et al.*, 1988; Beven and Carling, 1989; Carling and Petts, 1992). There is also a need for research which links theoretical studies of sediment deposition on floodplains with simulation models, of both floodplain topography and alluvial architecture.

The aim of the study reported was to incorporate a realistic model for deposition on floodplains into a simulation model for alluvial architecture derived by Crane (1982). An opportunity to test theoretical models of overbank deposition against textural analysis of data collected following a flood arose in early 1990, when severe flooding in the Tewkesbury area of the River Severn floodplain followed prolonged heavy rainfall in the North Wales and West Midlands catchments. The textural analysis of the flood deposits collected enabled the performance of two numerical models of overbank deposition to be tested (James, 1985; Pizzuto, 1987). The success of James' (1985) model in particular, in predicting the sediment distribution patterns for the Severn flood, led to the incorporation of a version of this model into Crane's (1982) computer simulation model of alluvial architecture. The results of including this more realistic model of deposition on floodplains are discussed below.

Floodplain Processes. Edited by Malcolm G. Anderson, Des E. Walling and Paul D. Bates.

3.2   BACKGROUND TO THE FLOOD EVENT

The River Severn catchment area is approximately 10 000 km^2 and includes extensive upland areas in North and Mid Wales. The catchment also includes parts of the West Midlands and Cotswolds escarpment drained by the Warwickshire Avon, a major tributary, which joins the Severn at Tewkesbury (Figure 3.1). The reach under study is the Lower Severn from the confluence with the River Teme, south of Worcester, to Gloucester (Figure 3.2). Gloucester is the present limit of tidal influence, although during southwesterly storms the influence of the tide may be recorded at the Mythe Bridge gauging station, north of Tewkesbury (J. Crabbe, 1990, personal communication).

   In the study reach, the Lower Severn has a low-sinuosity channel planform. The sinuosity is 1.15 between Worcester and Tewkesbury and 1.36 between Tewkesbury and Gloucester. The planform has changed little over the past 500 years (Brown, 1987). The channel is, on average, 60–70 m wide and has a relatively narrow floodplain (about six channel widths) bordered by Pleistocene terrace sediments and bedrock

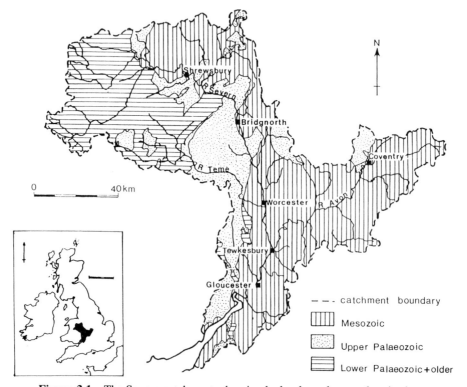

Figure 3.1   The Severn catchment, showing bedrock geology and main rivers

Figure 3.2 (*opposite*)   Details of the study area, including sampling sites

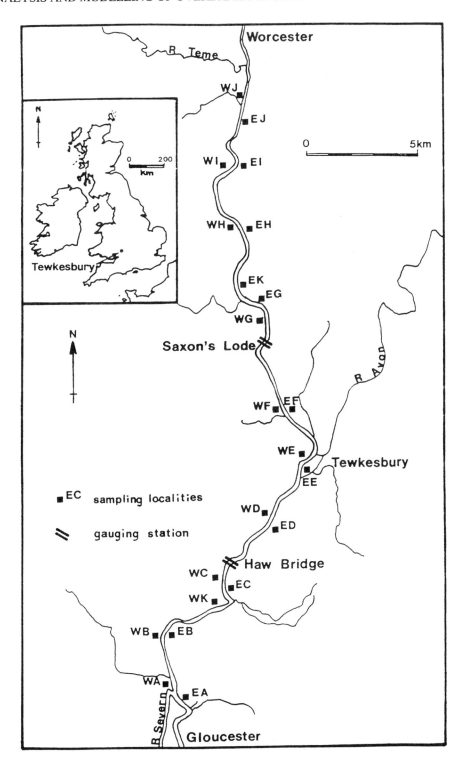

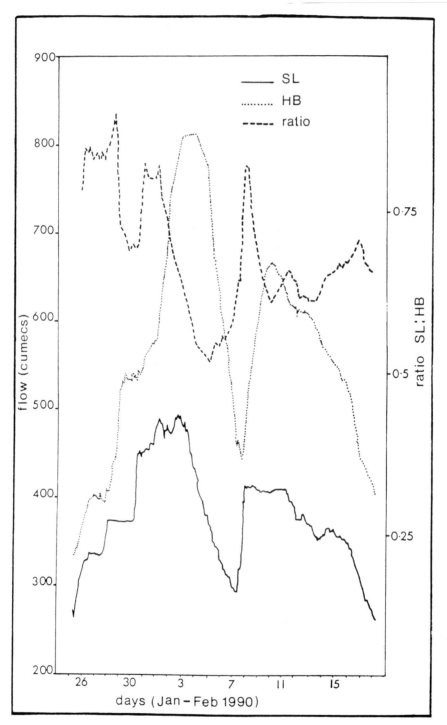

Figure 3.3 Graphs of flow at Saxon's Lode (SL) and Haw Bridge (HB) gauging stations for the period 25 January to 18 February 1990. The dashed curve is the ratio between the two indicating the influence of input from the Warwickshire Avon (data supplied by NRA)

outcrops, as at Wainlode Hill (British National Grid Reference SO 845 258). The river's stability is probably due to relative incision of the channel within the last 3000 years (Brown, 1983) and has been enhanced by embanking which has been progressive from the eighteenth century onward, and by channelization for flood control (i.e. the channel is straightened, steepened and deepened). Much of the study reach has tree-lined banks, which also aid stability.

During late January and February 1990, prolonged heavy rainfall across the Severn and Avon catchments resulted in severe flooding in the Tewkesbury area; the worst since 1960. Figure 3.3 shows the flows recorded by the National Rivers Authority (NRA) at the Saxon's Lode and Haw Bridge gauging stations, located respectively above and below the confluence of the Severn and the Warwickshire Avon (Figure 3.2). The ratio curve (Figure 3.3) shows the influence of water entering the system between the two stations, notably from the Avon.

Overbank flow persisted from 25 January until 18 February. As a consequence of the embankments being overtopped and water from tributaries entering low-lying areas, flood water was ponded over extensive areas (e.g. Hasfield Ham and Ashleworth Ham around SO 835 265) until well into March. The presence of embankments prevented the direct return of flood waters to the main channel.

3.3 FIELD AND LABORATORY METHODS

After the flood had largely subsided, sediment samples were collected along traverses normal to the channel at 18 localities (Table 3.1; Figure 3.2). Localities and traverses were chosen in order to provide a set of samples representative of the floodplain in the reach of the Severn under study. The localities were about 3.5 km apart and, where possible, sections of the river without embankments were selected. This was not always possible at all localities, and in several cases only short traverses extending from the channel bank to the embankment were involved. The samples were collected soon after the flood had subsided to avoid the effects of wind drying and deflation, mixing by bioturbation and contamination by agricultural activities. A further four localities were visited a week later to provide samples from more distant low-lying areas which were still under water during the previous visit. A total of 265 samples was collected.

Samples of approximately 200 g were collected near the channel bank where the deposited sediment consisted of up to 150 mm of sandy material lying over waterlogged and decaying vegetation. In some places the sand had a veneer of silty clay and at sites EE, EK, WA and WE (Figure 3.2) the thick sandy sediment had current ripple marks and small (wavelength ~40 mm) sinuous-crested dunes on the surface. In the pasture away from the channel, the deposit consisted mainly of a film of fine-grained material which was collected from grass and leaves. Where water had been ponded for longer periods a thicker deposit (up to 5 mm) of fine sediment could be sampled.

In the laboratory, the samples were air dried at 40°C, before coning and quartering to give small subsamples of approximately 25 g. The coarser sandy samples were weighed and then dry-sieved and the fine samples were passed through a Coulter Multisizer (Marriott, 1992).

Table 3.1 British National Grid References and sample numbers
for site codes shown on Figure 3.2

Site code	Grid references on transect (all SO)		Number of samples
	Start	Finish	
EA	8223 2134	8235 2134	11
EB	8173 2471	8212 2450	21
EC	8445 2664	8448 2668	10
ED	8659 2972	8660 2972	9
EE	8822 3280	8870 3263	12
EF	8678 3677	8680 3677	13
EG	8570 4061	8570 4063	10
EH	8506 4407	8509 4410	10
EI	8448 4727	8452 4723	10
EJ	8479 5051	8506 5048	12
EK	8469 4138	8505 4128	12
WA	8199 2166	8201 2167	10
WB	8151 2453	8146 2459	10
WC	8442 2730	8440 2729	8
WD	8665 2999	8663 3001	10
WE	8837 3297	8835 3299	10
WF	8670 3683	8660 3679	11
WG	8615 4030	8560 4021	24
WH	8475 4453	8462 4451	10
WI	8391 4705	8389 4707	9
WJ	8500 5101	8495 5103	10
WK	8400 2563	8358 2675	23

3.4 TEXTURAL PATTERNS

Mean, standard deviation, skewness and kurtosis were calculated for all samples
together with the weight per cent of sand, silt and clay (Marriott, 1993). At about 20 m
from the channel bank, the pattern of variation of mean and standard deviation with
distance (Figure 3.4a,b,c). shows a marked change from sediment with a mean grain
size in the fine to very fine sand range, to medium silt. This change is also shown on
the weight per cent sand–silt–clay variation diagrams, where there is a distinct
decrease in the proportion of sand present, again at a distance of around 20 m (Figure
3.5 a,b,c). The sand–silt–clay diagrams, however, show that sand is present
throughout the transect and that the proportion of sand increases slightly at the edge of
the floodplain. This is particularly clear at sites EB (Figure 3.5b) and WG (Figure 3.5a)
and might be due to slope wash from terrace edges.

    The standard deviation shows a similar marked change in value at 20 m from the
channel bank. The spread of values around the mean is lower for the samples collected
near to the channel than for those collected at a distance. This implies that the sandier
samples were better sorted than the finer-grained samples.

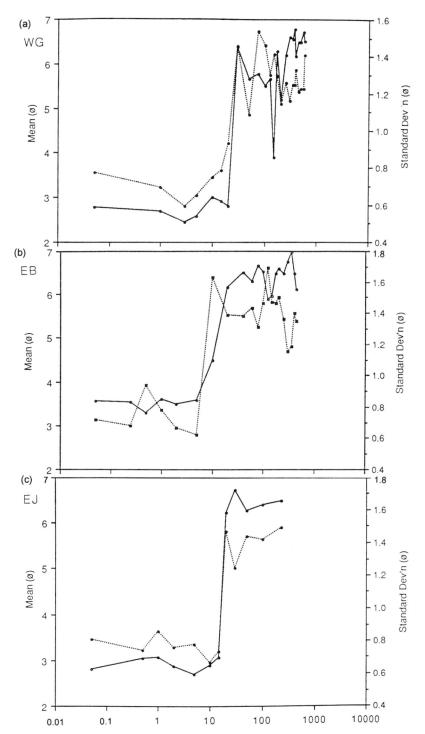

Figure 3.4 The variation of grain size mean and standard deviation with distance from the channel bank for three localities (a) WG; (b) EB; (c) EJ. Distance is given on a logarithmic scale. The mean is represented by the solid line

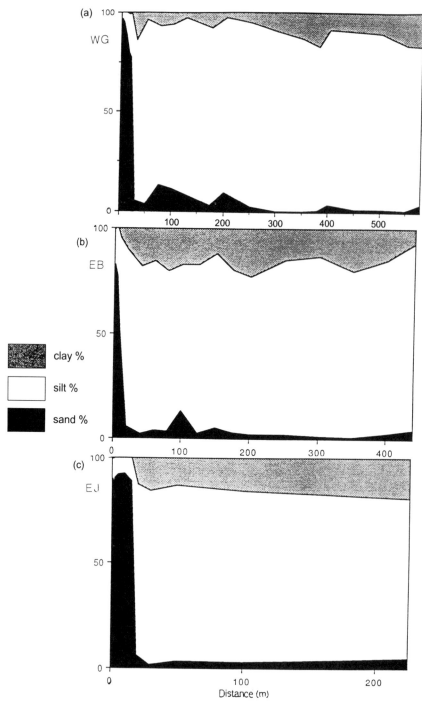

Figure 3.5 For the same localities as in Figure 3.4, these graphs show the variation in sand, silt and clay with distance from the channel bank: (a) WG; (b) EB; (c) EJ

The weight per cent sand, silt and clay were also plotted on a ternary diagram to demonstrate the textural relationships (Figure 3.6). The samples appear to be divided into two populations; in the sand to loamy sand, and silty clay loam to silt loam classes. The dual populations are also evident on cross-plots of standard deviation and skewness against mean for all samples (Figures 3.7a,b). The coarse and fine-grained sediment means form two separate groups with their associated standard deviation and skewness. The variation in mean values for all samples (Figure 3.8) also shows two populations, clustering around 3.250 ϕ and 6.50 ϕ respectively.

In summary, the mean grain size decreases sharply at around 20 m from the channel bank, where the sediment consists mainly of sand and appears to be better sorted than sediment collected at greater distances from the channel. A small proportion of sand is present in the flood deposit across the whole width of the floodplain. The textural relationships indicate that two distinct populations are present with characteristic means and standard deviations.

3.5 APPLICATION OF NUMERICAL MODELS TO THE FLOOD EVENT

Modes of transport and deposition are difficult to evaluate during flood events. Some indications as to the processes operating within flood waters can be given by observing the behaviour of flows in flume experiments, and by basing numerical models on the

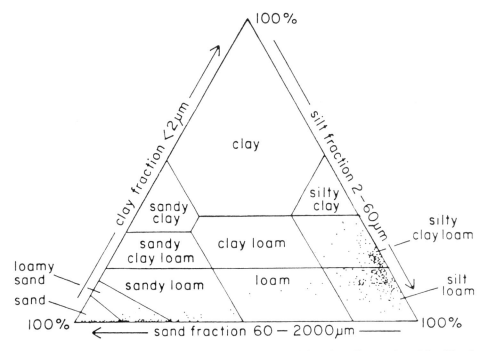

Figure 3.6 Ternary diagram showing textural relationships for all samples (classification according to Soil Survey of England and Wales)

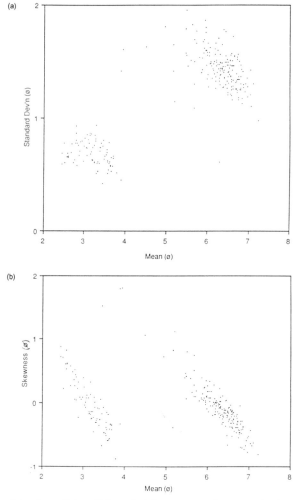

Figure 3.7 Cross-plots for all localities of (a) mean against standard deviation, (b) mean against skewness

observations. However, very few numerical models have been proposed which attempt to predict grain size distribution for overbank events. Most hydraulic research has concentrated on the hydraulics of flood channels and the nature of flow between the channel and its floodplain. Two models do, however, consider overbank deposition; those of James (1985) and Pizzuto (1987).

When overbank flow occurs, sediment may be transferred from the channel to the floodplain by different mechanisms (see review in Marriott, 1993). Coarse sediment deposited near to the channel may have been transported by traction as bedload, and fine-grained sediment as suspended load. During floods, flow is fast and deep in the channel and relatively slow and shallow over the floodplain. A strong reaction takes place between the two flows resulting in a transfer of momentum from the channel to

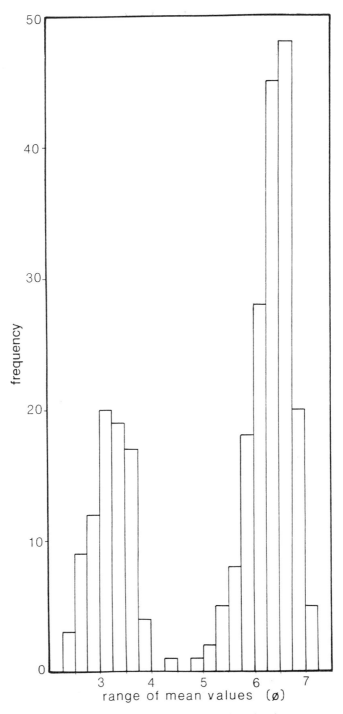

Figure 3.8 Histogram of central tendencies

the floodplain as the flow velocity is decreased in the channel and increased over the floodplain adjacent to the channel. This is evident, at least on a laboratory scale, in turbulent eddies along the interface between the channel and floodplain (Sellin, 1964).

The concentration of suspended sediment is greater in the channel than in the floodplain flow due to the greater transporting capacity of faster-flowing water (Walling, 1974). The interaction at the interface between the channel and floodplain would involve transfer of suspended sediment as well as momentum. This transfer of sediment by turbulent eddies is analogous to a diffusion process (Allen, 1985, pp. 124–129).

Sediment transfer by convection may also occur where there is a component of flow normal to the channel, i.e. where the flow in the channel is inclined to the direction of flow on the floodplain. This can result in asymmetrical deposition, with the floodplain on one side of the channel receiving significant accumulations of sediment while deposition on the other side is inhibited (James, 1985).

3.5.1 Assessment of the quantitative model of Pizzuto (1987)

Pizzuto (1987) considered deposition from steady flow using a diffusion analogy, and without varying diffusivity (i.e. coefficients included in the model which describe the vertical and lateral transport of suspended sediment by eddy exchange) with distance from the channel. The model produces curves representing the thickness of sediment of a particular grain size deposited at particular distances from the channel for various flood durations. According to Pizzuto's (1987) model, for floods of short duration (half an hour), the thickness of sediment deposited drops off rapidly within a short distance from the channel. As modelled flood duration is increased, the curve becomes increasingly concave upward and, for floods lasting longer than 10 hours, the difference in curve shape from that for a 10-hour flood is negligible. From this, Pizzuto (1987) assumed that the sediment concentration had reached a steady state, and he used a "steady-state" version of the model (i.e. one which did not include a time-dependent variable) to predict floodplain topographic profiles. The topographic curves and predicted grain size patterns were compared with deposition from flood events that have occurred over the past 250 years in the Brandywine Creek floodplain, USA. While the predictions seem to agree fairly well with findings on small floodplains, Pizzuto (1987) found that sand-sized sediment was actually carried further across the floodplain than his model predicted.

Computer programs in FORTRAN 77 were written (Marriott, 1993) for both the time dependent and steady-state solutions in order that Pizzuto's (1987) model could be applied to other floodplains. The equation for the steady-state model was given by Pizzuto (1987) as follows:

$$H = \frac{V_s^2}{e_z} Z_0 t \, \frac{\sinh(Gn)\mathrm{e}^{-G}}{\cosh(Gn)} + \mathrm{e}^{-Gn} \tag{3.1}$$

where H = thickness of sediment deposited during flood, V_s = sediment settling velocity, e_z = vertical sediment diffusivity, Z_o = suspended sediment concentration, n = distance across floodplain, and G represents a dimensionless variable relating the width of the floodplain W, sediment settling velocity and the vertical and horizontal diffusivities as:

$$G = WV_s/(e_y e_z)^{1/2} \qquad (3.2)$$

where e_y = horizontal sediment diffusivity.

The program for the steady-state model was checked with Pizzuto's (1987, his Figure 4) test data (Figure 3.9). Pizzuto (1987) had tested his model by comparison with grain size analysis of samples collected from the Brandywine Creek floodplain (see above). The steady-state model appeared to provide a useful indication of floodplain topography and might, therefore, be a realistic deposition model for incorporation into Crane's (1982) computer simulation model for alluvial architecture.

However, on using the floodplain width of the River Severn at site WG (see Table 3.4) as input for the steady-state program, the predicted thickness of sediment of grain size 4.75 ϕ (fine silt) shown on Figure 3.10 appears to be very unrealistic when compared with the actual grain size distribution (compare with the sand–silt–clay ratio diagram for WG, Figure 3.5a). The value of G calculated in the program was investigated and found to be 57.3, which is considerably larger than the values for Pizzuto's test data (Figure 3.9). This finding agrees with Pizzuto's (1987) conclusion that as the value of G increases, the curve becomes more concave upwards. The widths of the floodplains that correspond to Pizzuto's test values of G are very small (shown in Table 3.2). The widths in Table 3.2 were calculated using Equation 3.2 above and with values for diffusivity e_y and e_z as 0.013 and 0.00746 m^2/s respectively (Pizzuto's values).

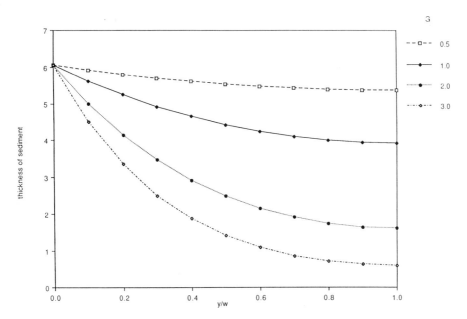

Figure 3.9 The output from the computer program of Pizzuto's (1987) steady-state model. y/w = distance from channel divided by floodplain width

Table 3.2 Widths of floodplains represented in Pizzuto's (1987) test data (see Figure 3.9)

G	Width (m)
0.5	0.62
1.0	1.2425
2.0	2.485
3.0	3.727

As the profiles in the test graph (Figure 3.9) were considered realistic, Pizzuto (1987) applied the steady-state model to localities on the Brandywine Creek floodplain. He showed consistency between the actual and predicted profiles (see his Figure 10). However, the widths of the floodplain sections examined on the Brandywine were between 18 and 216 m, so with values of G similar to that calculated for the experiment with Severn data, steep profiles as in Figure 3.10 might be expected. According to Pizzuto's data, very low values were chosen for G (between 0.005 and 0.05, see his Table 1) so as to achieve gently sloping profiles. This implies correspondingly large increases in the values for vertical and horizontal diffusivity (see Equation 3.2). The values for diffusivity were calculated for the sites in Pizzuto's (1987) Table 1 using the constant ratio $e_y = 1.71e_z$. This relationship was derived from work on diffusivities of channel flow by Ikeda and Nishimura (1985). The particle fall velocity V_s was calculated using the Stoke's Law equation. The results of these calculations are shown in Table 3.3.

The values for vertical and horizontal diffusivity of suspended sediment in overbank flows have not been measured and, as mentioned above, the relationship used was derived from work on channels. However, James (1985), in his model for sediment transfer to overbank sections, estimates vertical diffusivity and transverse diffusivity by calculations based on experimental work by Lau and Krishnappan (1977) and

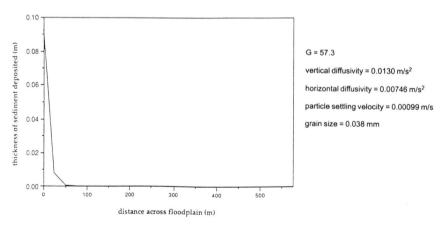

Figure 3.10 Prediction of floodplain topography using Pizzuto's (1987) steady-state model for the River Severn

Table 3.3 Horizontal and vertical diffusivities calculated for the floodplain widths and G values from Brandywine Creek (after Pizzuto, 1987, Table 1)

Site	Floodplain width (m)	Grain size (mm)	V_s(m/s)	G	e_z(m²/s)	e_y(m²/s)
TB	18.6	0.0195	0.00105	0.050	0.298	0.510
CF-E	150.0	0.0168	0.00078	0.011	8.119	13.884
CF-W	216.0	0.0168	0.00078	0.005	25.722	43.985
6S	76.2	0.0239	0.00157	0.011	8.302	14.197
6N	43.9	0.0137	0.00052	0.020	0.871	1.490
7S	33.5	0.0239	0.00157	0.025	1.606	2.746
7N	127.9	0.0195	0.00105	0.0095	10.791	18.453
8	51.8	0.0195	0.00105	0.027	1.538	2.630

Rajaratnam and Ahmadi (1981). The data for site WG on the River Severn (Table 3.4) were used in a program of James' (1985) model. The calculated vertical diffusivity ranged from 0.0137 to 0.0206 m²/s depending on the depth within the water column, and the transverse diffusivity was 2.98 m²/s. These values, for site WG on the Severn floodplain (575 m wide) and examining three grain sizes (0.038 mm, 0.180 mm, 0.355 mm), give values for G (using Equation 3.2) ranging from 2.817 to 247.582.

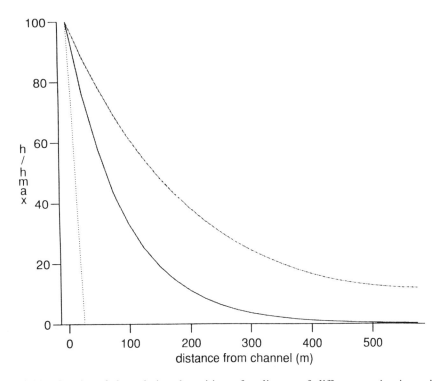

Figure 3.11 Graphs of the relative deposition of sediment of different grain size using the steady-state model of Pizzuto (1987) and a value of vertical diffusivity, e_z, of 0.0137 m²/s

Graphs of the relative deposition of sediment using the revised diffusivity values are shown in Figures 3.11 and 3.12. These predict that medium sand (0.355 mm) is deposited only within the vicinity of the channel, although fine sand (0.180 mm) is present to the outer margin of the floodplain. The modelled distribution of silt appears to be more realistic using these revised diffusivity values, although deposition drops off more sharply than found from the field data (see Figure 3.5a).

It would appear, therefore, that the calculations for thickness of sediment deposited made in Pizzuto's (1987) model may not be reliable and that the steady-state equation cannot be used effectively for large floodplains, due to the unrealistically large values of diffusivity that must be used in order to provide a realistic topographic profile. Pizzuto's (1987) steady-state model is considered unsuitable for substitution as the deposition model in Crane's (1982) program, since further calculations would be necessary to incorporate variations in diffusivity.

3.5.2 Assessment of the numerical model of James (1985)

The numerical model of James (1985) simulates the transfer of suspended sediment from a channel to an adjacent floodplain under steady, uniform flow conditions. It uses a diffusion analogy for sediment transfer and also the possibility of convective sediment supply. In this model the transverse diffusivity (e_y) is estimated with reference to

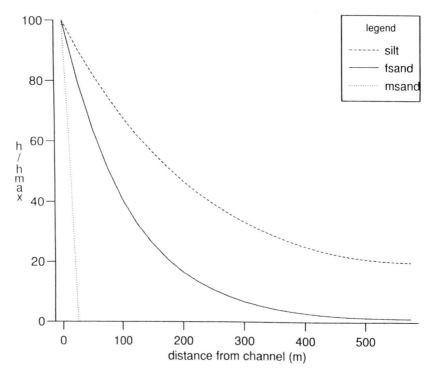

Figure 3.12 Graphs of the relative deposition of sediment of different grain size as in Figure 3.11, but using a value of vertical diffusivity, e_z, of 0.0206 m^2/s

distance across the floodplain using data obtained by Lau and Krishnappan (1977) and Rajaratnam and Ahmadi (1981). The vertical diffusivity (e_z) is estimated for different heights in the water column above the bed. The James (1985) model for the transfer of suspended material through a fluid by turbulence takes the form of an elliptic partial differential equation which, on integration, gives the distribution of sediment in the flow over the floodplain. Solution by finite difference approximations produces a table of suspended sediment concentration values predicted for a specific grain size throughout the water over the width of the floodplain. James (1985) compared the grain size distributions obtained from laboratory flume experiments with solutions derived from a computer program of the numerical model, and found that the model predicted the distributions reliably.

As James' (1985) numerical model predicts the distribution of deposited materials in a relative sense, it was thought that it would provide a satisfactory basis for examining sediment transport processes operating during the Severn flooding. Geometrical and hydraulic data (Table 3.4) from the west side of the Severn floodplain at locality WG (Upton Ham, SO 862 403), were used as input to a computer program of the model. Figure 3.13 shows the resulting curves of expected concentrations for three different grain sizes. The concentration is shown as a percentage of the maximum for each grain size, so that the curves can be plotted on the same diagram.

If Figure 3.13 is compared with the sand–silt–clay diagram for WG (Figure 3.5a), it can be seen that, in both Figures, sand-sized grains form a major proportion of the

Table 3.4 Geometric and hydraulic data for locality WG (SO 862 403) used as input to the computer program developed for James' (1985) model information supplied by NRA Severn/Trent region)

Channel			
Top width (m)	45.50		
Bottom width (m)	26.00		
Flood flow depth (m)	7.75		
Manning's n	0.018		
Floodplain			
Width (m)	575.00		
Hydraulic gradient	0.00262		
Flood flow depth (m)	1.25		
Manning's n	0.020		
Sediment			
Diameter (mm)	0.038	0.180	0.355
Fall velocity at 10°C (m/s)	0.00099	0.0223	0.0867
Relative density	2.65		
Finite difference grid parameters			
Number of vertical grid points	22		
Vertical grid spacing	25.00		
Number of horizontal grid points	6		
Horizontal grid spacing	1.25		

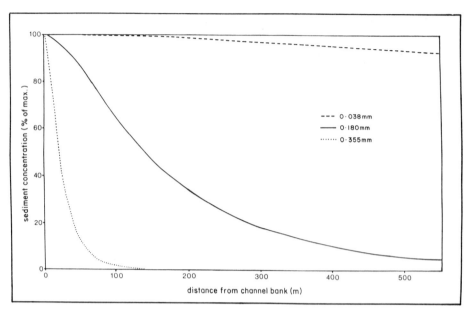

Figure 3.13 Output from the James (1985) numerical model, using different grain sizes, the River Severn channel cross-section and floodplain dimensions and depth of flow at site WG. The curves show relative values of predicted suspended sediment concentrations as distance from the channel bank increases. The sediment concentrations are shown as percentages of the maximum for each grain size from the values of the finite difference grid in the layer immediately above the floodplain surface

sediment closer to the channel, and sand is also present right across the floodplain. Silt retains a fairly consistent, but decreasing concentration across the whole width.

The findings from the sand–silt–clay diagrams also agree with Pizzuto's (1987) observations from Brandywine Creek, that sand is carried further than his model predicts. This feature is, however, satisfactorily predicted by the James (1985) model.

3.6   COMPUTER SIMULATION OF FLOODPLAIN DEVELOPMENT OVER TIME (ALLUVIAL ARCHITECTURE)

Crane (1982) developed a computer simulation model for the architecture of avulsion-controlled alluvial suites, as a contribution to a growing interest in fluvial successions due to their usefulness as sources of hydrocarbons, water and metalliferous reserves. The model was an attempt to obtain further insight into the distribution of channel sandstone bodies and thereby to aid prediction of the location and extent of possible reservoirs.

Crane's (1982) model was based on quantitative models of alluvial architecture developed by Allen (1978) and Leeder (1978), and a computer simulation model produced by Bridge and Leeder (1979). It provided a means of investigating possible responses to changes in the fluvial environment, such as increases in discharge or changes in rate of subsidence. It also provided an opportunity to investigate the influence of overbank deposition on floodplain construction and topography over (modelled) time.

3.6.1 Models for alluvial architecture

Allen's (1978) model considered deposition to be uniform over the entire floodplain and channel avulsion (i.e. switching to a new position) to occur at regular intervals. The effects of compaction were allowed for by specific avoidance factors for the location of the new channel belt. The model was used to demonstrate that the interconnectedness of channel sandbodies is low when they make up less than 50% of the suite and that successions with a wide range of sand–fine ratios can be generated in avulsion-controlled suites.

In Leeder's (1978) model, avulsion occurred randomly but both compaction and topography were ignored. Leeder's model was used to demonstrate the possible variations in one-dimensional sections across the width of the alluvial suite.

Bridge and Leeder (1979) developed a more complex model which incorporated compaction, location of the new channel belt at a topographic low and variation in deposition with distance across the floodplain. This model required solution by computer (Bridge, 1979), but still considered avulsion to occur at a fixed time interval, rather than including further calculations to evaluate the effects of floodplain topography on the occurrence of avulsion. The Bridge and Leeder (1979) model also investigated the effects of tectonic tilting at one margin of the basinal setting. The tilting had a strong effect on the position of the new channel belt after avulsion and appeared to mask the effects of other variables, including sedimentation. It is to be expected that tilting would also affect the regularity of avulsion, but this was not taken into account. Additionally, sedimentation was only calculated once at the occurrence of each avulsion, instead of allowing for a build-up of sediment with periodic flooding.

Crane's (1982) model involved a wider range of processes than the earlier models and attempted to reduce arbitrary effects to a minimum. For example, deposition on the floodplain was not fixed and new accretion on the floodplain was evaluated frequently; lithology of the overburden was taken into account when compaction was calculated; the occurrence of avulsion was determined by reference to floodplain topography and the mean flood flow velocity required to erode the floodplain sufficiently to cause channel relocation. Crane (1982) used his model to investigate sand–fine ratios and sandbody interconnectedness for alluvial suites. He also examined the influence of the palaeoenvironment on the simulated suites by changing input parameters such as mean annual flood discharge and subsidence rate. Crane (1982) considered that the preserved fluvial succession, particularly sandbody interconnectedness, was strongly influenced by the deposition rate.

3.6.2 Functions for deposition used in the models

The deposition function used in Crane's (1982) program is a pseudo-logarithmic function relating the deposition on the floodplain to that in the channel:

$$DZ = DMB.e^{aZ + b} \qquad (3.3)$$

where a and b are constants, Z = distance across floodplain from edge of channel belt, DZ = deposition on floodplain at distance Z and DMB = deposition in channel. By substituting for two known distances, zero (at the edge of the channel belt) and one channel belt width from the channel belt edge, where the rate of deposition is denoted by $RDWMB$, Crane (1982) found the values of a and b so that the deposition equation became:

$$DZ = DMB . RDWMB^{Z/WMB} \qquad (3.4)$$

where WMB = channel belt width.

Crane (1982) compared the deposition profile of his model with that of Bridge and Leeder (1979). They used an algebraic function:

$$DZ = DMB . (Z+1)^{\log e\, RDWMB/\log e\, WMB\, +\, 1} \qquad (3.5)$$

which produces an apparent rapid decrease in sedimentation near to the channel belt and a roughly constant rate across the floodplain. The function used by Bridge and Leeder (1979) was based on an equation presented by Kesel *et al.* (1974) derived from the study of a single flood. Crane's (1982) deposition function produces a variety of profiles dependent on the value given to the rate of deposition at one channel belt width from the channel belt edge (Figure 3.14).

3.6.3   Substitution of the numerical model of James (1985) for the deposition function in Crane's program

In order to incorporate James' (1985) model within Crane's program, the sediment concentrations produced by the model had to be converted to thicknesses deposited. The model computes suspended sediment concentration values throughout the flood water above the floodplain. These are presented in the form of a grid of concentration values relative to a given reference concentration for the channel. The thickness of deposition is time dependent and is not specifically predicted by James' (1985) model.

It is suggested in the Flood Studies Report (NERC, 1975) that a flood hydrograph can be approximated by a simple triangle, as in Figure 3.15. The discharge Q takes time t_p to reach a peak (Q_p). Time t_1, is the total duration of the flood (i.e. the total time discharge has exceeded bankfull discharge $Q_0 = Q_1$). The change in discharge with time is represented by the line t_0, Q_0-t_p, Q_p-t_1, Q_1. Comparison with Figure 3.3 indicates that (apart from "noise" created by irregularities in a natural system) it is not unrealistic to assume a simple triangle for a flood hydrograph.

The change in suspended sediment concentration within the channel over the duration of the flood is also required in order to use James' (1985) model, since a reference concentration is required as a basis for the calculations. As the model was tested by relative comparisons with test data, a reference concentration of 100 was used by James (1985) and in the test with Severn flood data above. However, to calculate thickness of sediment deposited by floods, a more realistic reference concentration is required. Walling (1974) examined the relationship between discharge and suspended sediment concentration for some British catchments and found a proportional relationship:

$$C_s = kQ \qquad (3.6)$$

where C_s (in kg/m^3) is the suspended sediment concentration, Q (cumecs) is the discharge and k is the constant of proportionality. Similar results were obtained by Carson *et al.* (1973) on the Eaton River, western United States. By observation from Walling's (1974) graphs and those of Carson *et al.* (1973), a value of 10 for k seems reasonable. However, if different geomorphic effects were to be included in the model, the value of k could be altered to reflect higher or lower suspended sediment yields in the catchment. Walling (1974) observed that suspended sediment concentration was controlled by season, with storms in early summer resulting in the highest concentrations. Bogen (1980) examined the suspended sediment loads of glacial meltwaters and observed similar changes with season and from year to year. He suggested that the concentration depended on the rate of increase in discharge and, in particular, on the length of time that had elapsed since water had flowed over the area exposed to fluvial erosion by the event measured. A flood which exceeds the level of

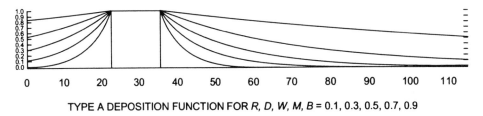

TYPE A DEPOSITION FUNCTION FOR *R, D, W, M, B* = 0.1, 0.3, 0.5, 0.7, 0.9

Figure 3.14 Examples from Crane's (1982) program for the deposition rate at various distances across the floodplain, see text for explanation of symbols

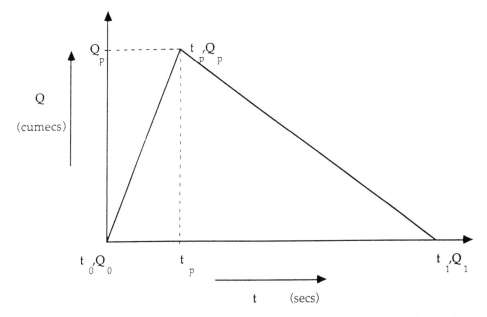

Figure 3.15 Flood hydrograph as a simple triangle, see text for explanation of symbols

erosion after a long low water period carries large amounts of sediment. Thus climatic effects could also be modelled by varying the value of k.

As, for the purpose of this simulation, a simple relationship is assumed between discharge and suspended sediment concentration, the triangle in Figure 3.15 can be used to show the change in sediment concentration over time. From this, the thickness of sediment deposited can be calculated (Marriott, 1993):

$$\text{thickness} = (w \cdot C_{sp} \cdot t_1)/(2 \cdot pf) \qquad (3.7)$$

where w = particle fall velocity, C_{sp} = suspended sediment concentration at peak discharge, t_1 = duration of flood and pf = porosity factor.

An estimate for the duration of a flood can be calculated from comparison of the average flood discharge for a given return period (Qd) and the peak discharge (Qp) (NERC, 1975). Thus the duration of a flood can be shown by (Marriott, 1993):

$$t_1 = (1 - r^2)/2r^2 \qquad (3.8)$$

where r represents the reduction ratio Q_d/Q_p.

With adjustments made to include the calculations for thickness of sediment

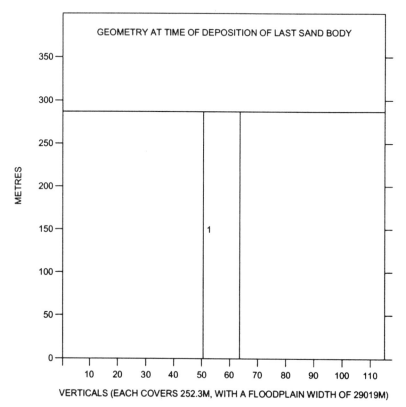

Figure 3.16 Output diagram from Crane's (1982) program using a high value for *RDWMB* (0.9) and a low value for *YFA* (0.7). See text for explanation of symbols

deposited, the program of James' (1985) numerical model was incorporated into
Crane's (1982) program.

3.6.4   Effects of the new deposition function on the simulation model

Crane (1982) suggested that the deposition function had a significant effect on the
simulated alluvial architecture derived from his program, and it can be seen from
the profiles obtained using his deposition function (Figure 3.14) that the value given
to the variable *RDWMB* (the rate of deposition at one channel belt width from the edge
of the channel belt) would influence the resulting overall sand–fine ratio and interconnectedness of sandbodies. With a high value for *RDWMB* (0.9), thick deposits are
built up near to the channel and for a considerable distance across the floodplain. If a
low value (0.7) is given to the other key variable, *YFA* (representing the erosive power
of the flood, see below), the river is unable to migrate or avulse and a stable system
results (Figure 3.16). Using Crane's (1982) program incorporating a deposition function
based on James' (1985) model, with the same low value for *YFA*, the simulated alluvial
succession (Figure 3.17) shows that for some periods the channel system was stable,

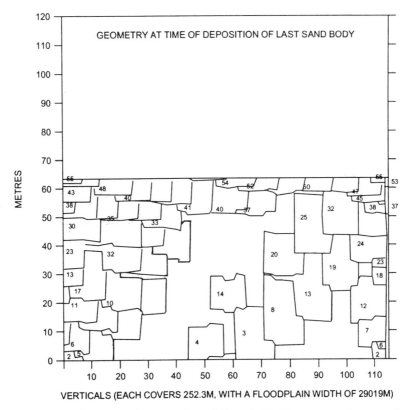

Figure 3.17   Output diagram using a version of Crane's (1982) program incorporating James'
(1985) numerical model. The value of *YFA* is 0.7

allowing thick sandbodies to develop, and at other times the channel avulsed and migrated. This has resulted in complex alluvial architecture with thick, interconnected sandbodies. In this amended program, deposition over the floodplain is not governed by an operator-set variable (i.e. the variable *RDWMB* is not used); instead, the thickness of deposition is influenced by the magnitude of periodic flooding calculated within the hydrological sections of Crane's (1982) program.

Within Crane's (1982) original program, deposition is calculated according to the maximum flood occurring in each 50-year period. The peak discharge for each 50-year flood is calculated using a random number generating subprogram. This is the only random element introduced into the program.

Using a low value for *RDWMB* (0.4), the deposition on the floodplain is considerably less near to the channel (see Figure 3.14), giving distinctive topography and increasing the (likelihood that avulsion will occur. Crane (1982) included the provision that avulsion could occur due to the following: (1) the base of the channel had built up above the level of the floodplain; (2) a severe flood caused sufficient erosion to result in relocation of the channel; (3) a pre-set period of (modelled) time had elapsed. Provision (3) is operator selected and was not used in the examples presented here. In

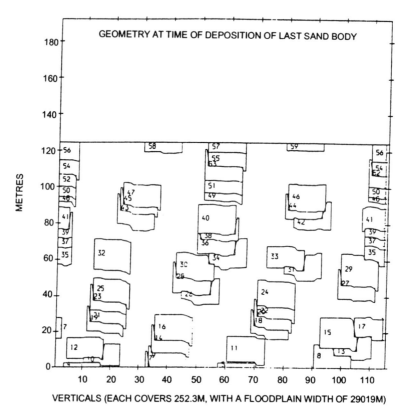

Figure 3.18 Output diagram from Crane's (1982) program using a low value for RDWMB (0.4) and a high value for *YFA* (1.0)

the event of avulsion occurring with situations (1) or (2), Crane (1982) allowed for topographic effects and proposed that the new channel would be located at the lowest point on the floodplain; if two localities were equal in level then the one nearer to the original channel position was chosen. This allows for the effects of compaction of different sediment types and the influence of underlying lithologies, in particular, old abandoned (buried) channel belts. It also follows the examples in Allen's (1978) exploratory work.

Figure 3.18 shows the simulated succession with values of 0.4 and 1.0 for *RDWMB* and *YFA* respectively, using the original program. It shows that avulsion occurred frequently and also that old channel belts were eroded by severe floods and therefore made available as sites for the new channel after avulsion. Using the revised program, avulsion is also shown to have occurred frequently. The channel also appears to migrate easily, combing the floodplain (see Figure 3.19). The difference in sandbody type, from isolated channel sandbodies (e.g. numbers 10 and 16 on Figure 3.19) to the interconnected series (e.g. numbers 17–22 on Figure 3.19) is probably related to the magnitude of flooding during the modelled period covered. A prolonged period of moderate floods would allow a relatively thick accumulation of floodplain sediment and result in an isolated channel sandbody, whereas a period of successive severe floods would result in very frequent avulsion due to the lack of time available for

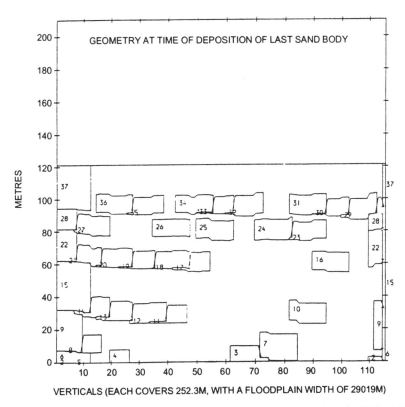

Figure 3.19 Output diagram from the revised version. The value of *YFA* is 1.0

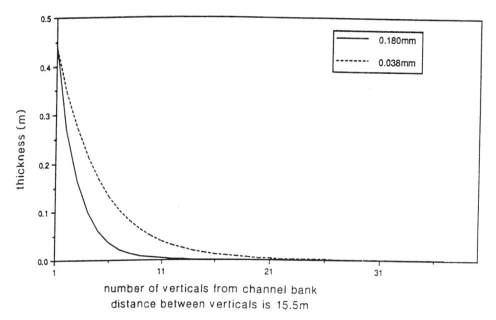

Figure 3.20   Thickness of deposition for different grain sizes for average 50-year floods using James' (1985) model in Crane's (1982) program

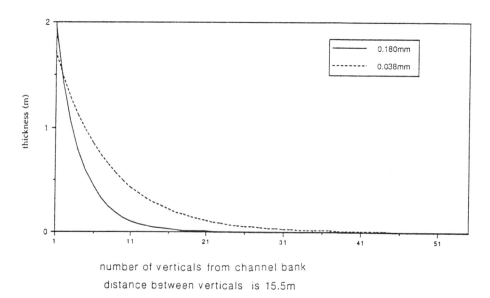

Figure 3.21   Thickness of deposition for the extraordinary 1500-year flood

build-up of sediment and because of the high value set for *YFA*. Additionally, the channel has to avulse to a vicinity very near its previous position (see numbers 17–18 on Figure 3.19) as a previously abandoned old channel (number 16) is still forming a topographic high on the floodplain The lowest point on the floodplain (and, as defined above, the most suitable for the position of the new channel) is near to position 17.

From these examples, it appears that the revised program may be providing more realistic patterns than the original. The deposition pattern for the revised program was examined for the first ten 50-year floods and plotted on a graph of thickness against distance across the floodplain (Figure 3.20). Two different grain sizes were investigated, namely, fine sand (0.180 mm) and fine silt (0.038 mm). It can be seen that, associated with these average 50-year floods, there is very little deposition of either fine sand or silt further than 250–300 m from the channel. In fact, it was only from an extraordinarily severe flood which occurred after 1500 years of model time, that sediment was deposited on distal floodplain areas (Figure 3.21).

3.7   DISCUSSION

It is generally assumed that floodplain deposits decrease in mean grain size as distance from the channel increases and that sand is deposited near to the channel, with fine sediment at greater distances (Bridge and Leeder, 1979). The results presented here support the first assumption but do not show that sand is absent from distal floodplain deposits. The graphs of mean and standard deviation against distance from the channel (Figure 3.4 a,b,c) record a sharp decrease in mean grain size within 20 m from the channel bank, but the patterns produced by the sand–silt–clay diagrams (Figure 3.5 a,b,c) indicate that sand is present throughout.

During flooding, the flow is relatively slower and shallower over the floodplain than in the channel, so its competence to carry suspended sediment is lower. The turbulent transfer of such sediment from the channel may therefore overload the flow on the floodplain near to the bank and result in deposition. The turbulent eddies are intense at the boundary between the channel and floodplain, but flow velocity declines outward away from the channel (Sellin, 1964; Rajaratnam and Ahmadi, 1979; Allen, 1985, p. 129). This leads to the coarser-grained sediment falling out of suspension near to the channel and a gradual decrease in mean grain size of deposited material, as distance from the channel increases. The distance over which particles are transported by turbulent diffusion depends on the characteristics of the channel and floodplain and the depth and velocity of flow (James, 1984). The difference in relative depth between the two flows influences the extent of the interaction zone (i.e. the zone of mixing of the two flows). It is possible that the sharp decrease in mean grain size at about 20 m from the channel bank (Figure 3.5a,b,c) relates to the areal extent of the interaction zone for the Severn flood. The turbulence associated with an interaction zone has been observed experimentally (e.g. Sellin, 1964), but field observations have been hampered by the adverse conditions experienced during flooding.

The sharpness of the decrease in average grain size is apparent from the histogram of central tendencies (Figure 3.8), which indicates a distinct lack of sediment in the coarse silt size range. As the Upper Severn and Teme drain a catchment of Palaeozoic and older rocks, and the Middle and Lower Severn and Warwickshire Avon flow over Mesozoic rocks (Figure 3.1), the clue to the lack of coarse silt size sediment may lie in these areas, since the contributions from these catchments to the flood differed (see ratio curve on Figure 3.3).

An alternative explanation may be indicated by the cross-plot diagrams (Figure 3.7a, b). These show two populations with distinct mean/standard deviation and mean/ skewness relationships which may reflect different modes of sediment transport. Near to the channel, turbulent diffusion would result in sediment of a range of grain sizes being deposited (see above); at distance, and at the margins of the floodplain in topographically lower-lying areas, water was ponded for several weeks. Fine-grained sediment was able to settle out from suspension during this time. Prior to ponding, flood water flows over the floodplain in a down-valley direction, so a proportion of the sediment transported and deposited from this flow would be derived from entrainment of existing floodplain material, i.e. sediment deposited by previous floods, colluvium from terrace edges and wind-blown sediment. However, since fine sand is present as well as silt in these more distal areas (see Figure 3.5a,b,c), it must be assumed that coarse silt would also be transported by the mechanisms in operation. In addition, the separation of the sediment grain size parameters into two distinct clusters could be an artifact of the two methods of grain size analysis undertaken. However, both methods covered a range of grain sizes from medium sand to clay, so the paucity of sediment of coarse silt size in the source areas is more likely to be the reason for the absence of significant quantities of coarse silt in the sediment deposited by the flood event under study.

At localities where flow in the channel is inclined to flow on the plain, the additional process of convection is involved in sediment transfer (James, 1985). The effects of this at the channel margins were shown by Toebes and Sooky (1967), who demonstrated the distribution of velocity vectors during overbank flow in a meandering river (see also Jackson, 1975). Convective transport often results in crevasse splays (e.g. Ritter, 1975). Toebes and Sooky (1967) also found that secondary, clockwise and anticlockwise helicoidal currents are established in a transverse direction at the meander bends. As discharge rises the strength of the helicoidal currents also increases, with a corresponding increase in the sediment transported from the outer to the inner bank (Hooke, 1975; Kikkawa et al., 1976). At bankfull stage, considerable thicknesses of sediment can be deposited at the inner bank (Bridge and Jarvis, 1982).

During flooding on the Severn, bankfull stage was exceeded for three weeks. Samples were collected after discharge receded and the water had retreated into the channel. The thick sandy deposits with ripple marks and dunes found at the channel bank at some localities, e.g. EE, EK, WA and WE (Figure 3.2), may be the result of transfer from the river bedload by secondary helicoidal currents acting at meanders as discharge fell.

The inclusion of a version of the numerical model of James (1985) as a deposition subroutine in Crane's (1982) simulation model does appear to influence the resulting alluvial suite. The location of the channel following avulsion, according to the assumptions built into the original program (see above), depends mainly on the floodplain topography. The topography, in turn, is influenced, during the operation of the program, by deposition, erosion and compaction. The mechanisms of overbank deposition modelled by James (1985) are turbulent diffusion and convection. Convective transport of sediment is particularly influential where the flow on the floodplain is at an angle to that in the channel (James, 1984, 1985) (see above). Consequently, as the original model produces a flow-normal section, diffusion is the only process considered in the version incorporated in the computer program.

The graphs in Figures 3.20 and 3.21 show that sediment is transported only a limited distance by diffusion and areas distant from the channel are only affected by extraordinarily severe floods. Other processes such as wash from terraces, colluvial deposits and aeolian transport are probably responsible for deposition of sediment at distal floodplain positions. Bates *et al.* (1992) (also see Bates and Anderson, 1993, and Bates *et al.*, Chapter 7, this volume) have developed a model for flow of floodwater on floodplains which shows the influence of floodplain topography on flood flowpaths. Their model indicates the likely importance of sediment transport from up-valley floodplain areas.

Pizzuto (1987) found that fine-sand-sized sediment was transported further across the floodplain than his model predicted and suggested that bedload transport might be responsible. James' (1985) model is shown to predict satisfactorily that fine sand can be transported by a diffusion process fairly large distances across the floodplain (up to 500 m, in the case of the River Severn). The values for vertical and horizontal diffusivity of suspended sediment used by Pizzuto (1987) and derived in James' (1985) model were based on work done in experimental channels (Lau and Krishnappan, 1977; Rajaratnam and Ahmadi, 1981; Ikeda and Nishimura, 1985), due to the fact that studies examining or modelling diffusivity values in overbank flows are lacking. It was shown that the diffusivity values have a strong influence on the predicted deposition profiles using Pizzuto's (1987) equation (Figures 3.11 and 3.12). Investigations into suspended sediment diffusivity during overbank flow would be worthwhile.

The large scale of the modelled channel/floodplain system of Crane's (1982) simulation for alluvial architecture seems to require the inclusion of further process models to allow for sediment deposition at the outer margins of the floodplain. However, it is not unreasonable to suggest that James' (1985) model does predict accurately what might occur during a period of 1500 years, since the average 50-year flood may not cover the whole extent of the large floodplain involved in the model. James (1990, personal communication) has suggested that the process of turbulent diffusion alone probably does not account for transport of sediment to a great distance across a floodplain.

3.8 SUMMARY

Examination of sediment collected following January–February 1990 flooding in the Tewkesbury area of the Severn floodplain indicates that the mean grain size of the flood deposits falls rapidly within 20 m of the channel bank, although fine sand is present in the samples across the entire width of the floodplain. The sharp decrease in mean grain size may indicate a shortage of coarse silt size sediment in the source areas. Alternatively, different sediment transport and deposition mechanisms may have been responsible for sediment distribution in different parts of the floodplain. This may refer especially to the influence of the zone of turbulent eddies where the channel and floodplain flows mix.

Transfer of sediment during overbank events may be modelled by analogy with a diffusion process. Models by Pizzuto (1987) and James (1985) were examined and tested using data from the River Severn. Pizzuto's model was found to predict the floodplain topographic profile unreliably, although, when revised values for vertical and transverse diffusivity were introduced, the predicted distribution of fine sand, in particular, was improved.

James' (1985) numerical model was, however, considered to predict more satisfactorily the distribution of sediment deposited during overbank flow. When James' (1985) model was used with geometric and hydraulic data from the Severn flooding, the relative distributions of sediment obtained were fairly consistent with the grain size distribution of the samples of flood sediment.

James' (1985) numerical model was incorporated into Crane's (1982) computer model for the simulation of alluvial architecture and appears to produce more realistic floodplain topography. The process of turbulent diffusion alone will not account entirely for the deposition on floodplains.

ACKNOWLEDGEMENTS

I acknowledge the assistance of NRA Severn Trent region for details of channel cross-sections and hydraulic data and Professor. C.S. James and Dr R.C. Crane for permission to use their computer programs.

REFERENCES

Allen, J.R.L. (1978) Studies in fluviatile sedimentation: an elementary geometrical model for the connectedness of avulsion related channel sandbodies. *Sedimentary Geology*, 24, 253–267.

Allen, J.R.L. (1985) *Principles of Physical Sedimentology*. George Allen & Unwin, London.

Baker, V.R., Kochel, R.C. and Patton, P.C. (eds) (1988) *Flood Geomorphology*. Wiley, New York and Chichester.

Bates, P.D. and Anderson, M.G. (1993) A two-dimensional finite element model for river flow inundation. *Proceedings of the Royal Society of London A*, **440**, 481–491.

Bates, P.D., Anderson, M.G., Baird, L., Walling, D.E. and Simm, D. (1992) Modelling floodplain flows using a two-dimensional finite element model. *Earth Surface Processes and Landforms*, *17*, 575–588.

Beven, K. and Carling, P.A. (eds) (1989) *Floods: Hydrological, Sedimentological and Geomorphological Implications*. Wiley, Chichester.

Bogen, J. (1980) The hysteresis effect of sediment transport systems. *Norsk. Geografisk Tidsskrift*, **34**, 45–51.

Bridge, J.S. (1979) A FORTRAN IV program to simulate alluvial stratigraphy. *Computers and Geosciences*, 5, 335–348.

Bridge, J.S. and Jarvis, J. (1982) The dynamics of a river bend: a study in flow and sedimentary processes. *Sedimentology*, 29, 499–541.

Bridge, J.S. and Leeder, M.R. (1979) A simulation model of alluvial stratigraphy. *Sedimentology*, **26**, 617–644.

Brown, A.G. (1983) Floodplain deposition and accelerated sedimentation in the Lower Severn basin. In: Gregory, K.J. (ed.), *Background to Palaeohydrology*, Wiley, Chichester 375–397.

Brown, A.G. (1987) Holocene floodplain sedimentation and channel response of the lower River Severn, UK. *Zeitschrift für Geomorphologie, Neue Folge*, **31**, 293–310.

Carling, P.A. and Petts, G.E. (eds) (1992) *Lowland Floodplain Rivers*. Wiley, Chichester, 302 pp.

Carson, M.A., Taylor, C.H. and Grey, B.J. (1973) Sediment production in a small Appalachian watershed during spring runoff: the Eaton Basin, 1970–1972. *Canadian Journal of Earth Sciences*, **10**, 1707–1734.

Crane, R.C. (1982) A computer model for the architecture of avulsion controlled suites. Unpublished PhD thesis, University of Reading.

Hooke, R. Le B. (1975) Distribution of sediment transport and shear stress in a meander bend. *Journal of Geology*, **83**, 543–565.

Ikeda, S. and Nishimura, T. (1985) Bed topography in bends of sand–silt rivers. *Journal of Hydraulic Engineering ASCE*, **111**, 1397–1411.

Jackson, R.G. II (1975) Velocity–bedform–texture patterns of meander bends in the lower Wabash River of Illinois and Indiana. *Geological Society of America Bulletin*, **86**, 1511–1522.

James, C.S. (1984) The distribution of fine sediment deposits on floodplains. *International Symposium on Urban Hydrology, Hydraulics and Sediment Control*, University of Kentucky, July 23–26, 1984, 247–254.

James, C.S. (1985) Sediment transfer to overbank sections. *Journal of Hydraulic Research*, **23**, 435–452.

Kesel, R.H., Dunne, K.C., McDonald, R.C., Allison, K.R. and Spicer, B.E. (1974) Lateral erosion and overbank deposition on the Mississippi River in Louisiana caused by 1973 flooding. *Geology*, 2, 461–464.

Kikkawa, A., Ikeda, S. and Kitagawa, A. (1976) Flow and bed topography in curved open channels. *Journal of the Hydraulics Division ASCE*, **102**, 1327–1342.

Lau, Y.L. and Krishnappan, B.G. (1977) Transverse dispersion in rectangular channels. *Journal of the Hydraulics Division ASCE*, **103**, 1173–1189.

Leeder, M.R. (1978) A quantitative stratigraphic model for alluvium, with special reference to channel deposit density and interconnectedness. In: Miall, A.D. (ed.), *Fluvial Sedimentology, Memoirs of the Canadian Society of Petroleum Geologists*, **5**, 587–596.

Marriott, S.B. (1992) Textural analysis and modelling of a flood deposit, River Severn, UK. *Earth Surface Processes and Landforms*, **17**, 687–697.

Marriott, S.B. (1993) Floodplain processes, palaeosols and alluvial architecture: modelling and field studies. Unpublished PhD thesis, University of Reading.

NERC (1975) *Flood Studies Report, vol. 1, Hydrological Studies*. Whitefriars Press, Institute of Hydrology, Wallingford.

Pizzuto, J.E. (1987) Sediment diffusion during overbank flows. *Sedimentology*, **34**, 301–317.

Rajaratnam, N. and Ahmadi, R.M. (1979) Interaction between main channel and floodplain flow. *Journal of Hydraulics Division* ASCE, **105**, 573–m588.

Rajaratnam, N. and Ahmadi, R.M. (1981) Hydraulics of channels with floodplains. *Journal of Hydraulic Research*, **19**, 43–60.

Ritter, D.F. (1975) Stratigraphic implications of coarse-grained gravel deposited as overbank sediment, Southern Illinois. *Journal of Geology*, **83**, 645–650.

Sellin, R.H.J. (1964) A laboratory investigation into the interaction between the flow in the channel of a river and that over its floodplain. *La Houille Blanche*, **7**, 793–802.

Toebes, G.H. and Sooky, A.A. (1967) Hydraulics of meandering streams with floodplains. *Journal of Waterways and Harbors Division ASCE*, **93**, 213–236.

Walling, D.E. (1974) Suspended sediment and solute yields from a small catchment prior to urbanisation. In: Gregory, K.J. and Walling, D.E. (eds), *Fluvial Processes in Instrumented Watersheds*, Institute of British Geographers Special Publication 6, 169–192.

Wolman, M.G. and Leopold, L.B. (1957) River floodplains: some observations on their formation. US *Geological Survey Professional Paper*, **282C**, 87–107.

4   Floodplain Palaeoenvironments

A.G. BROWN

Department of Geography, University of Exeter, UK

4.1   APPROACHES TO THE RECONSTRUCTION OF FLOODPLAIN PALAEOENVIRONMENTS

Floodplains form probably the most marked ecotones in the landscape. They are also unique among terrestrial landforms in having a well-preserved record of both their physical and biological history. There are a number of reasons for reconstructing floodplain environments:

1. To infer past processes of floodplain formation/evolution.
2. To elucidate the effects of past environmental changes on floodplain processes and evolution.
3. To define intrinsic controls on floodplain processes and to define boundary conditions for hydraulic models of floodplain inundation and sedimentation.
4. To reconstruct past bioclimatic conditions.
5. To reconstruct past ecologies and ecological change.
6. To reconstruct past land-use history and settlement history.

The first three of these justifications are primarily geomorphological, and form the main focus of this chapter, while the fourth is climatic, the fifth is botanical and the last is archaeological. This illustrates the interdisciplinary nature of studies of floodplain palaeoenvironments and it follows that work done for other purposes, especially by archaeologists, can be extremely valuable for studies of changing floodplain conditions and geomorphology (Needham and Macklin, 1992; Brown, 1996) and so no apology is made for including much recent alluvial geoarchaeological work in this chapter.

4.2   STRATIGRAPHY AND FLOODPLAIN EVOLUTION

Floodplains are complex assemblages of landforms which, as shown in Figure 4.1 include: channel features such as bars and bedforms, channel edge features such as banks benches and levees; and floodplain features such as old channels (oxbows), old levees (scroll bars), backswamps and crevasse-splays. The existence, development and arrangement of these features is in effect a record of the past history of the river and its

Floodplain Processes. Edited by Malcolm G. Anderson, Des E. Walling and Paul D. Bates.
© 1996 John Wiley & Sons Ltd.

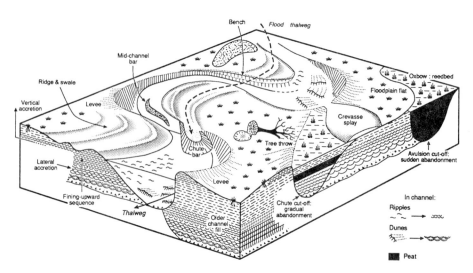

Figure 4.1 Block diagram of landforms associated with a meandering river and its floodplain (adapted and modified from Allen, 1965)

current activity including floodplain formation. Our ability to "read" the history of a floodplain from its sediments has come from experimental studies of river sedimentation and observations of contemporary rivers in many different environments.

As early as 1906 it was realized that floodplains were not solely created by floods (Fenneman, 1906), but that deposits formed by river migration were also an important component. Although well established in the literature, the terms vertical sediments, as used to refer to overbank deposits, and lateral sediments, as used to refer to within-channel deposits, are not ideal, as within-channel deposits may have an important vertical element from deposition at the shoulder of the point bar and on the bank (Taylor and Woodyer, 1978). Silt and clay may also be deposited in dead-water zones in the channel or at the entrance to cutoffs creating a clay plug (Figure 4.2, Erskine *et al.*, 1982). Indeed, an irregular, sinuous river prone to cutoff meanders and channel bifurcation can deposit a wide range of grain sizes from clay to gravel within the channel zone. Similarly, overbank sediments may not be laterally synchronous, as the location of the river may shift over time, as may crevasse-splays which are caused by levee failures. Sediments other than those produced by floods or channel migration are also common in floodplain sequences: these include lacustrine sediments deposited in floodplain lakes; wind-blown sediment including loess (usually reworked, cf. Burrin and Scaife, 1984); and colluvium from adjacent hillslopes. In very active geomorphic regions (i.e. high slopes and/or flashy hydrometeorological regimes) another set of sediments may be dominant, including landslide, debris-flow and mudflow deposits. These can also be important in temperate regions, even in the Holocene fill, and they can destroy or bury all archaeological evidence leaving little from which to construct the palaeoenvironment of a site or reach.

Many sedimentology textbooks (e.g. Reineck and Singh, 1973) contain idealized floodplain cross-sections in order to illustrate the different types of floodplain sediments and their spatial relationships. Facies models have been refined using

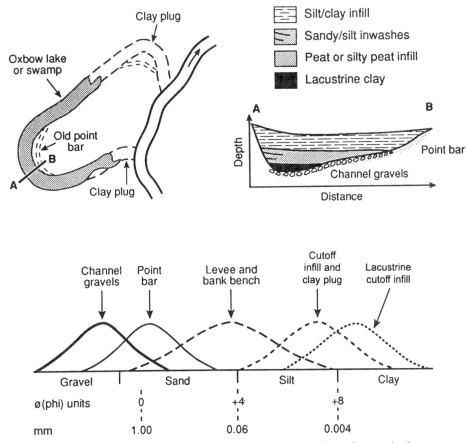

Figure 4.2 A chute cutoff and its generalized grain size distributions from point bar, concave-bank bench (also known as a slough) and the cutoff fill. Adapted from a Murrumbidgee example of Erskine *et al.* (1982)

statistical models of grain size (Brown, 1985), and chemical and biological analyses. However, there is no universal model and the occurrence of, or relative proportions of, each sediment type is controlled by river type and behaviour and is itself a measurable parameter of a floodplain fill. Figure 4.1 shows a segment of floodplain created by a meandering channel and Figure 4.3 shows three different but typical multiple-channel floodplains, each associated with different dominant modes of sedimentation: (a) a low-sinuosity sandy braided (unstable multiple channel) system; (b) an intermediate sinuosity anastomosing (stable multiple channel) system; and (c) a sinuous avulsion-dominated system. In reality these types are not always discrete, but may involve combinations, such as meandering multiple channels and intermediate types. Each of these channel–floodplain types, including anastomosis (Knighton and Nanson, 1993), can be seen as part of a continuum which may operate spatially, from one reach to another, or temporally, when the dominant mode of floodplain sedimentation changes due to changing flow, sediment calibre or sediment availability. This idealized pattern is further complicated by the fact that rivers very rarely if ever start off with a "clean

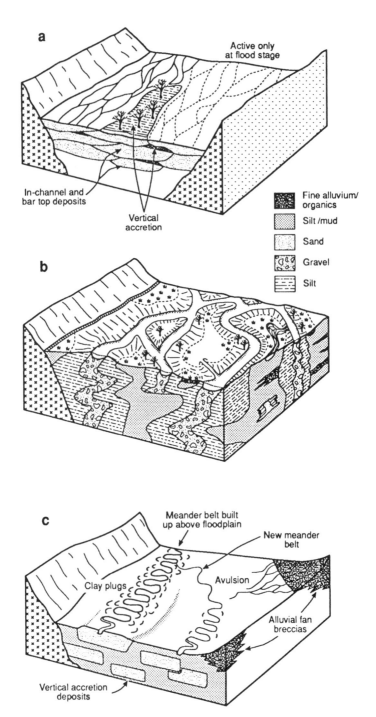

Figure 4.3 Idealized and simplified floodplain sedimentary systems: (a) a low-sinuosity sandy-braided floodplain; (b) an intermediate sinuosity anastomosing floodplain; and (c) a sinuous avulsion-dominated floodplain. Adapted and redrawn from diagrams in Smith and Smith (1980) and Allen (1965)

slate"; some forms inherited from a preceding regime will remain and influence contemporary processes of floodplain formation. This is the antecedence described by Croke and Nanson (1991), or inheritance, as described by Brown (1989) from the River Perry in Shropshire. In the Perry valley, the distribution of Holocene sediment types and rates of deposition were largely controlled by inherited downstream changes in slope (a basin-gorge topography) which have caused persistent relative downstream variation in stream power.

We know more about small meandering streams and their associated floodplain sediments than other channel types. This is partly due to the classic study by Wolman and Leopold (1957). They proposed a model of floodplain formation based on geomorphologic reasoning and observations of a small stream in the USA, Watts Branch Creek. In their model overbank, flood deposits are limited, because as each flood layer is deposited it requires successively higher, and therefore more infrequent floods to overtop the banks, leading in time to a decrease in the rate of overbank sedimentation. They argued that this self-limiting system restricted the occurrence of overbank sediments to approximately 20% of the sedimentary fill. This argument was supported by the observation that rivers had a constant recurrence interval of bankfull discharge. However, as Kennedy (1972) has pointed out, even the data they used, does, in fact, show that bankfull recurrence interval can vary by a factor of four. We now know that, within broad limits, bankfull capacity is very variable, both reach-to-reach and within reaches, and that it can change, even over short periods of time. There are two important assumptions that underpin the Wolman and Leopold (1957) model. The first is that there is no deposition on the channel bed as this would facilitate a constant, or even increasing, rate of overbank deposition (i.e. both bed and floodplain aggradation) and the second is that the channel migrates across the entire width of the floodplain. Most channels do not do this, especially those of larger rivers; instead they oscillate across a relatively small belt or belts of the floodplain. A good example of this is the Brahmaputra in Bangladesh where the meander belt itself displays a meandering pattern (Coleman, 1969) and has migrated due to tectonic activity. Indeed, we can say that rivers commonly display a preferred position within floodplains, which is one possible cause of floodplain and valley cross-sectional asymmetry (Palmquist, 1975). Also, some rivers are remarkably stable over long periods of time despite having a meandering planform. For these reasons, exactly the opposite ratio between lateral and vertical sediments to that predicted by the Wolman and Leopold model is possible, as for example is the case with the floodplain of the Delaware River in the USA (Ritter et al., 1973). The Wolman and Leopold model remains valid for the type of system they studied, which is a relatively small floodplain with a relatively high stream power. Their model also highlighted the importance of bed aggradation and degradation for the control of channel form and the mode of floodplain sedimentation.

Since overbank sedimentation is also controlled by the flood-series and sediment concentrations, there can also be continuing floodplain aggradation with no change in bed height, given increased flood magnitudes or sediment availability. This mechanism has been referred to as the stable-bed aggrading-banks (SBAB) model by Brown and Keough (1992a) and is illustrated in Figure 4.4. In this model the channel bed remains at approximately the same elevation while overbank deposition raises the height of the banks and floodplain. The increase in channel discharge required for continued

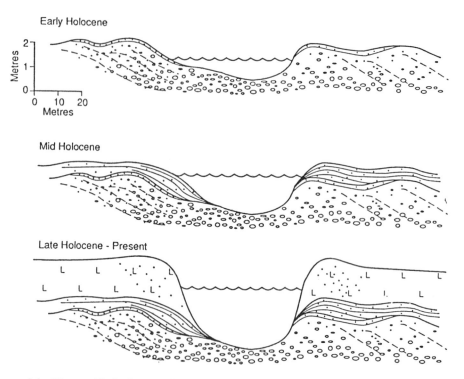

Figure 4.4 The stable-bed aggrading-banks (SBAB) model of floodplain and channel evolution, which facilitates a continued linear rate of increase in floodplain elevation and a constant recurrence interval of overbank events due to the increase in flow passing down the primary channel as secondary channels silt up. For more details see Brown *et al.* (1994)

overbank deposition as the bank height is raised can be the result of a number of mechanisms including a reduction in the number of live-water channels (due to the abandonment or siltation of channels in a multiple-channel system); a change in flow partitioning in a multiple-channel system (Harwood and Brown, 1993); a reduction in channel width; or possibly external factors such as climate or land-use change. The first two mechanisms are ways in which discharge can be accommodated as a system changes from an anastomosing pattern to a single-thread meandering pattern due to increased sedimentation or channel management. In reality most floodplain fills show evidence for a variety of processes including, point-bar sedimentation, overbank deposition, subaerial processes (soil formation and bioturbation), bed aggradation/ incision and channel abandonment. The discrimination of some of these facies and the inference of processes of formation can often only be made from open sections; however, it may be possible from large-diameter cores and thin-section and scanning electron microscopy. A classic example of a compound stratigraphy is that of the Duck River, Tennessee (Brackenridge, 1984), which shows a series of lateral units decreasing in age across the floodplain but all overlapped by a fine unit of recent age (Figure 4.5). This overlapping is caused by lack of incision of the channel into the bedrock and an increase in discharge and sediment loading due to land-use change associated with

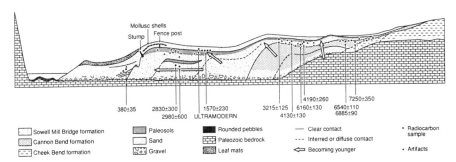

Figure 4.5 The generalized stratigraphy, artifact distribution and dating of a cross-section of the Duck River floodplain, Tennessee, USA. Adapted from Brackenridge (1984)

European colonisation. The existence of floodplain units or "parcels" of very different ages at approximately the same elevation and the time-transgressive overlapping of older units by overbank silts is typical of medium–low energy temperate floodplains.

The inference of past pathways and rates of floodplain evolution is highly dependent upon dating control. While there is not room here to review recent advances in dating alluvial sediments (see Brown, 1996) it should be noted that, in addition to radiocarbon and dendrochronology, recent success has been achieved using palaeomagnetics (Brown and Ellis, 1995) and optical stimulation luminescence (Macklin, personal communicator; Price, personal communication) although this does depend on the degree of bleaching that has occurred (Rhodes and Pownall, 1994).

4.3 FLOODPLAIN BIOGEOMORPHOLOGY AND ECOLOGICAL PROCESSES

As with all attempts to reconstruct palaeoenvironments, the present is one important key to the past. Floodplains are more productive than the land that surrounds them, due to the availability of water and nutrients. However, their productivity still varies with latitude and local climatic factors. The net above-ground primary productivity of a forested fen in Minnesota has been measured as $746 \, \mathrm{g\,m^{-2}} \, \mathrm{year^{-1}}$, while that of a Louisiana bottomland hardwood forest is $1374 \, \mathrm{g\,m^{-2}} \, \mathrm{year^{-1}}$ (Mitsch and Gosselink, 1986) Net productivity and the difference between floodplain productivity and that of surrounding land is controlled by water supply and is proportional to the difference between potential evapotranspiration and actual evapotranspiration. In subtropical climates, the difference is at its maximum because of the greater year-round supply and storage of water in floodplains. In cool temperate zones the contrast is smaller and differences in soil type and agricultural potential become more important. In the boreal zone, disturbance of vegetation by highly active rivers complicates the situation.

4.3.1 Hydrogeomorphological controls on floodplain vegetation

Temperate and boreal floodplains commonly contain several nested wetland systems which can be classified according to their position on a hydrodynamic energy gradient

(Gosselink and Turner, 1978). They vary in hydrological regime and in their water–plant–sediment interactions. The single most important regulator in alluvial wetlands is usually the hydroperiod, which reflects the duration, frequency, depth and season of flooding and the depth of the water table (Lugo *et al.*, 1990). Floodplains are therefore especially sensitive to changes in catchment conditions, including land use; so, even if they are unmanaged, their ecology will be affected by human activity. They are also more changeable over time than most other terrestrial ecosystems, due to both autogenic and allogenic factors. Under certain circumstances floodplains may record their own ecological and hydrological history in organic and mineral sediments and this provides us with an invaluable tool in the reconstruction of past environments, hydrological conditions and resources.

The primary controls on floodplain ecology are the water regime and the stability of each patch of floodplain surface. Secondary controls include soil type and fertility. All of the, factors are intrinsically related to the pattern and processes of floodplain formation. The water regime of floodplain soils varies from free-draining, drought-prone soils on levees to permanently waterlogged hydromorphic soils in backswamps. Soils also vary in texture from sands and gravels in high energy zones to silts and clays in low. energy zones, with organic matter content ranging from medium to very high levels (over 60% in some backswamps) The water regime of floodplain soils is a function of the flow regime of the river, adjacent slope inputs and the hydrological properties of the floodplain soils and sediments (Bradley and Brown, 1992, 1995). The most obvious example of the effect of the sediments themselves is the existence of perched aquifers. These are common in many temperate floodplains and are formed by impermeable layers in depressions sealing them off from the strata below, so that when the regional water table falls the site remains waterlogged (Brown and Bradley, 1995). The most common case is a cutoff with clays deposited in the oxbow lake, prior to

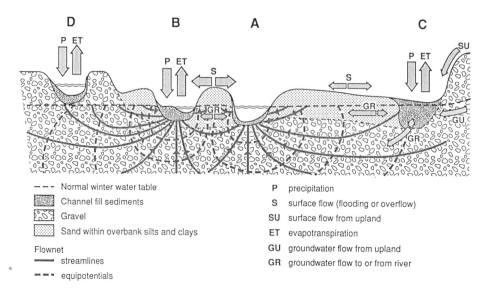

Legend	
--- - Normal winter water table	P precipitation
Channel fill sediments	S surface flow (flooding or overflow)
Gravel	SU surface flow from upland
Sand within overbank silts and clays	ET evapotranspiration
Flownet	GU groundwater flow from upland
streamlines	GR groundwater flow to or from river
equipotentials	

Figure 4.6 A simplified representation of the hydrology of a temperate floodplain

channel incision and a fall in the floodplain water table. Using a general model of the hydrological characteristics of floodplain soils, we can generate a simplified model of floodplain hydrology (Figure 4.6). Inputs are from rainfall, throughflow and channel flow.

The floodplain water table will reflect the hydraulic conductivity of floodplain sediments. For a given hydraulic gradient the water table will follow the ground topography more closely in soils with low hydraulic conductivity (e.g. silts and clays) and less closely in sediments with higher hydraulic conductivities (e.g. sands and gravels). This means that it may be quite irregular across the floodplain, reflecting the different rates of lateral and vertical water transmission. The hydrology of floodplains also provides the natural controls for floodplain ecology, as the patchwork of soils with different water regimes, stabilities and nutrient contents allows plants with different adaptive strategies to co-exist on floodplains. For temperate floodplains, it is possible to divide floodplain environments into three broad hydrogeomorphological categories: (a) relatively free-draining and coarse-textured soil environments; (b) levees and floodplain flats; and (c) backswamps and floodplain hydroseres (e.g. cutoffs). An additional vegetation-dominated environment described below is that of the organic debris dam and associated dead-water zones (d).

(a) Sand and gravel environments

These are not only free-draining, but frequently disturbed, and have undergone little or no weathering *in situ*; they thus have little soil development and low nutrient levels, especially nitrogen. Most nutrients must be derived either from allochthonous mud and particulate organic matter or directly from the river water. This niche is occupied by plants which have relatively short life cycles, expend a relatively high proportion of energy on reproduction, are tolerant of disturbance and frequent inundation, and are shade intolerant. The life-form of these plants tends to be herbaceous or shrubby. Typical sedimentary environments include: point bars, braid bars, mid-channel bars and crevasse-splays.

(b) Levee and intermediate flat environments

Natural levees, which tend to be most pronounced in mixed load systems where there is sufficient sand, frequently carry rather different vegetation from the rest of the floodplain. Natural levee vegetation does, however, grade into the floodplain flat community. The lack of natural floodplains in Northwest Europe and much of North America has meant that most studies relate to subtropical and tropical environments or boreal environments, and very little is known about the natural levee communities in temperate floodplain systems. The levee and channel edge, including the intermediate zone (Petts, 1987), is in reality a microgradient composed of a series of zones which include the water's edge, the bank edge, the levee summit and a slope grading into the floodplain flat (Figure 4.7). The ecology in each of these zones typically varies from emergent aquatics such as bulrushes (*Typha latifolia*) to dryland herbs. It is also not clear to what extent this variation has changed since the removal of floodplain forest. An excellent example of the relationship between vegetation associations and

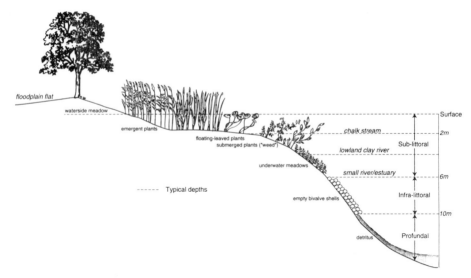

Figure 4.7 Channel, edge ecotones and the vegetation gradient. Adapted from Naiman *et al.* (1989)

floodplain microtopography is given by Wharton *et al*'s. (1982) studies of bottomland hardwood forests in southeastern USA (Table 4.1). The microrelief and the vegetation does not show a smooth change from one zone to the next and communities vary due to fluctuations in the elevation of any geomorphological feature.

(c) Backswamp environments

More is known about the ecology and palaeoecology of these environments of temperate floodplains, because they were the last areas to be cleared and have in some locations survived to the present day. The high organic matter content of backswamp soils is caused by a lack of free oxygen (anaerobic conditions) due to waterlogging, which results in low decomposition rates. Plants in this environment face two principal problems: firstly, a lack of available nitrogen due to the rate and pathway of decomposition and, secondly, a lack of oxygen, producing reducing conditions. These conditions favour many of the monocotyledon species (rushes, reeds and grasses) and some of tree species, especially those able to fix nitrogen directly, such as the alders (*Alnus* spp.).

In addition to these environments, the land–water ecotones are extremely important in floodplain systems as they are characterized by higher biological diversity than adjacent patches and can occur in practically all floodplain environments (Naiman *et al.*, 1989).

(d) Organic debris, dams and beavers

Vegetation can also create hydrological and sedimentary environments. The cleared and tamed floodplains of North America and Northwest Europe are far removed from the

Table 4.1 The general relationship between vegetation associations and floodplain topography

Floodplain zone	Vegetation association
river channel – edge	submergent – emergent aquatics – black willow – cottonwood
natural levee	sycamore – sweetgum – American elm
backswamp of first terrace flat	sugarberry – American elm – green ash or overcup oak – water hickory
low first terrace ridge	sweetgum – willow oak
high first terrace ridge	upland forest: white oak, blackgum, white ash, hickories, winged elm, loblolly pine
oxbow	bald cypress – tupelo
second terrace flats	sweetgum – willow oak
low second terrace ridge	swamp chestnut – cherrybark oak
high second terrace ridge	upland forest
upland	upland forest

largely forested floodplains from which they evolved. Along with ecological change, some important channel environments have largely disappeared. Fortunately, studies in some of the few remaining forests large enough to have natural floodplains have illustrated the potential role of these processes. In forested floodplains, trees fall into and across rivers; there is also a high input of organic material of all sizes from trunks to the microscopic. McDonald *et al.* (1982) have shown that in the Redwoods, northern California, USA, this organic debris can exert a fundamental control on channel pattern and channel change. Obviously, the Redwoods are rather an extreme case, but studies in other ecosystems have shown similar processes operating, if not so dramatically. Even in small floodplains (with small trees), such as in the New Forest in Hampshire, UK, organic debris forms small dams and tree falls can cause channel diversion (Davis and Gregory, 1994). These retain fine sediment and, when they burst, send small flood-waves down the channel (Gurnell and Gregory, 1984). Studies of a small multipie-channel (anastomosing) river in southwestern Ireland have shown how organic debris dams are one of the causes of channel diversion and the maintenance of an intricate and diverse fluvial system (Harwood and Brown, 1993; Brown, *et al.*, 1995). In boreal environments trees much smaller than the channel widths are still important because they can from a nucleus for bar formation (Nanson, 1981). Even in arid and semi-arid environments organic debris can be important. The lower Murray in Australia is now completely blocked by river red gums which used to be regularly cleared out for the paddlesteamer traffic.

When considering the impact of organic debris on past channels in cleared landscapes, there are two points worth remembering. Firstly, trees of the wildwoods were not of the same shape and size as those of the same species we see today; for example, both birch and alder are commonly thought of as rather short, almost scrubby trees, but this need not be the case as both can grow to over 20 m with straight single trunks and form a canopy crown. This obviously makes a considerable difference in terms of their potential hydrological effects. Secondly, in some regions, mammals such as the beaver, can have a major effect on fluvial processes. The European beaver

(*Castor fiber*), which became extinct in the British Isles early in the Medieval period, can have a considerable impact upon forest growth, local groundwater tables and river conditions. A mature European beaver is about 1 m long and weighs 15–30 Kg. It will fell just about any tree species, up to a diameter of about 1 m, for food and shelter, although it prefers poplar (*Populus*), willow (*Salix*) and maple (*Acer*). The lodges beavers construct across streams can be up to 3 m high and are constantly repaired. The trees within them may also sprout. The effect of beavers can be extremely important because they will attempt to dam all the watercourses in an area, often the only limit being set by existing dams. The rise in the local water tables kills trees by paludification, and once dead the trees are easily blown down by wind. It is not known how long the dams and lodges generally last, but some are known to have existed for over 1000 years (Coles, 1992). The result will be a stepped stream profile with very effective sediment traps and the accumulation of lake-like sediments in small, steep, stream valleys, a phenomenon hard to explain without some local control on baselevels. Coles (1992) has suggested that beavers may have been a major cause of late Mesolithic and early Neolithic clearance in the UK, through the paludification of river and stream-side forest. The North American Beaver (*Castor canadiensis/Castor fiber*) still has a major impact on Canadian fluvial systems, especially in protected areas such as the Algonquin National Park.

The pattern of these environments and habitats, including those caused by beaver is fundamentally controlled by the geomorphic system including the size, slope and regime of the river, its sediment load and its history of channel movement. The species composition of these environments is also the result of adaptive evolution and adjustment to climate change, including the return-migration of species in areas glaciated during the Pleistocene. These ecological factors operate through colonization and succession.

4.3.2   Colonisation and succession

Newly created landsurfaces on floodplains can be colonized by plants whose seeds can arrive by wind, animal vectors or water. Water is also an effective means of vegetative colonization through the transport of living gametes along with organic debris. The colonization of surfaces is rapid because of these different mechanisms and the proximity of seed sources. However, seedling mortality is often very high. Floodplains are not simple hydroseres, but have strong components of hydroseral succession within their pattern of vegetation change. Whereas strongly deterministic patterns can be shown in the vegetation change of cutoffs (open-water species → floating aquatics → attached aquatics → reedbeds → willow or alder scrub), this is just one component of floodplain vegetation and all these successional stages can occur in close proximity at all times. Neither is the floodplain just a simple patchwork of parcels of different successional stages, as events such as floods, channel change and changes in the sedimentation rate will affect all parcels to some extent, and these are normal components of the system. Because the pattern of vegetation colonization is so closely tied to floodplain development, Everitt (1968) was able to construct maps of the age of the floodplain of the Little Missouri river in North Dakota, USA, from coring and ring-counts of cottonwood trees (*Populus sargentii*). The floodplain forest is composed of a series of even-aged stands which exhibit a progression in age up-valley and away from the channel, providing a complete record of the historical migration of the channel. This allowed Everitt to calculate floodplain deposition and erosion rates. He also noted how flood-training of saplings could occur whereby a sapling bent parallel to the flow is buried and shoots produce linear clumps of trees as a result.

Our ideas concerning succession have been profoundly changed by studies of wetlands. The classical view of hydroseral succession where waterbodies are terrestrialized by the build-up of inorganic and organic material cannot fully explain either the pattern of communities on floodplains or observations of floodplain vegetation change. The end-point of dry land is highly unlikely without hydrological change. Walker (1970) found from pollen and bog stratigraphy that successional sequences in northern peatlands were variable, with reversals and skipped stages, and these were influenced by the dominant species first reaching a site. A bog rather than terrestrial forest was the most typical end-point. Wetlands may also exist virtually unchanged for 1000 years or more, while some respond to decadal shifts in the precipitation/evapotranspiration ratio (Barber, 1976; Barber *et al.*, 1994). All alluvial wetland successions are influenced by allogenic factors, including flooding, groundwater changes, climatic change subsidence and vegetation disturbance. Van der Valk (1981) has therefore replaced the classical autogenic succession model by a Gleasonian model in which the presence and abundance of each species depends upon its life-history and its adaptation to the environment of a site. The environmental factors make up the "environmental sieve" in this model and, as the environment changes, so does the sieve and hence the species present. In reality, both autogenic and allogenic forces act to change, alluvial vegetation.

The floodplain is also part of the river continuum (*sensu* Vannote *et al.* 1980). The reason why zonation is so sharp in many floodplains is that environmental gradients are ecologically steep and groups of species have similar tolerances and so tend to group on these gradients. The principal gradients are those already identified: hydroperiod, water-table depth, and soil fertility. The systematic nature and pattern of floodplain sedimentation creates environments on these gradients, so creating chronosequences of vegetation. Chronological or developmental patterns of floodplain vegetation are the result of both autogenic and allogenic forces. Considerable emphasis is therefore put upon allogenic factors in the continuum model of wetland and floodplain vegetation. The river/floodplain ecosystem can also be regarded as a continuum from headwaters to the sea. Organic matter is the linking and driving component of the river continuum model of fluvial ecology (Vannote *et al.*, 1980). The model, which is analogous to the model of downstream physical energy expenditure devised by geomorphologists (Langbein and Leopold, 1966), proposes that energy input and organic matter transport, storage and use by macroinvertebrate functional feeding groups (e.g. shredders, grazers, collectors and predators) is largely regulated by fluvial geomorphological processes (Vannote *et al.*, 1980). The continuum model provides a standard against which changes in organic matter inputs can be compared both for present, and theoretically, for past hydrological systems. However, it is not a universally applicable model and it is biased towards those processes that predominate in forested landscapes.

4.4   FLOODPLAIN PALAEOECOLOGY

Floodplains of large rivers generally offer a range of potential sites that can yield palaeoecological data. These range from floodplain lakes, through much smaller dead-water bodies to lateral peat deposits and even in some areas raised bogs (i.e. ombrogenous or rain-fed bogs). Many palaeoecological studies on floodplains have made implicit use of th fact that floodplains present a variety of different successional habitats in different stages o development at any one time and also that some of these habitats preserve a record of past ecological change. This has been made more explicit in the model of synchronic and diachronic analysis used by Amoros *et al.* (1987) and illustrated in Figure 4.8. Most of the techniques used in the diachronic analysis of floodplain palaeoecology have been adapted from lake and peat bog studies. Despite many practical and interpretive disadvantages, floodplains do have one great advantage for palaeoecological work and that is their frequency. Even in areas barren of lakes, they permit palaeoecology to be studied from the uplands and glaciated regions to the lowland plains. One disadvantage is frequent disturbance that may produce hiatuses in the record. The ability to infer ecological conditions outside this zone is a function of the palaeoecological method or data source and the size and location of the palaeoecological site.

The preservation of organic remains is very variable in floodplain environments and largely determined by the water regime (Brown and Bradley, 1995). Chemical conditions are of secondary importance and, as long as decay is slowed down or inhibited by waterlogging, organic remains, including non-woody tissue and pollen, can survive in both acidic and alkaline environments. This means that, for *in situ* macroscopic remains, there is a bias towards the wetter floodplain environments. Permanent waterlogging can occur in non-organic and coarse deposits; therefore, if wood and other organic materials are deposited contemporaneously with gravels and

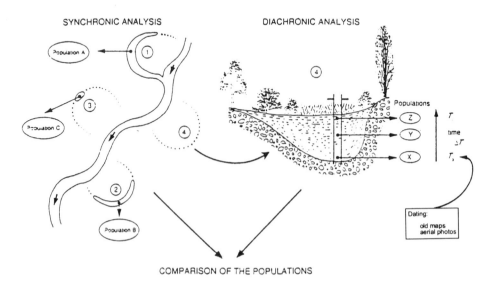

Figure 4.8   Synchronic and diachronic analysis from Amoros *et al.* (1987). Reprinted by permission of John Wiley & Sons Ltd

they remain waterlogged, they will be preserved. Beyond these general taphonomic statements, more specific constraints have to be examined organism by organism.

In practice, no method really provides a comprehensive picture of past ecological conditions, only a fragmentary picture of some parts of the ecosystem from which the rest is inferred. Often, it is only a part of the system that is under investigation, such as water chemistry or forest resources. The choice of palaeoecological method is inevitably a compromise between the questions most in need of being answered, and the opportunities offered by the environment. The questions asked of palaeoecological data vary from those requiring relatively little inference from the data to those requiring far too much. Problems arise because the death assemblage may not mirror the living assemblage in, for example, age structure, and many animals do not expire in random locations but tend to die in "cemetery" locations. Living assemblages are themselves not always easy to describe and classify, as witnessed by the vast literature on the subject, since the result largely depends upon the method used, e.g. life-form versus floristic classification in plants. We also have to assume that the ecophysiology of the species has not changed in the time period being studied. An example of the practical problems this poses is given by coleoptera: they are excellent indicators of past temperatures (Atkinson *et al.*, 1986) until the late Holocene, when human activity, producing new habitats and internal climates, partially invalidates the climatic-range method.

Ecological change can occur when there is no environmental change; this is because not all populations are in equilibrium with their environment, especially climate, because of the lags produced by the processes of colonization and succession. Since life on the Earth is structured in trophic levels, with each level dependent upon the level below, there will be a lagged response to any external change and this lag will generally increase with increasing trophic status. This problem is at its most acute in glaciated latitudes of the Old World and North America, where vast areas have been in climatic disequilibrium for a significant part of the Pleistocene. If we wish to use palaeoecology to reconstruct past environmental controls on floodplains, then we must be able to isolate and quantify these environmental controls. This is not easy, as is shown by the volume of modern ecophysiological studies. In particular, problems have arisen because it is now clear that the law of limiting factors (Leibig's law) as originally proposed is inadequate, as it is not necessarily the one factor which is in least supply in relation to demand that controls productivity (e.g. water, light, nitrogen and other nutrients), but complex interactions between factors involving inhibition, blocking and soil micro-organisms (Bannister, 1976).

4.4.1 Plant macrofossils

Plant macrofossils, which are plant remains visible to the naked eye, are both common and varied in alluvial deposits in the temperate zone. They are variable both in type and provenance, the latter being a matter of interpretation from the depositional setting and evidence of transport. Wood and woody tissue can preserve well in waterlogged conditions, with the internal structure or wood anatomy allowing identification to the genus and often species level. Although few experiments have been undertaken to confirm this, observation suggests that bark removal is caused by fluvial transport and exposure, so no bark removal at all suggests very little transport and limited exposure prior to burial. Whether the wood is allochthonous or autochthonous is not easy to prove, except where it is in its life-position. An example of life-position is where the tree-bole is sitting upright or near upright with the roots penetrating the underlying gravel strata. Apart from palaeoenvironmental information wood now provides the highest resolution dating through the use of regional dendrochronological curves extending back over most of the Holocene, which now exist for most of Northwest Europe and much of North America. Large tree trunks are particularly common in gravels in Central Europe and the Netherlands. These are most commonly oaks (*Quercus*, so-called black oaks or ranen), but occasionally other species occur, especially pine (*Pinus*). From such ranen a detailed record of fluvial activity has been constructed for the Main and Weser rivers (Becker, 1975; Becker and Schirmer, 1977) and they have been used to date sedimentary units in the Rhine, Danube and Vistula (Starkel, 1991). They are rare in the British Isles, the only significant accumulation being in the Middle and Lower Trent valley (Salisbury *et al.*, 1984; Salisbury, 1992), due to the lack of gravel units of middle and late Holocene age, although black oaks are more common in organic units especially around the coast and in large wetland basins. The difference between the British Isles and mainland Europe is probably due to the maritime fluvial regime of British rivers, the deposition of peats, silts and clays rather than sand and gravel and early deforestation of terraces and valley sides.

Cutoffs often contain *in situ* plant macrofossils which reflect the seral stages of cutoff infilling. There may also be an allochthonous input by wind, tributary streams or floods. For example, alder catkins are commonly found some distance from alder trees and in accumulations of flood debris. There are many pathways by which macrofossils may be incorporated into the floodplain fill, as well as problems associated with differential production and differential preservation (Watts, 1978). It is possible for very resistant macrofossils to be reworked into later deposits, and this may produce anomalous radiocarbon dates. At least in glaciated regions, this problem is limited to Holocene materials, and will not influence estimates of the antiquity of the earliest occurrence of any particular species in the palaeoecological history of a floodplain. In practice, detailed stratigraphic work, attention to the condition of the macrofossil fractions and the identification of any anomalous components will minimize the risks of dating reworked macrofossils.

4.4.2   Plant microfossils: pollen, spores and diatoms

Pollen analysis and macrofossil analysis are complementary techniques. Pollen analysis can provide a picture of the off-site as well as site vegetation, but it is often taxonomically less precise. Although in general macrofossil-rich sediments contain pollen, this is not always the case, as demonstrated by two examples, one obvious, the other not: in coarse deposits trees and organic debris may be deposited in "natural traps", but pollen will not. Secondly, many of the chalk valleys of southern England contain substantial thicknesses of reed-peat but sometimes contain little or no pollen. In this case, the most likely cause is the water table which fluctuates relatively little but is slightly higher during the autumn and winter, allowing the preservation of dead plant material, but falls during the late spring and summer, causing enough oxidation and microbial activity to destroy the pollen but not the macrofossils (Barber, personal communication). Pollen may also be preserved in non-organic deposits such as loess and alluvial silts and clays.

Recent improvements in techniques have allowed pollen to be recovered from silts and clays where the concentration is relatively low. Many pollen studies of alluvial sediments have been reported as just lists of types and percentage occurrence; however, it has now been accepted for some time that from suitable sites (the most waterlogged), coherent pollen diagrams can be constructed (Brown, 1982; 1988; 1996). There are several floodplain facies from which percentage diagrams have been produced, including: cutoff infills, buried lateral peat beds, terrace depressions (if they are perched aquifers) and backswamp peats. If the site has no, or few, hiatuses and a reasonably smooth accumulation curve, it is possible to construct; influx diagram (using concentration × accumulation rate, in grains cm^{-2} yr^{-1}). Influx diagrams allow ecological processes such as succession and competition to be investigated using pollen analysis (Bennett, 1983; Brown, 1988).

The interpretation of floodplain and terrace depression pollen diagrams faces the same problems as other environments, along with some others. The universal problems which are well known and discussed in many texts can be summarized by the three differentials; differential preservation, differential productivity and differential transport/contributing area. Differential plant pollen productivity is not known to vary in alluvial as compared with non-alluvial sites. There is, however, evidence that pollen preservation varies with environment and that some alluvial environments may have biased spectra as a result. From the work of Sangster and Dale (1961) we know that some pollen types, such as poplar and bulrush, are more likely to be destroyed in ponds which are rather similar to cutoffs, than in lakes or bogs. Many alluvial diagrams have parts of the spectra where concentrations are very low and the number of types is reduced to those known to be more resistant, such as ferns (*Filicales*), other spores and *Compositae ligulflora* type (daisy family) (Lewis and Wiltshire, 1992). In addition, typically more degraded and damaged pollen is found in alluvial contexts due to abrasion by inorganic particles and partial oxidation. Some floodplain environments are not permanently waterlogged and exposure to continued evaporation is one of the most important causes of pollen degradation (Holloway, 1981); so the interpretation of pollen from these sites must pay due attention to differential degradation as well as the mechanisms of transport. Indeed, one of the most significant complicating factors in alluvial sites is the transport/contributing source area differential. Much work has been done on the wind-blown component and Tauber's (1965) model is applicable to floodplain sites such as oxbow lakes. The atmospheric pollen input can be divided into the rain-out component (largely long distance), recanopy component, the trunk-space component and the local component from plants growing on the site. The relative proportions of these components will vary with oxbow size. Several other workers have looked at the decrease in wind-blown pollen away from a vegetation boundary (Salmi, 1962) and found an exponential rate where the b term was affected by the prevailing wind direction and pollen grain size. Dispersal equations similar to those used for smoke plumes have been used to model this process (Prentice, 1985). The most widely quoted model, produced by Jacobsen and Bradshaw (1981), would suggest that from an oxbow lake of about 30 m in diameter (ignoring any shape effects) approximately 8% of the pollen would be of regional origin, 13% extra-local and 79% local. Their model does not have any site input, but cutoffs will have an important site input as they easily become choked with vegetation as succession proceeds. Figure 4.9 takes the Tauber model and adapts it for a typical cutoff, which also carries some inflow from a minor tributary. The lateral peat bed, which has accumulated as peat under a canopy of wet woodland, can be viewed as similar to a woodland soil with site size varying from zero (i.e. no canopy break) to the dimensions of a small forest hollow or clearing. In both cases, the main source areas are relatively small and this is why, in pre-deforestation floodplain sites, pollen diagrams generally reflect the local and extra-local zones. This is an advantage rather than a disadvantage if the floodplain vegetation is the focus of interest. However, in many investigations it is not and there are two approaches to "seeing through the floodplain corridor vegetation to the surrounding landscape" including terrace vegetation. One approach is to assume, preferably on the basis of

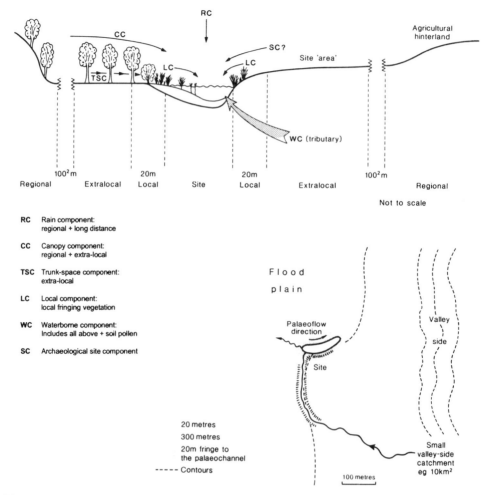

Figure 4.9 A modified version of the Tauber model for a small oxbow lake with a small stream input

macrofossil analysis, that all the inputs of certain pollen types are exclusively of floodplain origin and remove them from the sum (Janssen, 1959). Apart from the vast number of grains which may have to be counted to obtain a reasonable pollen sum, there are two other problems with this approach. Firstly, as succession occurs, local vegetation will change and so therefore should the sum, making the resultant diagram very difficult to interpret. Secondly, the assumption that all of a species such as alder comes from the floodplain will effectively render it invisible in the rest of the landscape, where it could be an important component. For these reasons, and to facilitate the comparison of diagrams, most workers have moved to a total land pollen sum which commonly only excludes aquatics, spores and sometimes Cyperaceae (sedges). The second approach is the comparison of floodplain and non-floodplain sites from the same region.

One reason why alluvial deposits were avoided in the past, even when they were adjacent to archaeological sites, was the fear that pollen would be selectively transported by water either into or out of the site. The ecological coherence of floodplain diagrams suggests this is not an insurmountable problem. This is supported by some laboratory experiments by Brush and Brush (1972) which suggest that differential settling out of flood waters is unlikely. Due to the constant turbulence (including thermally induced) and the low specific gravity of pollen, either all types will be suspended or all will settle out, whereas a substantial component of differential settling may occur with wind. So, while flood waters often contain high concentrations of pollen (Brown, 1985), flood deposits often contain very little. Indeed, Brown (1985) has recorded concentrations as high as 230 grains ml^{-1} during a flood on the Highland Water in the New Forest, United Kingdom. However, a silt flood layer sampled after the event, but before it had dried-out, had a very low concentration of under 10 grains cm^{-3}. As has been shown by studies of the pollen budgets of small lakes, most of this waterborne pollen enters lakes and ponds (Crowder and Cuddy, 1973; Peck, 1973). It should also be noted from these studies that the waterborne pollen spectra probably reflects the vegetation of the basin more accurately than the wind-blown component.

Pollen diagrams from floodplains and lakes are one of the prime data sources for studying the processes of floodplain colonization, competition, succession, natural disturbance and anthropogenic influence. Anthropogenic activity has received far more attention than the other factors mentioned, but Bennett (1983) has calculated the doubling times of the early Holocene rise in tree taxa around Hockham Mere, in the eastern United Kingdom, from an influx diagram. We now know that vegetation response significantly lagged behind climate change in the early Holocene and the composition of the flora varied due to the distance of refuges, differential migration rates, successional processes and local soil variations. This is indicated by significant differences between pollen diagrams from adjacent areas, such as the delayed appearance and rise of taxa, or even reversals in the order of dominance. Palaeovegetation maps have been produced in Europe (Huntley and Birks, 1983) and in the United States, using pollen analysis (Delcourt and Delcourt, 1981), and Bennett (1989) has attempted to map the forest types present in the United Kingdom at 5000 BP. At this scale, floodplains are invisible except for large low-lying areas such as the East Anglian Fens or Somerset Levels. At the regional scale, the role of floodplains becomes more evident, especially for wetland taxa. In this context the debate as to the processes involved in the immigration, rise and eventual dominance of alder (*Alnus*) and hazel (*Corylus*) in southern Britain is particularly interesting as it has geomorphological implications due to increased bank resistance and turbulence induced in overbank flows.

Because very different soils can lie adjacent to each other on floodplains, it is possible to add a spatial aspect to the pollen diagram interpretation taking into account the ecophysiology of the taxa. For example, Brown (1988) has shown that the very high lime pollen found in some floodplain sites comes from lime forest on adjacent terraces. This is confirmed by a decline in the values of pollen away from the terrace edge. Other factors such as natural vegetation disturbance on floodplains due to windthrow and floods are difficult to isolate from anthropogenic activity. Janssen (1972) has compared pollen diagrams from a small floodplain surrounded by sandy acidic uplands, the Dommel Valley, in North Brabant in the Netherlands, with extensive raised bogs 30–40 km to the east. This revealed that in the Sub-Boreal the uplands were covered by grassland, heath and oak–birch woods, while the valley was covered by alder and reed swamp. In order to evaluate the role of local effects, Janssen (1986) has proposed the use of contrast diagrams where the curves of selected taxa of different ecological affinities are superimposed for sites in the river valley and adjacent uplands. Traditionally, palaeoecologists have attempted to infer the spatial pattern by ecophysiological reasoning, in a similar manner to the identification of floodplain and non-floodplain components of a spectra. This is often reinforced by a knowledge of the local flora derived from macrofossil analysis. This procedure is ultimately unsatisfactory, as it assumes that the soil and water characteristics (of the soils) on different lithologies were similar to those today. There is here an important role for palaeosol pollen analysis in the determination of local soil/vegetation conditions.

In the course of identifying pollen and common spores, other far less well-known microfossils are frequently encountered. These have been extensively studied by Van Geel (1986) and are mostly fungal spores. Despite the problems of identification, some have been shown to have quite specific ecological affinities. Two that are common on floodplains include *Gaumanomyces* spores from a parasitic fungi which grows on Cyperaceae (sedges) and another is the ascospore *Mougeotia* which is only found in pools and puddles which dry out in summer. It is generally assumed that these non-pollen microfossils do not travel far and so may add extra information to the pollen analysis of local conditions.

Diatoms are unicellular algae which may be preserved in sediments because of a strong exoskeleton or frustule made of biogenic silica. They have a variety of life-forms and be found in a variety of habitats including large rivers, cutoffs, marshes and soils. They a relatively easy to extract (Batterbee, 1986) and can be identified below species level. Work on their ecology by Hustedt and others has led to a series of environmental spectra int which species may be allocated. These include a salt-tolerance spectra (the Halobian system), a nutrient status spectra (several methods including the Araphidineae : Centrales ratio), a pH spectra (Hustedt, 1930) and a thermal spectra. These systems all have considerable problems in their accuracy and applicability, although salinity and pH are the most reliable (Batterbee, 1986). Diatoms have as yet been used relatively little in the reconstruction of floodplain environments, despite occurring in high concentration in some backswamp deposits (especially epipelic and epiphytic species). Occasionally, bands of pure white diatomite can be found in these sediments. One reason for their neglect is that less is known about the environmental controls of attached species in comparison to planktonic species, although work has been done in rivers in the United Kingdom (Swale, 1969), the United States and elsewhere (Foged, 1947–1948). They may also be derived from a variety of environments upstream. However, Brown and Barber (1985) have shown how they can indicate a change in floodplain hydrological conditions, even prior to this being reflected in the sedimentary record. Diatom analysis has been applied to samples from the Medieval town of Svendborg in Denmark (Foged, 1978), illustrating the poor water quality of many of the town ditches, dykes and moats, and the proximity of saline water from the coast. Similar work has been undertaken on other waterfront sites, and in London diatom work has been undertaken in order to test the hypothesis that the Roman Thames might have been freshwater whereas the present river is tidal (supposedly due to tectonic subsidence). The diatom evidence from Roman waterfronts in the City of London clearly shows that the river was as tidal then as it is now (Batterbee, 1988).

4.4.3 Palaeohabitats; snails and insects

Much recent reconstruction of floodplain environments has used animal rather than vegetation data. Freshwater, and to a lesser extent, dryland mollusca are extremely common in alluvial deposits, especially the basal units of palaeochannel infills. They can even make up a significant proportion of the sediment. They are relatively easily extracted, although with clay-rich sediments disaggregation may be difficult without crushing the shells. In the United Kingdom, there are only some 50 native freshwater species and so identification to species level is not tremendously difficult. They have been extensively used in both interglacial and glacial contexts as palaeoclimatic indicators. In the Holocene, the climatic linkage is largely replaced by a habitat linkage associated with human activity. The percentage of species from dry ground can be highly significant, indicating direct soil input from valley sides (Shotton, 1978) and the freshwater species may be allocated to one of four environmental groups following Sparks (1961; slum, catholic, ditch and moving water). Work by Evans *et al.* (1992) has shown that variations in wet ground taxa can reflect changes in floodplain vegetation, although there are several problems caused by taphonomic mixing, behavioural mixing and the lack of modern comparative data. Evans *et al.* (1992) recommend the use of autochthonous deposits, such as fine-grained overbank deposits which lack river species or fine-grained river deposits which lack land species, in order to define past "taxocenes". Mollusca can also give sedimentological information, as death assemblages of freshwater mollusca can indicate dead channel areas or even extensive flood events. Brown (1983b) reported that after a flood of the River Stour in Dorset, United Kingdom, the floodplain was covered with mollusca and sediment flushed from the channel bed. The most common mollusca were the swan mussel (*Anodonta cygnea*) and duck mussel (*Anodonta anatina*); these species can also occasionally be found in life-positions in lower point-bar sediments.

In general, insects have several advantages for palaeoenvironmental work. These include: the vast number of species that exist, their penetration into almost all ecological niches, their strong relationships to environmental conditions and their high mobility. Of all the insects, beetles (coleoptera) are particularly useful, because their strong exoskeletons are preserved in a wide variety of sedimentary environments and because there is a long record of taxonomic work. Samples can be collected from exposures, disaggregated, sieved and coleoptera separated from other organic remains by paraffin flotation. While some species clearly indicate changes in climate (Atkinson *et al.*, 1986; Coope, 1986), others can indicate the presence of grazing animals (e.g. dung beetles), the presence of different types of waterbodies, from still to fast-flowing, and human activity including crops and buildings. In a study of the contemporary floodplain of the River Trent in the UK, three different habitats were differentiated by Greenwood *et al.* (1991): riparian damp grassland, fen (including cutoffs) and fen/carr (willow with seasonally standing water). It was also noted in this and another study (Shotton and Osborne, 1986) that flood debris typically contained species derived from all alluvial environments, but not species adapted to fast flow, which successfully held on! The ability of coleoptera to indicate specific past conditions has been used by Osborne (1988) to show that during the Bronze Age the River Avon in Worcestershire in the UK had a clean stony bed without the covering of mud and silt that is seen today. This is inferred from the large numbers of beetles from the family Elmidae in the sample. The beetles also showed that the floodplain was already cleared of trees by this time and covered with open grassland maintained by large grazing animals, presumably domestic cattle. There seems little doubt that the full potential of coleoptera for palaeoenvironmental analysis of floodplains has yet to be realized.

The insect remains most commonly recovered from alluvial contexts, other than coleoptera, are water-fleas or cladocera (Frey, 1986) and to a lesser extent water midges or chironomids (Hoffman, 1986). Cladocera are relatively easily extracted by methods similar to those used for pollen analysis, but without the use of acetylation which can damage the chitin exuvia (Frey, 1986). Over 90 species occur in European inland waters (Amoros and Van Urk, 1989), but the most useful for palaeo-environmental inference are found in still waters, including oxbow lakes, pools, marshes and backwaters. Cladoceran analysis of a core from a cutoff on the upper Rhône in France has provided evidence of succession caused by the embankment of the river in the 1880s (Roux *et al.*, 1989).

Chironomids, or in practice the head capsules, can be preserved in large numbers and have been shown to be good indicators of past ecological conditions (Amoros and Van Urk, 1989). These distinctive head capsules allow identification, although not always to species level: details of the methods of sampling and analysis can be found in Walker and Paterson (1985) and Hoffman (1986). Several workers have identified characteristic habitats which include: mobile sand, snag habitats produced by branches and trees in the river, the hyporheic zone (interstitial water-flow through gravels), and eroding clay banks (Barton and Smith, 1984). Klinke (1989) has shown that chironimids characteristic of shifting sand habitats were common in the Lower Rhine prior to the eighteenth century and their decline is probably due to channel constriction and deforestation of the floodplain. This study seems to show that past river entomology can be reconstructed accurately from vertically accreted alluvial sediments, allowing the possibility of extending the use of biological pollution indicators back into the past.

There are many other fossils that can occur in floodplain sediments and which can yield significant palaeoenvironmental information. These include mammals (especially small mammals), fish remains (Casteel, 1976) and rhizopods (Tolonen, 1986). However, most are rare, although some are more common on archaeological sites themselves. A possible exception to this are ostracods which are small arthropods (0.15–15 mm long), with a carapace composed of two chitinous or calcareous valves. They occupy nearly all marine and freshwater environments, and much is known about their rather restricted ecological preferences. Details of extraction and identification can be found in Brasier (1980). Although they have generally been used in the study of ocean cores and coastal environmental sequences, as they are excellent indicators of salinity, they have also been used in lacustrine (Frey, 1964) and alluvial environments where leaching has not occurred. In a study of alluvial peats from the Rivers Avon (Worcestershire) and Thames in the UK, Siddiqui (1971) identified 19 species. All the species are still living in Northwest Europe and their ecology suggested that the River Avon peat was deposited in a pool of standing water while the Thames peat accumulated in a channel or pool with slow-flowing water and abundant macrophyte growth.

4.5 HOLOCENE FLOODPLAIN PALAEOENVIRONMENTS IN NORTHWEST EUROPE

4.5.1 Disequilibrium, colonization and early Holocene floodplains

Dramatic change in fluvial regime occurred during the Lateglacial–Holocene transition. We now know that these changes were extremely abrupt, indeed the approximately 7°C warming at the end of the Younger Dryas may have occurred in as little as 50–100 years (Dansgaard *et al.*, 1989), a rate well in excess of the possible adjustment of either ecology or fluvial landforms. Studies of the Gipping by Rose *et al.* (1980) and the Nene and Soar by Brown (1995) show a sequence of change from braided to meandering and back to braided conditions. The result of these rapid changes was early Holocene floodplains in disequilibrium with discharge and climate; abundant palaeochannels undergoing hydroseral succession and an open floodplain being invaded by birch (*Betula*), hazel (*Corylus*) and possibly pine (*Pinus*).

During the recolonization of northern European floodplains, a new vegetation pattern emerged. This is well illustrated by Van Leeuwaarden and Janssen's (1987) analysis of the pollen record from oxbow lakes and upland pingo melt-holes in North Brabant, the Netherlands, during the Lateglacial and early Postglacial. They showed that the successional pathway varied between the two environments between 12 000 BP and 7300 BP, not so much in the order of appearance of the river-forest trees, but in the time of appearance due to edaphic differences between the more fertile and wetter soils in the valleys and the nutrient-poor upland soils. These can be described by the following successional notation:

valley $B_{PO} \rightarrow P_O \rightarrow P_{BC} \rightarrow P_C \rightarrow A_T$
upland $B \rightarrow B_C \rightarrow C_Q \rightarrow C_{QU}$

where the upper case is the dominant (subscript denotes also present) and A is *Alnus* (alder), B is *Betula* (birch), C is *Corylus* (hazel), P is *Pinus* (pine), Q is *Quercus* (oak), T is *Tilia* (lime), O is *Populus* (poplar) and U is *ulmus* (elm). Brown (1988) has also illustrated the emergence of the ecotone between the floodplain and terrace slope environments during the early to mid-Holocene in the West Midlands of Britain. The early rise in alder (*Alnus*) in some valleys, which still, however, occurred well after it first arrived in the region (assuming it had not survived through the Devensian) may well be due to out-competition of willow (*Salix*) which had itself out-competed birch (*Betula*), due to changing floodplain conditions and soil development. There are therefore complex reciprocal linkages between vegetation and sedimentation on Holocene floodplains. Irrespective of whether a native inoculum may have persisted through the Devensian, the most probable agencies of the spread and then the expansion of alder are: dispersal by birds and disturbance by beavers and human activity (Chambers and Elliott, 1989).

An important second consideration in the early Holocene is the strong role of inheritance in fluvial systems. This includes the filling of natural stores and changes in the drainage network and channel pattern which can produce changes in sediment conveyance and output with no change in climate or catchment controls (Brown, 1987). By the Atlantic period, floodplains had evolved to be highly interrelated biophysical forest systems upon which Neolithic and later peoples were to have a tremendous impact.

An important and remarkably rich early Holocene alluvial site has been excavated at Noyen-sur-Seine, northern France (Mordant and Mordant, 1992). The Mesolithic of the Paris basin is relatively well known, but most of the sites have come from gravel "islands" and organic channel and valley bottom sites are rare. The excavation of several over the last few years suggests this is due to factors controlling discovery (e.g. lowered water tables), rather than avoidance of valley bottoms by Mesolithic people. During the Pre-Boreal, the Seine valley at Noyen contained wide shallow channels which filled with gravel and organic silts. During the Boreal, there is evidence of additional small channels which changed rapidly and filled with gravel and organic debris. It is in association with these channels that the Mesolithic fish-traps were found, suggesting that they had been placed in small, shallow channels independent of the main river in a system that was probably transitional between braiding and anastomosing. During the Atlantic, the area reverted to grassy marshland and the channel was confined to approximately its present position in the early Bronze Age. The site has provided a high-resolution pollen diagram covering 4000 years which, interestingly, shows no evidence of Mesolithic clearance in a landscape clearly heavily utilized by humans (Leroyer, 1989, cited in Mordant and Mordant, 1992). The first evidence of clearance and fire is from the late Neolithic. The diagram does, however, reflect the variety of vegetation types in the valley: gravelly islands covered with grass and bushes, swamps and river channels bordered with alder woodland and loessic terraces supporting hazel, oak and elm forest.

4.5.2 Equilibrium, stability and middle Holocene floodplains

The late Boreal and early Atlantic periods saw profound changes in the floodplain environments of many northwest European rivers valleys. There is evidence from predominantly non-alluvial sites of a rise in water tables and from alluvial sites of either a lack of sedimentation or the initiation of cut-and-fill cycles (Robinson and Lambrick,

1984; Brown and Keough, 1992a; Macklin and Lewin, 1994). This and the changes that occurred in floodplain vegetation have been attributed to greater precipitation caused by increased oceanicity of the climate associated with the Holocene sea-level rise. Rivers may well have had more equable river regimes with higher summer flow and clear-water floods, although the increase in precipitation may, to some extent, have been offset by the higher evapotranspiration and interception losses associated with a forest cover (Lockwood, 1979). Although there is abundant evidence that alder was present in northern France and Britain before 10 000 BP, it was around 2000 years before it became a major forest component. The expansion of alder, although rather abrupt at most sites, was diachronous in Britain by at least 2300 years. These characteristics are consistent with a model of alder migration, probably with two or more European foci, with expansion being controlled by local factors including competition. It was not a simple response to climate change but probably related to the increase in fine and organic-rich fine soils covering and developing on coarser gravelly bars and banks (Brown, 1988). It is also possible that the spread of alder was facilitated by vegetation disturbance by beavers and Mesolithic gatherer–fisher–hunters (Smith, 1984). Once alder had invaded northwest European floodplains, it was not easily removed and had profound effects on channel processes and floodplain ecology. In particular, the dense alder–hazel woodlands inhibited channel change and were probably responsible for the maintenance of anastomosing channel systems into the Atlantic, for which there is evidence from the Thames (Needham, 1992); East Midland valleys (Brown and Keough, 1992b); the Gipping (Rose et al., 1980) and the chalk valleys (Evans et al., 1994). A similar co-evolution of vegetation and channel pattern seems to have also occurred in northern Germany (Hagedorn, personal communication) and can be inferred in northwestern France (Morzadec-Kerfourn, 1974).

Many studies of floodplain alluviation in Britain have shown considerable thicknesses of Lateglacial sediments and Post-Boreal sediments, but frequently relatively little sediment from the early–mid Holocene (Boreal and early Atlantic). This is illustrated by Macklin and Lewins's (1994) survey of dated alluvial units in lowland and upland Britain (Needham and Macklin, 1992) and longitudinal work on valleys in southern Britain (Burrin and Scaife, 1984; Scaife and Burrin, 1992). In the cases where several palaeochannels have been dated from one reach, there is often a conspicuous lack of evidence of channel abandonment during the period 9000–5000 BP. There are two possible causes. Firstly, floodplain erosion with incision and a net loss of sediment from the reach may have removed the evidence. However, in many lowland river valleys there is little evidence of significant early Holocene erosion, unlike the uplands where early Holocene terraces can be found (Macklin and Lewin, 1986). An alternative explanation is channel stability, relatively little channel change and low rates of, or no, overbank deposition across much of the morphological floodplain. This is in agreement with the vegetation evidence previously discussed derived from floodplain peats. The floodplain soils developed on Glacial and Lateglacial gravels were also in some cases stripped-off or truncated by overbank floods in the mid to late Holocene (Brown and Keough, 1992a).

In the larger, and more active lowland and piedmont rivers of Central Europe, which have a continental regime, we see more evidence of aggradation and incision cycles in the early Holocene. Studies on the Wisloka and other rivers in Poland (Mamakowa and

Starkel, 1974; Starkel, 1981) and in the former Czechoslovakia (Rybnicek and Rybnickova, 1987), have shown that the Boreal–Atlantic was characterized by fluvial incision and the erosion of floodplain peats. Not surprisingly, the lower reaches of many large continental European rivers display a tendency towards aggradation by fine rather than coarse sediments and peat growth (Vandenberghe and de Smedt, 1979).

Mid-Holocene (Neolithic) landsurfaces have been excavated from most of the major river valleys of Europe. Particularly well-documented evidence of Neolithic valley settlement comes from Pleszow in the Vistula valley, Poland. Godlowska *et al.* (1987) have examined the impact on the valley vegetation using pollen cores from oxbow lakes close to the Neolithic settlements on a loess terrace (Figure 4.10). The palaeochannels date from the late Mesolithic (before 6300 BP). Clearances are associated with each of the seven settlement phases from the Musik phase (B, around 4200 BC) through to the Funnel Beaker culture (G, 2800 BC). Rather complex changes in the floodplain environment occurred during the Zeliezovce cultural phase, as the valley floor became drier with a decline in water plants and increase in dryland plants. This may have restricted the growth of alder and promoted cereal cultivation on the lower terrace slopes. The relative importance of hydrological change and anthropogenic activity in the decline of alder at this time is unknown; however, alder recovered towards the end of the phase, when there was a period of flooding as recorded by a thin flood silt. The flooding forced abandonment of the lower fields and encouraged reed growth. As well as illustrating the dramatic changes that occurred on the floodplain during the Holocene, this study highlights an interesting problem; the extent to which the

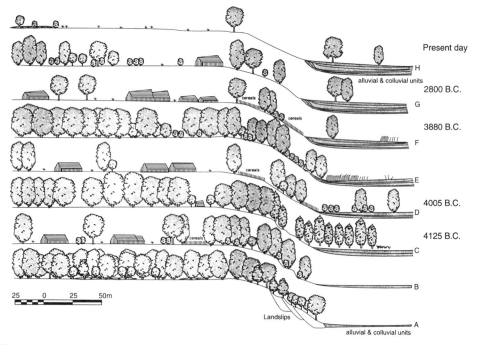

Figure 4.10 A visualization of the settlement phases at Pleszow, adapted from Godlowska *et al.* (1987)

expansion and contraction of arable land on floodplains was caused by fluctuations in the floodplain water table and flooding, before humans constructed drainage systems and flood defences.

During the mid-Holocene the floodplains of Europe were probably at their most diverse with regional variations in climate, soils and groundwater conditions largely unmitigated by human impact.

4.5.3 Floodplain deforestation

The pollen and macrofossil evidence indicates that by 6000 BP probably all lowland floodplains in Britain and northern France were dominated by types of alder–hazel–oak woodland, although there would undoubtedly have been regional variations in species composition with more oceanic elements in the west and continental elements in the east. In floodplain pollen diagrams, alder (the only native British species being *Alnus glutinosa*) is the dominant mid-Holocene taxon, frequently reaching 60% TLP (total land pollen). As Janssen (1959) pointed out, this can present problems, if it is the non-local vegetation that is of interest to the palynologist or archaeologist and in the underestimation of trees that are believed to have been characteristic of floodplains such as black poplar (*Populus nigra*, Peterken, personal communication) but which produce relatively little, very small or fragile pollen. In addition to pollen, there is wood evidence from many floodplains and even *in situ* remains of alder carrs (on peat) and alder wet woodland (on mineral soils), preserved by burial by sediment in many valleys, including the Thames at Runnymede (Needham, 1992). Brown and Keough (1992a) have excavated three alder woodland sites from the Middle Nene valley in the East Midlands, as shown in Figure 4.11. The stability of the alder woodland suggests that the community became non-invadable by other species probably due to waterlogging and substrate stability. There is very little evidence of the cyclic changes in vegetation that one would expect if there had been successional waves associated with channel migration. Alder also has the ability to survive flooding and a rise in the local water table, through the production of adventitious roots (McVean, 1954). The decrease in the number of pollen types that are recorded accompanying high alder values is probably not entirely due to the "drowning-out" effect, but also the limited diversity of alder woodland. The alder decline seen in nearly all pollen diagrams is highly diachronous. If we look at sites just within the River Severn catchment in western Britain, we can see that it varies from as early as 3100 BP to 2500 BP (Figure 4.12), and at one site in the Nene it occurs as early as 5200 BP. Many of these sites show a two-stage decline, with an initial fall, a recovery or standstill, and a second decline to very low values; at a few sites there are three stages (Brown, 1988). The step-like declines have been ascribed to management and abandonment, but they may simply be caused by the spatial pattern of deforestation in relation to the core location, i.e. block deforestation at some distance from the core which, for obvious reasons, is likely to be the last site deforested. The observation that this phenomena is less common or pronounced with other tree types (Brown, 1988) would fit, as they would rarely form such pure stands. There is ample evidence of the management of alder in later Prehistory; in fact, coppicing of alder is known from as early as the early Neolithic from the Somerset Levels (Coles and Coles, 1986), and Brown and Keough (1992a)

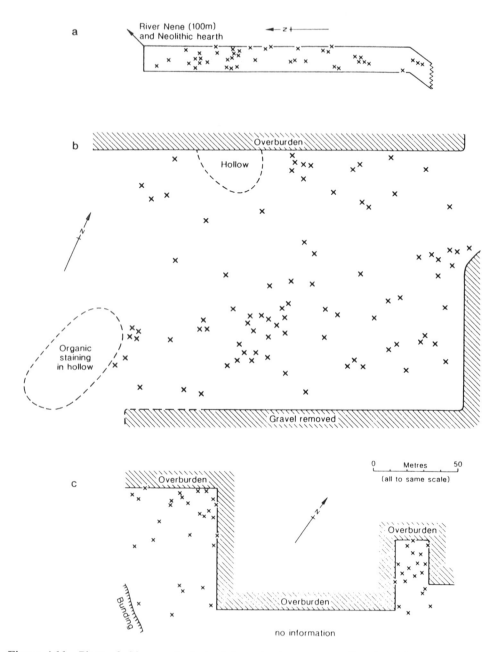

Figure 4.11 Plans of alder woods derived from *in situ* tree roots. Reprinted by permission of John Wiley & Sons Ltd from Brown and Keough (1992a)

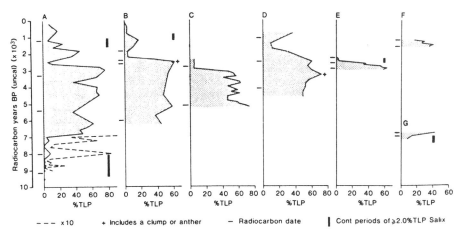

Figure 4.12 The alder decline in the River Severn valley. Comparison of alder percentage of TLP at Widen Marsh (A), Ashmoor Common (B), Longney (C), Callow End (D), (Brown, 1982), Ripple Brook (E) and two sites in the Perry catchment; Ruyton-IX-Towns (F) and Tetchill Brook g Junction (G). Reproduced with the permission of the Trustees of *New Phytologist* from Brown (1988)

have suggested that the seemingly multiple-stemmed alder root clusters found in the Nene valley may have been coppiced. Several sites, including four in the West Midlands, show an increase in willow after the alder decline, which may be at least partly natural, as alder had out-competed both birch and willow on floodplains during the early Holocene. In historical times willow species were planted for wood and to line rivers and ditches, and osiers (*Salix viminalis, Salix triandra* and *Salix purpea*) were extensively planted for the basket industry.

4.5.4   The managed palaeoenvironment: late Holocene floodplains

From the Bronze Age onwards in the lowlands of northwest Europe, floodplain and fen-margin sites reveal evidence for increased flooding and alluviation. The picture is much less clear for upland sites (Richards *et al.*, 1987). This is partly due to the lack of preserved archaeological sites on the valley floors and the more dynamic response of upland rivers to individual storms. Before describing some of the evidence for lowland alluviation, it is worth considering the types of data may be available, other than a dated increase in overbank silt deposition. Firstly, pedological data: a lack of soil development and soil structure, caused by a lack of bioturbation and soil development. Soil micromorphology frequently reveals a decrease in pedological fabric and an increase in un-bioturbated sedimentary microfeatures (Limbrey, 1992). This could be caused by an increase in flood frequency and/or an increase in the sediment loading of overbank flows. The former is probably more important than the latter because fine sediment is supply rather than transport-limited. A second, more ambiguous data type is the existence of an *in situ* peat horizon or tree stumps, which may result from a rise in the floodplain water table, which may, or may not, be accompanied by increased alluviation (Robinson, 1992). Third, a coarsening-up sequence, especially a band of fine gravel or coarse sand, traceable laterally across a site, and lastly, the partial destruction of a site by channel change or catastrophic flooding. Just as the nature of the evidence varies, so does the nature of the hydrological change implied; from the lumped hydroclimatic response of water-table variations to the episodic response exemplified by floods. In this context, we have to remember that one flood does not imply hydrological change, in contrast to evidence of a permanent rise in floodplain water-tables which does. Variations in the frequency of flooding and rates of alluviation change the floodplain environment; for example, floodplain "islands" or terrace remanents may disappear – as alluviation raises the floodplain around them, it may even bury them completely.

The Thames basin provides two sites which show good evidence of floodplain change in Later Prehistory. These are the Neolithic–Late Bronze Age sites at Runnymede, near Egham (Needham, 1992), and further upstream, the Iron Age site at Farmoor near Oxford (Lambrick and Robinson, 1979). The Runnymede site is particularly interesting, as the interstratified alluvial and settlement record is unusually detailed and well recorded. The site consists of an occupation floor with post-holes, cultural material, and a midden under a metre of alluvium. Waterlogging of the entire site most probably caused abandonment after a relatively short period of time. In 1978, some 80 m of excavations revealed a silted-up river channel with a double row of pile-driven timbers and a Prehistoric landsurface behind it. This was interpreted by the excavators as a wharfage, which fits well with the evidence of long-distance trade deduced from the Bronzes discovered at the site (Needham, 1985). The position of the wharf and the evidence of channel erosion may be reconciled by the existence in this reach of an anastomosing channel or simply a main channel with some smaller secondary channels. The Runnymede sites have produced abundant environmental evidence including mollusca and pollen. The mollusca point to damp conditions during the period of occupation and a cleared landscape, but with some light woodland in the vicinity. The Runnymede sites display evidence of increased flooding at the beginning and towards the end of the Bronze Age. There is evidence from an overbank gravel spread of a large flood at $c.\ 4000$ BP. Similar flood layers have been dated to $c.\ 3000$ BP from the River Stour in the West Midlands (Brown, 1988). The identification of discrete events is relatively rare in predominantly silt and clay alluvial systems, probably because they result from a combination of an unusually large flood event (*cf.* mega-flood) and continued floodplain aggradation, preventing the pedological reworking of the flood layer.

The Iron Age and Roman sites at Farmoor lie on the first gravel terrace of the Upper Thames and also on the Holocene floodplain close to an infilled cutoff (Lambrick and Robinson, 1979). The excavators of this site employed a wide range of environmental techniques, including analysis of pollen, seeds, wood, plant macrofossils, mollusca and insects. Pollen, from all the contexts sampled, showed a cleared floodplain with very low tree percentages, but a high percentage of grasses and herbs. The herbs were indicative of both grazing, or pasture and arable land. The arable indicators included cereal pollen, some of which may have been transported to the site on cereal grains themselves. The non-cereal seeds indicated a variety of habitats, including marshland, grassland, woodland, disturbed ground and scrub. Not surprisingly, in the light of the pollen analytical results, the wood discovered on the site was mostly from dry land trees, especially the cherry family (*Prunus* sp.). The Early Iron Age deposits at

Farmoor contained relatively few mollusca, mostly those with terrestrial affinities; however, the Middle Iron Age deposits produced a large number of aquatic mollusca, including the common bithynia (*Bithynia tentaculata*) and the flat valve snail (*Valvata cristata*), both indicative of flooding. The molluscan evidence for the Roman flooding is less strong. Taken along with the archaeological remains, the environmental evidence indicates that the farmstead was located in open wet grassland which was used for pasture and was liable to flooding. The rich floodplain grassland was probably being seasonally utilized, perhaps in spring. From both environmental and stratigraphic evidence, there seems to have been a marked increase in flooding after the Middle Iron Age, with seeds of aquatic plants and aquatic mollusca brought onto the site, along with increased sedimentation. At this time, flooding did not reach the lower terrace; however, by the Roman period it had. This increased flooding may well have been instrumental in the abandonment of the site. The gravel terrace at Farmoor is bisected by a palaeochannel only some 250 m from the main archaeological site. Although it is difficult to date, the infilling of Late Devensian peat stratified with sands suggests a Devensian date for channel excavation and, on archaeological evidence, it must have been abandoned by the river by the Iron Age. However, as Lambrick and Robinson (1979) point out, it probably functioned as a flood channel for a considerable period of time after the Iron Age and it remained a marshy depression providing valuable summer grazing. In the Kennet valley, a tributary of the Thames, environmental work has revealed two episodes of accelerated alluviation: one Late Neolithic/Early Iron Age, the other Post-Medieval. Snail analysis indicates that the valley bottom was cleared in the Neolithic to provide dry grassland and at various times the valley was dry enough for occupation (Evans *et al.* 1988).

The majority of *lowland* floodplains in Britain show remarkably little *channel change* during the Roman and Medieval periods. The location of structures such as settlements quays, bridges and weirs and the distribution of ridge and furrow provides ample evidence of stability from floodplains of large rivers such as the Thames and Severn. This is not to imply that there has been no channel change, especially since there are obvious problems with using the survival of constraining structures as evidence of little channel change. However, change seems to have been restricted to alterations in the number and dominance of channels where there was originally a multichannel system, siltation, occasional cutoffs, and locally active reaches. The Severn shows all of these features. At Gloucester, a map dated 1455 AD shows a third, easterly, channel which seems to have progressively silted-up, and then been built over in the eighteenth century. A Medieval bridge over this channel was excavated and indicated a channel width of about 40 m (Hurst, 1974). In general, the history of channel change at Gloucester has been one of the maintenance of the anastomosing channel pattern, although reduced to two channels, with siltation of the smaller branches. A similar pattern of development seems to have occurred at Runnymede on the Thames (Needham, 1992). The changes that have occurred on the floodplain are often relatively minor, and involve little or no reworking of the floodplain. For example, at Powick on the Lower Teme, a meander loop was abandoned by a chute-cutoff across its neck some time after the establishment of the Medieval parish boundary, but before 1740 AD (Brown, 1983a), but a grossly overgrown meander under a kilometre downstream has remained almost identical in planform for at least the last 400 years. The entire Lower Severn has been plotted (Brown, 1983a) using first edition Ordnance Survey maps of the 1830s and 1840s and compared to modern maps using fixed ground control points. The conclusion reached was that very little lateral channel change has occurred over the majority of the reaches during the last 150 years. At some sites, there has been limited lateral channel shifting; for example, the Roman quay at Wroxeter stands 16.5 m from the present channel bank and at Aylburton, near Gloucester, the Roman quay is approximately 40 m to the east of the present channel; but even these sites suggest very low rates of change. Even just upstream of Ironbridge Gorge, between Leighton and Buildwas, where the river displays a classic highly meandering planform, there has been remarkably little channel change over the last 150 years. However, as soon as we move into the piedmont zone, close to the Welsh mountains, the Severn shows evidence of being far less laterally stable. Lewin (1987) has shown significant channel change, some constrained, at sites such as Welshpool, Penstrowed, Llandinam, Lanidloes and Maesmawr during the last 90 years. A tributary of the Severn, the Vyrnwy, also displays evidence of post-Roman incision, and active migration over the last 200 years, although, as Lewin (1992) shows, there is also a reach of lower gradient which has been laterally stable in the historical period, with overbank sedimentation dominant. Although, as the Vyrnwy shows, the gross pattern, at the catchment scale is fundamentally related to stream power (Ferguson, 1981; Brown, 1987; Lewin, 1987), it does not explain the amount of change at the local level without the addition of a resistance factor. As Lewin (1987) points out, the pattern is complex, since different reaches, bends and cutoffs have responded differently to the same event. Comparable studies of smaller piedmont or lowland rivers initiated on recently glaciated terrain (e.g. last Glacial outwash), such as the Dane (Hooke *et al.*, 1990) and the Bollin (Mosley, 1975), both in Cheshire, have revealed considerable channel change over the last 150 years. Hooke *et al.* (1990) have been able to date much of the floodplain and the terraces of the River Dane at Swettenham; and a considerable proportion of the floodplain area has been reworked since 1840. They have also shown that there was a major phase of aggradation caused by soil erosion in the Medieval period, followed by dissection in the last 300 years which created a low terrace approximately 2 m above the modern floodplain.

4.6   FLOODPLAIN PALAEOENVIRONMENTS AND MODELLING FLOODPLAIN EVOLUTION

Over the last few years there has been a convergence between geomorphological modelling and landscape/GIS modelling and visualization. Both use digital elevation models (DEMs), either in the case of geomorphological models to derive slopes and flow paths (e.g. floodplain models of Bates *et al.*, 1992) or in the case of landscape models to provide the topography onto which vegetation can be draped or visualized (e.g. Haynes-Young *et al.*, 1993; the IDRISI system, etc.). The recent advent of landscape models with some dynamic capabilities (e.g. Vistamorph) and hydraulic models with real-time visualisation (e.g. HEC-2) brings these two areas of research even closer together. Figure 4.13 provides a foretaste of this approach using the visualization package Vistapro, and a base DEM derived from a USGS digital map of a reach of the Washington river (USA).

Figure 4.13 Synthesized visualizations of a Northwest European floodplain: (a) at the beginning of the Holocene, (b) in the middle Holocene. For more details see text

Table 4.2 Morpho-ecological classes of vegetation on floodplains and some geomorphological implications

Vegetation/ land-use type/ structures	Geomorphological context	Historical/geographical occurrence (UK)	Geomorphological effects
Bare ground, weeds	Bar surfaces especially point bars, eroding banks: typical of actively migrating braided and wandering rivers	Restricted in modern times, by channelization but much more common in northern/upland rivers and probably in the early Holocene and possibly Medieval to Little Ice age	Limited increase in substrate protection and strength, soil formation, little effect on flow hydraulics
Weed/grass/ lichen/ metalophoyte communities	Gravel bars	Northern/upland high energy rivers, metalophyte communities probably only common in Lateglacial/early Holocene and after metal mining	Limited increase in substrate protection and strength, soil formation, little effect on flow hydraulics
Lawns	Grazed areas adjacent to channels on fine alluvium	Lower energy rivers, typical of semi-natural or protected areas, in the past restricted to deforested or unforested area with high grazing	Significant increase in surface shear resistance, low roughness: low rates of overbank erosion or deposition
Wet meadow/ rough pasture/hay meadow	Floodplain flat, hams and backswamps	Common in Iron Age floodplains, managed grazing	Turf resistant to erosion and probably with a relatively high sediment trap efficiency
Reedbeds	Backswamps, abandoned channels	Both natural and managed for thatch and wildfowl	By analogy with estuarine reedbed resistant to erosion and associated with high rates of sediment accretion
Osier beds	Backswamps and abandoned channels	Planted in rows/strips for basket making	Unlikely to be eroded by lateral migration, and generally away from high sedimentation zones
Arable	On levees, on floodplain flat if floodplain drained sufficiently	Common in the high/late Medieval periods, and again since drainage, crops include barley and potatoes	Low resistance to overbank erosion, Ap horizon easily lost, relatively low trap efficiency
Willow/alder scrub	Upper levels of point bars and island bars, abandoned channels, backswamps	Common as a seral stage to alder woodland	Increases bar/island stability and bar accretion

(continued)

Table 4.2 *Continued*

Vegetation/ land-use type/ structures	Geomorphological context	Historical/geographical occurrence (UK)	Geomorphological effects
Alder/willow/ hazel/oak woodland	Backswamps and floodplain flat	Before deforestation common over large areas of floodplain, has been managed for timber and charcoal (gunpowder)	Stabilizes banks but creates hollows through windthrows, creates an uneven microtopography favouring avulsion and anastomosing channels
Alder carr	Backswamps, abandoned channels on peat	Common where the water table is at the surface and there is a relatively low input of suspended sediment, managed in the past for timber and charcoal	Generally away from channels, dense root network, effect on deposition not known
Poplar plantations	Floodplain flat and drier backswamps	Planted to dry-out land and produce timber (matches particularly)	As for rough grassland
Watermeadows/ floating meadows	Floodplain flat, particularly chalk valleys, utilize multiple channels	Methods for controlling flood inundation and maximize deposition fo silt, pre-Medieval to present	Channel is managed and high rate of deposition of silt and clay
Drainage ditches/ rhynes/dykes	From floodplain flat/ backswamp to channel	Greatest effect nearer to channel/ levee, first systematic drainage during the Roman period	Backflooding, reduced residence time of water on the floodplain
Embankments	Along channel margins and surrounding settlements	Comprehensive in coastal zone	Prevent return flow to river, reduce frequency of overbank inundation crevasse splay/avulsion caused by failure
Fords/bridges	Normally located on riffles	Bridges built with piers at ford locations on major routes	Upstream aggradation, flood obstruction
Causeways	Across hams and floodplain flats	From late Prehistory to present	Obstruct overbank flows and alters flow patterns, may increase sedimentation
Fish ponds	Dammed secondary channel or from scratch	Associated with monastic establishments, were regularly cleaned out	Increase floodplain sedimentation
Watermills	Use secondary channels or constrained reaches and/or small gravel islands	Least and mill dam construction, good evidence from the Roman period onwards, density can be limited by river fall	Flow obstruction, flood obstruction, settling in mill pond

The occurrence of water and vegetation have been manipulated in order to illustrate a very generalized representation of a hypothetical Northwest European floodplain during the early and middle Holocene. The ultimate aim of this approach will be interactive modelling of changing floodplain processes (e.g. overbank inundation and sedimentation) and changing floodplain topography, channel patterns and vegetation, thereby producing more realistic models of floodplain evolution.

The partial dependence of floodplain sedimentation on vegetation-related overbank deposition and the well-established effects of vegetation on lateral deposition through lateral migration both imply that vegetation can have a highly significant role in the evolution of floodplains. Coarse-fraction overbank deposition can be induced by large roughness elements on the floodplain, such as hedges, tree stumps and tussocks (Brown, 1983a), and there is increasing evidence that different plant communities have different trap efficiencies for fin sediment (Brown and Brookes, in preparation). It follows that changes in floodplain vegetation whether caused by human activity or climate change, can influence floodplain formation independently of catchment controls. Although not explicitly included in recent attempts to model floodplain formation, models such as that of Howard (1992; Chapter 2, this volume) include several vegetation-sensitive parameters such as bank erosion (which is taken to be proportional to near-bank velocity perturbation) and the rate of point-bar accretion. There are two fundamental ways in which vegetation can be included in hydraulic floodplain models. The first is by increasing or decreasing the ease with which channel banks and the floodplain are eroded. The protective effect of vegetation can be modelled by increasing shear resistance of units with age, simulating succession and increasing vegetation cover. Observations suggest that turf generally protects the floodplain surface, except where scour is induced by some macroroughness element. However, vegetation can increase the rate of erosion, particularly through failure and increased roughness caused by tree throwing. In forested floodplains this is a significant geomorphological process and is responsible for the creation of the floodplain topography and the location of new channels. The effect of vegetation on deposition can be modelled by increasing roughness of the floodplain surface and the creation of dead-water zones where all sand and gravel will settle instantaneously. While this simulates the deposition of coarse sediments, close to the channel and downstream of obstructions, it does not work for silt and clay. Recent models have used diffusion theory (James, 1985; Pizzuto, 1987; Marriott, 1992, Chapter 3, this volume) but this ignores the effect of vegetation, by regarding it as constant, and so is only applicable to grassed floodplains. The deposition of silt and clay is complex and controlled by several factors including aggregation (both in the water column and on plant leaves), impact sedimentation and possibly electrochemical forces, and it is not sufficiently understood for deterministic modelling. A feasible short-term alternative is to define a trap efficiency for different types of vegetation, which could be spatially distributed in a floodplain model. More studies are required of flood sedimentation under different land uses. An alternative could involve the determination of typical sedimentation rates for different land uses from caesium-137 cores (Walling et al., 1992; Chapter 12, this volume). In the long run, the incorporation of vegetation and avulsion may be prerequisites for stratigraphic floodplain models or dynamic facies models. Some of the main land uses of British floodplains during the Holocene and comments on their probably geomorphological implications are given in Table 4.2.

While it is not possible at present to specify the floodplain conditions which influenced Holocene floodplain evolution, some general trends can be postulated. Early Holocene floodplains offered little resistance to channel or floodplain erosion, irrespective of stream power, but this resistance increased until by c. 6000 BP lowland floodplains were highly resistant to channel migration and channel change could only take place by avulsion. These floodplains also had a high sediment trap efficiency adjacent to the channel (and a steep decline with distance from the channel) and high water tables including organic accumulation in backswamps. Deforestation of floodplains probably decreased bank resistance and increased the sedimentation zone across the floodplain, a process which has been partially reversed by channelization and embanking.

The study of palaeoenvironments provides us with some important general points concerning floodplain processes and evolution:

1. Floodplains have changed dramatically during the Holocene, due to both natural and anthropogenic factors.
2. Floodplain palaeoenvironments both reflect and induce geomorphological change.
3. Floodplain models are only valid for defined channel and vegetation conditions and must be specified/determined from palaeoenvironmental studies for long-term floodplain modelling or models of past floodplains.
4. Due to their wealth of resources, floodplain corridors have long attracted human activity and this has had an important effect both directly through management of the number and stability of channels and indirectly through floodplain vegetation.
5. The rates and pathways of floodplain evolution reflect not only inputs to the floodplain but also floodplain conditions, particularly vegetation type and cover.

The uncoupled (river–floodplain) nature of modern and recent floodplains allows them to be modelled without a vegetation component or with vegetation included through an adjustment to a hydrodynamic function, whereas the co-evolution of river and floodplain in the early Holocene might require vegetation to be a driving variable along with sediment supply and discharge. The study of floodplain palaeoenvironments also highlights interesting theoretical questions with contemporary significance, one of the most obvious being what are the short-term and long-term effects of floodplain deforestation on floodplain evolution?

REFERENCES

Allen, J.R.L. (1965) A review of the origin and characteristics of recent alluvial sediment *Sedimentology*, **5**, 89–191.

Amoros, C. and Van Urk, G. (1989) Palaeoecological analysis of large rivers: some principles and methods. In: G.E. Pett, H. Moller and A.L. Roux (eds), *Historical Change of Large Alluvial Rivers*, Wiley, Chichester, 143–165.

Amoros, C., Roux, A.L., Reygrobellet, J.L., Bravard, J.P. and Pautou, G. (1987) A method for applied ecological studies of fluvial hydrosystems. *Regulated Rivers*, **1**, 17–36.

Atkinson, T.C., Briffa, K.R., Coope, G.R., Joachim, M.J. and Perzy, D.W. (1986) Climatic calibration of coleopteran data. In: B.E. Berglund (Ed.), *Handbook of Palaeoecology and Palaeohydrology*, Wiley, Chichester, 851–858.

Bannister, P. (1976) *Introduction to Physiological Plant Ecology*. Blackwell, Oxford.

Barber, K.E. (1976) History of vegetation. In: P.D. Moore and S.B. Chapman (eds), *Methods in Plant Ecology*, Blackwell, Oxford.

Barber, K.E., Chambers, F.M., Maddy, D., Stoneman, R. and Brew, J.S. (1994) A sensitive high resolution record of late Holocene climatic change from a raised bog in northern England. *The Holocene*, **4**, 198–205.

Barton, D.R. and Smith, S.M. (1984) Insects of extremely small and extremely large aquatic habitats. In: V.H. Resh and D.M. Rosenberg (Eds), *The Ecology of Aquatic Insects*, Praeger, New York, 456–483.

Bates, P.D., Anderson, M.G., Baird, L., Walling, D.E. and Simm, D. (1992) Modelling floodplain flows using a two-dimensional finite element model. *Earth Surface Processes and Landforms*, **17**, 575–588.

Batterbee, R.W. (1986) Diatom analysis. In: B.E. Berglund (ed.), *Handbook of Holocene Palaeoecology and Palaeohydrology*, Wiley, Chichester, 527–570.

Batterbee, R.W. (1988) The use of diatoms in archaeology: a review. *Journal of Archaeological Science*, **15**, 621–644.

Becker, B. (1975) Dendrochronological observations on the postglacial river aggradation in the southern part of Central Europe. *Biuletyn Geologiczny*, **19**, 127–136.

Becker, B. and Schirmer, W. (1977) Palaeoecological study on the Holocene valley development of the river Main, Southern Germany. *Boreas*, **6**, 303–321.

Bennett, K.D. (1983) Postglacial population expansion of forest trees in Norfolk, U.K. Nature (*London*) **303**, 164–167.

Bennett, K.D. (1989) A provisional map of forest types for the British Isles 5000 years ago. *Journal of Quaternary Science*, **4**, 141–144.

Brackenridge, G.R. (1984) Alluvial stratigraphy and radiocarbon dating along the Duck river, Tennessee. Implications regarding floodplain origin. *Bulletin of the American Geological Society*, **95**, 9–25.

Bradley, C. and Brown, A.G. (1992) Floodplain and palaeochannel wetlands: geomorphology, hydrology and conservation. In: C. Stevens, J.E. Gordon, C.P. Green and M.G. Macklin (eds), *Conserving Our Landscape*, English Nature, Peterborough, 117–124.

Bradley, C. and Brown, A.G. (1995) Modelling of hydrological processes in a floodplain wetland. In: J. Gidert (ed.), *Groundwater/Surface Water Ecotones*, Cambridge University Press, Cambridge.

Brasier, M.D. (1980) *Microfossils*. Allen & Unwin, London and Boston.

Brown, A.G. (1982) Human impact on the former floodplain woodlands of the Severn. In: M. Bell and S. Limbrey (eds), *Archaeological Aspects of Woodland Ecology*, British Archaeological Reports, International Series No. 146, 93–105.

Brown, A.G. (1983a) Late Quaternary palaeohydrology, palaeoecology and floodplain development of the lower River Severn. Unpublished PhD thesis, University of Southampton.

Brown, A.G. (1983b) An analysis of overbank deposits of a flood at Blandford-Forum, Dorset, England. *Revue de Geomorphologie Dynamique*, **32**, 95–99.

Brown, A.G. (1985) The potential of pollen in the identification of suspended sediment sources. *Earth Surface Processes and Landforms*, **10**, 27–32.

Brown, A.G. (1987) Holocene floodplain sedimentation and channel response of the lower River Severn, United Kingdom. *Zeitschrift für Geomorphologie*, NF, **31**, 293–310.

Brown, A.G. (1988) The palaeoecology of Alnus (alder) and the Postglacial history of floodplain vegetation. Pollen percentage and influx data from the West Midlands, United Kingdom. *New Phytologist*, **110**, 425–436.

Brown, A.G. (1989) Holocene floodplain diachronism and inherited downstream variations in fluvial processes: a study of the river Perry, Shropshire, England. *Journal of Quaternary Science*, **5**, 39–51.

Brown, A.G. (1995) Lateglacial–Holocene sedimentation in lowland temperate environments: floodplain metamorphosis and multiple channel systems. *Palaeoclimate Research/ Paläoklimforschung*, Special Issue 9, **14**, 1–15.

Brown, A.G. (1996) *Alluvial Geoarchaeology*. Cambridge University Press, Cambridge (in press).

Brown, A.G. and Barber, K.E. (1985) Late Holocene palaeoecology and sedimentary history of a small lowland catchment in Central England. *Quaternary Research*, **24**, 87–102.

Brown, A.G. and Bradley, C. (1995) Past and present alluvial wetlands and the archaeological resource: implications from research in East Midland valleys, U.K. In: M. Cox and V. Straker (eds), *Wetlands: Nature Conservation and Archaeology*, HMSO London, 189–206.

Brown, A.G. and Ellis, C. (1995) People, climate and alluviation: theory, research design and new sedimentological and stratigraphic data from Etruria Italy. *Papers of the British School in Rome* 63, 45–73.

Brown, A.G. and Keough, M.K. (1992a) Palaeo-channels, palaeo-landsurfaces and the 3-D reconstruction of floodplain environmental change. In: P.A. Carling and G.E. Petts (eds), *Lowland Floodplain Rivers: A Geomorphological Perspective*, Wiley, Chichester, 185–202.

Brown, A.G. and Keough, M.K. (1992b) Palaeochannels and palaeolandsurfaces: the geoarchaeological potential of some Midland (U.K.) floodplains. In: S. Needham and M.G. Macklin (Eds), *Archaeology under Alluvium*, Oxford, 185–196.

Brown, A.G., Keough, M.K. and Rice, R.J. (1994) Floodplain evolution in the East Midlands, United Kingdom: the Lateglacial and Flandrian alluvial record from the Soar and Nene valleys. *Philosophical Transactions of the Royal Society Series A*, 348, 261–293.

Brown, A.G., Stone, P. and Harwood, K. (1995) *The Biogeomorphology of a Wooded Anastomosing River: The Gearagh on the River Lee in County Cork, Ireland*. Discussion Papers in Geography, University of Leicester.

Brush, G.S. and Brush, L.M. (1972) Transport of pollen in a sediment laden channel: a laboratory study. *American Journal of Science*, **272**, 359–381.

Burrin, P.J. and Scaife, R.G. (1984) Aspects of Holocene valley sedimentation and floodplain development in southern England. *Proceedings of the Geologists Association*, **95**, 81–96.

Casteel, R.W. (1976) *Fish Remains in Archaeology and Palaeoenvironmental Studies*. Academic Press, London.

Chambers, F.M. and Elliott, L. (1989) Spread and expansion of *Alnus* Mill. in the British Isles: timing, agencies and possible vectors. *Journal of Biogeography*, **16**, 541–550.

Coleman, J.M. (1969) Brahmaputra river: channel processes and sedimentation. *Sedimentary Geology*, **3**, 129–239.

Coles, B. (1992) The possible impact of beaver on valleys big and small in a temperate landscape. In: S. Needham and M.G. Macklin (eds), *Archaeology under Alluvium*, Oxbow Books, Oxford, 93–99.

Coles, B. and Coles, J. (1986) *Sweet Track to Glastonbury*, Thames and Hudson, London.

Coope, G.R. (1986) Coleopteran analysis. In: B.E. Berglund (ed.), *Handbook of Holocene Palaeoecology and Palaeohydrology*, Wiley, Chichester, 703–714.

Croke, J. and Nanson, G.C. (1991) Floodplains; their character and classification on the basis of stream power, sediment type, boundary resistance and antecedency. *Geomorphology*, 20–35.

Crowder, A.A. and Cuddy, D.G. (1973) Pollen in a small basin: Wilton Creek Ontario. In: H.J.B. Birks and R.G. West (eds), *Quaternary Plant Ecology*, Cambridge University Press, Cambridge, 61–76.

Dansgaard, W., White, J.W.C. and Johnson, S.J. (1989) The abrupt termination of the Younger Dryas climate event. *Nature*, **339**, 532–533.

Davis, R.J. and Gregory, K.J. (1994) A new distinct mechanism of river bank erosion in a forested catchment. *Journal of Hydrology*, **157**, 1–11.

Delcourt, P.A. and Delcourt, H.R. (1981) Vegetation maps for eastern North America: 40,000 B.P. to the present. In: Romans, R.C. (ed.), *Geobotany II*. Plenum Publishing Corporation, 123–165.

Erskine, W., Melville, M., Page, K.J. and Mowbray, P.D. (1982) Cutoff and oxbow lakes. *Australian Geographer*, **15**, 174–180.

Evans, J.G., Limbrey, S., Mate, I. and Mount, R.J. (1988) Environmental change and land-use history in a Wiltshire river valley in the last 14000 years. In: J.C. Barrett and I.A. Kinnes (eds), *The Archaeology of Context in the Neolithic and Bronze Age: Recent Trends*, University of Sheffield, Department of Archaeology and Prehistory, Sheffield, 97–104.

Evans, J.G., Davies, P., Mount, R. and Williams, D. (1992) Molluscan taxocenes from Holocene overbank alluvium in southern central England. In: S. Needham and M. G. Macklin (eds), *Archaeology under Alluvium*, Oxbow Books, Oxford, 65–74.

Everitt, B.L. (1968) The use of cottonwood in an investigation of the recent history of a floodplain. *American Journal of Science*, 266, 417–439.

Fenneman, N.M. (1906) Floodplains produced without floods. *Bulletin of the American Geographical Society*, 38, 89–91.

Ferguson, R.L. (1981) Channel form and channel changes. In J. Lewin (ed.) *British Rivers*, Allen and Unwin, London, 90–125.

Foged, N. (1947–1948) Diatoms in watercourses in Funen. I–VI. *Dansk. Bot. Ark.*, 12(5), (12(6), 12(9), 12(12).

Foged, N. (1978) *Diatom Analysis. The Archaeology of Svendborg, Denmark*, No. 1. Odense University Press, Odense.

Frey, D.G. (1964) Remains of animals in Quaternary lake and bog sediments and their interpretation. *Arch. Hydrobiol. Beih.*, 2, 1–114.

Frey, D.G. (1986) Cladoceran analysis. In: B.E. Berglund (ed.), *Handbook of Holocene Palaeoecology and Palaeohydrology*, Wiley, Chichester, 667–692.

Godlowska, M., Kozlowski, J., Starkel, L. and Wasylikowa, K. (1987) Neolithic settlement at Pleszow and changes in the natural environment in the Vistula valley. *Przeglad Archeologiczny*, 34, 133–159.

Gosselink, J.G. and Turner, R.E. (1978) The role of hydrology in freshwater wetland ecosystems. In: R.E. Good, D.F. Whigham and R.L. Simpson (eds), *Freshwater Wetlands: Ecological Processes and Management Potential*, Academic Press, London and New York, 63–78.

Greenwood, M.T., Bickerton, M.A., Castella, E., Large, A.R.G. and Petts, G.E. (1991) The use of Coleoptera (Arthropoda: Insecta) for floodplain characterisation of the river Trent. *Regulated Rivers*, 6, 321–332.

Gurnell, A.M. and Gregory, K.J. (1984) The influence of vegetation on stream channel processes. In: T.P. Burt and D.E. Walling (eds), *Catchment Experiments in Geomorphology*, Geo Books, Norwich,

Harwood, K. and Brown, A.G. (1993) Changing in-channel and overbank flood velocity distributions and the morphology of forested multiple channel (anastomosing) systems. *Earth Surface Processes and Landforms*, 18, 741–748.

Haynes-Young, R., Green, D.R. and Cousins, S.H. (1993) *Landscape Ecology and GIS*. Taylor and Francis, London.

Hoffman, W. (1986) Chironomid analysis. In: Berglund, B.E. (ed.), *Handbook of Holocene Palaeoecology and Palaeohydrology*, Wiley, Chichester, 715–728.

Holloway, R.G. (1981) Preservation and experimental diagenesis of pollen exine. Unpublished PhD thesis, Texas A & M University.

Hooke, J.M., Harvey, A.M., Miller, S.Y. and Redmond, C.E. (1990) The chronology and stratigraphy of the alluvial terraces of the river Dane valley, Cheshire. *Earth Surface Processes and Landforms*, 15, 717–737.

Howard, A. (1992) Modelling floodplain–channel interactions. In: P. Carling and G.E. Petts (eds), *Lowland Floodplain Rivers: A Geomorphological Perspective*, Wiley, Chichester, 1–42.

Huntley, B. and Birks, H.J.B. (1983) *An Atlas of Past and Present Pollen Maps for Europe: 0–13000 Years Ago*. Cambridge University Press, Cambridge.

Hurst, H. (1974) Excavations at Gloucester, 1971–1973. Second interim report. *The Antiquarians Journal*, 54, 8–51.

Hustedt, F. (1930) *Die Susswasser-Flora Mitteleuropas. Heft 10: Bacillarriophyta (Diatomaceae)*. Jena, Verlag Von Gustav Fischer.

Jacobsen, G.L. and Bradshaw, R.H.W. (1981) The selection of sites for palaeovegetational study. *Quaternary Research* 11, 80–96.

James, C.S. (1985) Sediment transfer to overbank sections. *Journal of Hydraulic Research*, **23**, 435–452.

Janssen, C.R. (1959) Alnus as a disturbing factor in pollen diagrams. *Acta Botanica Neerlandica*, **8**, 55–58.

Janssen, C.R. (1972) The palaeoecology of plant communities in the Dommel valley, North Brabant, Netherlands. *Journal of Ecology*, **60**, 411–437.

Janssen, C.R. (1986) The use of pollen indicators and of the contrast between regional and local pollen values in the assessment of the human impact on vegetation. In: K.E. Behre (ed.), *Anthropogenic Indicators in Pollen Diagrams*. Balkema, Rotterdam, 203–208.

Kennedy, B.A. (1972) "Bankfull" discharge and meander forms. *Area*, **4**, 209–212.

Klinke, A. (1989) The Lower Rhine: palaeoecological analysis. In: Petts, G.E., Moller, H. and Roux, A.L. (eds), *Historical Change of Large Alluvial Rivers*. Wiley, Chichester and New York, 183–202.

Knighton, A.D. and Nanson, G.C. (1993) Anastomosis and the continuum of channel pattern. *Earth Surface Processes and Landforms*, **18**, 613–625.

Lambrick, G.H. and Robinson, M.A. (1979) *Iron Age and Roman Riverside Settlement a Farmoor, Oxfordshire*. Council for British Archaeology Report 32, York.

Langbein, W.B. and Leopold, L.B. (1966) River meanders – theory of minimum variance *United States Geological Survey Professional Paper*, 422H, 15 pp.

Lewin, J. (1987) Historical river channel changes. In: K.J. Gregory, J. Lewin and J.B Thornes (eds), *Palaeohydrology in Practice: A River Basin Analysis*. Wiley, Chichester, 161–170.

Lewin, J. (1992) Alluvial sedimentation style and archaeological sites: the Lower Vyrnwy, Wales. In: S. Needham and M.G. Macklin (eds), *Archaeology under Allavium*, 103–110, Oxbow Books, Oxford, 103–110.

Lewis, J.S.C. and Wiltshire, P.E.J. (1992) Late glacial and early Flandrian archaeology in the Colne valley (Middlesex); possible environmental implications. In: S. Needham and M.G. Macklin (eds), *Archaeology under Alluvium*, Oxbow Books, Oxford, 235–248.

Limbrey, S. (1992) Micromorphological studies of buried soils and alluvial deposits in a Wiltshire river valley. In: S. Needham and M.G. Macklin (eds), *Archaeology under Alluvium*, Oxbow Books, Oxford.

Lockwood, J.G. (1979) Water balance of Britain 50,000 BP to the present. *Quaternary Research*, **12**, 297–310.

Lugo, A.E., Brinson, M.M. and Brown, S. (1990) *Forested Wetlands*. Ecosystems of the World 15, Elsevier, Amsterdam.

Macklin, G.M. and Lewin, J. (1986) Terraced fills of Pleistocene and Holocene age in the Rheidol valley, Wales. *Journal of Quaternary Science*, **1**, 21–34.

Macklin, G.M. and Lewin, J. (1994) Holocene river alluviation in Britain. *Zeitschrift für Geomorphologie*, **88**, 109–122.

Mamakowa, K. and Starkel, L. (1974) New data about the profile of Quaternary deposits at Brzeznica in Wisloka valley. *Studie Geomorph. Carpatho-Balcanica*, **8**, 47–59.

Marriott, S. (1992) Textural analysis and modelling of flood deposits: river Severn, U.K. *Earth Surface Processes and Landforms*, **17**, 687–698.

McDonald, A., Keller, E.A. and Tally, T. (1982) The role of large organic debris on stream channels draining redwood forests in North-west California. In: J.K. Hardin, D.C. Marron and A. McDonald (eds), *Friends of the Pleistocene 1982 Pacific Cell Fieldtrip Guidebook. Late Cenozoic History and Forest Geomorphology of Humbolt County, California*, 226–245.

McVean, D.N. (1954) Ecology of *Alnus glutinosa* (L.) Gaertn. Postglacial history. *Journal of Ecology*, **44**, 331–333.

Mitsch, W.J. and Gosselink, J.G. (1993) *Wetlands*. Second edition, Van Nostrand Rheinhold, New York.

Mordant, D. and Mordant, C. (1992) Noyen-Sur-Seine: a mesolithic waterside settlement. In: B. Coles (ed.), *The Wetland Revolution in Prehistory*, Wetland Archaeological Research Project Occasional Paper 6, Exeter, 55–64.

Morzadec-Kerfourn, M.T. (1974) Variations de la ligne de rivage Armorican au Quaternaire. *Memoirs Society Geologic et Mineral de Bretagne*, **17**, 1–208.

Mosley, M.P. (1975) Channel changes on the river Bollin, Cheshire, 1872–1973. *East Midlands Geographer*, **6**, 185–199.

Naiman, R., Decamps, H. and Fournier, F. (1989) *Role of Land/Inland Water Ecotones Landscape Management and Restoration*. MAB Digest 4, UNESCO, Paris.

Nanson, G.C. (1981) New evidence of scroll-bar formation on the Beatton river. *Sedimentology*, **28**, 889–891.

Needham, S. (1985) Neolithic and Bronze Age settlement on the buried floodplain of Runneymede. *Oxford Journal of Archaeology*, **4**, 125–137.

Needham, S. (1992) Holocene alluviation and interstratified settlement evidence in the Thames valley at Runnymede Bridge. In: S. Needham and M.G. Macklin (eds), *Archaeology under Alluvium*, Oxbow Books, Oxford, 249–260.

Needham, S. and Macklin, M.G. (1992) *Archaeology under Alluvium*. Oxbow Books, Oxford.

Osborne, P.J. (1988) A Late Bronze Age fauna from the river Avon, Warwickshire, England: its implications for the terrestrial and fluvial environment and for climate. *Journal of Archaeological Science*, **15**, 715–727.

Palmquist, R.C. (1975) Preferred position model and subsurface valleys. *Bulletin of the Geological Society of America*, **86**, 1392–1398.

Peck, R.M. (1973) Pollen budget studies in a small Yorkshire catchment. In: H.J.B. Birks and R.G. West (eds), *Quaternary Plant Ecology*, Blackwell, Oxford, 43–60.

Petts, G.E. (1987) Ecological management of regulated rivers: a European perspective. *Regulated Rivers*, **1**, 363–369.

Pizzuto, J.E. (1987) Sediment diffusion during overbank flows. *Sedimentology*, **34**, 304–317.

Prentice, I.C. (1985) Pollen representation, source area and basin size: towards a unified theory of pollen analysis. *Quaternary Research*, **23**.

Reineck, H-E. and Singh, I.B. (1973) *Depositional Sedimentary Environments*. Springer-Verlag, New York, 230 pp.

Rhodes, E.J. and Pownall, L. (1994) Zeroing of the OSL signal in quartz from young glaciofluvial sediments. *Radiation Measurements*, **23**, 581–585.

Richards, K.S., Peters, N.R., Robertson-Rintoul, M.S.E. and Switsur, V.R. (1987) Recent valley sediments in the North York Moors: evidence and interpretation. In: V. Gardiner (ed.), *International Geomorphology 1986*, Part I, Wiley, Chichester, 869–883.

Ritter, D.F., Kinsey, W.F. and Kauffman, M.E. (1973) Overbank sedimentation in the Delaware river valley during the last 6000 years. *Science*, **179**, 374–375.

Robinson, M. (1992) Environment, archaeology and alluvium on the river gravels of the South Midlands. In: S. Needham and M.G. Macklin (eds), *Archaeology under Alluvium*, Oxbow Books, Oxford, 197–208.

Robinson, M. and Lambrick, G.H. (1984) Holocene alluviation and hydrology in the Upper Thames basin. *Nature*, **308**, 809–814.

Rose, J., Turner, C., Coope, G.R. and Bryan, M.D. (1980) Channel changes in a lowland river catchment over the last 13,000 years. In: R.A. Cullingford, D.A. Davidson and J. Lewin (eds), *Timescales in Geomorphology*, Wiley, Chichester, 159–176.

Roux, A.L., Bravard, J-P., Amoros, C. and Pautou, G. (1989) Ecological changes of the French Upper Rhone river since 1750. In: G.E. Petts, H. Moller and A.L. Roux (eds), *Historical Change of Large Allavial Rivers*, Wiley, Chichester and New York, 323–350.

Rybnicek, K. and Rybnickova, E. (1987) Palaeogeobotanical evidence of middle Holocene stratigraphic hiatuses in Czechoslovakia and their explanation. *Folia Geobot. Phytotax Praha*, **22**, 313–327.

Salisbury, C.R. (1992) The archaeological evidence for palaeochannels in the Trent valley. In: S. Needham and M.G. Macklin (Eds), *Archaeology under Alluvium*, Oxbow Books, Oxford, 155–162.

Salisbury, C.R., Whitley, P.J., Litton, C.D. and Fox, J.L. (1984) Flandrian courses of the river Trent at Colwick, Nottingham. *Mercian Geologist*, **9**, 189–207.

Salmi, M. (1962) Investigations on the distribution of pollens in an extensive raised bog. *Bulletin of the Commission for Geology of Finland*, **204**, 152–193.

Sangster, A.G. and Dale, H.M. (1961) A preliminary study of differential pollen grain preservation. *Canadian Journal of Botany*, **39**, 35–51.

Scaife, R. and Burrin, P.J. (1992) Archaeological inferences from alluvial sediments: some findings from southern England. In: S. Needham and M.G. Macklin (eds), *Archaeology Under Alluvium* Oxbow Books, Oxford, 75–92.

Shotton, F.W. (1978) Archaeological inferences from the study of alluvium in the lower Severn–Avon valleys. In: S. Limbrey and I.G. Evans (eds), *Man's Effect on the Landscape: The Lowland Zone*. Council for British Archaeology Research Report 21, 27–32, York.

Shotton, F.W. and Osborne, P. (1986) Faunal content of debris left by an exceptional flood of the Cuttle Brook at Temple Balsall Nature Reserve. *Proceedings of the Coventry Natural History Society*, **10**, 359–363.

Siddiqui, Q.A. (1971) The palaeoecology of non-marine ostracoda from Fladbury, Worcestershire and Isleworth, Middlesex. In: H.J. Oertli (ed.), *Paleoecologie Ostracodes*, Bull. Centra Rech. Pau – SNPA, 5 suppl., 331–339.

Smith, A.G. (1984) Newferry and the Boreal–Atlantic transition. *New Phytologist*, **98**, 35–55.

Smith, D.G. and Smith, N.D. (1980) Sedimentation in anastomosing river systems: examples from alluvial valleys near Banff, Alberta. *Journal of Sedimentary Petrology*, **50**, 157–164.

Sparks, B.W. (1961) The ecological interpretation of Quaternary non-marine mollusca. *Proceedings Linnaen Society*, **172**, 71–80.

Starkel, L. (ed.) (1981) The evolution of the Wisloka valley near Debica during the late Glacial and Holocene. *Folia Quaternaria*, **53**, 1–91.

Starkel, L. (1991) The Vistula valley: a case study for Central Europe. In: Starkel, L., Gregory, K.J. and Thornes, J.B. (eds), *Temperate Palaeohydrology*, Wiley, Chichester, 171–188.

Swale, E.M.F. (1969) Phytoplankton in two English rivers. *Journal of Ecology*, **57**, 1–23.

Tauber, H. (1965) Differential pollen dispersal and the interpretation of pollen diagrams. *Danm. Geol. Unders. Ser. II*, **89**, 1–69.

Taylor, G. and Woodyer, K.D. (1978) Bank deposition in suspended-load streams. In: A.D. Miall, (ed.), *Fluvial Sedimentology, Memoirs Canadian Society of Petroleum Geologists*. **5**, 257–275.

Tolonen, K. (1986) Rhizopod analysis. In: B.E. Berglund (ed.), *Handbook of Holocene Palaeoecology and Palaeohydrology*, Wiley, Chichester, 645–666.

Van der Valk, A.G. (1981) Succession in wetlands: a Gleasonian approach. *Ecology*, **62**, 688–696.

Vandenberghe, J. and de Smedt, P. (1979) Palaeomorphology in the eastern Scheldt basin (Central Belgium) *Catena*, **6**, 73–105.

Van Geel, B. (1986) Applications of fungal and algal remains and other microfossils in palynological studies. In: B.E. Berglund (ed.), *Handbook of Holocene Palaeoecology and Palaeohydrology*, Wiley, Chichester, 497–506.

Van Leeuwaarden, W. and Janssen, C.R. (1987) Differences between valley and upland vegetation development in eastern Noord-Brabant, the Netherlands, during the Late Glacial and early Holocene. *Review of Palaeobotany and Palynology*, **52**, 179–204.

Vannote, R.L., Minshall, G.W., Cummins, K.W., Sedell, J.R. and Cushing, C.E. (1980) The river continuum concept. *Canadian Journal of Fisheries and Aquatic Science*, **37**, 130–137.

Walker, D. (1970) Direction and rate in some British post-glacial hydroseres. In: D. Walker and R.G. West, (eds) *Studies in the Vegetational History of the British Isles*, Cambridge University Press, 117–139.

Walker, I.R. and Paterson, C.G. (1985) Efficient separation of subfossil Chironimidae from lake sediments. *Hydrobiologia*, **122**, 189–192.

Walling, D.E., Quine, T.A. and He, Q. (1992) Investigating contemporary rates of floodplain sedimentation. In: P.A. Carling and G.E. Petts (eds), *Lowland Floodplain Rivers*, Wiley, Chichester, 165–184.

Watts, W.A. (1978) Plant macrofossils and Quaternary palaeoecology. In: D. Walker and J.C. Guppy (eds), *Biology and Quaternary Environments*, Australian Academy of Science, Canberra, 53–68.

Wharton, C.H., Kitchens, W.M., Pendleton, E.C. and Sipe, T.W. (1982) The ecology of bottomland hardwood swamps of the southeast: a community profile. In: J.R. Clark and J. Benforado (eds), *Wetlands of the Bottomland Hardwood Forests*, Elsevier, Amsterdam, 87–100.

Wolman, M.G. and Leopold, L.B. (1957) River floodplains: some observations on their formation. *United States Geological Survey Professional Paper*, 282C, 87–107.

5 River Channel and Floodplain Hydraulics

D.W. KNIGHT* AND K. SHIONO†

* *School of Civil Engineering, University of Birmingham, UK*

† *Department of Civil and Building Engineering, Loughborough University of Technology, UK*

5.1 INTRODUCTION

Rivers do not generally "burst their banks" once the discharge and water level reach certain critical values. This commonly used idiomatic phrase, equally popular with the media and the general public, has undoubtedly served to reinforce a misunderstanding of basic fluvial processes. Rivers by their very nature flow in valleys with the more frequently occurring discharges flowing inbank and hence form visually identifiable river channels, usually of a single thread but sometimes occasionally braided, with a multiplicity of interconnected channels. Periodically in the hydrological time-scale, higher discharges will occur that cause the channel to flow in an overbank condition, thereby increasing the flow area, depth and width. Although the channel cross-section is now somewhat different from the inbank channel, incorporating the main river channel and the adjacent floodplains, it is still none the less an open channel albeit with a more complex geometry, roughness and planform. The floodplains are therefore an integral part of the whole river system. Furthermore, floodplain flows are a natural consequence of the hydrological flow regime, even if they do occur infrequently. However, it should be recognized that for overbank flow, not only does the cross-section shape change significantly from its inbank shape, but also the streamwise pathways for flow may alter considerably, as for example when the original main river channel is of a meandering nature contained within a valley of a more uniform shape and planform. There is therefore in nature a continuum of hydraulic processes that occur within a river as the discharge increases, with additional processes coming into action above the bankfull level. Inevitably there is also a significant increase in the complexity of the flow behaviour once overbank flow starts. Whereas inbank flows may be treated as if they were predominately one-dimensional flows in the streamwise direction, despite known three-dimensional mechanisms being present in all flows, overbank flows must be treated differently as certain three-dimensional processes begin to be especially important, particularly the main channel/floodplain interaction. It is this interaction among others that makes the analysis of floodplain flows inherently difficult.

Floodplain Processes. Edited by Malcolm G. Anderson, Des E. Walling and Paul D. Bates.
© 1996 John Wiley & Sons Ltd.

This review seeks to highlight some recent studies in river channel/floodplain flows and to identify those processes that are important for modellers to consider. Following a description of river channel hydraulics, the particular issue of floodplain hydraulics is addressed in some detail, beginning with an overview of recent studies and then considering the governing equations, turbulence structures, boundary shear stress and velocity distributions and modelling strategies. The practical implications of overbank flow are then considered with respect to conveyance capacity, hydraulic resistance parameters, sediment transport, flood routing and model calibration procedures. A short reference list is included to allow the reader to follow up certain key issues. Of necessity the list of references is brief, but it is hoped that these will provide suitable coverage of this fascinating topic.

5.2   RIVER CHANNEL HYDRAULICS

The flow of water in channels is governed by the Navier–Stokes equations (Schlichting, 1979). One-dimensional versions of these equations are known as the St Venant equations (see, for example, Cunge et al., 1980). The resistance laws which are generally adopted for open-channel flow are those based on steady flow and include the Darcy–Weisbach (1857), Manning (1891) and Chezy (1769, see 1921) formulae, with the Colebrook–White equation as the all-important ancillary equation for the Darcy–Weisbach coefficient (1937, 1939). These resistance laws essentially relate the conveyance capacity of the channel to the cross-sectional shape, longitudinal bed slope and resistance parameters. Tables and charts exist for those cases in which the channel boundary is fixed (Barr and HR Wallingford, 1993), and further laws exist for those cases in which the channel is formed in sediment and hence free to adjust its shape and roughness to the prevailing discharge and sediment load. In addition to the traditional empirically based regime type equations developed for irrigation purposes (Ackers, 1992a), it is now possible to develop rational theories for equilibrium channels based on a combination of empirical equations (White et al., 1982; Chang, 1988). Tables exist for the geometrical shapes of alluvial channels in a state of equilibrium, based on the combination of laws relating sediment transport, sediment roughness and some extremal hypothesis such as minimization of the stream power (Bettess and White, 1987). Commercial software based on one-dimensional considerations exists in many forms, and most notable among these are ISIS (HR Wallingford and Halcrow) and MIKE11 (Danish Hydraulic Institute), which deal with the movement of water and pollutants in river channels. Flood routing at a catchment scale may be simulated via RIBAMAN (HR Wallingford), which deals with river basin management issues. Consideration is now given to these three topics of conveyance capacity, flood routing and sediment equilibrium in a little more detail before considering them again in Section 5.3 with the additional processes arising from overbank flow.

5.2.1   Conveyance capacity

The essential difficulty in analysing even simple steady uniform flow in a prismatic channel is shown in Figure 5.1. The Navier–Stokes equations apply at a single point in

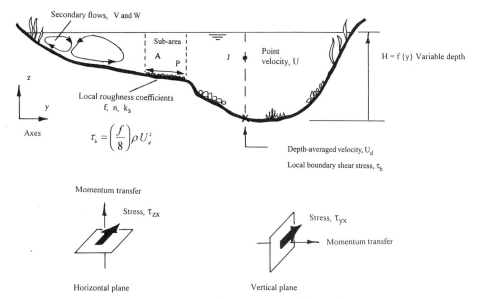

Figure 5.1   Flow in a natural channel

the fluid such as at point J. At this point the governing equation for the streamwise motion of a small element within the cross-section of an open channel, with a plane bed inclined in the streamwise direction, is:

$$\rho\left[V\,\frac{\partial U}{\partial y} + W\,\frac{\partial U}{\partial z}\right] = \rho g\,\sin\theta + \frac{\partial\tau_{yx}}{\partial y} + \frac{\partial\tau_{zx}}{\partial z} \qquad (5.1)$$

$$\begin{array}{ccc} \text{Secondary flows} & = \text{Weight force} + & \text{Reynolds stresses} \\ \text{(vorticity)} & & \text{(lateral)} + \text{(vertical)} \end{array}$$

where $\{UVW\}$ are velocity components in the $\{xyz\}$ directions {i.e. streamwise, lateral and normal to the bed}, ρ is fluid density, θ is channel bed slope, g is gravitational acceleration and τ_{yx} and τ_{zx} are Reynolds stresses on planes perpendicular to the y and z directions respectively. The driving gravity force is thus expended not only against two Reynolds stress terms which control the vertical and lateral shearing processes arising from friction forces on the channel bed and sides, but also against maintaining the secondary flows transverse to the mean streamwise direction of flow with velocity components V and W. Equation (5.1) thus requires some turbulence closure equations before it can be solved to give a three-dimensional simulation of the flow field. Turbulence models which might be considered for this problem are the $k-\varepsilon$, algebraic stress, Reynolds stress or large eddy simulation variety. See Rodi (1980), Naot and Rodi (1982), Nezu (1993), Cokjlat and Younis (1994), Lin & Shiono (1994), Thomas and Williams (1995) and Younis (1996, Chapter 11, this volume) and other chapters in this book for further details of these approaches. The modelling is thus at a three-dimensional level.

Usually river engineers are only concerned with the parameters at the boundaries and therefore equation 5.1 has to be integrated over the depth, width or area before being of much practical use. If lateral distributions are of importance, as they often are in rivers, then integration over the depth must be undertaken. The depth-averaged form of equation 5.1 as given by Shiono and Knight (1988, 1991) is:

$$\rho g H S_0 - \frac{1}{8}\rho f U_d^2 \left(1 + \frac{1}{s^2}\right)^{1/2} + \frac{\partial}{\partial y}\left[\rho \lambda H^2 \left(\frac{f}{8}\right)^{1/2} U_d \frac{\partial U_d}{\partial y}\right] = \frac{\partial}{\partial y}[H(\rho UV)_d] \quad (5.2)$$

where s is the channel side slope $(1:s$, vertical : horizontal) and U_d is the depth mean velocity defined by

$$U_d = \frac{1}{H}\int_0^H U\,dz \quad (5.3)$$

Defining Γ as the right hand side of equation 5.2, then the terms f, λ and Γ are the local friction factor, dimensionless eddy viscosity and secondary flow parameters respectively, defined and discussed more fully following equation 5.21 in Section 5.3.3. Further details are given in Knight and Shiono (1990) and Shiono and Knight (1991). This equation is essentially a two-dimensional one giving lateral distributions of U_d or τ_b across the channel width. Some examples of this model are given by Knight (1989), Shiono and Knight (1990, 1991) and Abril (1995a,b).

Integration over the area should lead to a one-dimensional equation. However, it is customary in one-dimensional analysis to ignore individual terms in three-dimensional equations and to lump all hydraulic effects into a simple bulk flow resistance law such as the Manning and Darcy–Weisbach laws:

$$Q = (AR^{2/3}S_f^{1/2})/n \quad (5.4)$$

$$Q = (8g/f)^{1/2}AR^{1/2}S_f^{1/2} \quad (5.5)$$

where Q is the discharge, A is the area, R is the hydraulic radius, S_f is the friction slope and n and f are resistance coefficients. These coefficients are known to vary with depth or discharge in most channel flows. The variation of the resistance coefficient, f, with Reynolds number and relative roughness is normally defined by the Colebrook–White equation (1937, 1939). It should be remembered that this resistance law strictly only applies to flow in circular pipes, and that although it may be used for certain prismatic channels the equation is not suitable for flow in very complex cross-sections. In this one-dimensional approach therefore the shape of the cross-section is not explicitly included and has to be accounted for by some other means (Keulagan, 1938; Engelund, 1964; Kazemipour and Apelt, 1979; Knight et al., 1994, Rhodes and Knight, 1994a,b). Those cases in which there is a heterogeneous roughness distribution around the wetted perimeter likewise have to rely on ancillary equations (Task Force, 1963; Alhamid, 1991; Knight et al. 1992a), since again the Colebrook–White equation is only valid for uniform roughness patterns. It should be clear from this brief description and from a comparison of equations 5.1 to 5.5 that the roughness coefficients f or n are essentially somewhat crude measures of the net effect of the three influences of vertical shear, lateral shear and secondary flow that occur in straight open-channel flow, as indicated by the various terms in equation 5.1. Streamwise curvature is another order of complexity and is sometimes dealt with along similar lines by simply enhancing the resistance coefficient (Wark et al., 1994).

For a particular cross-section therefore equations 5.4 and 5.5 link the depth, h, via the hydraulic radius, R, to the discharge, Q, to give a stage–discharge curve, h versus Q. For steady flow this is a unique relationship, given certain channel properties, and is of particular interest to river engineers since it relates the two primary parameters in any flood problem, namely water level and discharge. Typical stage–discharge curves are shown in Figure 5.2 for a laboratory and a natural channel, flowing both inbank and overbank, in order to illustrate some general features. It may be seen from Figure 5.2a that in general Q increases with depth but that once bankfull level is reached, under certain circumstances there may be an actual reduction in Q despite a larger flow area. This is usually more apparent in laboratory channels than in nature (Knight and Demetriou, 1983; Knight and Hamed, 1984, 1990), but even in the field very shallow inundations on the floodplains are problematic to analyse and model (Knight et al., 1989). However once above this small floodplain depth region, at higher depths, Q increases significantly due to the increased flow areas, with the slope of the h versus Q curve decreasing as the width of the floodplain increases. There is therefore a continuum of depths linked to discharges for channels flowing both inbank and overbank, albeit complicated by the interaction process between the main river and the floodplains or by storage effects (see Figure 5.2(b)). For unsteady flow the curve is not unique and is further complicated by dynamic effects. These are now dealt with in the next section.

5.2.2   Flood routing

For unsteady one-dimensional flow in an open channel, the principles of mass and momentum conservation lead to the well-known St Venant equations (Cunge et al., 1980):

$$\frac{\partial A}{\partial t} + \frac{\partial Q}{\partial x} = q^* \tag{5.6}$$

$$\frac{\partial Q}{\partial t} + \frac{\partial}{\partial x}\left(\beta\,\frac{Q^2}{A}\right) + gA\left(\frac{\partial h}{\partial x} + S_f - S_0\right) = 0 \tag{5.7}$$

where $q^* =$ lateral inflow/outflow per unit length. For a momentum correction coefficient, β, equal to 1.0, the momentum equation may be expressed in terms of the section mean velocity, u, to give the friction slope, S_f, as

$$S_f = S_0 - \frac{\partial h}{\partial x} - \frac{u}{g}\frac{\partial u}{\partial x} - \frac{1}{g}\frac{\partial u}{\partial t} \tag{5.8}$$

$$\text{steady uniform flow} \longrightarrow$$
$$\text{steady non-uniform flow} \longrightarrow$$
$$\text{unsteady non-uniform flow} \longrightarrow$$

Various categories of flow may therefore be attributed to the various terms in the equation 5.8 as indicated above. Using equation 5.4 with $S_f = S_0$, and denoting Q_n as the

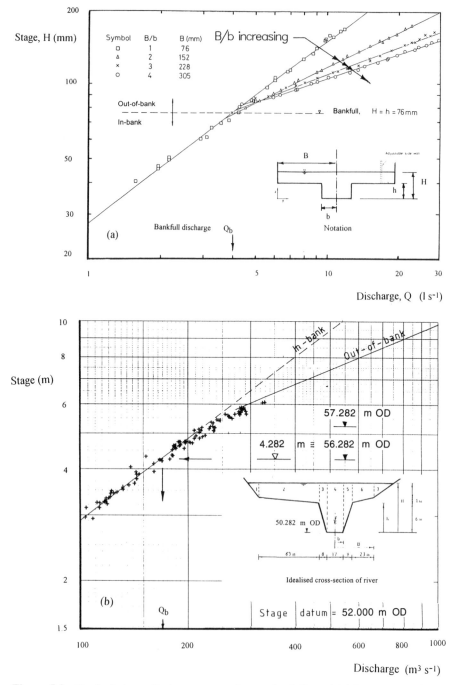

Figure 5.2   Typical stage–discharge curves for overbank flow: (a) laboratory, (b) field

steady discharge at uniform flow at normal depth, $h = h_n$, then combining equations 5.4 and 5.8 gives

$$Q = Q_n \left[1 - \frac{\partial h}{S_0 \partial x} - \frac{u}{gS_0} \frac{\partial u}{\partial x} - \frac{1}{gS_0} \frac{\partial u}{\partial t} \right]^{1/2} \tag{5.9}$$

kinematic wave $\longrightarrow$

diffusion analogy $\longrightarrow$

full dynamic wave $\longrightarrow$

where the terms are again grouped to indicate different levels of flood routing model. Of interest here is the diffusion model (Price, 1973, 1985; NERC, 1975) which results from equations 5.6 and 5.9 and gives the convective-diffusion equation

$$\frac{\partial Q}{\partial t} + c \frac{\partial Q}{\partial x} = D \frac{\partial^2 Q}{\partial x^2} \tag{5.10}$$

where D = diffusion coefficient = $Q/2BS_0$, and the wave speed, c, is given by

$$c = \frac{1}{B} \frac{dQ}{dh} \tag{5.11}$$

It follows from equation 5.10 that the discharge in a channel during a flood event has the characteristics of a wave that translates and attenuates. The reduction of the St Venant equations to the convective-diffusion equation implies that the relationship between stage and discharge is no longer uniquely defined from simple steady flow formulae such as the Manning equation assuming $S_f = S_0$, but is of a more complex looped nature arrived at by inserting equation 5.8 into equation 5.4 (Henderson, 1966; Knight, 1989). The gradient of the stage discharge curve is related to the kinematic wave speed by equation 5.11, and thereby indicates that during a flood c will vary as Q, dQ/dh and B change with time. For a simple rectangular channel the wave speed may be given by

$$c = \left[\frac{5}{3} - \frac{4h}{3(B + 2h)} \right] \tag{5.12}$$

indicating that for a very wide channel $c = 5/3 \times$ velocity (NERC, 1975).

The routing of a flood down a one-dimensional channel may be accomplished by solving either equations (5.6 and 5.7) or equation (5.10). One relatively simple and effective solution procedure to the latter equation is the variable parameter Muskingum–Cunge method (VPMC) which allows for the variation of the travel time constant, K, and the distance weighting parameter, ε, in the equations

$$I - O = \frac{dS}{dt} \tag{5.13}$$

$$S = K[\varepsilon I + (1 - \varepsilon)O] \tag{5.14}$$

where I = inflow, O = outflow and S = storage. Cunge (1969) showed that K and ε may be related to the wave speed, c, and the attenuation parameter, α, by

$$K = \frac{\Delta x}{c} \qquad (5.15)$$

$$\varepsilon = \frac{1}{2} - \frac{D}{c\Delta x} = \frac{1}{2} - \frac{\alpha \bar{Q}_p}{Lc\Delta x} \qquad (5.16)$$

where $\bar{Q}_p$ = mean peak discharge and L = reach length, and the diffusion coefficient, D, is linked to α by

$$D = \frac{\alpha \bar{Q}_p}{L} \qquad (5.17)$$

Provided c and α are known for all Q, then the values of K and ε may be obtained for all times throughout a hydrograph and a flood routed explicitly through a river system. Further details on flood routing may be found in Price (1973, 1978, 1985, 1986). Typical wave speed and attenuation parameters are shown in Figure 5.3 and some actual data in Figure 5.4, taken from the UK *Flood Studies Report* (NERC, 1975). Natural rivers do not exhibit a simple relationship as expressed by equation 5.12 but, as shown, the wave speed typically reaches a maximum value at around two thirds of bankfull flow, Q_b, before decreasing to a minimum at a particular overbank flow with a shallow floodplain depth, probably associated with the maximum main channel/floodplain interaction effect. The overbank case is considered later in more detail, but Figures 5.3 and 5.4 are presented here to illustrate again a continuum of events that occur between inbank and overbank flows.

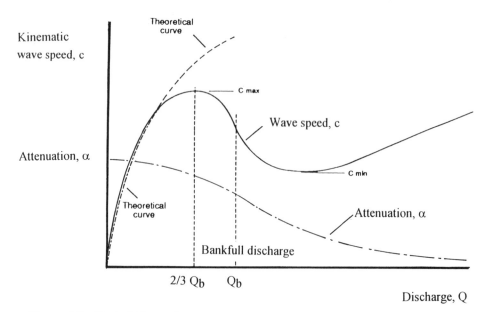

Figure 5.3   Typical kinematic wave speed–discharge and attenuation–discharge curves

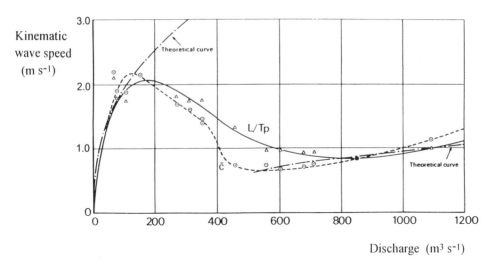

Figure 5.4 Example of wave speed–discharge curve for River Wye, Erwood to Belmont reach (after NERC, 1975)

5.2.3 Equilibrium sediment channels

Natural rivers possess more degrees of freedom than rigid boundary channels in which the cross-section shape and planform geometry of the channel do not change with discharge. In alluvial channels composed of sediments the channel will seek to respond to a given discharge, Q, and sediment load, X, by adjustment of its breadth, B, depth, D, velocity, U, and slope, S. In this section the symbols B, D, U and S are used since the cross-sectional shape of an equilibrium channel is as yet undefined and involves linking the surface width, B, and mean depth, D, to other parameters. In general for a given sediment size the six variables

$$\{Q, X, B, D, S, U\} = 0 \tag{5.18}$$

may be linked by four equations, based on sediment transport, sediment resistance, continuity and one extremal hypothesis. This means that two variables can be selected to determine the remaining four. For example

$$\{Q, X\} = f\{U, B, D, S\} \tag{5.19}$$

For inbank flows the design of stable or equilibrium channels may be based on either the rational equilibrium approach given above (White *et al.*, 1982) or on the empirical regime "laws" obtained from observation of natural rivers and irrigation canals (Ackers, 1992a). See also Cao (1995) and Cao and Knight (1995, 1996a,b), Various extremal hypotheses exist (Bettess and White, 1987), but most are based on the principle that "for an alluvial channel the necessary and sufficient condition of equilibrium occurs when the stream power per unit length of channel, $\rho g Q S_f$, is a minimum, subject to certain constraints" (Chang, 1980). The implications of this are illustrated in Figure 5.5 where for a given sediment load the slope is shown to reach a minimum at the equilibrium breadth that produces the maximum sediment discharge

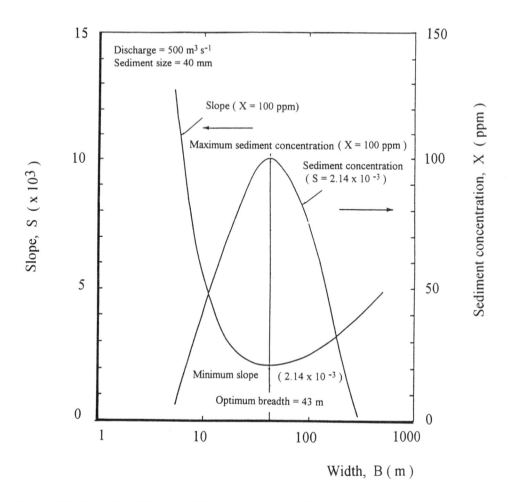

Figure 5.5 Typical equilibrium width relationship for minimum stream power and maximum sediment transport rate (after White *et al.*, 1982)

illustrated in Figure 5.5 where for a given sediment load the slope is shown to reach a minimum at the equilibrium breadth that produces the maximum sediment discharge rate, X, for a given slope, S. An alluvial channel in equilibrium is thus one that maximizes the sediment carrying capacity for given external constraints. It is therefore possible to construct a theoretical sediment rating curve, X versus Q, in a similar way to the stage–discharge curve, h versus Q. Such a curve is shown in Figure 5.6 taken from an example by Ackers (1992a). In this hypothetical river, the sediment motion is assumed to occur only in the main channel of a compound section with two floodplains.

It can be seen from Figure 5.6 that for discharges up to bankfull of $16.9 \text{ m}^3 \text{ s}^{-1}$, X increases with Q as would be expected. At higher discharges, whereas X might be

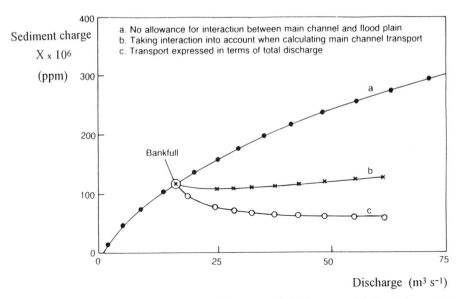

Figure 5.6   Sediment transport in a hypothetical river with 0.25 mm sand (after Ackers, 1992a)

the main river channel in such a way that the sediment load may actually decrease once overbank flow occurs and only possibly recover to the same rate at a much higher discharge. In a somewhat similar way to there being two possible depths for the same discharge in Figure 5.2, Figure 5.6 shows that there are now two or three possible Q's for the same sediment transport rate, X. The same general relationship between X and Q is also confirmed by calculations by Knight and Abril (1996) for a different hypothetical case.

In natural rivers the presence of sediment in the main river, in bankside regions and on the floodplains further complicates the simplified approach given here. This simplified analysis has only been presented to illustrate the possible influence of floodplains in straight channels on one important sediment parameter, namely the bedload sediment transport rate. When dealing with alluvial channels many other hydraulic and geomorphological aspects need to be considered, such as the details of sediment deposition and erosion, meandering and braiding behaviour, possible width adjustment, variations in bed forms with discharge (and hence variations in that part of the flow resistance due to form drag), differential bedload rates associated with graded sediments, possible armouring when dealing with gravel-bed rivers, cohesive soil properties when dealing with silts, and the influence of bankside vegetation, to name but a few. The range of issues that need to be considered in river mechanics is extensive, as recently summarized by Knight (1996).

The three topics presented here serve to show how the variation of water level, wave speed and sediment transport rate vary from inbank to overbank flow. In each case there is a departure from a simpler relationship for inbank flow to a more complex relationship for overbank flow. These more complex relationships are the result of additional three-dimensional processes present in overbank flow and are dealt with more thoroughly in the next section.

5.3 FLOODPLAIN HYDRAULICS

5.3.1 Overview of recent studies

It is important to have a good conceptual model of the interaction process that occurs between the main channel and any floodplain flows before attempting any theoretical or experimental work. Although sketches of perceived behaviour may be helpful, they may also cloud one's perspective and mislead one into a misunderstanding of the real physics of the flow. In the same way, too precipitate a judgement about any governing equations, or which terms in that equation might be thought to be trivially small, may also prejudge the real situation. Although sketches of the process given in Figures 5.7 and 5.8, and photographs by Sellin (1964) have been helpful, they may also have coloured our imagination. In these figures various three-dimensional structures are illustrated, each with their own length and time-scale. In the zone of immediate interest, i.e. at the top of the main river channel bank at the edge of the floodplain, there are large interrelated vortex structures which are individually and corporately important.

One dominant feature, first photographed at the water surface by Sellin (1964), are the vortices with vertical axes which develop in any highly sheared zone between two

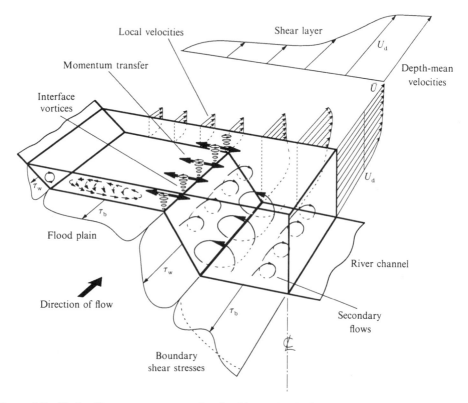

Figure 5.7 Hydraulic parameters associated with overbank flow (after Shiono and Knight, 1991)

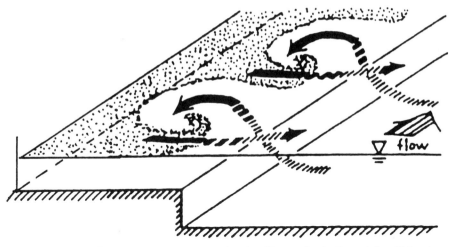

Figure 5.8 Large-scale vortex structure associated with overbank flow (after Fukuoka and Fujita, 1989)

co-flowing streams at different velocities. These are indicated in Figure 5.7 as a bank of vertical "interface vortices" and in Figure 5.8 as somewhat larger vortices or eddies convecting high momentum fluid from the main channel onto the floodplain. It is the periodic nature of these vortices (Knight and Shiono, 1990; Meyer and Rehme, 1994) that accounts for some of the momentum transfer between the deep and shallow regions. In addition to vorticity generated along vertical axes, Figure 5.7 also shows some helical secondary flows in the longitudinal streamwise direction. It is well known that longitudinal vorticity is present in all turbulent flows in non-circular sections (Einstein and Li, 1958, Liggett *et al.*, 1965; Tracy, 1965; Melling and Whitelaw, 1976; Perkins, 1979; Imamoto *et al.*, 1993) and that these cause perturbations in any lateral distribution of boundary shear stress, as shown in Figure 5.7, and affect the pattern of isovels (Perkins, 1979; Knight and Patel, 1985; Knight, *et al.*, 1994). Secondary flows are usually directed inwards towards channel corners and outwards at re-entrant corners as shown by Perkins (1979). The floodplain cell appears to be significant for reasons to be explained later with regard to Figures 5.9–5.11. Recent research has highlighted both the three-dimensional nature of all the vortex structures (Fukuoka and Fujita, 1989; Imamoto and Ishigaki, 1990; Imamoto and Kuge, 1974; James and Wark, 1994; Lai, 1986; Muto, 1995; Nezu, 1993; Rhodes, 1991; Tominaga and Nezu, 1991; Yuen, 1989) and also the complexity of the turbulence in the interaction zone. Since these are still not yet fully understood, it is not surprising that there is no simple method of analysing overbank flow.

From a practical point of view the effect of vorticity and turbulence on velocity and boundary shear stress is illustrated in Figure 5.7. At low depths on the floodplain, the faster moving water in the main river channel and the slower moving water on the floodplain produce a shear layer laterally across the system. The depth mean velocity, U_d, is seen to decrease from one value in the river to a lower value on the floodplain. The boundary shear stress likewise responds in a similar manner, although perturbations may occur wherever longitudinal secondary flows are strong. Narrow channels are

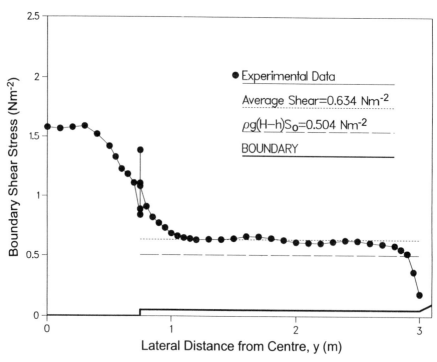

Figure 5.9 Lateral distribution of boundary shear stress from Flood Channel Facility test (FCF experiment 080501)

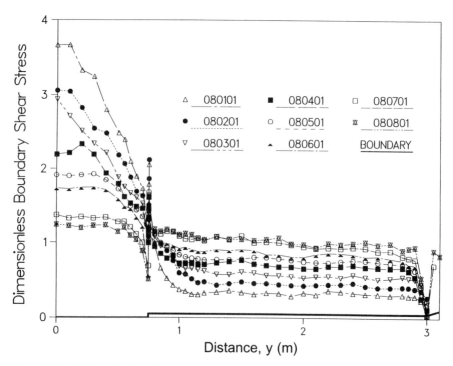

Figure 5.10 Dimensionless boundary shear stress distributions from FCF tests (Series 8)

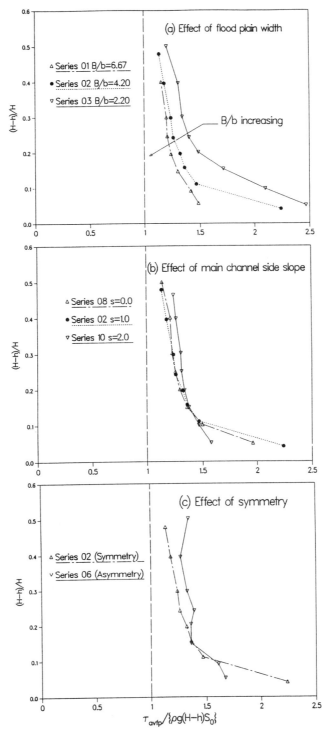

Figure 5.11 Ratio of measured to "two-dimensional" boundary shear stresses from FCF tests (Series 1–10), showing: (a) effect of floodplain width, (b) effect of main channel side slope, (c) effect of symmetry

particularly susceptible to such secondary flows (Knight and Lai, 1985). Figure 5.7 also shows that the shear layer width extends both into the river channel and on to the floodplain. The width of this shear layer is composed of two parts, one from Reynolds stresses and the other from secondary flows/vorticity, and each spread laterally at different rates (Shiono and Knight, 1991). The combined effect of both vortical structures and turbulence on boundary shear stress is shown in Figures 5.9–5.11, taken from a representative data set from the UK Flood Channel Facility (FCF) research programme (Series 8 with vertical sides to the main river channel). In Figure 5.9 the local boundary shear stress, τ_b, varies from 1.6 N m^{-2} in the river channel to 0.634 N m^{-2} on the floodplain, with an obvious shear layer transition from one to the other, ignoring shear stresses on the side wall. Another distinctive feature of floodplain flows is shown by the difference between the "two-dimensional" value of boundary shear stress, 0.504 N m^{-2}, based on the local depth and bed slope, $\tau_b = \rho g (H - h) S_o$, compared with the actual measured value of 0.634 N m^{-2} that occurs on the floodplain, some 26% higher. The difference arises because of the right hand side of equation 5.2 and the influence of secondary flows. The flow is not in fact "two-dimensional" in the usual sense, despite a constant plateau of values of both τ_b and U_d on the floodplain. Figure 5.11 illustrates the effect of this for a number of series of experiments in which the channel side slope and floodplain geometry were varied. In each case it is clear that at low floodplain depths the increase in the boundary shear stress is sometimes over twice that calculated on local depth and energy slope values. See Knight and Sellin (1987), Shiono and Knight (1991) and Knight and Cao (1994) for further details.

Another general feature of floodplain flows is their strong dependence on the floodplain depth or relative depth, $((H - h)/H)$. The "relative depth" is used to describe the ratio between the floodplain depth, $(H - h)$, and the main river depth, H, where $h =$ river bank height. The maximum interaction between the river and the floodplain usually occurs at relative depths of around 0.10–0.30, typically at 0.20, and consequently the main focus of experimental work should be in this region. However, this is practically difficult because at laboratory scales the bank height, h, is inevitably small making floodplain depths at low relative depths extremely small and difficult to set up and undertake measurements in. Turbulence and secondary flow measurements at such small floodplain depths are obviously exceptionally difficult. The FCF experiments in the UK were all conducted for $(H - h)/H$ values between 0.05 and 0.50. Examples of the difficulties of floodplain measurements are highlighted by Rhodes and Knight (1994b) and Knight and Demetriou (1983). It is important to realize the strong depth dependence of all parameters or physical effects so as to interpret any description or experimental data accordingly. Most authors recognise this and functional relationships for the interaction process are usually expressed in depth dependence terms (Imamoto and Kuge 1984; Knight *et al.*, 1983, 1984, 1990, 1991; Tominaga and Nezu 1991; Ackers, 1992b, 1993a,b). Figures 5.10 and 5.14 illustrate why this is so, using Series 8 data for boundary shear stress and coherence data for Series 2 from Ackers (1992b). At low depths (test 080101) Figure 5.10 shows that the main channel value is three to four times the overall channel average and the floodplain is correspondingly much less than the channel average. As the relative depth changes from 0.05 to 0.5 (tests 080101 to 080801) the lateral distribution changes so that at the maximum depth tested the main channel and floodplain values are of a comparable order of magnitude. This arises because, as Figure 5.10 shows, the main channel and floodplain velocities tend to equalize, thus removing a strong lateral shear influence. However, at these large depths the vortical structures may still influence the flow considerably. It should be noted that in Figures 5.10 and other data (Knight and Demetriou, 1983; Lai, 1986), the aspect ratio of the main channel is significantly different, which highlights another variable in any general analysis.

All the previous figures 5.1–5.11, and particularly Figures 5.7 and 5.8, have demonstrated the complexity of the floodplain hydraulics. It is clear that three-dimensional flow structures are present and that one or two-dimensional methods of analysis are likely to be incomplete. However, for practical purposes such methods of analysis are still required since no general three-dimensional model or experimental turbulence data are currently available. The next few sections therefore deal further with floodplain flows treated from a one, two or three-dimensional point of view.

5.3.2   One-dimensional analysis

One traditional approach to the hydraulics of flows in complex cross-sections has been to subdivide the channel into a number of discrete sub-areas as shown in Figure 5.12. The conveyance capacities are then calculated via equations 5.4 or 5.5, using the appropriate sub-area values for A, P, R, f and n, and the individual conveyances are then summed to give the total discharge in the whole channel. One example of the difficulty in applying this approach is shown for the River Severn at Montford in Figure 5.13 Using measured friction slope and overbank velocity data in the field, Knight *et al.* (1989) calculated the conveyance in three sub-areas and their corresponding roughness values. The variation of Manning n with depth for the whole channel is shown to decrease sharply just above bankfull level due to abrupt changes in the hydraulic radius. On the other hand the sub-area values for the main channel had to be increased in order to obtain the correct sub-area conveyance capacity. The sub-area values for the floodplain had to be correspondingly reduced, to levels well below their actual values. Figure 5.13 therefore serves to show the difficulties in arriving at the correct stage–discharge curve, which for this particular site has been shown in Figure 5.2b.

Alternatively, a composite roughness coefficient may be adopted for the whole channel based on aggregating values from the sub-areas. Although the difficulties in using this approach for channels with complex variations in hydraulic radius have already been demonstrated in Figure 5.13, this method is still advocated in various textbooks. In this approach, however, certain simplifying assumptions have still to be made in order to obtain a composite roughness value. Historically these have ranged from making pairs of complementary assumptions, such as that: (i) the total shear force equals the sum of the constituent sub-area shear forces and that the sub-area velocities vary in proportion to the depth to a one sixth power law (Pavlovskii, 1931), (ii) the velocities are equal in each sub-area and the friction slope is the same for all sub-areas (Horton, 1933; Einstein, 1934), or (iii) that the total discharge equals the sum of the constituent discharges and that the friction slope is the same for all sub-areas (Lotter, 1933). These are summarized in Chow (1959), and discussed by Knight *et al.* (1983, 1984) in relation to flood channels. These early division and composite roughness

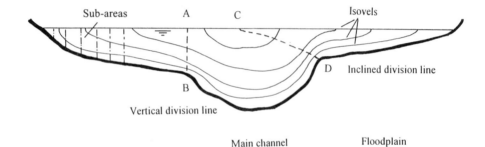

Figure 5.12   Channel subdivision methods for calculation of discharge

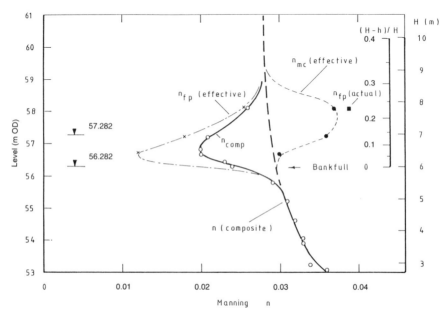

Figure 5.13   Variation of Manning "n" resistance coefficient for overbank flow at Montford, River Severn (after Knight *et al.*, 1989)

methods are essentially flawed, since the simple addition of the individual flows in each sub-area does not necessarily equal the total flow due to the main channel/ floodplain interaction process and the invalidity of most of assumptions listed above. However four variations on the channel division methods have been suggested in order to simulate the interaction process more accurately.

The first is based on altering the sub-area wetted perimeters, typically excluding the length AB in the calculation of P for the floodplain (see Figure 5.12), but including it in the value of P for the main channel. This is intended to have the effect of retarding the flow in the main channel sub-area and enhancing it in the floodplain sub-area. However since the length of the line AB is small at low depths of submergence on the floodplain, when the interaction effect is very high, this approach will inevitably fail (Wormleaton *et al.*, 1982, Knight *et al.*, 1983, 1984). Although many variants on this kind of approach have been tried, it is now recognized that it is better to adjust the discharges in each sub-area by some appropriate method later as a separate step in the calculation procedure. This is in fact the basis of the "coherence" method which is discussed as the second variation.

The second one-dimensional approach is based on the "coherence" concept by Ackers (1993a). Where the differences in velocity and depth between the various sub areas are large, then significant interaction between the sub-areas would be expected to occur. "Coherence" is defined as the ratio of the basic conveyance calculated by treating the channel as a single unit, with perimeter weighting of the friction factor, to that calculated by summing the basic conveyances of the separate zones. Thus the coherence *COH* is defined as

$$COH = \frac{\displaystyle\sum_{i=1}^{i=n} A_i \sqrt{\left[\sum_{i=1}^{i=n} A_i \middle/ \sum_{i=1}^{i=n} (f_i P_i)\right]}}{\displaystyle\sum_{i=1}^{i=n} [A_i \sqrt{(A_i/(f_i P_i))}]} \qquad (5.20)$$

The closer to unity the COH approaches, the more appropriate it is to treat the channel as a single unit, using the overall geometry. Where the coherence is much less than unity then discharge adjustment factors are required in order to correct the individual discharges in any sub-area. It appears from experimental studies of overbank flow (Ackers, 1993b) that different adjustment factors are required in at least four distinct regions of depth. A typical set of data are shown in Figure 5.14. Further details of this approach are given by Ackers (1992b, 1993a,b).

The third one-dimensional approach which has been popular is to quantify the apparent shear stresses (ASSs) or forces (ASFs) on the sub-area division lines, usually again on the basis of experimental evidence. Many authors have attempted to develop empirical equations for these ASSs. These can then be included in the first method to give the effective shear force or resistance for each sub-area and hence the correct sub area conveyance capacity. Most of the equations which have been published use the ratio of the floodplain depth to main channel depth, already referred to as the "relative depth", as a primary variable with the floodplain width and relative roughness as subsidiary variables (Knight and Hamed, 1984). Although most of the equations produced to date fit particular experimental data sets well, they only apply within the range of parameters tested and are not generally applicable. It is not surprising therefore that multiple regressions of data, covering a wide range of depths, floodplain widths, roughnesses, aspect ratios, etc., are not very effective, given the obvious subtleties of the interaction process, as illustrated by Figure 5.14, in which it is necessary to specify at least four regions of depth in order to obtain accurate discharge adjustment equations for use in the Ackers coherence approach.

A fourth one-dimensional approach, which follows directly on from the third, is to consider drawing the division lines along lines of zero shear stress as indicated by line CD in Figure 5.12, rather than at predetermined positions dictated by the analyst. However, the three-dimensional nature of the velocity field makes it extremely difficult to generalize the position of these division lines for all types of channel shape, flow

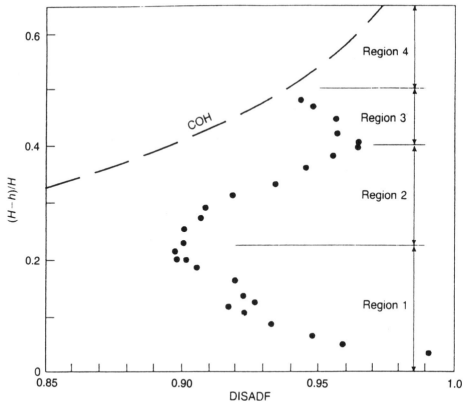

Figure 5.14   Discharge adjustment factors, DISADF, for FCF tests, Series 2 (after Ackers, 1993a)

depth and roughness configuration (Knight and Hamed, 1984). Moreover, it is known from three-dimensional turbulence considerations that orthogonal lines to the isovels do not necessarily imply lines of zero shear stress (Tracy, 1965; Melling and Whitelaw, 1976).

Certain features are apparent from this summary of one-dimensional approaches. Firstly, it is clear that the prediction of the conveyance capacity in the form of a stage discharge curve is the most fundamental test or requirement of any predictive computational method. However, it should be noted that even though a particular method performs well in predicting the total conveyance capacity for a given stage, it does not necessarily imply that that method is a good one. It is imperative to check that the division of flow between sub-areas is also correct, as the two adjustment procedures for adjusting the main channel discharge downwards and the floodplain discharge upwards, are clearly self compensating. This is best achieved by checking the proportion of the total flow in the main channel against a validated data set. As many authors have shown (e.g. Wormleaton *et al.*, 1982; Knight and Hamed, 1984; Ackers, 1992b), it is quite possible to have a moderately successful h versus Q prediction method based on division lines, but it may lead to gross errors in the internal distribution of the flow.

Secondly, it should be noted that all these methods concentrate on obtaining the correct discharge in the various sub-areas, whereas for sediment transport or other processes it is the sub-area boundary shear stresses that may be equally of importance. The boundary shear is a particularly difficult parameter to predict accurately due to its sensitivity to secondary flows (Knight and Patel, 1985; Knight *et al.*, 1994). Any one-dimensional method should therefore be capable of predicting not only the discharge but also the mean shear stress or force in each sub-area. None of the methods mentioned previously explicitly give adjustment factors for the boundary shear stress, although Ackers (1992a) does suggest using the square of the discharge adjustment factor for the main channel to give the correct boundary shear stress there. This follows from the often assumed square law relationship between velocity and shear stress. The boundary shear stress is however a sensitive parameter and more sophisticated methods are normally required in order to accurately predict its distribution around the wetted perimeter. This topic is now dealt with in the next section.

5.3.3 Two-dimensional analysis

A logical development of the one-dimensional approaches described earlier would be to divide the channel cross-section into not only many more sub-areas, but also to include analytically in those sub-areas those processes which are known to be important and which are absent from simple bulk flow equations like equations 5.4 and 5.5. Equation 5.2 forms the basis of such an approach since it includes lateral shear and secondary flows in addition to the usually dominant bed friction. It should therefore provide a more accurate representation of the flow physics present in overbank flow. The equation is based on depth-averaged parameters of velocity, U_d, Reynolds stress, $\bar{\tau}_{yx}$, and secondary flow, $H(\rho UV)_d$. These in turn are governed by three calibration coefficients, f, λ and Γ, concerned with bed friction, lateral shear (via depth-averaged eddy viscosity) and secondary flow (via lateral gradient of $H(\rho UV)_d$) respectively, where

$$\tau_b = \left(\frac{f}{8}\right)\rho U_d^2; \; \bar{\tau}_{yx} = \rho \bar{\varepsilon}_{yx} \frac{\partial U_d}{\partial y}; \; \bar{\varepsilon}_{yx} = \lambda U_* H; \; \frac{\partial (H\rho UV)_d}{\partial y} = \Gamma \tag{5.21}$$

Unlike one-dimensional methods, which just have one lumped parameter such as f or n, there are now three parameters and hence the possibility arises of simulating the lateral variations of U_d, τ_b, $\bar{\tau}_{yx}$ and Γ. The method is still however approximate, on account of the depth-averaging procedure, but has three degrees of freedom. Since depth-averaged models are frequently used in engineering practice, and lateral distributions of key parameters are useful in design and an improvement on single discharge values in sub—areas, there are obvious advantages in adopting this method of approach. Furthermore lateral integration of key parameters within any sub area or zone will of course yield mean sub-area values required for any one-dimensional analysis or model.

The analytical solution to equation 5.2 may be obtained by writing it in the form of second-order linear ordinary differential equations, one for a constant depth domain and the other for a variable depth domain.

The solution given by Shiono and Knight (1991) for a sub area with a constant depth, H, is

$$U_d = \left\{ A_1 e^{\gamma y} + A_2 e^{-\gamma y} + \frac{8 g S_0 H}{f} (1 - \beta) \right\}^{1/2} \tag{5.22}$$

and for a sub-area with a channel side slope of $1 : s$ (vertical : horizontal) is

$$U_d = \{ A_3 \xi^{\alpha_1} + A_4 \xi^{-\alpha_1 - 1} + \omega \xi + \eta \}^{1/2} \tag{5.23}$$

where

$$\left.\begin{aligned}
\alpha_1 &= -\frac{1}{2} + \frac{1}{2} \left\{ 1 + \frac{s(1 + s^2)^{1/2}}{\lambda} (8f)^{1/2} \right\}^{1/2} \\[2mm]
\beta &= \frac{\Gamma}{\rho g H S_0} \\[2mm]
\gamma &= \left(\frac{2}{\lambda}\right)^{1/2} \left(\frac{f}{8}\right)^{1/4} \frac{1}{H} \\[2mm]
\eta &= -\frac{\Gamma}{\dfrac{(1 + s^2)^{1/2}}{s} \rho \left(\dfrac{f}{8}\right)} \\[2mm]
\omega &= \frac{g S_0}{\dfrac{(1 + s^2)^{1/2}}{s} \left(\dfrac{f}{8}\right) - \dfrac{\lambda}{s^2} \left(\dfrac{f}{8}\right)^{1/2}}
\end{aligned}\right\} \tag{5.24}$$

and ξ is the local depth function on any side slope domain, given by

$$\xi = H - \left(\frac{y - b}{s}\right) \tag{5.25}$$

By dividing the channel into any number of elements, equations 5.21–5.25 give the lateral variation of U_d and τ_b, assuming that any channel can be discretized by a finite number of linear elements, and that the three coefficients, f, λ and Γ, are known or prescribed for each sub-area. The analytical solution also serves as a closed form check on direct numerical simulations of equation 5.2 using finite difference or finite element methods.

Four examples of the use of this model are shown in Figures 5.15–5.18. In each case experimental data from the Flood Channel Facility (FCF) was chosen to test a finite element model, and comparisons are made for two channels and three methods of calibration. Figure 5.15 shows a symmetric channel and how a calibration based on selecting appropriate f, λ and Γ values gives good agreement between the experimental data and the model. Figure 5.16 shows an asymmetric channel calibrated in the same way. Figure 5.17 shows the same data as Figure 5.15 calibrated with constant λ values

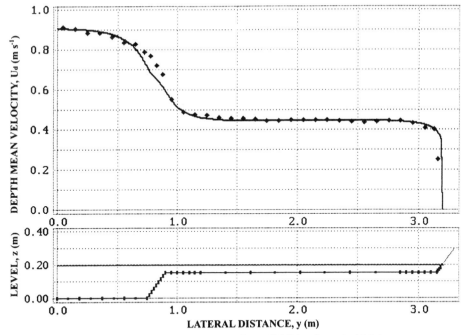

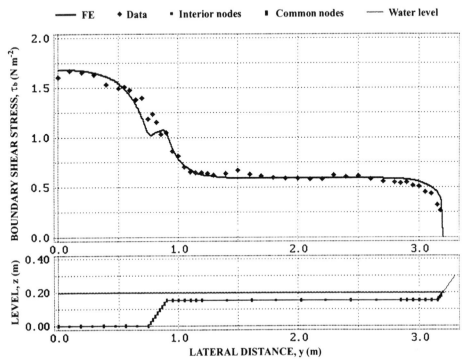

Figure 5.15   Example calibration of two-dimensional model (FCF experiment 020501)

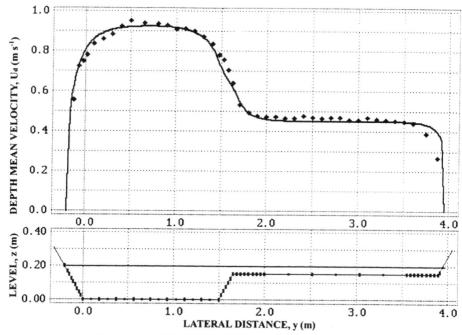

Simulation of SERC-FCF experiment 060501, using actual f data,
$\lambda = 0.07$ on main channel and equation for λ on floodplain
FINITE ELEMENT SOLUTION FOR DEPTH MEAN VELOCITY, Ud (m s⁻¹)

—— FE   ♦ Data   ▪ Interior nodes   ▪ Common nodes   —— Water level

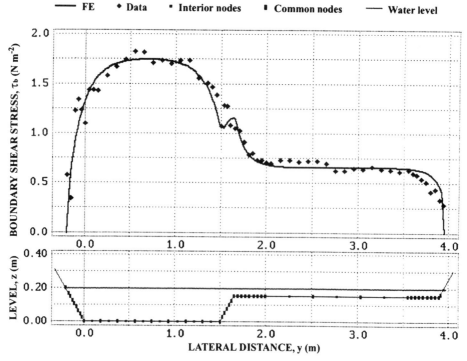

Figure 5.16   Example calibration of two-dimensional model (FCF experiment 060501)

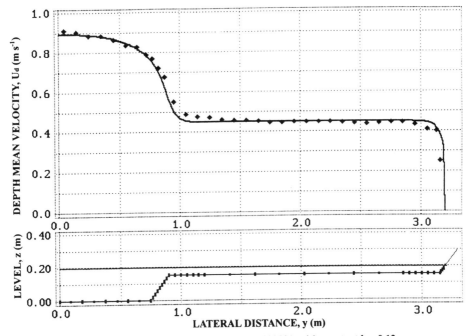

Simulation of SERC-FCF experiment 020501, with constant λ = 0.13 on
main channel and floodplain, and actual f data
FINITE ELEMENT SOLUTION FOR DEPTH MEAN VELOCITY, Ud (m s⁻¹)

—— FE   ◆ Data   ▪ Interior nodes   ▪ Common nodes   —— Water level

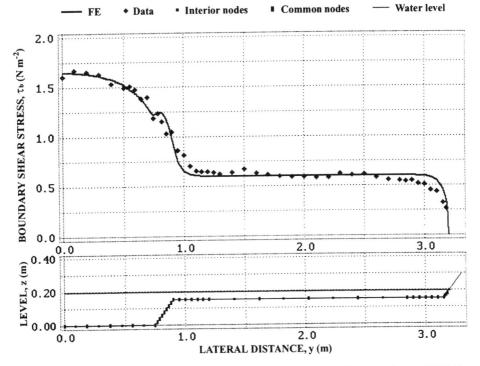

Figure 5.17   Example calibration of two-dimensional model (FCF experiment 020501)

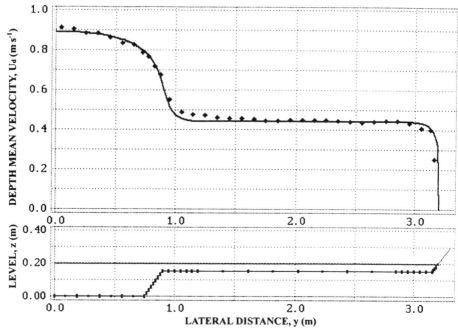

**Simulation of SERC-FCF experiment 020501, with constant λ = 0.13 on
main channel and floodplain, and neglecting secondary flow Γ=0**
FINITE ELEMENT SOLUTION FOR DEPTH MEAN VELOCITY, Ud (m s⁻¹)

—— FE   ◆ Data   ▪ Interior nodes   ▮ Common nodes   —— Water level

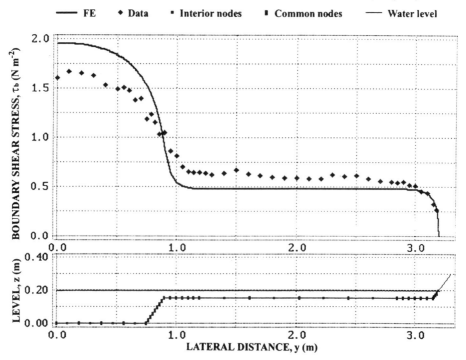

Figure 5.18   Example calibration of two-dimensional model (FCF experiment 020501)

($\lambda = 0.13$) and variable Γ values. Figure 5.18 shows the same data as Figure 5.17 with different f and the same λ values (λ again held constant at 0.13), but with Γ values set to zero. This demonstrates the importance of the Γ term in equation 5.1, particularly with regard to calibrating a model to give the correct boundary shear stress. The good agreement between model and data in Figures 5.15–5.17 and the poor agreement in Figure 5.18 indicate the need to consider carefully the calibration philosophy and the number of variables required in any calibration procedure. Full details of the model are given in Abril (1995a,b). Similar examples are shown in Shiono and Knight (1990), Wark et al. (1990), Shiono (1993), and Knight and Abril (1996). Clearly the two-dimensional model is superior to a one-dimensional model since more use may be made of the lateral output than bulk values. However for a complete analysis of the overbank problem a fully three-dimensional model is required.

5.3.4 Three-dimensional analysis

A description of the procedures required for a full three-dimensional analysis is beyond the scope of this chapter. Reference should be made to Chapter 11 and also the work of Rodi (1980), Naot and Rodi (1982), Lin and Shiono (1992, 1994), Shiono and Lin (1992), Cokljat and Younis (1994) and Thomas and Williams (1995) to name but a few. It is however appropriate to highlight certain three-dimensional features that should be of concern to modellers, based on the authors experimental work. Four particular issues are now considered.

Firstly it should be noted that the logarithmic velocity law is not necessarily valid in the complex region where the interaction process between the river and its floodplain is most intense. The vertical and lateral distributions of Reynolds stress were derived from equation 5.1 by Knight et al. (1990) in the form

$$\tau_{zx} = \rho g(H - z)\sin\theta + \int_z^H \frac{\partial \tau_{yx}}{\partial y}\, dz + \int_z^H \frac{\partial(\rho UV)}{\partial y}\, dz + \rho UW \qquad (5.26)$$

$$\bar{\tau}_{yx} = (\rho UV)_d - \frac{1}{H}\int_0^y \left[\rho g H S_0 - \tau_b \sqrt{\left(1 + \frac{1}{s^2}\right)} \right] \qquad (5.27)$$

At the river/floodplain interface the individual terms in equations 5.26 and 5.27 change significantly. The net effect of all the terms is illustrated by some experimental data shown in Figures 5.19 and 5.20, taken from the FCF Series 02, at a relative depth (($H - h)/H$) of 0.152. Figure 5.19 shows some lateral distributions of τ_{yx} at different depths, z, within the flow for various y values. At the channel centreline, $y = 0$, τ_{yx} is zero, as would be expected from the symmetry of the flow. Between $0.5 < y < 1.5$ m, there is a strong shear layer region, in which these Reynolds stresses are significant. The maximum stresses occur at the $y = 0.90$ m position, i.e. at the top of the main channel bank at the beginning of the floodplain. Figure 5.20 shows the corresponding τ_{zx} values for the same experiment. It can be seen that the vertical distribution of τ_{zx} is approximately linear at the centreline of the channel, $y = 0$, as would be expected from the first term in equation 5.26, and when derivatives with respect to y are zero. In the vicinity of the main river channel bank ($y = 0.81$ m), the distributions are highly

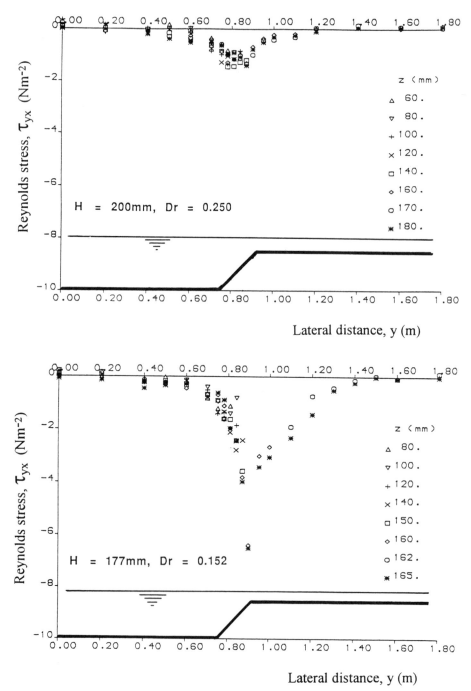

Figure 5.19 Lateral variation of Reynolds stresses, τ_{yx}, near the main channel/floodplain interaction zone (FCF experiments 020301 and 020501)

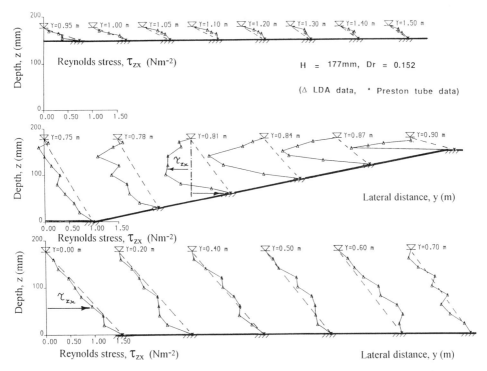

Figure 5.20 Vertical variation of Reynolds stresses, τ_{zx}, near the main channel/floodplain interaction zone (FCF experiment 020301)

non-linear in agreement with the influence of the various terms in equation 5.26. At large values of y ($y > 1.50$ m), i.e. outside the region in which τ_{yx} is non-zero (see Figure 5.20), but still within the region where vorticity and secondary flows are influential, the vertical distributions return to a linear form. This is discussed further in Knight *et al.* (1990) and Shiono and Knight (1991) where three-dimensional spatial plots of these stresses may be found. The very high quality experimental closed duct data of Meyer and Rehme (1994) also confirm these spatial variations. It is evident that there are strong three-dimensional effects occurring in this region and that care needs to be taken in predicting boundary shear stress from assumed logarithmic velocity profiles.

Secondly, the relative contributions of the depth-averaged Reynolds stress term, $\bar{\tau}_{yx}$, and the secondary flow term, $(\rho UV)_d$, to the apparent shear stress (ASS) acting on any vertical interface at a given lateral position is important. Prior to the detailed experimental work on the FCF the authors were of the opinion that the right hand side of equation 5.2 was small in comparison with the third term on the left hand side. The experimental evidence shows otherwise, and the data presented by Shiono and Knight (1990, figure 3, and 1991, figures 9 and 10) indicate that these two terms are roughly comparable in magnitude. Recent numerical work based on a large eddy simulation model (LES) tends to confirm this finding, although further analysis of FCF test series 2 is required before it can be confirmed. See, for example, equation 5.21 and figure 9 of Thomas and Williams (1995) using data from test series 22 from the FCF.

Thirdly it appears from the experimental evidence that the lateral spread or influence of the Reynolds stresses is considerably less than the lateral influence of the secondary flow/vortical structures. For example Shiono and Knight (1990, figure 2 and 1991, figures 9 and 10) show that the lateral penetration onto the floodplain of the former is around 1 m, whereas for the latter it is around 3 m, using data taken from series 2 of the FCF tests. Quantifying the lateral spread of these influences could be useful in devising simpler subdivision methods of analysis based on regions of the flow where parameters have known values or none (see Rhodes and Knight, 1995). In this context it is interesting to note from Figure 5.19 that the maximum depth-averaged Reynolds stress, $\bar{\tau}_{yx}$, occurs at approximately the top edge of the main river channel, i.e. at $y = 0.90$ m, making it a suitable position to locate a vertical division line in any of the channel subdivision methods. This is precisely where many authors have chosen to locate such a line, and possibly accounts for its continued use and any efficacy that it might have.

Fourthly it is recognized that spectral analysis of the pulsating flow should be undertaken in the interface zone in order to quantify the quasi-periodic motion that occurs there. Temporal averaging of the governing equations may in fact obscure certain phenomena which have been highlighted by Tamai *et al.* (1986), Shiono and Knight (1990) and Meyer and Rehme (1994). It should be noted that the stress terms in equation 5.1 include both mean and fluctuating components, but that once decomposed into Reynolds stress and secondary flow terms as in equation 5.2, the right hand side of equation 5.2 includes at least two effects. These have been represented in Figure 5.7 as both vorticity in the vertical direction at the floodplain/main channel interface and longitudinal vorticity on the floodplain. In the former there are some known frequency effects which still need to be fully understood. However, the term $(\rho UV)_d$ includes both vorticity effects and is therefore a convenient general "sink" term representing both influences. This appears to be a fortuitous way of representing the vorticity in river/floodplain processes.

5.4   PRACTICAL ASPECTS OF OVERBANK/FLOODPLAIN FLOWS

The foregoing paragraphs have highlighted the development of hydraulic analysis from simple one-dimensional models to three-dimensional models in the last 60 or so years. The overbank flow problem has done much to stimulate thinking about what were fairly straightforward concepts in the 1930s, such as hydraulic radius, friction laws, etc., and has led to a reappraisal of familiar one-dimensional hydraulic analyses. It has been demonstrated that vorticity and lateral shear are clearly important in these complex flows. The three topics selected for comment earlier in Section 5.2 are now briefly discussed.

The conveyance capacity problem is related to both the flood routing and the sediment transport problems mentioned earlier. In the flood routing problem, described in Section 5.2.2, equation 5.11 shows how the stage–discharge curve and top width are related to wave speed. It is therefore of paramount importance to be able to determine accurately the conveyance capacity of a compound channel at a given depth, with all the flow physics of the interaction process suitably modelled. This is especially true at very low submergence depths on the floodplain, when the gradient of the stage–discharge curve is at its most uncertain analytically, due to the difficulty of accurately modelling the interaction process. In the sediment problem, described in Section 5.2.3, it is the redistribution of the discharge and boundary shear stress between sub-areas that is important. The lateral distributions of U_d and τ_b are usually vital for good sediment resistance or transport predictions. The two-dimensional model described in Section 5.3.3 should be helpful here, in addition to the one-dimensional coherence method of Ackers. In both flood routing and sediment transport, Figures 5.3 and 5.6 indicate that there are significant effects on both phenomena during shallow overbank flow and that these need to be taken into account in any simulation process. Experimental validation of both figures to a benchmark standard is highly desirable.

The analytical solution to equation 5.2, given through equations 5.21–5.25 provides a useful closed form check on any numerical simulation. The calibration philosophy for this solution is still developing, but with three parameters, f, λ and Γ, different types of overbank flow problems can be simulated quite well, as Figures 5.15–5.17 indicate. The three parameters appear to control the three main elements which are often present in open-channel flow, namely bed shear, lateral shear and secondary flows. The practical implication of this approach is that it gives the lateral variation of the two parameters most commonly required by river engineers, U_d and τ_b. These distributions have been shown to be correctly simulated through the use of the local friction factors, which relate depth-averaged velocity and local shear stress together. It is clearly

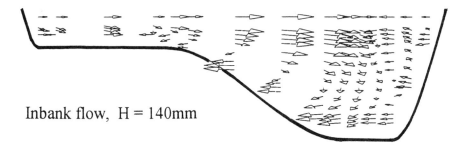

Inbank flow, H = 140mm

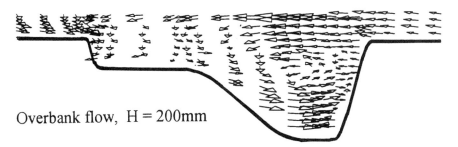

Overbank flow, H = 200mm

Figure 5.21 Secondary flow vectors at the apex of a bend in a 110° meander channel with natural cross-sections (inbank and overbank flows, FCF studies)

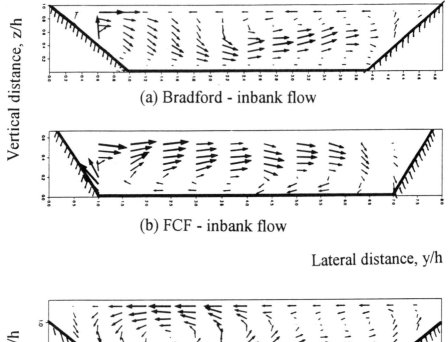

(a) Bradford - inbank flow

(b) FCF - inbank flow

Lateral distance, y/h

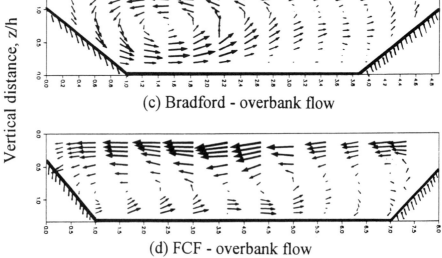

(c) Bradford - overbank flow

(d) FCF - overbank flow

Lateral distance, y/h

Figure 5.22   Secondary flow vectors at the apex of a bend in two 60° meander channels with trapezoidal cross-sections and different aspect ratios (inbank and overbank flows, FCF and Bradford studies)

important to include the secondary flow term and its parameter, Γ, rather than to rely on enhanced friction factors or eddy viscosities alone to calibrate a two-dimensional model. The overestimation of floodplain friction factors by use of the local depth and bed slope has been highlighted. The depth-averaged form of the flow physics still however represents a certain level of approximation and a three-dimensional model is required for more refined simulations. However, the level of the two-dimensional model is still adequate for many practical applications.

There are naturally many other issues requiring further analysis. Once a channel is no longer considered to be straight, which is invariably the case with natural rivers, the flow becomes even more complicated. Figures 5.21 to 5.28 illustrate some experimental data from a number of meandering channel studies with bankfull and overbank flow. At bankfull flow, Figure 5.21 shows that for a left hand bend in a meander channel the secondary flow at the bend apex is clockwise, as would be expected from curvilinear analysis. However, at higher discharges with overbank flow, the secondary flow circulation is reversed, due to the vorticity generated in the lower main channel in the crossover region immediately upstream of that particular bend. The mechanism is described more fully by Sellin *et al.* (1993). Figure 5.22 shows the same effect, observed in two different experimental channels, both with trapezoidal cross-sections, as distinct from a channel with natural cross-sections (as in a meandering channel)

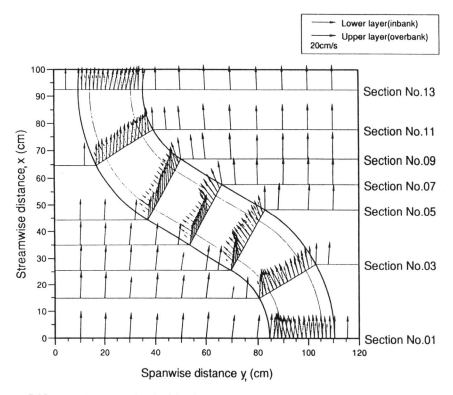

Figure 5.23 Depth-averaged velocities in two layers around a 60° meander bend for overbank flow with $(H - h)/H = 0.25$

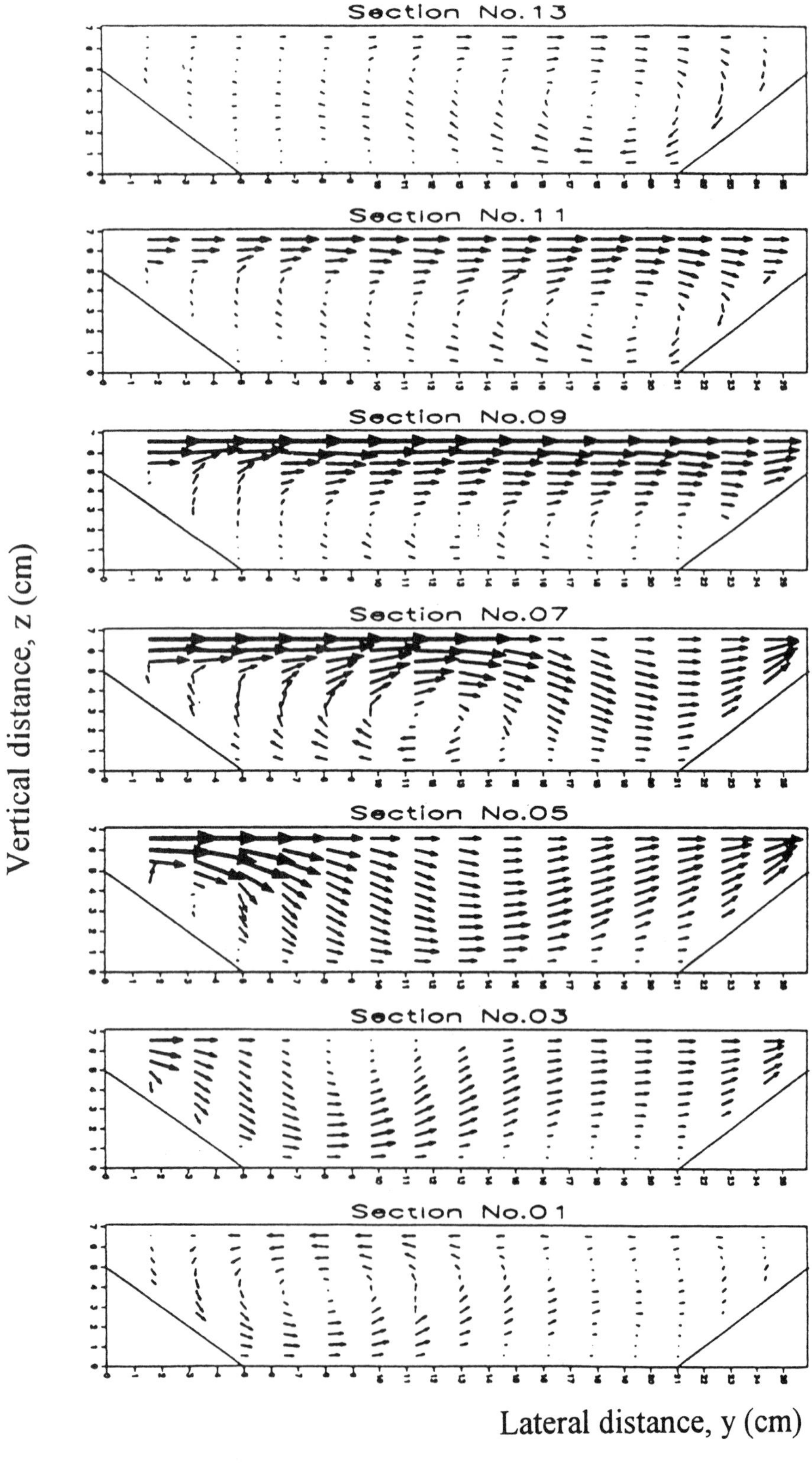

Section No.13

Section No.11

Section No.09

Section No.07

Section No.05

Section No.03

Section No.01

Vertical distance, z (cm)

Lateral distance, y (cm)

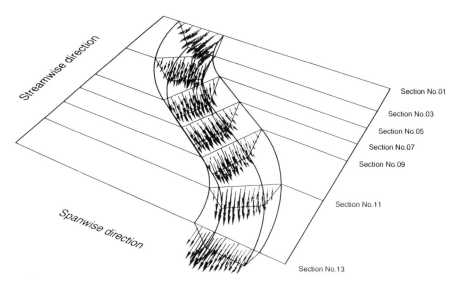

Figure 5.25 Velocity vectors for flow in a trapezoidal meandering channel at bankfull

observed in two different experimental channels, both with trapezoidal cross-sections, as distinct from a channel with natural cross-sections (as in a meandering channel) shown in Figure 5.21. These flow structures have a profound effect on the distribution of boundary shear stress (Knight *et al.*, 1992b). They also indicate the difficulty of applying sub-area division methods of analysis (Wark *et al.*, 1994). Figure 5.23 shows some data from a meandering channel in which the upper and lower flows have been depth-averaged in two layers, separated horizontally at the bankfull stage (Muto, 1995). The difference in the velocity vectors between the upper and lower flows is apparent. The variation of vorticity around the whole bend sequence is shown in Figure 5.24, where at the crossover region the faster upper flow is seen to create a clockwise vortex within the main channel (Shiono and Muto, 1993). Some flow is seen also to plunge downwards into the main channel from the upstream floodplain, and some is ejected upwards onto the downstream receiving floodplain. This interchange of fluid between the river and its floodplains may have significant effects on both sediment and pollution behaviour in meandering channels. This is the mechanism referred to earlier and described by Sellin *et al.* (1993). Figures 5.25 and 5.26 illustrate bankfull and overbank flow for the same planform geometry, with a trapezoidal meandering main channel, and an overbank flow region in which the outer edges of the floodplains boundaries are straight. Figures 5.27 and 5.28 illustrate two overbank flows in a meandering channel with floodplain edges that also meander. Full details of this study may be found in Muto (1995).

Figures 5.21 to 5.28 indicate that for meandering channels, even under steady flow conditions, there are many additional features to consider, such as the turbulence,

---

Figure 5.24 (*opposite*) Secondary flow vectors around a 60° meander bend for overbank with $(H - h)/H = 0.25$

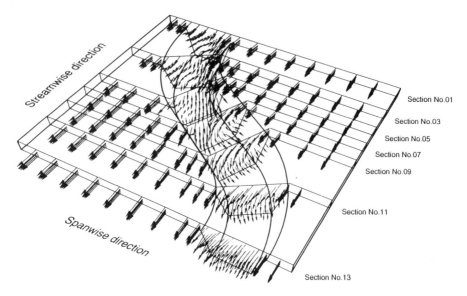

Figure 5.26   Velocity vectors for flow in a trapezoidal meandering channel with overbank flow and straight floodplain alignment with $(H - h)/H = 0.25$

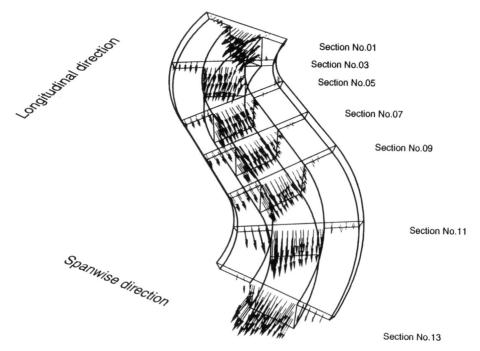

Figure 5.27   Velocity vectors for flow in a rectangular meandering channel with overbank flow and curved floodplain alignment with $(H - h)/H = 0.25$

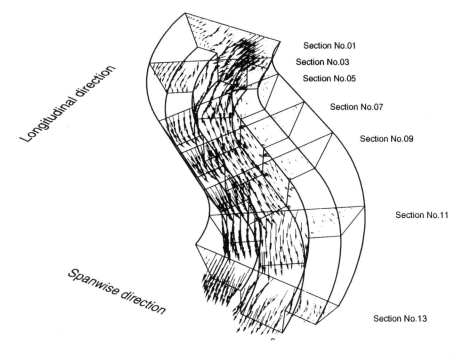

Figure 5.28 Velocity vectors for flow in a rectangular meandering channel with overbank flow and curved floodplain alignment with $(H - h)/H = 0.50$

even further issues requiring consideration. These might include the way in which the water progresses transversely across floodplains, the dynamic effects the rising and falling water levels have on turbulence and the overall effect of these on the speed and attenuation of the flood wave. It is hoped that this brief review, mainly of steady flow, may have served to highlight those topics which are important and those which are worthy of further study.

5.5 CONCLUSIONS

River channels and floodplains should be thought of as an integral system, albeit one in which significant changes occur in certain parameters governing conveyance capacity, sediment transport rate, flood wave speed and pollution dispersion at the bankfull stage. It is important to recognize that floodplains are an essential part of any river system. It is possible to analyse such a system from a one-dimensional point of view, and the coherence approach appears to be the most promising for the stage–discharge relationship, with discharge adjustment factors being used to obtain the correct division of flow between sub-areas. Equivalent boundary shear stress adjustment factors may be inferred from the discharge adjustment factors, or alternatively derived separately, to give the mean shear stresses on individual boundary elements.

An analytical solution has been given for two-dimensional overbank flow which is based on a depth-averaged form of the Navier–Stokes equations. Its application to rivers using finite difference and finite element methods has been demonstrated. The three coefficients f, λ and Γ have been shown to mimic the three basic processes of bed friction, lateral shear and secondary flow reasonably well. Functional relationships may be found between the relative depth and the ratios of these coefficients between the main river channel and the floodplain. It is important to appreciate that models cannot be calibrated on enhanced values of f and λ alone with Γ equal to zero. The use of correct local friction factors is vital if the local boundary shear stresses are to be predicted accurately.

As is often the case in turbulence and river engineering studies, there is a dearth of high quality data in sufficient spatial detail with which to benchmark or calibrate three-dimensional models accurately. There is therefore a need to fund, plan and execute further experimental studies aimed at specific gaps in our knowledge. It is important that future work be directed towards understanding the dynamics of the vortex structures that occur at the main channel/floodplain interface in straight channels. The focus of future studies, both at laboratory and field scales, should be directed towards shallow floodplain depths when the interaction process is at its most intense. Considerably more effort is required for the study of meandering channels, because the spatial variation of parameters inevitably means an increase in cost and time to collect data. Unsteady overbank flows should not be ignored since in natural flood events the discharge is rarely steady but varies continuously with time. However, it should be remembered that the variability imposed by the hydrology may be of more practical significance than a more detailed understanding of the three-dimensional hydrodynamics of overbank flow.

ACKNOWLEDGEMENTS

The authors would like to express their thanks to a number of people who have assisted them in the preparation of this chapter. Figures 5.9–5.11 containing the computer plots of the FCF data were prepared by Professor Shuyou Cao, Sichuan Union University, Chengdu, China, during his sabbatical leave at The University of Birmingham. Figures 5.15–5.18 were prepared by Boris Abril, Cuenca University, Cuenca, Ecuador, during his research work at the University of Birmingham, and Figures 5.23–5.28 were prepared by Yasunori Muto of Bradford University, England. The authors are grateful to these research workers and to those publishers who have made some other diagrams available. Finally both authors acknowledge the support of the Engineering and Physical Sciences Research Council (EPSRC), formerly the SERC, for their funding of the UK Flood Channel Facility and the various research programmes associated with it. The assistance of many of the staff at HR Wallingford is also gratefully acknowledged.

REFERENCES

Abril, B. (1995a) Numerical modelling of turbulent flow and sediment transport in compound channels by the finite element method. MPhil thesis, University of Birmingham, UK.
Abril, B. (1995b) Numerical modelling of turbulent flow in compound channels by the finite element method. *Hydra 2000, Proc. 26th IAHR Congress*, London, UK, September, Vol. 5, pp. 1–6.

Abril, B. (1995b) Numerical modelling of turbulent flow in compound channels by the finite element method. *Hydra 2000, Proc. 26th IAHR Congress*, London, UK, September, Vol. 5, pp. 1–6.

Ackers, P. (1992a) 1992 Gerald Lacey Memorial Lecture – canal and river regime in theory and practice: 1929–92. *Proc. Instn of Civil Engineers, Water, Maritime and Energy*, 96, Sept., Paper No. 10019, pp. 167–178.

Ackers, P. (1992b) Hydraulic design of two stage channels. *Proc. Instn Civil Engineers, Water, Maritime and Energy*, 96, Dec., Paper No. 9988, pp. 247–257.

Ackers, P. (1993a) Stage-discharge functions for two-stage channels: the impact of new research. *Journal Instn Water & Environmental Management*, 7(1), 52–61.

Ackers, P. (1993b) Flow formulae for straight two-stage channels. *Journal of Hydraulic Research*, IAHR, 31(4), 509–531.

Alhamid, A.A.I. (1991) Boundary shear stress and velocity distributions in differentially roughened trapezoidal channels. PhD thesis, University of Birmingham.

Barr, D.I.H. and Wallingford, H.R. (1993) *Additional Tables for the Hydraulic Design of Pipes, Sewers and Channels*. Thomas Telford, London, 229 pp.

Bettess, R. and White, W.R. (1987) Extremal hypotheses applied to river regime. In: *Sediment Transport in Gravel-Bed Rivers*, C.R. Thorne, J.C. Bathurst and R.D. Hey (eds), Wiley, Chichester, pp. 767–789.

Cao, S. (1995) Regime theory and a geometric model for stable alluvial channels. PhD thesis, University of Birmingham, UK.

Cao, S. and Knight, D.W. (1996a) Regime theory of alluvial channels based upon the concepts of stream power and probability. *Proc. Instn of Civil Engineers, Water, Maritime and Energy* (in press).

Cao, S. and Knight, D.W. (1996b) Design approach for hydraulic geometry of straight alluvial channels. *Journal of Hydraulic Engineering*, ASCE (in press).

Chang, H.H. (1980) Geometry of gravel streams. *Journal of Hydraulic Engineering*, ASCE, 106(HY9), 1443–1456.

Chang, H.H. (1988) *Fluvial Processes in River Engineering*. Wiley, Chichester.

Chézy, A. (1921) See "Antoine Chézy, histoire d'une formule d'hydraulique" by G. Mouret, *Annales des Ponts et Chaussees*, Vol. II.

Chow, V.T. (1959) *Open Channel Hydraulics*. McGraw-Hill, New York.

Cokjlat, D. and Younis, B.A. (1994) On modelling turbulent flows in non-circular ducts. *ASME Forum on Turbulent Flows*, ASME, Lake Tahoe, June, pp. 1–6.

Colebrook, C.F. (1939) Turbulent flow in pipes, with particular reference to the transition region between the smooth and rough pipe laws. *Journal of the Institution of Civil Engineers*, 11, 133–156.

Colebrook, C.F. and White, C.M. (1937) Experiments with fluid friction in roughened pipes. *Proc. Roy. Soc.*, A, 161, 367–381.

Cunge, J.A. (1969) On the subject of a flood propagation computation method (Muskingum method). *Journal of Hydraulic Research*, IAHR, 7(2), 205–230.

Cunge, J.A., Holly, F.M. and Verwey, A. (1980) *Practical Aspects of Computational River Hydraulics*. Pitman, London.

Darcy, H. (1857) Sur des recherches experimentales relatives au movement des eaux dans les tuyaux. *Computs rendus des seances de l'Academie des Sciences*.

Einstein, J.A. (1934) "The hydraulic or cross section radius", Schweizensche Bauzeitung, Zurich, 33(8), 89–91 (in German).

Einstein, H.A. and Li, H. (1958) "Secondary currents in straight channels". *Trans. American Geophysical Union*, 39, 1085–1088.

Engelund, F. (1964) "Flow resistance and hydraulic radius". *ACTA*, Ci 24, Copenhagen.

Fukuoka, S. and Fujita, K. (1989) "Prediction of flow resistance in compound channels and its application to design of river courses". *Proc. JSCE*, (in Japanese).

Henderson, F.M. (1966) *Open Channel Flow*. Macmillan, New York.

Horton, R.E. (1933) "Separate roughness coefficients for channel bottom and sides". *Engineering News Record*, 111(22), 652–653.

Imamoto, H, and Ishigaki, T. (1990) "On the hydraulics of an open channel flow in complex cross section (4)". *Kyoto University Disaster Prevention Research Annual Report No. 33B-2*, pp. 559–569 (in Japanese).

Imamoto, H. and Kuge, T. (1974) "On the basic characteristics of an open channel flow in complex cross section". *Kyoto University Disaster Prevention Research Annual Report No. 17*, pp. 665–679.

Imamoto, H., Ishigaki, T. and Shiono, K. (1993) "Secondary flow in a straight open channel". *Annals of the Disaster Prevention Research Institute*, Kyoto University, No. 36 B-2, April, pp1–9 (in Japanese).

James, C.S. and Wark, J.B. (1994) "Conveyance estimation for meandering channels". *Report SR 329*, HR Wallingford, December, pp. 1–86.

Kazemipour, A.K. and Apelt, C.J. (1979) "Shape effects on resistance to uniform flow in open channels". *Journal of Hydraulic Research*, **17**(2) 129–147.

Keulegan, G.H. (1938) "Law of turbulent flow in open channels", *Journal of Research of the National Bureau of Standards*, Vol. 21, No. 6, Research Paper 1151, December.

Knight, D.W. (1989) "Hydraulics of flood channels". In: *Floods: Hydrological, Sedimentological and Geomorphological Implications*, K. Beven (ed.), Wiley, Chichester, 83–105.

Knight, D.W. (1996) "Issues and directions in river mechanics – closure of sessions 2, 3 and 5". In: *Issues and Directions in Hydraulics: An Iowa Hydraulics Colloquium in Honor of Professor John F. Kennedy*, T. Nakato and R. Ettema (eds), Iowa Institute of Hydraulic Research, May, A.A. Balkena, Rotterdam, pp. 435–462.

Knight, D.W. and Abril, B. (1996) "Refined calibration of a depth averaged model for turbulent flow in a compound channel". *Proc. Instn of Civil Engineers, Water, Maritime and Energy* (in press).

Knight, D.W. and Cao, S. (1994) "Boundary shear in the vicinity of river banks". *Proc, ASCE National Conf. on Hydraulic Engineering*, Buffalo. New York, August, Vol. 2, pp. 954–958.

Knight, D.W. and Demetriou, J.D. (1983) "Flood plain and main channel flow interaction". *Journal of Hydraulic Engineering*, **109**(8), 1073–1092.

Knight, D.W. and Hamed, M.E. (1984) "Boundary shear in symmetrical compound channels". *Journal of Hydraulic Engineering*, ASCE, **110**(10), 1412–1430.

Knight, D.W. and Lai, C.J. (1985) "Turbulent flow in compound channels and ducts". *Proc. 2nd Int. Symp. on Refined Flow Modelling and Turbulence Measurements*, Iowa, USA Sept., Paper 121, pp. 1–10, Hemisphere Publishing Co., Iowa, USA.

Knight, D.W. and Patel, H.S. (1985) "Boundary shear stress distributions in rectangular duct flow". *Proc. 2nd Int. Symp. on Refined Flow Modelling and Turbulence Measurements*, Iowa. USA Sept., Paper 122, pp. 1–10, Hemisphere Publishing Co., Iowa, USA.

Knight, D.W. and Sellin, R.H.J. (1987) "The SERC Flood Channel Facility". *Journal of the Institution of Water & Environmental Management*, **1**(2), 198–204.

Knight, D.W. and Shiono, K. (1990), "Turbulence measurements in a shear layer region of a compound channel". *Journal of Hydraulic Research, IAHR.* **28**(2), 175–196. (Discussion in *IAHR Journal*, 1991, 29(2), 259–276.)

Knight, D.W., Demetriou, J.D. and Hamed, M.E. (1983) "Hydraulic analysis of rivers with floodplains". *Proc. Int. Conf. on Hydraulic Aspects of Floods*, British Hydromechanics Association, BHRA, Cranfield, Sept., Paper E1, pp. 129–144.

Knight, D.W., Demetriou, J.D. and Hamed, M.E. (1984) "Stage discharge relationships for compound channels". *Proc. Int. Conf. on Hydraulic Design of Channel Control Structures in Water Resources Engineering: Channels and Channel Control Structures* (Ed. K.V.H. Smith), Southampton, April, pp. 4.21–4.25, Springer-Verlag, Heidelberg.

Knight, D.W., Shiono, K. and Pirt, J. (1989) "Prediction of depth mean velocity and discharge in natural rivers with overbank flow". *Proc. Int. Conf on Hydraulic and Environmental Modelling of Coastal. Estuarine and River Waters*, (Ed. R.A. Falconer *et al.*), University of Bradford. Gower Technical Press, Paper 38, pp. 419–428.

Knight, D.W., Samuels, P.G. and Shiono, K. (1990) "River flow simulation: research and developments". *Journal of the Institution of Water and Environmental Management*, **4**(2), 163–175.

Knight, D.W., Alhamid, A.A.I. and Yuen, K.W.H. (1992a) "Boundary shear in differentially roughened trapezoidal channels", *Proc. 2nd Int. Conf. on Hydraulic and Environmental Modelling of Coastal, Estuarine and River Waters* (Ed. R.A Falconer *et al.*), University of Bradford, Gower Technical Press.

Knight, D.W., Yuan, Y.M. and Fares, Y.R. (1992b) "Boundary shear in meandering river channels". *Proc. Int. Symp. on Hydraulic Research in Nature and Laboratory*, Yangtze River Scientific Research Institute, Wuhan, China, Nov., Vol. 2, pp. 102–106.

Knight, D.W., Yuen, K W.H. and Alhamid, A.A.I. (1994) "Boundary shear stress distributions in open channel flow". In: *Physical Mechanisms of Mixing and Transport in the Environment*, K. Beven, P. Chatwin and J. Millbank (eds), Wiley, Chichester, 51–87.

Lai, C.J. (1986) "Flow resistance, discharge capacity and momentum transfer in smooth compound closed ducts". PhD thesis, the University of Birmingham, UK.

Lai, C.J. and Knight, D.W. (1988) "Distributions of streamwise velocity and boundary shear stress in compound ducts". *Proc. 3rd Int. Symp. on Refined Flow Modelling and Turbulence Measurements*, Tokyo, Japan, July, pp. 527–536.

Liggett, J.A., Chiu, C.L. and Miao, L.S., (1965) "Secondary currents in a corner". *Journal of the Hydraulics Division*, ASCE, **91**(HY6), 99–117.

Lin, B. and Shiono, K. (1992) "Prediction of pollutant transport in compound channel flows". *Proc. 2nd Int. Conf. on Hydraulic and Environmental Modelling of Coastal, Estuarine and River Waters* (Ed. R.A. Falconer *et al.*), University of Bradford, Gower Technical Press, pp. 373–384.

Lin, B. and Shiono, K. (1994) "Three dimensional numerical modelling of rectangular open channel flows". *Journal of Hydraulic Engineering*, Chinese Hydraulic Engineering Society, No. 3, pp. 47–56 (in Chinese).

Lotter, G.K. (1933) "Considerations on hydraulic design of channels with different roughness of walls". *Trans, All-Union Scientific Research Institute of Hydraulic Engineering, Leningrad*, **9**, 238–241 (in Russian).

Manning, R. (1891) "On the flow of water in open channels and pipes". *Trans. Inst. Civil Engineers of Ireland, Dublin*, **20**, 161–207.

Melling, A. and Whitelaw, J.H. (1976) "Turbulent flow in a rectangular duct". *Journal of Fluid Mechanics*, **78**(2), 289–315.

Meyer, L. and Rehme, K. (1994) "Large scale turbulence phenomena in compound rectangular channels". *Experimental Thermal and Fluid Science*, 286–304.

Muto, Y. (1995) "Turbulent flow in two stage meander channels". PhD thesis, University of Bradford, UK.

Naot, D. and Rodi, W. (1982), "Calculation of secondary currents in channel flow". *Journal of the Hydraulics Division*, ASCE, **108**(HY8), 948–968.

NERC (1975) *Flood Studies Report, Vols 1–5*, Natural Environmental Research Council, Swindon, UK.

Nezu, I. (1993) "Turbulent structures and the related environment in various water flows". *Scientific Research Activities, Department of Civil and Global Environment Engineering, Kyoto University*, Kyoto, Japan, December, pp. 1–140.

Pavlovskii, N.N. (1931) "On a design formula for uniform movement in channels with nonhomogeneous walls". *Trans, All-Union Scientific Research Institute of Hydraulic Engineering, Leningrad*, **3**, 157–164 (in Russian).

Perkins, H.J. (1979) "The formation of streamwise vorticity in turbulent flow". *Journal of Fluid Mechanics*, **44**(4), 721–740.

Price, R.K. (1973a) "Flood routing methods for British rivers". *Proc. Instn of Civil Engineers, London* **55**, Paper 7674, 913–930.

Price, R.K. (1978) "A river catchment flood model". *Proc. Instn of Civil Engineers, London*, **65**, Paper 8141, 655–668.

Price, R.K. (1985) "Hydraulics of floods". In: *Advances in River Engineering* (T.H.Y. Tebbut) (ed.), Proc. Int. Symp. at the University of Birmingham, July, Elsevier, pp. 302–310.

Price, R.K. (1986) "Flood routing". In: *Developments in Hydraulic Engineering3*, P. Novak (ed.), Elsevier, 129–173.

Rhodes, D.G. (1991) "An experimental investigation of the mean flow structure in wide ducts of simple rectangular and compound trapezoidal cross-section in particular zones of high lateral shear", 4 volumes. PhD thesis, University of Birmingham, UK.

Rhodes, D.G. and Knight, D.W. (1994a) "Distribution of shear force on boundary of smooth rectangular duct". *Journal of Hydraulic Engineering*, ASCE, **120**(7), 787–807.

Rhodes, D.G. and Knight, D.W. (1994b) "Velocity and boundary shear in a wide compound duct". *Journal of Hydraulic Research*, IAHR **32**(5), 743–764.

Rhodes, D.G. and Knight, D.W. (1995) "Lateral shear in a wide compound duct". *Journal of Hydraulic Engineering*, ASCE, **121**(11), 829–832.

Rodi, W. (1980) "Turbulence models and their application in hydraulics". *State of the art paper, Int. Assoc. for Hydraulic Research*, IAHR, Delft, Netherlands, pp. 1–104.

Schlichting, H. (1979) *Boundary Layer Theory*, McGraw-Hill, New York.

Sellin, R.H.J. (1964) "A laboratory investigation into the interaction between the flow in the channel of a river and that over its flood plain". *La Houille Blanche*, No. 7, 793–801.

Sellin, R.H.J., Ervine, D.A. and Willetts, B.B. (1993) "Behaviour of meandering two-stage channels". *Proc. Instn Civil Engineers Water Maritime and Energy*, 101, June, pp. 99–111.

Shiono K. and Knight, D.W. (1988) "Two dimensional analytical solution for a compound channel". *Proc. 3rd Int. Symp. on Refined Flow Modelling and Turbulence Measurements*, Tokyo, Japan, July, pp. 503–510.

Shiono K. and Knight, D.W. (1990) "Mathematical models of flow in two or multi stage straight channels". *Proc. Int. Conf. on River Flood Hydraulics*, (Ed. W.R. White), Wiley, Paper G1, pp. 229–238.

Shiono, K. and Knight, D.W. (1991) "Turbulent open channel flows with variable depth across the channel". *Journal of Fluid Mechanics*, **222**, 617–646 (and **231**, 693).

Shiono, K. and Lin, B. (1992) "Three dimensional numerical model for two stage open channel flows". *Hydrocomp '92*, 25–29 May, Budapest, Hungary, pp. 123–130.

Shiono, K. (1993) "Numerical flow modelling for inbank and overbank flows in natural rivers". *Proc. Int. Symp. on Hydroscience and Engineering* (Ed. S.Y. Wang, University of Mississippi), Washington, DC, USA, June, Vol. 2, pp. 1265–1270.

Shiono, K. and Muto, Y. (1993) "Secondary flow structure for inbank and overbank flows in trapezoidal meandering compound channel". *Proc. 5th Int. Symp. on Refined Flow Modelling and Turbulence Measurements*, Paris, France, September, pp. 645–652.

Tamai, N., Asaeda, T. and Ikeda, Y. (1986) "Study on generation of periodical large surface eddies in a composite channel flow". *Water Resources Research*, **22**(7), 1129–1138.

Task Force on friction factors in open channels (1963) "Friction factors in open channels". *Journal of the Hydraulics Division*, ASCE, **89**(HY2).

Thomas, T.G. and Williams, J.J.R. (1995) "Large eddy simulation of turbulent flow in an asymmetric compound open channel". *Journal of Hydraulic Research*, **33**(1), 27–41.

Tominaga, A. and Nezu, I. (1991) "Turbulent structure in compound open channel flows. *Journal Hydraulic Engineering*, ASCE, **117**(1), 21–41.

Tominaga, A. and Nezu, I. (1993) "Flows in compound channels. *Journal of Hydroscience and Hydraulic Engineering*, special issue, Research and Practice of Hydraulic Engineering in Japan, No S1–2. Fluvial Hydraulics, JSCE, pp. 121–140.

Tracy, H.J. (1965) "Turbulent flow in a three-dimensional channel". *Journal of the Hydraulics Division*, ASCE, **91**(HY6), 9–35.

Wark, J.B., Samuels, P.G. and Ervine, D.A. (1990) "A practical method of estimating velocity and discharge in a compound channel". In: *River Flood Hydraulics*, W.R. White (ed.), Wiley, Chichester, 163–172.

Wark, J.B., James, C.S. and Ackers, P. (1994) "Design of straight and meandering channels". *National Rivers Authority R&D Report No. 13*, pp. 1–86.

White, W.R., Paris, E. and Bettess R. (1980) "The frictional characteristics of alluvial streams: a new approach". *Proc. Instn Civil Engineers*, London, Part 2, Vol. 69, Sept., pp. 737–750.

White, W.R., Bettess, R. and Paris, E. (1982) "Analytical approach to river regime". *Journal of the Hydraulics Division*. ASCE. **108**(HY10), 1179–1193.

Wormleaton, P.R., Allen, J. and Hadjipanos, P. (1982) "Discharge assessment in compound channel flow". *Journal of the Hydraulics Division*, ASCE, **108**(HY9), 975–993.

Yuen, K.W.H. (1989) "A study of boundary shear stress, flow resistance and momentum transfer in open channels with simple and compound trapezoidal cross section". PhD thesis, University of Birmingham, UK.

6 Recent Advances in Numerical Methods for Fluid Flows

J.-M. HERVOUET and L. VAN HAREN
Laboratoire National d'Hydraulique, Chatou, Paris, France

6.1 INTRODUCTION

Sedimentology, long-term morphodynamics, water quality and contaminant dispersion, protection against floods, are some of the many topics relying on a clear understanding of hydrodynamics. The latter discipline is thus the corner-stone of many studies involving floodplain processes. To predict or explain the flow patterns in a river, to design bridges and dykes, engineers have long been doomed to resort to empirical or simplified formulas, or to build scale models. Despite the basic flaw of similitudes, scale models have been widely used up to now. Their range of application is still very important, though shrinking every day. Scale models are now gradually replaced by numerical simulations, but will still thrive for a long time, at least for very complex problems like the assessment of dyke stability, or simply for validating numerical models.

Computational Fluid Dynamics (CFD) is now rapidly evolving and we shall try in this chapter to show the state of the art in this domain: what can be done, and what is still beyond reach. We shall first briefly describe the basic equations, emphasizing their limitations and the underlying hypothesis. Then we shall detail, with the example of the TELEMAC system (see Hervouet *et al.*, 1994), the two-dimensional and three-dimensional computation of free surface flows. Examples of real studies will show the actual capacity of numerical tools to tackle the challenging problems raised by the study of floodplain processes.

6.2 BASIC EQUATIONS

6.2.1 Notation

The principal notations used in this chapter will be as follows

Latin alphabet

c	:	Celerity of waves
C	:	Chézy coefficient
Cf	:	Friction coefficient
DT	:	Time-step

Floodplain Processes. Edited by Malcolm G. Anderson, Des E. Walling and Paul D. Bates.
© 1996 John Wiley & Sons Ltd.

f_i	:	Volume forces in the momentum equation
Fr	:	Froude number
F_x, F_y	:	Depth-averaged source terms in the momentum equation
g	:	Gravity
h	:	Water depth
h_{prop}	:	"Propagation" depth
k	:	Turbulent kinetic energy
K	:	Strickler coefficient
n	:	Manning coefficient
p	:	Pressure
Q_x	:	Depth multiplied by the first component of the horizontal velocity u
Q_y	:	Depth multiplied by the first component of the horizontal velocity v
Re	:	Reynolds number
R_{ij}	:	Reynolds stress tensor
t	:	Time
T	:	Temperature
u	:	First component of the velocity in Saint-Venant equations
u_{conv}	:	First component of the advecting field
U_1, U_2, U_3	:	Components of the velocity in the Navier–Stokes equations
v	:	Second component of the velocity in Saint-Venant equations
v_{conv}	:	Second component of the advecting field
Z	:	Free surface elevation
Zf	:	Bottom elevation

Greek alphabet

β	:	Thermal expansion coefficient (per °C)
ε	:	Turbulent dissipation rate
Γ_t	:	Turbulent dispersion (in $m^2 s^{-1}$)
φ_i^h	:	Test function at point i in the continuity equation
φ_i^u	:	Test function at point i in the momentum equation
κ	:	Karman constant, equal to 0.41
ν	:	Kinematic viscosity (in $m^2 s^{-1}$), $\nu = \mu/\rho$, where μ is the molecular viscosity
ν_t	:	Turbulent viscosity (in $m^2 s^{-1}$)
τ_{ij}	:	Viscous part of the stress tensor
θ	:	Bottom slope
θ_h	:	Semi-implicitation coefficient of the depth
θ_u	:	Semi-implicitation coefficient of the velocity
ρ	:	Density
ψ_i^h	:	Basis function at point i for the depth
ψ_i^u	:	Basis function at point i for the velocity

6.2.2 Navier–Stokes equations

Unlike in other nearby disciplines, such as sedimentology and water quality, the basic equations are well known in hydrodynamics. The only and difficult problem is to solve them. All equations used in hydrodynamics stem from the three-dimensional Navier–Stokes equations that read:

Continuity

$$\frac{\partial(\rho U_i)}{\partial x_i} = 0 \tag{6.1}$$

Momentum

$$\frac{\partial(\rho U_i)}{\partial t} + \frac{\partial(\rho U_i U_j)}{\partial x_i} = -\frac{\partial p}{\partial x_i} + \frac{\partial \tau_{ij}}{\partial x_j} + \rho f_i + \rho g_i \tag{6.2}$$

where x_i denotes the space coordinates, U_i represents the three components of the velocity, ρ is the density and p is the pressure. Here the indices i and j are 1, 2 or 3 and the so-called Einstein tensorial notation has been used, which means that repeated indices represent sums. For instance:

$$\frac{\partial(\rho U_i U_j)}{\partial x_j} = \frac{\partial(\rho U_i U_j)}{\partial x_1} + \frac{\partial(\rho U_1 U_2)}{\partial x_2} + \frac{\partial(\rho U_i U_3)}{\partial x_3} \tag{6.3}$$

Furthermore, f_i represents all volume forces other than pressure and gravity (Coriolis force in large estuaries for example), g_i represents the three components of gravity ($g_1 = g_2 = 0$, $g_3 = -g$) and

$$\tau_{ij} = \mu \left(\frac{\partial U_i}{\partial x_j} + \frac{\partial U_j}{\partial x_i} \right) \tag{6.4}$$

is the viscous part of the tensor of constraints. The kinematic viscosity v, given in $m^2 s^{-1}$, is equal to μ/ρ.

The continuity equation expresses the mass conservation, while the momentum equation is actually the fundamental law of dynamics, written for fluids. The only assumption in these equations is that the fluid should be Newtonian and this is indeed an excellent approximation for water. The main difficulty of the Navier–Stokes equations stems from the non-linear terms that challenge the numerical algorithms and are also responsible for the flow turbulence. As a matter of fact it is important to note here that turbulence, with all its complexity, is contained in the Navier–Stokes equations.

We have given here the compressible form of the equations. Most of the time the density will be considered constant. Density effects will appear only in estuaries or also in the study of thermal plumes.

At the time being, no industrial tool for solving directly the 3D Navier–Stokes equations for free surface flows is available. One of the main difficulties is the free surface itself, which causes the computational domain to vary in time. Many kinds of simplifications have been proposed, the most popular and widely used being the Shallow Water equations given by Barré de Saint-Venant one century ago. As will be shown in section 6.2.4, the 3D Navier–Stokes equations for free surface flows may now be solved provided that we accept the hypothesis of hydrostatic pressure. Because this hypothesis is also shared by 2D Shallow Water equations, these 3D equations are sometimes called 3D Shallow Water equations.

6.2.3 Shallow Water equations

The 2D Shallow Water, or "Saint-Venant", equations are obtained by means of an averaging of the 3D Navier–Stokes equations over the depth. The new variables obtained are mean values over the depth. The two components of the horizontal depth-averaged velocity, u and v, will thus be:

$$u = \frac{1}{h} \int_{Zf}^{Z} U_1 \, dz \qquad \text{and} \qquad v = \frac{1}{h} \int_{Zf}^{Z} U_2 \, dz \qquad (6.5)$$

where Zf is the bottom elevation and Z is the free surface. The depth h is equal to $Z - Zf$ (see Figure 6.1) and is an unknown. Solving the equations will consist of finding the values of u, v and h everywhere in a domain, during a given lapse of time, as functions of the initial conditions and of the boundary conditions.

Giving a full derivation of the equations would be far too long, so we will only recall the basic assumptions and approximations that have to be made. First the vertical acceleration must be considered negligible, and this goes along with the hydrostatic pressure assumption, stating that:

$$-\frac{1}{\rho} \frac{\partial p}{\partial z} - g = 0 \qquad (6.6)$$

i.e. $p(x, y, z) = -\rho g z +$ a constant value. As a consequence the vertical velocity will be assumed to remain small and will have no specific equation. An often overlooked consequence is that steep slopes should be avoided when they are facing the flow. It is usually considered that slopes of $1 : 10$ are a reasonable limit. The computer programme

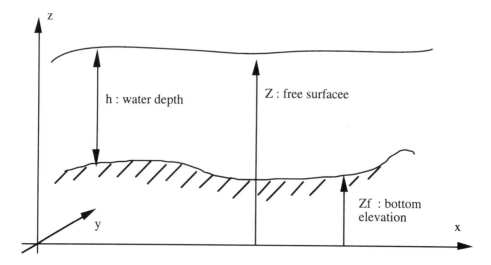

Figure 6.1 Definition of bottom topography and free surface

TELEMAC-2D presented below will be able to cope with much steeper slopes, but it should be kept in mind that this is out of the validity domain of the equations.

The impermeability of the bottom and of the free surface are widely used in the derivation; they are written:

$$\frac{\partial Z}{\partial t} + U^s \cdot n = 0 \quad \text{and} \quad \frac{\partial Zf}{\partial t} + U^f \cdot n = 0 \tag{6.7}$$

where U^s is the velocity at the free surface and U^f the velocity at the bottom. This means that a particle situated, for example, on the free surface will follow it in its movement. Furthermore, n is a vector normal to the surfaces with components $(\partial Z/\partial x, \partial Z/\partial y, -1)$. The derivation of the Shallow Water equations is mainly based on the Leibnitz rule, that reads:

$$\frac{\partial}{\partial x} \int_{Zf}^{z} F \, dz = \int_{Zf}^{z} \frac{\partial F}{\partial x} \, dz + F(x, y, Z) \frac{\partial Z}{\partial x} - F(x, y, Zf) \frac{\partial Zf}{\partial x} \tag{6.8}$$

Using the impermeability conditions, the non-integral terms (the last two in the above expression) will generally disappear. However, the averaging procedure of the non-linear terms will give rise to expressions still containing the original 3D variables. For instance, after some basic algebraic manipulations:

$$\frac{\partial}{\partial y} \int_{Zf}^{z} U_1 U_2 \, dz = \frac{\partial}{\partial y} (huv) + \frac{\partial}{\partial y} \int_{Zf}^{z} (U_1 - u)(U_2 - v) dz \tag{6.9}$$

The second part of this expression cannot be reduced to a simpler form unless U_1 and U_2 are constant on the vertical. These terms are generally called dispersion terms and will be treated as additional diffusion terms at the level of the turbulence model of the Shallow Water equations (see Section 6.2.4). The existence of dispersion is clearly a limitation of the Shallow Water equations: if the horizontal velocity varies too much along the vertical, the average value will have no physical meaning. For example a contaminant will not have the same velocity if it travels near the surface or near the bottom. It is a property of long waves to have a constant velocity along the vertical. For this reason the Shallow Water equations are well suited for the computation of floods, tides and tsunamis.

We eventually get the following set of equations:

Continuity

$$\frac{\partial h}{\partial t} + \text{div}(Q) = 0 \tag{6.10}$$

Momentum

$$\frac{\partial Q_x}{\partial t} + \text{div}(uQ_x) = -hg \frac{\partial Z}{\partial x} + hF_x + \text{div}(hv_t \, \mathbf{grad} \, u) \tag{6.11}$$

$$\frac{\partial Q_v}{\partial t} + \text{div}(uQ_y) = -hg \frac{\partial Z}{\partial y} + hF_y + \text{div}(hv_t \, \mathbf{grad} \, v) \tag{6.12}$$

Here $Q_x = hu$ and $Q_y = hv$, while the vector Q with components (Q_x, Q_y) is sometimes called discharge although its units are in $m^2 s^{-1}$. Furthermore, v_t is a diffusion coefficient which takes into account molecular viscosity, dispersion and turbulence (see Section 6.2.4). Here F_x and F_y now stand for a number of depth-averaged volume forces, such as the Coriolis force, and several supplementary source terms, like the effect of wind, friction, etc., that stem from the depth-averaging procedure.

The above equations are said to be in a "conservative form". The depth h acts here as the density in compressible Navier–Stokes equations and Saint-Venant equations behave basically as equations of a compressible flow. A very important consequence is the existence of discontinuities in the solutions, called "hydraulic jumps" (which is the equivalent of shock waves in compressible flow). These discontinuities are one of the challenging difficulties of these equations for numerical schemes.

By developing the derivatives in the conservative equations, the so-called "non-conservative" equations are obtained, which read:

Continuity

$$\frac{\partial h}{\partial t} + u \cdot \mathbf{grad}(h) + h \operatorname{div}(u) = 0 \tag{6.13}$$

Momentum

$$\frac{\partial u}{\partial t} + u \cdot \mathbf{grad}(u) = -g \frac{\partial Z}{\partial x} + F_x + \frac{1}{h} \operatorname{div}(hv_t \mathbf{grad}\, u) \tag{6.14}$$

$$\frac{\partial v}{\partial t} + u \cdot \mathbf{grad}(v) = -g \frac{\partial Z}{\partial v} + F_y + \frac{1}{h} \operatorname{div}(hv_t \mathbf{grad}\, v) \tag{6.15}$$

In theory, the non-conservative equations are different from the original conservative form when there are discontinuities. However, the difference is not that clear in Computational Fluid Dynamics and closely depends on the numerical treatment of equations.

We will now briefly discuss some of the important properties of Shallow Water equations.

Celerity of waves

When there is no current, long waves will travel at the speed $c = \sqrt{gh}$. This is the basic propagation speed of tides and tsunamis. Indeed, c is the hydrodynamic equivalent of the speed of sound. When the velocity is greater than c, the flow is said to be supercritical, when the velocity is equal to c the flow is critical and when it is lower than c it is fluvial, or subcritical. The value of the Froude number $Fr = \sqrt{u^2 + v^2}/c$ is an important feature of the flow.

Riemann invariants

The theory of characteristics applied to a simplified form of the equations (linearized and without friction) shows that certain quantities, called the Riemann invariants, remain constant along characteristic curves. To be more precise, in a 1D example, we can show that:

$$\text{the quantity } u + 2c \text{ will be constant on curves } \frac{dx}{dt} = u + c \qquad \text{and}$$

$$\text{the quantity } u - 2c \text{ will be constant on curves } \frac{dx}{dt} = u - c$$

A consequence is that for a supercritical flow, all the information will come from upstream. On the contrary, a fluvial flow will be influenced both by upstream and downstream conditions. This basic physical principle of the flow has a severe impact on numerical techniques. Skipping from a fluvial flow $(Fr < 1)$ to a supercritical one $(Fr > 1)$ is a dramatic change and will cause many numerical schemes to fail.

Bottom friction

A very important term in the equations, especially in river flows, is the bottom friction. The general 1D expression of friction is given by the formula $\tau = (1/2)\rho C f u^2$, with units in $kg/m/s^2$, where ρ is the density, u the velocity of the flow and Cf is the friction coefficient (without dimension). The 2D equivalent of this formula is $\tau = (1/2)\rho C f |u| u$.

The friction coefficient Cf is not frequently used in hydraulics and is generally replaced by the Chézy coefficient which is defined as $C = \sqrt{2g/Cf}$ (in $m^{1/2}/s$). According to the Chézy law, the friction force will then appear in the non-conservative momentum equation as a source term having the following form:

$$F_x = -\frac{1}{\cos\theta}\frac{g}{hC^2} u\sqrt{u^2 + v^2} \qquad \text{and} \qquad F_y = -\frac{1}{\cos\theta}\frac{g}{hC^2} v\sqrt{u^2 + v^2} \quad (6.16)$$

where θ is the bottom slope. Another formulation is given by the Manning–Strickler empirical law which expresses C as a function of the hydraulic radius R_H, that is $C = KR_H^{1/6}$ where K (in $m^{1/3} s^{-1}$) is the Strickler coefficient. In 2D flows, R_H is replaced by the water depth. A variation of the Strickler law is Manning's law, where K is replaced by n, with $n = 1/K$. According to Nikuradse, the Chézy coefficient may be related to the grain size k_s on the bottom:

$$C = 7.63 \ln\left(11\frac{h}{k_s}\right) \qquad (6.17)$$

Mostly, the value of the friction coefficients will be unknown and will have to be estimated by means of measurements or by using catalogues (see e.g. Guide for Selecting ... (1989)). The relation with the grain size of the sediments on the bottom is not obvious since the friction will often depend on the vegetation and on the presence of ripples resulting in a much higher value. Different friction coefficients will give different flood hydrographs, with considerable shifts in timing. An automatic estimation of friction with the help of series of measurements is thus an important issue in Computational Fluid Dynamics.

6.2.4   Extension to three-dimensional hydrostatic equations

In this section we will discuss in more detail the equations which were called in Section 6.2.3 the "three-dimensional shallow water equations". Indeed, since they also rely on a vertical hydrostatic pressure distribution, they are only suitable to calculate typical shallow water applications in which the vertical velocities are much smaller than the horizontal ones. However, they are very useful because they make it possible to take into account vertical density fluctuations, which makes them suitable for the treatment of applications with stratification or salinity effects.

In order to derive the 3D Shallow Water Equations, pressure and density are decomposed into mean values $p_0(z)$ and $\rho_0(z)$, and fluctuations around these mean values, $\Delta p(z)$ and $\Delta\rho(z)$, as follows:

$$p(z) = p_0(z) + \Delta p(z); \qquad \rho(z) = \rho_0(z) + \Delta\rho(z) \tag{6.18}$$

We suppose that the mean flow structure is in a state of hydrostratic equilibrium, so that $\partial p_0(z)/\partial z = -\rho_0(z)g$, and that the variations around the equilibrium state are negligible, which means that $\Delta p/p_0 \ll 1$ and $\Delta\rho/\rho_0 \ll 1$. These decompositions can then be substituted into the momentum equations which will eventually, after linearizing the resulting equations, lead to what is called the Boussinesq approximation of the Navier–Stokes equations.

However, as far as the Shallow Water approximation is concerned, we will only use the vertical momentum equation to derive an expression for the pressure fluctuations, due to density fluctuations. If we suppose that vertical velocities are sufficiently small and we neglect diffusion and source terms, then only the pressure and gravity terms remain in the vertical momentum equation. Substituting the above decompositions and linearizing (which means that products of fluctuating quantities are ignored), will give then the following result:

$$0 = \frac{1}{\rho}\frac{\partial p}{\partial z} + g = \frac{1}{\rho_0 + \Delta\rho}\left(\frac{\partial p_0}{\partial z} + \frac{\partial \Delta p}{\partial z}\right) + g$$

$$\cong \underbrace{\frac{1}{\rho_0}\frac{\partial p_0}{\partial z}}_{-g}\left(1 - \frac{\Delta\rho}{\rho_0}\right) + \frac{1}{\rho_0}\left(1 - \frac{\Delta\rho}{\rho_0}\right)\frac{\partial \Delta p}{\partial z} + g$$

$$\cong -g\left(1 - \frac{\Delta\rho}{\rho_0}\right) + \frac{1}{\rho_0}\frac{\partial \Delta p}{\partial z} + g$$

$$= \frac{1}{\rho_0}\frac{\partial \Delta p}{\partial z} + \frac{\Delta\rho}{\rho_0}g = 0 \tag{6.19}$$

The fluctuating part of the pressure can then be found to have the following form (in the TELEMAC-3D approach, the density $\rho_0(z)$ in front of the pressure term is replaced by a reference density, denoted here by $\tilde{\rho}$):

$$\Delta p(z) = \tilde{\rho}g\int_z^z \frac{\Delta\rho(z)}{\rho_0(z)}\,\mathrm{d}z \tag{6.20}$$

where the convention is that the pressure is zero at the free surface Z (see also Figure 6.1). The final set of equations used in TELEMAC-3D is then the following:

Continuity

$$\text{div}(U) = 0 \tag{6.21}$$

Momentum

$$\frac{\partial U_1}{\partial t} + U \cdot \text{grad}(U_1) = -\frac{1}{\tilde{\rho}}\frac{\partial p}{\partial x} + \text{div}(\nu_t \, \text{grad}(U_1)) + f_x \tag{6.22}$$

$$\frac{\partial U_2}{\partial t} + U \cdot \text{grad}(U_2) = -\frac{1}{\tilde{\rho}}\frac{\partial p}{\partial x} + \text{div}(\nu_t \, \text{grad}(U_2)) + f_y \tag{6.23}$$

$$p = \underbrace{\rho_0 g(Z - z)}_{p_0} + \underbrace{\tilde{\rho} g \int_z^z \frac{\Delta \rho}{\rho_0} \, dz}_{\Delta p} \tag{6.24}$$

Temperature

$$\frac{\partial T}{\partial t} + U \cdot \text{grad}(T) = \text{div}(\Gamma_t \, \text{grad}(T)) + Q \tag{6.25}$$

Equation of state

$$-\frac{1}{\rho}\frac{\partial \rho}{\partial T} = \beta \tag{6.26}$$

In the continuity equation, no density terms are taken into account since we suppose that density fluctuations will remain small compared to the mean equilibrium state of the flow. Furthermore, the so-called reference density that appears in the momentum equations can actually be seen to be dissimulated in a corrected pressure $p/\tilde{\pi}$. Also, in the latter equations, the source terms are practically limited to Coriolis forces, since bottom friction and wind shear are directly taken into account by means of the boundary conditions at the bottom and at the free surface. In the equation for the temperature T, turbulence is taken into account by means of a turbulent dispersion Γ_t, which is the thermodynamic equivalent of the turbulent viscosity ν_t, and source terms are represented as Q. Finally, the equation of state enables the expression of temperature in terms of density, by means of the thermal expansion coefficient β. For instance, $\beta = 1.5 \ 10^{-4} \, °C^{-1}$ for pure water at 15 °C, but takes a negative value in the case of salinity.

6.2.5 The problem of turbulence

Turbulence remains one of the fundamental unresolved problems of physics, the complexity of the problem being due essentially to the non-linearity of the convective terms of the Navier–Stokes equations. But as far as we are concerned here it is more of a practical problem, that is to say of how to model the influence of the flow turbulence on the mean structure of the flow. It is also the problem of which turbulence model to use in which situation.

In order to better understand the different approaches to turbulence modelling, and to be able to chose between them, it will prove useful to gain some insights into the fundamental concepts underlying turbulence theory. We will therefore start with a brief introduction into the problem of turbulence, after which the different turbulence models that can be found in literature will be briefly discussed. We will especially focus on the turbulence model used in the TELEMAC system.

Some fundamental concepts of turbulence theory

The overwhelming complexity of the problem can best be illustrated by a short sidewalk in spectral space. Suppose for the sake of simplicity that we deal with an infinite domain occupied by homogeneous and isotropic turbulence[1], that is not subjected to any body force (like rotation, gravity and so on). The spectrum of the kinetic energy of this turbulence, after a Fourier transform of the flow structure from physical space into spectral space, will then also be isotropic and can be integrated over a spherical shell of radius $\kappa = |\boldsymbol{\kappa}|$ and thickness $\delta\kappa$, where $\boldsymbol{\kappa}$ is the wavenumber vector. The time evolution of this integrated kinetic energy density, which we will call $E(\kappa)$, is then governed by the following (simple) equation:

$$\frac{\partial E(\kappa, t)}{\partial t} = T - 2\nu\kappa^2 E(\kappa, t) \tag{6.27}$$

In this equation, ν is the kinematic viscosity ($\nu = 1.0 \times 10^{-6} \text{ m}^2/\text{s}$ for pure water at 20 °C) and the second term at the right hand side represents the dissipation of kinetic energy by molecular viscosity. We notice especially the presence of the square of the wavenumber in this term, which means that dissipation is most effective at small scales (large wavenumbers). The first term at the right hand side, formally denoted by T, stems from the non-linear convective terms in physical space and will not be detailed here, given its complexity. We will however quote its most important property, that is:

$$\int_0^\infty T \, \mathrm{d}\kappa = 0 \tag{6.28}$$

It is for this reason that this term is called a transfer term, since it transports energy from one wavenumber to (several) others, its total contribution to the total energy balance being zero. Although the detailed energy transfer mechanisms in spectral space are very complicated, one can easily establish a general tendency. As we have argued above, kinetic energy will be primarily dissipated at small scales (high wavenumbers) and energy will thus be transferred from larger scales (small wavenumbers) to the smaller ones, to make up for this loss. This phenomenon is called the (spectral) kinetic energy cascade, in view of its counterpart in physical space, which describes how large

---

[1] Turbulence of which all statistical moments are independent of position (homogeneity) and orientation (isotropy).

eddies fall apart in smaller ones, as illustrated by the following little poem (anonymous):

Big whirls have little whirls,
That feed on their velocity,
And little whirls have lesser whirls,
And so on to viscosity.

Integration of the spectral energy balance over the whole spectral space leads to the following expression:

$$\underbrace{\frac{\mathrm{d}}{\mathrm{d}t} \int_0^\infty E(\kappa, t) \, \mathrm{d}\kappa}_{\frac{\mathrm{d}}{\mathrm{d}t} k(t)} = \underbrace{-2\nu \int_0^\infty \kappa^2 E(\kappa, t) \, \mathrm{d}\kappa}_{\varepsilon(t)} \qquad (6.29)$$

which states that, in this simple configuration, the decay of total (integrated) kinetic energy k is equal to its dissipation rate ε (which can be seen to be always positive). In this simple energy balance, $\mathrm{d}k(t)/\mathrm{d}t = -\varepsilon(t)$, the kinetic energy will continue to decline, since we did not take into account any source terms.

In real applications however, turbulence will be continuously created by flow inhomogeneities like shear, recirculation zones, breaking waves, etc. Moreover this production will have a spatial distribution determined by the flow geometry and spatial diffusion terms will come into play. But even so, the general mechanism of the energy cascade will remain the same: kinetic energy production will primarily inject at the large scales, of the order of the geometry that determines the flow, and will then be cascaded down towards the small energy dissipating scales. The large scales are inhomogeneous and determined by the geometry of the problem; the small scales are nearly homogeneous and isotropic.

It is important to have some idea of the size of the small dissipating scales, compared to the large-scale flow structures. If we suppose that the scales at which dissipation takes place are are very much smaller than the large scales (this is called "the separation of scales" in turbulence theory), then their dynamics can only depend on the kinematic viscosity ν and the rate at which they are supplied with energy, that is approximately ε in an equilibrium state. By means of a dimensional analysis (ν is in $\mathrm{m}^2\,\mathrm{s}^{-1}$ and ε is in $\mathrm{m}^2\,\mathrm{s}^{-3}$) we then find the following size for the smallest eddies: $\eta \sim (\nu^3/\varepsilon)^{1/4}$, which is called the Kolmogorov length scale.

In order to estimate the Kolmogorov length scale η, we need an estimation for the dissipation rate ε. We have already argued that the rate at which energy is dissipated at the Kolmogorov scales will be approximately equal to the rate at which it is injected at the large scales. So if the large eddies have a typical velocity U then their kinetic energy is proportional to U^2, and if their size is typically L, their turnover time will be of the order of L/U, and their dissipation rate can then be estimated as $\varepsilon \propto U^3/L$ (the latter statement is usually referred to as the first law of turbulence theory). Thus, for a flow with a velocity of $U = 1$ m/s and large scales of the order of 1 m, and with $\nu = 1.0 \times 10^{-6}$ m^2/s, we will find that the Kolmogorov scale is only $\eta \sim 0.03$ mm!

Generally speaking, the ratio between large scales and small scales is found to be:

$$\frac{L}{\eta} \sim Re^{3/4}, \quad \text{with } Re = \frac{UL}{\nu} \tag{6.30}$$

where Re is called the Reynolds number. The larger the Reynolds number is, the more important will be the separation of scales. For a given flow geometry, that is L fixed, increasing the turbulence intensity by increasing the velocity, will lead to smaller Kolmogorov scales, and thus finer scale turbulence. Indeed, the Kolmogorov scales then move to smaller scales, where viscosity is more effective, thus re-establishing the balance between production and dissipation.

In our simple example, the Reynolds number of the flow is approximately 10^6 so that a complete numerical simulation of the turbulence in 3D would require $Re^{9/4} = 10^{13.5}$ grid points, which is far beyond reach at the moment and probably for some time to come (nowadays it is possible to treat at best 10^7 grid points). Of course, this goes along with, or is partly exchangeable with, the problem of limited CPU time. Notice also that in general, the Reynolds number will be much larger than 10^6.

Luckily, for practical applications we are normally not interested in knowing all the details of the flow turbulence. We "only" want to model the impact of the turbulence on the mean flow properties, or we want to model the interaction between all different eddy sizes and the mean flow pattern.

Turbulence modelling

Turbulence modelling is in general based on the Reynolds decomposition, in which the local velocity U_i is decomposed into a mean[1] flow $\overline{U}_i$ and a fluctuating velocity U_i' (in the same way we decompose the pressure field p in a mean part $\bar{p}$ and a fluctuating part p' and the body force f_i into $\overline{f}_i$ and f_i'):

$$U_i = \overline{U}_i + U_i'; \quad p = \bar{p} + p'; \quad f_i = \overline{f}_i + f_i' \tag{6.31}$$

Substituting these decompositions into the Navier–Stokes equations and average leads to the so-called Reynolds equations that describe the behaviour of the mean flow:

$$\frac{\partial(\rho\overline{U}_i)}{\partial t} + \frac{\partial(\rho\overline{U}_i\overline{U}_j)}{\partial x_j} = \frac{\partial}{\partial x_j}\left(-\bar{p}\delta_{ij} + \mu\left(\frac{\partial \overline{U}_i}{\partial x_j} + \frac{\partial \overline{U}_j}{\partial x_i}\right) - \underbrace{\overline{\rho u_i' u_j'}}_{R_{ij}}\right) + \rho\overline{f}_i + \rho g_i \tag{6.32}$$

Here the term R_{ij} stems from the decomposition treatment of the non-linear convective terms and is called the Reynolds stress tensor. It is a new term that formally describes the impact of the turbulence on the mean flow field and for which an expression in terms of mean flow quantities has to be found in order to close the equations.

There are basically two ways to obtain a closure of the Reynolds equations. The first and most elaborated method consists of developing explicit equations for the Reynolds stress tensor; we will briefly discuss this technique further on in this section. Most

---

[1] Strictly speaking this is an ensemble average, but in practice we use a time averaging, which supposes that an ergodicity criterion is fulfilled.

turbulence models, however, apply the so-called Boussinesq eddy-viscosity concept (not to be confused with the Boussinesq approximation, presented in Section 6.2.4), which expresses the Reynolds stress tensor in terms of the local velocity gradients, and introduces a "turbulent viscosity" v_t, as follows:

$$\frac{1}{\rho} R_{ij} = v_t \left(\frac{\partial \overline{U}_i}{\partial x_j} + \frac{\partial \overline{U}_j}{\partial x_i} \right) - \frac{2}{3} k \delta_{ij} \tag{6.33}$$

The second term on the right hand side ensures that the trace of the Reynolds stress tensor is equal to the turbulent kinetic energy. This term acts as a dynamic pressure and is therefore usually dissimulated in the normal pressure term. However, it is also normally a very small term that can often be neglected (as is the case in the TELEMAC system). We should bear in mind that the *turbulent viscosity is a property of the flow*, so is time and place dependent, whereas the *kinematic viscosity is a property of the liquid*.

Upon substituting the Boussinesq expression into the Reynolds equation, we obtain the following result:

$$\frac{\partial \overline{U}_i}{\partial t} + \frac{\partial \overline{U}_i \overline{U}_j}{\partial x_j} = -\frac{1}{\rho} \frac{\partial \bar{p}}{\partial x_i} + \frac{\partial}{\partial x_j} \left(\underbrace{(v + v_t)}_{v_e} \left(\frac{\partial \overline{U}_i}{\partial x_j} + \frac{\partial \overline{U}_j}{\partial x_i} \right) \right) + \bar{f}_i + g_i \tag{6.34}$$

Here, the combination of kinematic and turbulent viscosity is often referred to as an effective viscosity $v_e = v + v_t$. In the TELEMAC system, the term $\partial \overline{U}_j / \partial x_i$ is neglected, which greatly simplifies the equations since then no direct coupling exists between the different velocity components. Experience has shown that this coupling can generally be neglected. Turbulence modelling now consists of finding an expression for the turbulent viscosity.

The latter equations immediately lead to the 3D Shallow Water equations, given in Section 6.2.4, for i and j equal to 1,2 (and $g_1 = g_2 = 0$), while the vertical momentum equation will give the pressure term. For notational simplicity, however, the overbars, representing the Reynolds averaging, have been omitted whereas v_t, stands for an effective viscosity.

As far as the depth-averaged equations are concerned, the vertical averaging procedure is applied to the Reynolds equations rather than directly to the Navier–Stokes equations (again, overbars are omitted in Section 6.2.3 for the sake of simplicity) and the Boussinesq approximation is applied in its vertically averaged form. The vertically averaged R_{11}, $R_{12} = R_{21}$ and R_{22} components of the Reynolds stress tensor are expressed in terms of the vertically averaged horizontal velocity gradients, while the vertically averaged R_{13} and R_{23} Reynolds stress terms will give rise to the bottom friction terms R_{13}^{bot} and R_{13}^{bot} as well as to their surface equivalents R_{13}^{surf} and R_{13}^{surf}, that could represent the action of wind. This finally results in the following depth-averaged equation:

$$\frac{\partial (hu_i)}{\partial t} + \frac{\partial (hu_i u_j)}{\partial x_j} = -hg \frac{\partial Z}{\partial x_i} + \frac{\partial}{\partial x_j} \left(\underbrace{(v + v_t)}_{v_e} \left(\frac{\partial u_i}{\partial x_j} + \frac{\partial u_j}{\partial x_i} \right) \right) + \underbrace{h\bar{f}_i + (R_{i3}^{surf} - R_{i3}^{bot})}_{hF_i} \tag{6.35}$$

In this equation the subscripts i and j can again only take the values 1 or 2, so that the depth-averaged mean horizontal velocity is $u = u_1$ and the depth-averaged mean lateral velocity is $v = u_2$. We should also realize that the effective viscosity now contains a supplementary contribution, that comes from the vertical inhomogeneity of the horizontal velocity profiles and which was called turbulent dispersion in Section 6.2.3. However, due to a lack of fundamental knowledge about the behaviour of the dispersion terms, generally no explicit modelling is proposed. The term hF_i now contains both body forces and friction forces, as stated in Section 6.2.3.

Notice finally that from a turbulence point of view, there is no formal difference between the 3D equations and the depth-averaged equations. The difference is indeed only in the value of the turbulent viscosity, for which we will now briefly discuss the most important models (a more detailed description can be found in Rodi, 1993).

Constant eddy-viscosity model   The most rudimentary turbulence model, though often used in hydraulic engineering, consists of specifying a constant turbulent viscosity for the whole flow field. For depth-averaged calculations, one often expresses the turbulent viscosity as $v_t \propto u_* h$, where h is the water depth and u_* is the bottom friction velocity. The proportionality constant in the latter case not only takes into account the real turbulence but also the vertical flow inhomogeneities; a value of 0.0765 has been given in Rastogi and Rodi (1978) based on experimental channel flow data.

However, the constant eddy viscosity model is in general of little practical use. In the simple flow situations where it is valid, namely the calculation of quasi-2D large waterbodies, there is always an equilibrium state between the main driving force of the flow and the bottom friction, in which turbulence only plays a minor role. However, when the turbulence model becomes important, in more complex flow situations, the constant eddy viscosity model is far too simple. This kind of model is actually only useful for the calculation of tracers in simple configurations, since the turbulent terms are always dominant in the convection-diffusion equation.

Mixing length model   The first proper turbulence model is Prandtl's mixing length model, that has been inspired by kinetic gas theory, the mixing length l_m being the equivalent of the mean free path of the molecules[1]. By means of a dimensional analysis one obtains the following expression:

$$v_t = l_m^2 \left| \frac{\partial \overline{U}}{\partial y} \right| \tag{6.36}$$

Much work has been done to determine appropriate values of the mixing length for different flow situations. For instance for the turbulent flow in the vicinity of a rigid wall the mixing length is equal to the distance from the wall times the von Karman constant $\kappa \approx 0.4$. This also expresses, of course, the lack of generality of the mixing length model.

---

[1] However, momentum exchange between molecules only takes place on average at a distance of the mean free path, whereas a turbulent eddy exchanges momentum at intermediate scales.

One equation model Another turbulence model that is worth brief mention is the so-called one equation model that reads:

$$\nu_t = c'_\mu \sqrt{k}L \tag{6.37}$$

Here one has to solve a supplementary modelled (the complete equation contains unknown terms) equation for the turbulent kinetic energy k, that reads:

$$\underbrace{\frac{\partial k}{\partial t} + \overline{U}_j \frac{\partial k}{\partial x_j}}_{} = \underbrace{\frac{\partial}{\partial x_i}\left(\frac{\nu_t}{\sigma_k}\frac{\partial k}{\partial x_i}\right)}_{\text{transport}} + \underbrace{\nu_t\left(\frac{\partial \overline{U}_i}{\partial x_j} + \frac{\partial \overline{U}_j}{\partial x_i}\right)\frac{\partial \overline{U}_i}{\partial x_j}}_{\text{production}} - \underbrace{C_D \frac{k^{3/2}}{L}}_{\varepsilon} \tag{6.38}$$

The level of kinetic energy in the flow primarily depends on the constant C_D, and the turbulent viscosity is determined by the value of $c'_\mu C_D$ (a value 0.08 is recommended in literature and $\sigma_k \approx 1$).

The k–ε model The $k-\varepsilon$ model is the most commonly used turbulence model, because it is sufficiently general to treat many practical flow applications, while not being too complicated. The $k-\varepsilon$ model is implemented in TELEMAC-3D and in its depth-averaged version in TELEMAC-2D (the model does not take into account thermal effects). Because of its importance we will discuss this model in more detail than the other turbulence models.

In the $k-\varepsilon$ model, the turbulent viscosity is expressed as a function of the turbulent kinetic energy and its dissipation rate:

$$\nu_t = c_\mu \frac{k^2}{\varepsilon} \tag{6.39}$$

Here, the equation for k is the same as in the one-equation turbulence model, except for the dissipation rate, for which we now have to resolve an extra equation. The full set of the $k-\varepsilon$ model equations reads as follows:

$$\frac{\partial k}{\partial t} + \overline{U}_i \frac{\partial k}{\partial x_i} = \frac{\partial}{\partial x_i}\left(\frac{\nu_t}{\sigma_k}\frac{\partial k}{\partial x_i}\right) + P - \varepsilon \tag{6.40}$$

$$\frac{\partial \varepsilon}{\partial t} + U_i \frac{\partial \varepsilon}{\partial x_i} = \frac{\partial \varepsilon}{\partial x_i}\left(\frac{\nu_t}{\sigma_\varepsilon}\frac{\partial \varepsilon}{\partial x_i}\right) + \frac{\varepsilon}{k}(c_{1\varepsilon}P - c_{2\varepsilon}\varepsilon) \tag{6.41}$$

where P is the production of kinetic turbulent energy by shear:

$$P = \nu_t\left(\frac{\partial \overline{U}_i}{\partial x_j} + \frac{\partial \overline{U}_j}{\partial x_i}\right)\frac{\partial \overline{U}_i}{\partial x_j} \tag{6.42}$$

The constants of the $k-\varepsilon$ model are calibrated on two classical experiments, the free decay of grid turbulence and the turbulent flow that develops between two parallel walls. The first experiment permits us to find a value for the constant $c_{2\varepsilon}$, while the second one gives values for the constants c_μ and $c_{1\varepsilon}$. Finally, the constants c_k and c_ε have been "optimized", considering the performance of the $k-\varepsilon$ model for both test cases with different values for these constants. The constants that are generally employed are listed in Table 5.1.

The depth-averaged form of this standard $k-\varepsilon$ model has been implemented in TELEMAC-2D, following the ideas of Rodi (1993). The vertical-averaging procedure gives rise to supplementary production terms in both equations, P_{kv} for the kinetic energy and $P_{\varepsilon v}$ for the dissipation rate, due to bed shear:

$$P_{kv} = c_k \frac{u_*^3}{h} \quad \text{with:} \quad c_k = \frac{1}{\sqrt{Cf}} \tag{6.43}$$

$$P_{\varepsilon v} = c_\varepsilon \frac{u_*^4}{h^2} \quad \text{with:} \quad c_\varepsilon = 3.6 \frac{c_{\varepsilon 2}\sqrt{c_\mu}}{Cf^{3/4}} \tag{6.44}$$

where Cf is the friction coefficient and u_* is the bottom friction velocity, that has been defined earlier in this chapter. It is of course obvious that all quantities involved here are depth averaged.

Reynolds stress modelling   An approach that is totally different from the Boussinesq concept is Reynolds stress modelling or higher order closure modelling. In this case one directly develops evolution equations for all terms of the Reynolds stress model by substracting the Reynolds equation from the Navier–Stokes equation and upon combining the result for the different components:

$$\frac{\partial}{\partial t}\overline{u_i' u_j'} = \overline{u_i' \frac{\partial u_j'}{\partial t} + u_j' \frac{\partial u_i'}{\partial t}}$$
$$= \overline{(U_i - \overline{U_i})\frac{\partial (U_j - \overline{U_j})}{\partial t}} + \overline{(U_j - \overline{U_j})\frac{\partial (U_i - \overline{U_i})}{\partial t}} \tag{6.45}$$

This modelling is generally said to be of second order since it gives expressions for the second statistical moments of the turbulence. In the resulting equations there will now appear several new and unknown terms that contain triple moment of the form $u_i' u_j' u_k'$, which are then expressed in already known mean flow and/or statistical quantities. One can also derive evolution equations for the triple moments itself, using the same procedure as described above, and to model the quadruple moments. This hierarchy of higher order closure models clearly illustrates what is called "closure problem of turbulence".

Second-order Reynolds stress modelling leads to six equations (three for the trace and three for off-diagonal elements, since $u_i' u_j' = u_j' u_i'$), which makes the system very

Table 6.1   The constants of the $k-\varepsilon$ model

c_μ	$c_{1\varepsilon}$	$c_{2\varepsilon}$	σ_k	σ_ε
0.09	1.44	1.92	1.0	1.3

expensive, while moreover the numerical treatment is not simple. A somewhat less restricting method is the algebraic Reynolds stress model, in which some supplementary model assumptions to the full Reynolds stress model make it possible to express them in terms of the kinetic energy k and its dissipation rate ε.

Large-eddy simulation A completely different modelling approach that should be briefly mentioned here because of its increasing importance, is the Large-Eddy Simulation (LES) technique. The LES is conceptually speaking situated between the direct numerical simulation of all turbulence scales (which is not feasible for practical flow applications as we have seen before) and classical turbulence modelling in which only the mean flow is calculated. In a LES, the time evolution of all turbulent scales that can be represented on the numerical grid is explicitly simulated, which is the reason why LES is sometimes called a numerical experiment. Indeed, like in a real experiment, one has to continue the simulation for several integral timescales of the flow in order to calculate statistics like the mean flow profiles. The modelling part of LES concerns all turbulence scales that cannot be represented by the numerical grid, and is therefore called subgrid-scale modelling.

The LES technique has two major advantages over classical turbulence modelling. Firstly, since the turbulence modelling in a LES only concerns the subgrid scales the overall flow pattern is much less sensitive to the turbulence model. Secondly, since the subgrid scales are much less problem dependent and thus more isotropic than the large scales, it is much easier to define a sound turbulence model. As a consequence, the turbulence models in LES are generally much more simple than the classical turbulence models that need to represent all flow aspects. The major inconvenience of LES is its cost, since we need to pursue the simulation until stable statistics can be obtained, whereas classical turbulence models are formulated in terms of overall statistical quantities of the flow. For this reason, until now LES has been considered primarily as a research tool, to calculate relatively simple flow situations and to validate classical turbulence models.

The most well-known subgrid-scale model for LES is the Smagorinsky model, which can be seen to be a generalized form of Prandtl's mixing length model. The Smagorinsky model reads:

$$v_t = l^2 S; \quad \text{with:} \quad l = C_S \Delta$$

and:

$$S = \left[\frac{1}{2} \left(\frac{\partial \overline{U_i}}{\partial x_j} + \frac{\partial \overline{U_j}}{\partial x_i} \right) \left(\frac{\partial \overline{U_i}}{\partial x_j} + \frac{\partial \overline{U_j}}{\partial x_i} \right) \right]^{1/2} \quad (6.46)$$

For the constant C_S the values found in literature range from 0.17 to 0.23 The principal difference with the mixing length model is the use of the grid space Δ instead of a physically based length scale.

The one-equation model discussed earlier is also often employed as subgrid-scale model. We thus have the following expression for the subgrid turbulent viscosity:

$$v_t = c'_\mu \sqrt{k}\, \Delta \quad (6.47)$$

where k is now the subgrid kinetic energy. Also the expression for the dissipation rate is built from the grid spacing and the subgrid kinetic energy: $\varepsilon = C_D k^{3/2}/\Delta$.

We notice finally that Reynolds stress modelling and LES are meant to take into account the full 3D structure of the flow and they are therefore practically restricted to 3D applications.

6.3   FINITE ELEMENT IMPLEMENTATION

6.3.1   The choice of non-structured meshes

The flexibility offered by non-structured meshes is of utmost importance if high accuracy is to be obtained close to important features such as outfalls, groynes and weirs. As a consequence we favour Finite Element techniques. We do not infer however that other techniques such as Finite Differences or Finite Volumes are of a lesser interest. Actually those techniques offer extra possibilities for designing numerical schemes, especially for advection equations. For example, efforts for adapting Finite Volume schemes to non-structured meshes are a very interesting approach. As a matter of fact, Finite Elements offer a valuable theoretical framework and Finite Volumes are sometimes included in this (see Idelsohn and Onate, 1994).

6.3.2   The algorithms in TELEMAC-2D

TELEMAC-2D solves the non-conservative Shallow Water equations with the finite element technique. The fundamental operation in Finite Elements is the variational formulation. A complete description of this operation will be found in Goutal (1987). A good choice of approximation spaces for the unknowns is very important to ensure the existence and uniqueness of the solutions. However, theoretical proofs are often valid only for very simple types of boundary conditions. We shall give here only the practical aspects of the variational formulation:

- Every equation of the form $E = 0$ is treated as: $\int_\Omega E\varphi_j \, d\Omega = 0$ where φ_j is called a test function and Ω is the computational domain.
- We only consider a discrete number of time-steps denoted t^n and equal to $t^0 + n\,DT$ where t^0 is the starting time and DT is called the time-step. The derivative in time of a function is written: $\partial f/\partial t = (f^{n+1} - f^n)/DT$ where f^{n+1} is the function f at time t^{n+1} (new time-step), and f^n is f at time t^n (old time-step).
- Basically we use Galerkin finite elements and every unknown f is discretized on a non-structured mesh, in the form $f = \sum_{i=1}^{n} f_i \psi_i$, where n is the number of points in the mesh, ψ_i is a basis function and f_i the value of f at the point number i. Up to now linear basis functions on triangles have been chosen and each vertex is associated with a basis. Quadratic functions or other types of elements could be chosen. According to the Galerkin technique, test functions and basis functions are identical. However, for generality we shall denote them differently in this paragraph. We shall also assume that different discretizations may be applied to different variables. Thus:
  ψ_i^h will be the basis function for the depth at point i,
  φ_i^h will be the test function at point i for the continuity equation,
  ψ_i^u will be the basis function for the two components of velocity at point i,
  φ_i^u will be the test function at point i for the momentum equations.

The number of basis functions for the depth is denoted as nph and the number of basis functions for the velocity is denoted as npu.

We then have to solve:

$$\frac{h^{n+1} - h^n}{DT} + \boldsymbol{u} \cdot \mathbf{grad}(h) + h \operatorname{div}(\boldsymbol{u}) = 0 \tag{6.48}$$

$$\frac{u^{n+1} - u^n}{DT} + \boldsymbol{u} \cdot \mathbf{grad}(u) = -g \frac{\partial Z}{\partial x} + F_x + \frac{1}{h} \operatorname{div}(hv_t \, \mathbf{grad} \, u) \tag{6.49}$$

$$\frac{v^{n+1} - v^n}{DT} + \boldsymbol{u} \cdot \mathbf{grad}(v) = -g \frac{\partial Z}{\partial y} + F_y + \frac{1}{h} \operatorname{div}(hv_t \, \mathbf{grad} \, v) \tag{6.50}$$

The terms of the form $\boldsymbol{u} \cdot \mathbf{grad}(f)$ with f equal to h, u or v are called the advection terms and deserve a special treatment. Applying a standard discretization to these terms would give poor results. Special methods may be applied, for example the SUPG technique (Brookes and Hughes, 1982), or numerical schemes based on the method of characteristics coupled with a fractional step approach (see Hervouet, 1986). When the latter option is selected, the following equations are first solved:

$$\frac{\tilde{h} - h^n}{DT} + \boldsymbol{u} \cdot \mathbf{grad}(h) = 0 \tag{6.51}$$

$$\frac{\tilde{u} - u^n}{DT} + \boldsymbol{u} \cdot \mathbf{grad}(u) = 0 \tag{6.52}$$

$$\frac{\tilde{v} - v^n}{DT} + \boldsymbol{u} \cdot \mathbf{grad}(v) = 0 \tag{6.53}$$

and $\tilde{h}$, $\tilde{u}$, and $\tilde{v}$ are then used in a second step, called the propagation step, where the advection terms are removed.

The advection terms are still present in the propagation step when the method of characteristics is not used (e.g. when SUPG-based methods are used). We shall thus describe this propagation step with the advection terms.

The time discretization of the functions h, u and v has not yet been fully detailed. To get a good accuracy in time, they should be treated in a semi-implicit way, for example h should be $(h^n + h^{n+1})/2$. However, this would lead to non-linear terms and an unstable numerical scheme. In fact, we write for every function f:

$$f = \theta f^{n+1} + (1 - \theta) f^n, \quad \text{with } \theta > 0.5 \text{ but close to it.} \tag{6.54}$$

To avoid non-linear terms, expressions such as $h \operatorname{div}(\boldsymbol{u})$ are written in the form:

$$h \operatorname{div}(\boldsymbol{u}) = h_{\text{prop}} \operatorname{div}(\theta_u u^{n+1} + (1 - \theta_u) u^n) \tag{6.55}$$

where θ_u is called the implicitation coefficient of the velocity. Logically h_{prop} should be also of the form $\theta_h h^{n+1} + (1 - \theta_h) h^n$ but it is actually equal to $\theta_h h'^{n+1} + (1 - \theta_h) h^n$, h'^{n+1} being only an approximation of h^{n+1} (θ_h is the implicitation coefficient of the depth). h'^{n+1} is only h^n in simple cases but if one needs greater accuracy then it is necessary to solve the same time-step a number of times (these are called sub-iterations); the value of h'^{n+1} can be updated with the result of a previous sub-iteration. h_{prop} is called the "propagation depth".

In the same way $\boldsymbol{u}.\mathbf{grad}(f)$, for any function f, will be written $\boldsymbol{u}_{\text{conv}}.\mathbf{grad}(\theta_f f^{n+1} + (1 - \theta_f)f^n)$ where $\boldsymbol{u}_{\text{conv}}$ is the advecting field equal to $\theta_u \boldsymbol{u}'^{n+1} + (1 - \theta_u)\boldsymbol{u}^n . \boldsymbol{u}'^{n+1}$ is again an approximation of $\boldsymbol{u}^{n+1}$.

Further features of the time discretization in TELEMAC-2D are the following:

- The diffusion terms are treated in a fully implicit way.
- The unknown h^{n+1} is conveniently replaced by the increment $h^{n+1} - h^n$.
- The friction terms are fully implicit.

Our equations have now the form:

Continuity

$$\frac{h^{n+1} - h^n}{DT} + \boldsymbol{u}_{\text{conv}}.\mathbf{grad}(h) + h_{\text{prop}} \operatorname{div}(\boldsymbol{u}) = 0 \qquad (6.56)$$

Momentum

$$\frac{u^{n+1} - u^n}{DT} + \boldsymbol{u}_{\text{conv}}.\mathbf{grad}(u) = -\theta_h g \frac{\partial(h^{n+1} - h^n)}{\partial x} - g \frac{\partial Z^n}{\partial x} + F_x + \frac{1}{h} \operatorname{div}(hv_t \, \mathbf{grad} \, u^{n+1})$$

$$(6.57)$$

$$\frac{v^{n+1} - v^n}{DT} + \boldsymbol{u}_{\text{conv}}.\mathbf{grad}(v) = -\theta_h g \frac{\partial(h^{n+1} - h^n)}{\partial y} - g \frac{\partial Z^n}{\partial y} + F_y + \frac{1}{h} \operatorname{div}(hv_t \, \mathbf{grad} \, v^{n+1})$$

$$(6.58)$$

We have now to multiply these equations by the test functions and integrate them on the computational domain. This gives, for every point of the mesh:

$$\int_\Omega \left(\frac{h^{n+1} - h^n}{DT}\right) \varphi_i^h \, d\Omega + \int_\Omega \boldsymbol{u}_{\text{conv}}.\mathbf{grad}(h)\varphi_i^h \, d\Omega + \int_\Omega h_{\text{prop}} \operatorname{div}(\boldsymbol{u})\varphi_i^h \, d\Omega = 0 \qquad (6.59)$$

$$\int_\Omega \left(\frac{u^{n+1} - u^n}{DT}\right) \varphi_i^u \, d\Omega + \int_\Omega \boldsymbol{u}_{\text{conv}}.\mathbf{grad}(u)\varphi_i^u \, d\Omega \qquad (6.60)$$

$$= -\int_\Omega g\theta_h \frac{\partial(h^{n+1} - h^n)}{\partial x} \varphi_i^u \, d\Omega - \int_\Omega g \frac{\partial Z^n}{\partial x} \varphi_i^u \, d\Omega + \int_\Omega F_x \varphi_i^u \, d\Omega$$

$$+ \int_\Omega \frac{1}{h} \operatorname{div}(hv_t \, \mathbf{grad} \, u^{n+1})\varphi_i^u \, d\Omega$$

$$\int_\Omega \left(\frac{v^{n+1} - v^n}{DT}\right) \varphi_i^u \, d\Omega + \int_\Omega \boldsymbol{u}_{\text{conv}}.\mathbf{grad}(v)\varphi_i^u \, d\Omega \qquad (6.61)$$

$$= -\int_\Omega g\theta_h \frac{\partial(h^{n+1} - h^n)}{\partial y} \varphi_i^u \, d\Omega - \int_\Omega g \frac{\partial Z^n}{\partial y} \varphi_i^u \, d\Omega + \int_\Omega F_y \varphi_i^u \, d\Omega$$

$$+ \int_\Omega \frac{1}{h} \operatorname{div}(hv_t \, \mathbf{grad} \, v^{n+1})\varphi_i^u \, d\Omega$$

In the continuity equation we apply an integration by parts on the following term:

$$\int_\Omega h_{\text{prop}} \operatorname{div}(\boldsymbol{u})\varphi_i^h \, d\Omega = \int_\Gamma h_{\text{prop}} \boldsymbol{u} \cdot \boldsymbol{n}\varphi_i^h \, d\Gamma - \int_\Omega \boldsymbol{u} \cdot \operatorname{grad}(h_{\text{prop}}\varphi_i^h) \, d\Omega \quad (6.62)$$

In this form the flux across the boundaries appears as:

$$\int_\Gamma h_{\text{prop}} \boldsymbol{u} \cdot \boldsymbol{n}\varphi_i^h \, d\Gamma \quad (6.63)$$

This is a boundary condition that has to be given. On solid boundaries these terms will be, for example, set to zero to express the impermeability.

The diffusion terms are also integrated by parts and give new boundary terms standing for the friction on the solid boundaries. For example:

$$\int_\Gamma \varphi_i^u \nu_t \operatorname{grad}(u^{n+1}) \cdot \boldsymbol{n} \, d\Gamma \quad (6.64)$$

in which $\operatorname{grad}(u^{n+1}) \cdot \boldsymbol{n}$ is equal to $\partial u^{n+1}/\partial n$, and a friction law will read $\partial u^{n+1}/\partial n = \alpha u^{n+1}$, where α is the friction coefficient.

After the integration by parts, there remains to apply the discretization in space of our functions, namely:

$$h = \sum_{i=1}^{nph} h_i \psi_i^h, \qquad u = \sum_{i=1}^{npu} u_i \psi_i^u, \qquad v = \sum_{i=1}^{npu} v_i \psi_i^u$$

Our set of equations then becomes:

Continuity

$$\sum_{j=1}^{nph} \left(\frac{h_j^{n+1} - h_j^n}{DT} \right) \int_\Omega \psi_j^h \varphi_i^h \, d\Omega + \sum_{j=1}^{nph} h_j \int_\Omega \boldsymbol{u}_{\text{conv}} \cdot \operatorname{grad}(\psi_j^h)\varphi_i^h \, d\Omega$$

$$+ \int_\Gamma h_{\text{prop}} \boldsymbol{u} \cdot \boldsymbol{n}\varphi_i^h \, d\Gamma - \sum_{j=1}^{npu} u_j \cdot \int_\Omega \psi_j^u \operatorname{grad}(h_{\text{prop}}\varphi_i^h) \, d\Omega = 0 \quad (6.65)$$

Momentum

$$\sum_{j=1}^{npu} \left(\frac{u_j^{n+1} - u_j^n}{DT} \right) \int_\Omega \psi_j^u \varphi_i^u \, d\Omega + \sum_{j=1}^{npu} u_j \int_\Omega \boldsymbol{u}_{\text{conv}} \cdot \operatorname{grad}(\psi_j^u)\varphi_i^u \, d\Omega$$

$$= -\sum_{j=1}^{nph} (h_j^{n+1} - h_j^n) \int_\Omega g\theta_h \frac{\partial \psi_j^h}{\partial x} \varphi_i^u \, d\Omega - \int_\Omega g \frac{\partial Z^n}{\partial x} \varphi_i^u \, d\Omega + \int_\Omega F_x \varphi_i^u \, d\Omega$$

$$+ \int_\Omega \varphi_i^u \nu_t \operatorname{grad}(u^{n+1}) \cdot \boldsymbol{n} \, d\Gamma - \sum_{j=1}^{npu} u_j^{n+1} \int_\Omega \nu_t \operatorname{grad}(\psi_j^u) \cdot \operatorname{grad}(\varphi_i^u) \, d\Omega \quad (6.66)$$

$$\sum_{j=1}^{npu} \left(\frac{v_j^{n+1} - v_j^n}{DT} \right) \int_\Omega \psi_j^u \varphi_i^u \, d\Omega + \sum_{j=1}^{npu} v_j \int_\Omega \boldsymbol{u}_{\text{conv}} \cdot \operatorname{grad}(\psi_j^u)\varphi_i^u \, d\Omega$$

$$= -\sum_{j=1}^{nph} (h_j^{n+1} - h_j^n) \int_\Omega g\theta_h \frac{\partial \psi_j^h}{\partial y} \varphi_i^u \, d\Omega - \int_\Omega g \frac{\partial Z^n}{\partial y} \varphi_i^u \, d\Omega + \int_\Omega F_y \varphi_i^u \, d\Omega$$

$$+ \int_\Gamma \varphi_i^u \nu_t \operatorname{grad}(v^{n+1}) \cdot \boldsymbol{n} \, d\Gamma - \sum_{j=1}^{npu} v_j^{n+1} \int_\Omega \nu_t \operatorname{grad}(\psi_j^u) \cdot \operatorname{grad}(\varphi_i^u) \, d\Omega \quad (6.67)$$

We now have a linear system where the unknowns are all the values h_i^{n+1}, u_i^{n+1}, and v_i^{n+1}. This system is written in the form: $AX = B$. Here X is a vector containing all the unknowns that can be represented in the form of a block of three vector δH, U and V respectively containing the unknowns $h_i^{n+1} - h_i^n$, u_i^{n+1} and v_i^{n+1}.

$$X = \begin{pmatrix} \delta H \\ U \\ V \end{pmatrix} \tag{6.68}$$

In the same way, B is a right hand side consisting of three vectors called $CV1$, $CV2$ and $CV3$:

$$B = \begin{pmatrix} CV1 \\ CV2 \\ CV3 \end{pmatrix} \tag{6.69}$$

A is a matrix composed of nine other matrices. Among these nine matrices two are zero because we have no coupling between the two components of the velocity. The matrix A can thus be written as:

$$A = \begin{pmatrix} AM1 & BM1 & BM2 \\ -CM1^{\mathrm{T}} & AM2 & 0 \\ -CM2^{\mathrm{T}} & 0 & AM3 \end{pmatrix} \tag{6.70}$$

The reason why we have chosen the form $-CM1^{\mathrm{T}}$ will not be detailed here but stems from symmetry considerations. The seven matrices in the block A are composed of some well-known (and some other less common) matrices in finite elements:

Mass matrices

$$M^h(i,j) = \int_\Omega \psi_j^h \varphi_i^h \, d\Omega \quad \text{and} \quad M^u(i,j) = \int_\Omega \psi_j^u \varphi_i^u \, d\Omega \tag{6.71}$$

Advection matrices

$$T^h(i,j) = \int_\Omega \mathbf{u}_{\text{conv}} \cdot \mathbf{grad}(\psi_j^h) \varphi_i^h \, d\Omega \quad \text{and} \quad T^u(i,j) = \int_\Omega \mathbf{u}_{\text{conv}} \cdot \mathbf{grad}(\psi_j^u) \varphi_i^u \, d\Omega \tag{6.72}$$

Diffusion matrix

$$D^u(i,j) = \int_\Omega \nu_t \, \mathbf{grad}(\psi_j^u) \cdot \mathbf{grad}(\varphi_i^u) \, d\Omega \tag{6.73}$$

"Gradient" or "divergence" type matrices

$$B_x^{uh}(i,j) = -\int_\Omega \psi_j^u \frac{\partial}{\partial x} (h_{\text{prop}} \varphi_i^h) \, d\Omega \quad \text{and} \quad B_y^{uh}(i,j) = -\int_\Omega \psi_j^u \frac{\partial}{\partial y} (h_{\text{prop}} \varphi_i^h) \, d\Omega \tag{6.74}$$

$$C_x^{uh}(i,j) = -\int_\Omega \frac{\partial \psi_i^h}{\partial x} \varphi_j^u \, d\Omega \quad \text{and} \quad C_y^{uh}(i,j) = -\int_\Omega \frac{\partial \psi_i^h}{\partial y} \varphi_j^u \, d\Omega \tag{6.75}$$

Friction matrix, on the boundary

$$F^u(i,j) = \int_\Gamma v_t a \psi_j^u \varphi_i^u \, d\Gamma \tag{6.76}$$

Bottom friction matrices (e.g. the Chézy law)

$$F_u = -\int_\Omega \frac{1}{\cos\theta} \frac{g}{hC^2} u^{n+1} \sqrt{(u^n)^2 + (v^n)^2} \varphi_i^u \, d\Gamma \tag{6.77}$$

According to the linear system obtained, our seven matrices in the block A are:

$$AM1 = \frac{M^h}{DT} + \theta_h T^h$$

$$AM2 = AM3 = \frac{M^u}{DT} + \theta_u T^u + D^u - G^u + F^u$$

$$BM1 = \theta_u B_x^{uh}, \qquad BM2 = \theta_u B_y^{uh}$$

$$CM1 = g\theta_h C_x^{uh}, \qquad CM2 = g\theta_h C_y^{uh} \tag{6.78}$$

The terms T^h and T^u only appear if the method of characteristics is used. The right hand sides are:

$$CV1 = (\theta_h - 1)T^h H^n + (1 - \theta_u)(B_x^{uh}U^n + B_y^{uh}V^n) + M^h SCE + TB1 \tag{6.79}$$

where $TB1$ is

$$-\int_\Gamma h_{\text{prop}} \boldsymbol{u} . \boldsymbol{n} \varphi_i^n \, d\Gamma$$

With the method of characteristics: $(\theta_h - 1)T^h H^n$ would have to be replaced by $(M^h/DT)(\tilde{H} - H^n)$.

$$CV2 = \frac{M^u}{DT} U^n + (\theta_u - 1)T^n U^n + C_x^{uh} Z^n + M^u F_x \tag{6.80}$$

With the method of characteristics: $(M^u/DT)U^n + (\theta_u - 1)T^n U^n$ would have to be replaced by $(M^u/DT)\tilde{U}$.

$$CV3 = \frac{M^u}{DT} V^n + (\theta_u - 1)T^n V^n + C_y^{uh} Z^n + M^u F_y \tag{6.81}$$

With the method of characteristics: $(M^u/DT)V^n + (\theta_u - 1)T^n V^n$ would have to be replaced by $(M^u/DT)\tilde{V}$.

The majority of our matrices depend on time and must be consequently rebuilt at every time-step. The efficiency of the algorithms used for building these matrices is thus of utmost importance. This is why the so-called Element By Element technique (EBE, see Hughes *et al.*, 1987 and Hervouet, 1991) is used in the TELEMAC system. The matrices are not fully assembled (only the diagonal is assembled) and this spares considerable time. Furthermore, the Gauss quadrature generally used to work out the matrices in finite elements have been replaced by exact formulas obtained directly in FORTRAN with a symbolic computation software. These formulas are more concise and yield much faster algorithms.

The linear system is then solved by iterative techniques. This is done in four steps:

• Treatment of Dirichlet boundary conditions (when some unknowns are given on a boundary).
• Preconditioning, i.e. transforming the linear system for a better efficiency of the iterative techniques.
• Use of an iterative solver such as the conjugate-gradient method. The main operations are matrix-vector products and dot products.
• Recovering the original unknowns (preconditioning sometimes changes the unknowns).

Again the EBE techniques prove to be very efficient when solving the linear system. As a matter of fact, the algorithms obtained for the matrix-vector product are naturally accelerated on vector or parallel machines, and also on recent workstations.

6.3.3   Extension to three-dimensional implementations

Space discretization, variational formulation and solution procedures are broadly the same in 2D and 3D and we shall only emphasize the specific features arising in 3D.

The finite element which has been chosen is the prism with a linear interpolation. As a matter of fact, 3D meshes in TELEMAC-3D are obtained by means of a superimposition of 2D triangular meshes, as shown in Figure 6.2.

The lower 3D mesh follows the bottom topography and the upper one is the free surface. The elevation of a point in a 3D mesh is given by the formula:

$$z = Zf(x, y) + \theta(Z(x, y, t) - Zf(x, y)) \text{ with } 0 \leqslant \theta \leqslant 1 \qquad (6.82)$$

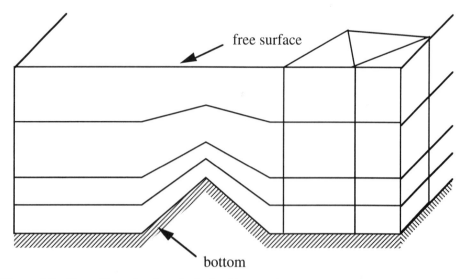

Figure 6.2   Three-dimensional mesh obtained by superimposition of two-dimensional meshes

In finite elements, θ may be a function of x and y and the intermediate 2D meshes are not mandatorily evenly spaced. The difficulty is the fact that this 3D mesh evolves with time. To overcome this problem, the so-called σ-transformation is used, i.e. the elevation z is changed into z^*, such that:

$$z = Zf(x,y) + \theta(Z(x,y,t) - Zf(x,y)) \qquad \text{with} \qquad 0 \leq \theta \leq 1 \qquad (6.83)$$

where $\bar{Z}$ is a given constant assumed to be 1 from now on. This σ-transformation is responsible in finite differences for "creeping diffusion" effect, where iso-value surfaces of vertical stratification do not remain horizontal but tend to follow the 2D meshes. This problem vanishes in finite elements when the diffusion matrices are rigorously computed.

In TELEMAC-3D, a fractional step approach is used where the first two steps deal with the advection and diffusion terms and give provisional values of the horizontal velocity denoted u_D and v_D. The main difference with 2D Shallow Water equations is the treatment of the propagation step where the remaining terms of the momentum equation and the continuity equation are solved. The equations in this last step read:

$$\left\{ \begin{array}{l} \text{div}(U) = 0 \\[2mm] \dfrac{U_1 - u_D}{\Delta t} = -\dfrac{1}{\tilde{\rho}} \dfrac{\partial p}{\partial x} + f_x \\[4mm] \dfrac{U_2 - v_D}{\Delta t} = -\dfrac{1}{\tilde{\rho}} \dfrac{\partial p}{\partial y} + f_y \end{array} \right. \qquad (6.84)$$

Averaging the latter equations on the vertical naturally gives Saint-Venant equations devoid of advection and diffusion terms (actually contained in u_D and v_D). These averaged equations are then solved by TELEMAC-2D and yield the mean velocity and the free surface elevation, thus enabling computation of an actualised 3D mesh. Using the fact that

$$p = \rho_0 g(Z - z) + \tilde{\rho}g \int_z^Z \frac{\Delta\rho}{\rho_0} \, dz \qquad (6.85)$$

which gives:

$$\frac{\partial p}{\partial x} = \rho_0 g \frac{\partial Z}{\partial x} + \tilde{\rho}g \frac{\partial}{\partial x}\left(\int_z^Z \frac{\Delta\rho}{\rho_0} \, dz \right) \qquad (6.86)$$

and a similar equation for y, U_1 and U_2 are obtained from the momentum equation. In the transformed mesh, the continuity equation reads:

$$\left(\frac{\partial h}{\partial t} \right)_{x,y,z^*} + \left(\frac{\partial(hu)}{\partial x} \right)_{y,z^*,t} + \left(\frac{\partial(hv)}{\partial y} \right)_{x,z^*,t} + h\left(\frac{\partial w^*}{\partial z^*} \right)_{x,y,t} = 0 \qquad (6.87)$$

where * denotes the transformed coordinates and variables. This equation is integrated along the vertical, on one hand from the bottom to an elevation z, and then from z to the free surface (with $\bar{Z} = 1$). This yields:

$$z^* \left(\frac{\partial h}{\partial t}\right)_{x,y,z^*} + \int_0^{z^*} \left[\left(\frac{\partial(hu)}{\partial x}\right)_{y,z^*,t} + \left(\frac{\partial(hv)}{\partial y}\right)_{x,z^*,t}\right] dz^* + h[w^*(z^*) - w^*(0)] = 0 \quad (6.88)$$

and:

$$(z^* - 1)\left(\frac{\partial h}{\partial t}\right)_{x,y,z^*} + \int_1^{z^*} \left[\left(\frac{\partial(hu)}{\partial x}\right)_{y,z^*,t} + \left(\frac{\partial(hv)}{\partial y}\right)_{x,z^*,t}\right] dz^* + h[w^*(z^*) - w^*(1)] = 0$$

$$(6.89)$$

Noting that $w^*(0)$ and $w^*(1)$ are known, both equations are combined to obtain:

$$hw^*(z^*) = h(1 - z^*)w^*(0) + hz^* w^*(1) + z^* \int_0^1 \left[\left(\frac{\partial(hu)}{\partial x}\right)_{y,z^*,t} + \left(\frac{\partial(hv)}{\partial y}\right)_{x,z^*,t}\right] dz^*$$

$$- \int_0^{z^*} \left[\left(\frac{\partial(hu)}{\partial x}\right)_{y,z^*,t} + \left(\frac{\partial(hv)}{\partial y}\right)_{x,z^*,t}\right] dz^* \quad (6.90)$$

The seemingly complex right hand side only contains known quantities.

This last step giving the vertical velocity is a critical one in 3D Shallow Water computations. As a matter of fact, as the vertical velocity is obtained from the continuity equation and the result u_D and v_D of the preceding steps, the inaccuracies will be supported by the vertical velocity. A consequence is that, unlike in 2D, trespassing the hydrostatic assumption in 3D results in degraded solutions with spurious vertical velocities.

6.3.4 Flooding, drying and tidal flats

Many of the recent studies performed with TELEMAC-2D involve wetting and drying areas and it seems a general trend linked to the increasing number of environmental problems resorting to hydraulics, in rivers or estuaries. The ability to cope with such situations is thus of utmost importance. The problem is the behaviour of the numerical algorithms in the dry zones, e.g. some terms in the equations are divisions with h in the denominator and they tend to infinity as h tends to 0.

Two kinds of solutions have been implemented:

- solving the equations everywhere and coping with spurious terms,
- removing the dry zones from the computational domain.

The first one is the simplest but corrections must be applied in the wetting and drying zones, to avoid infinite terms and spurious values of the free surface gradient. As a matter of fact, in dry areas, the free surface gradient is equal to the gradient of the bottom topography and in that case must not act as a driving force in the momentum equation. This problem is exemplified in 1D in Figure 6.3.

The second option, often referred to as the "moving boundary technique", consists of removing the drying zones from the domain. This can be a very cumbersome task and might require the definition of new meshes, leading to computationally demanding

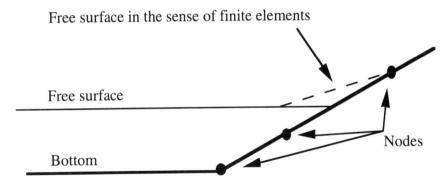

Figure 6.3 Spurious values of the free surface gradient

operations to be performed at every time-step, for example the renumbering of the unknowns, thus spoiling the efficiency. To avoid this in TELEMAC-2D, dry elements are kept in the mesh, but their occurrence in the computations is cancelled by an array that is set to 0 for dry elements and 1 for the others. A thorough study of all these algorithms showed that one could restrict this operation to a very small number of routines, mainly those with an assemblage.

Both techniques are used at the present time. In very difficult cases with very steep slopes, removing the dry elements is better. In river computations, and if there is a very small number of elements in a cross-section, solving the equations everywhere is preferable. The latter method worked in some cases with only two elements in a cross-section, but this is obviously out of any theoretical background and this type of simulation is not recommended.

In 3D computations, an additional difficulty is the fact that the volume of elements is zero when there is no water. Only the removal of dry elements is thus practicable.

6.3.5 Limitations and further developments

We are certainly not at the end of the ongoing process of improvement of the algorithms, especially in the field of advection schemes. Mesh adaptation may also be of interest to cope with hydraulic jumps and uneven bottom topography. Parameter estimation applied to the friction coefficients will also be very useful to speed up the calibration of river models. Whatever the improvements, however, some limitations of the 2D equations will remain. The helicoïdal currents in a bend are beyond reach and they are very important for sediment transport applications. Salt wedges and stratification obviously cannot be reproduced in 2D. An evolution towards 3D computations is thus necessary.

6.4   EXAMPLES

6.4.1 Two-dimensional examples

Two examples are presented here. The first one concerns a classical study of hydrodynamics in a river with sand banks, and the second one describes the propagation of a wave on a floodplain after a dam break.

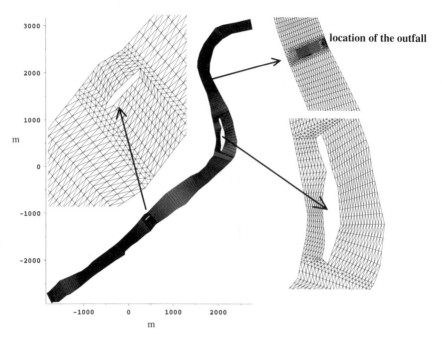

Figure 6.4   10 000 elements mesh representing a 8 km reach of the River Loire at Saint-Laurent

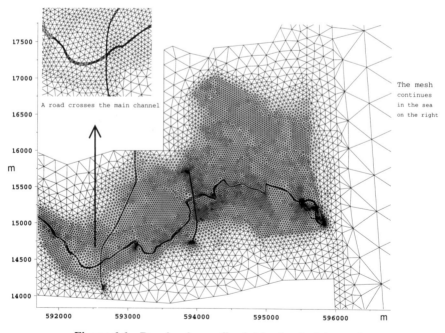

Figure 6.6   Dam break on a floodplain: detail of the mesh

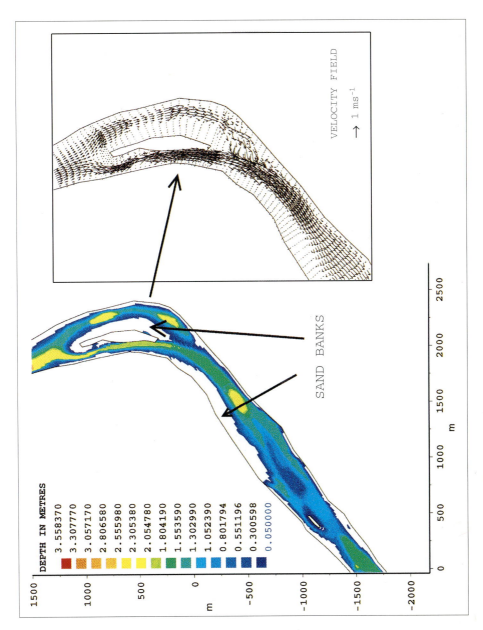

Plate I Figure 6.5   River Loire at Saint-Laurent: sand banks and velocity field

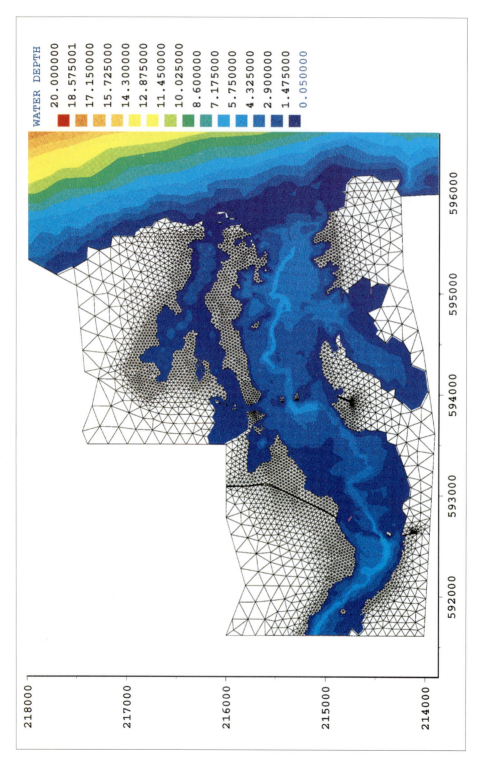

Plate II Figure 6.7 After a dam break, a flood wave reaches the sea

River Loire at Saint-Laurent

This model was set up for studying the thermal plume of a nuclear power plant. The River Loire is well known for its sand banks and a treatment of dry zones is mandatory here. This kind of study falls well within the range of applications of our equations because of the shallow waters and also because the outfall is designed to obtain a full mixing of the hot water along the vertical. Figure 6.4 shows the finite element mesh, with close-up views around the outfall and two permanent islands. The location of the

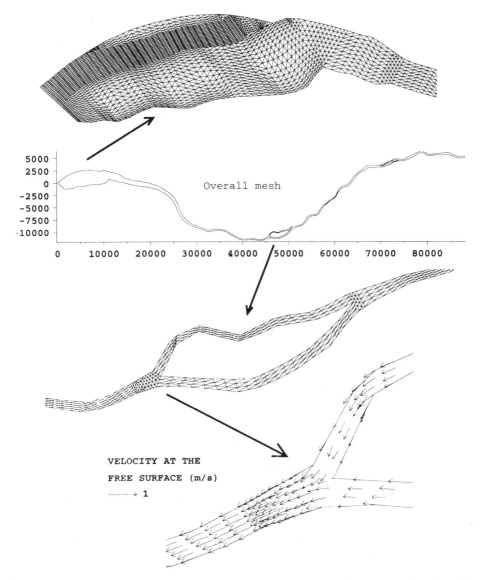

Figure 6.8 Three-dimensional study of hydrodynamics in the Loire estuary. Details of the mesh and of the velocity field

sand banks is given at the left part of Figure 6.5 (Plate I), whereas the right side shows
the velocity field.

Dam break on a flood plain

In France, as in most other countries, this kind of study is required by law for every
important dam. In the present case, the flood in the upper part of the river has been
simulated by a 1D computation. The result is then used as an upstream boundary

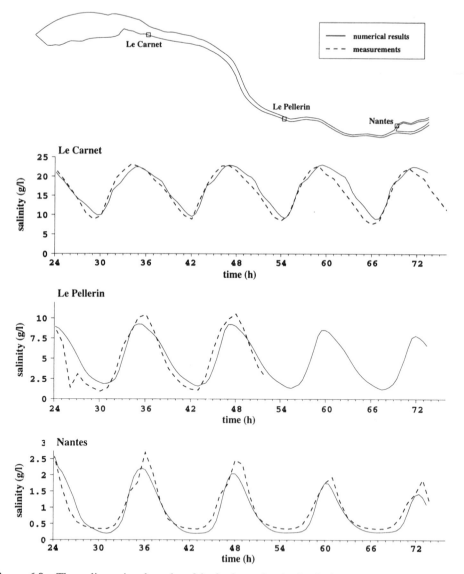

Figure 6.9   Three-dimensional study of hydrodynamics in the Loire estuary. Comparison of
computed and measured salinity at the free surface

condition in the 2D computation. The computation includes a part of the sea, where the flood finally ends up. Figure 6.6 shows details of the mesh containing 20 000 elements, with main channel, roads and dykes. Figure 6.7 (Plate II) is the water depth at the instant when the flood reaches the sea.

6.4.2 Three-dimensional examples

This computation was performed within the framework of a European MAST (MArine Science and Technology) project on the 3D numerical modelling of physical processes of cohesive sediment transport in estuarine environments. The main goal was the simulation of the salt wedge and of the turbidity maximum. Figure 6.8 shows the 80-km long reach discretization, along with close-up views of the mesh and of the velocity field. The 3D mesh of about 60 000 prisms was obtained by superimposition of ten 2D meshes. Figure 6.9 compares the computed and measured salinity at the free surface.

6.5 CONCLUSIONS

We have tried here to show the important role played by Computational Fluid Dynamics for studying floodplain processes. However, the idea of a rampant technology rapidly evicting other techniques would certainly be deceitful. Many obstacles have been overcome, but many are still ahead. Some basic numerical problems are not solved and very common situations such as a free surface flow followed by a pressure flow (a reservoir and a turbine) are not satisfactorily dealt with, even with very sophisticated tools. Sometimes a precise knowledge of physical phenomena, bedload transport in supercritical situations for example, is still lacking.

All these difficulties must be kept in mind. The rapid development of user-friendly interfaces, made available to non-specialists, is certainly a necessary evolution but is at risk of misuse and must not be done at the expense of the algorithms hidden behind. The wealth of nice-looking animations, with the argument of helping decision makers and obtaining public approval is a very good thing if and only if limitations and uncertainties are also presented. At the end of this chapter it is thus very important to recall that the present tools are far from fully satisfactory. Improvement will come, on the one hand, from the ongoing progress of computer technology that will enable higher and higher accuracy; the problem of turbulence could for instance be solved with a Large-Eddy Simulation approach. On the other hand, relying only on the progress of machines would be a dead end for hydrodynamics and an important research effort must be dedicated to enhance the basic algorithms.

The present trends are mesh adaptation, parameter estimation, and new approaches coupling finite elements and finite volumes, together with the adaptation to new computer architectures such as parallel machines. To avoid an uncontrolled radiation into many different programmes, a general trend is now the building of software systems sharing pre- and post-processors and libraries, and including a number of modules from different disciplines: water quality, wave modelling, sedimentology and so on. Controlling the overall quality of results becomes a difficult task for such systems and validation appears to be a major issue. Computer programmes should now be presented along with portfolios of test-cases including comparisons with analytical solutions and with measurements. Guidelines for such documents have been published by the International Association for Research in Hydraulics. The evaluation of the error interval of a free surface flow computation, however, is still one of the many open problems that we have to face. This simple fact gives us an idea of the remaining task, and is a commitment to humility.

REFERENCES

Brookes, A.-N. and Hughes, T.J.R. (1982) Streamline Upwind Petrov Galerkin formulations for convection dominated flows with particular emphasis on the Navier–Stokes equations. *Computer Methods in Applied Mechanics and Engineering*, **32**, 199–259.

Goutal, N. (1987) Résolution des équations de Saint-Venant en régime transcritique par une méthode d'éléments finis. Application aux bancs découvrants. Thèse de doctorat de l'Universite Paris VI.

Guide for selecting Manning's roughness coefficients for natural channels and flood plains. (1989) *United States Geological Survey. Water Supply Paper* 2339.

Hervouet, J.-M. (1986) CARAC, module de convection en éléments finis, par la méthode des caractéristiques. *Rapport EDF* HE41/86.21.

Hervouet, J.-M. (1991) Vectorisation et simplification des algorithmes en éléments finis. *EDF Bulletin de la Direction des Etudes et Recherches*. Série C, Mathématiques, Informatique no. 1, 1–37.

Hervouet, J.-M., Hubert, J.-L., Janin, J.-M., Lepeintre, F. and Peltier, E. (1994) The computation of free surface flows with TELEMAC. An example of evolution towards hydroinformatics. Special issue of the *Journal of the International Association for Research in Hydraulics*, **32**.

Hughes, T.J.R, Ferencz, R.M. and Hallquist, J.O. (1987) Large-scale vectorized implicit calculations in solid mechanics on a CRAY X-MP/48 utilizing EBE preconditioned conjugate gradients. *Computer Methods in Applied Mechanics and Engineering*, **61**, 215–248.

Idelsohn, S.R. and Onate, E. (1994) Finite Volumes and Finite Elements: two good friends. *International Journal for Numerical Methods in Engineering*, **37**, 3323–3341.

Rastogi, A.-K. and Rodi, W. (1978) Predictions of heat and mass transfer in open channels. *Journal of the Hydraulics Division, ASCE*, **104** (HY3), 397–420.

Rodi, W. (1993) *Turbulence Models and their Application in Hydraulics, A State-of-the-art Review. AIRH Monograph Series*, third edition, A.A. Balkema, Rotterdam.

7 Analysis and Development of Hydraulic Models for Floodplain Flows

P.D. BATES, M.G. ANDERSON, D.A. PRICE, R.J. HARDY and C.N. SMITH
Department of Geography, University of Bristol, UK

7.1 INTRODUCTION

Current developments in terms of floodplain topographic evolution models, computational fluid dynamics (CFD) and floodplain sedimentation models are beginning to offer the prospect of significant advances in process representation within the floodplain environment. Additional to the prospects in these individual fields are the opportunities that now present themselves for *integrating* these advances at medium river reach scales (10–60 km). Central to this integration is the development of a suitable hydraulic model platform.

For such applications a number of investigations have already been made using two-dimensional hydraulic models (see, for example, Gee *et al.*, 1990; Baird and Anderson, 1992; Baird *et al.*, 1992; Anderson and Bates, 1994). These applications show the feasibility of applying such schemes and in particular provide evidence for achieving a high level of physical process representation. This is essential if critical aspects such as the time/space domain of inundation extent are to be accurately portrayed. This issue is important in what is generally regarded as a data-sparse environment and guidance will be increasingly sought for field monitoring strategies from model results. This point is reinforced by Baird *et al.* (1992) who illustrate the inflow for a two-dimensional hydraulic model being provided by an ungauged catchment hydrology model for a 25 km river reach. In this application, the hydrology–hydraulic model combination provided a better prediction of the reach outflow hydrograph than that provided by the application of the ungauged hydrology model to the entire catchment.

CFD developments offer increased opportunities to examine three important areas relevant to floodplain processes:

- as a platform for integrating advances in long-term channel change and sedimentation;
- providing improved physical representation of process;
- providing a powerful addition to hydrological models, especially in relation to boundary conditions of the CFD schemes (upstream inflow, hillslope inflow and floodplain infiltration).

Floodplain Processes. Edited by Malcolm G. Anderson, Des E. Walling and Paul D. Bates.
© 1996 John Wiley & Sons Ltd.

The advances in numerical methods outlined in Chapter 6 can thus be furthered by applications, the results of which can be assessed in terms of their contribution to these three areas.

Thus, a key task in the analysis and development of hydraulic models for floodplain flow problems is the simulation of actual flood events (Section 7.2). Such studies fulfil a number of objectives. Firstly, they enable us to ascertain if the models correctly reproduce known physical processes. Secondly, we may determine whether model response to parameter variation accords with expectations. Lastly, and most importantly, we may compare model predictions to observed flows in order to begin model validation. By such studies it becomes possible to comment on model capabilities and limitations and so begin to define an agenda for future research. In this chapter we explore two specific areas where current hydraulic model studies in reach-scale floodplain environments indicate a need for further research. These concern development of model sensitivity analysis which can control for such attributes as topography and mesh resolution (Section 7.3) and the relaxation of previously impermeable hydraulic model boundary conditions (Section 7.4).

7.2   APPLICATION OF TELEMAC-2D

Suitable hydraulic models for reach-scale (10–60 km) floodplain flow problems need to fulfil a number of criteria:

- A two-dimensional, or higher, representation of the flow field should be implemented to capture known process dynamics (Knight and Shiono, Chapter 5, this volume).
- Efficient numerical solution algorithms should be chosen to enable high space/time resolution models to be constructed.
- The model should be capable of representing complex topography, such as a channel meandering within a wider floodplain belt, with a minimum number of computational points. This consideration gives a significant advantage to numerical methods, such as the finite element technique, which are based on unstructured grids.
- Given that simulation of river flood flows must represent a dynamically moving flow field boundary some consideration should be given to the performance of the chosen numerical algorithms in response to wetting and drying processes.
- Characteristics of the turbulence field relevant to the scale of application should be simulated realistically.

The two-dimensional finite element model TELEMAC-2D described by Hervouet and Van Haren (Chapter 6, this volume) meets all the above criteria and therefore forms the basis of the research reported in this chapter. Although other codes exist, which could be applied to this problem, considerable validation of TELEMAC-2D against analytical solutions as well as other fluvial flow problems has been undertaken (see, for example Hervouet, 1989, 1993; Hervouet and Janin, 1994; Hervouet and Van Haren, Chapter 6, this volume). Such studies increase confidence that the code can be extended to consider river channel/floodplain applications.

In order to illustrate the practical application of the TELEMAC-2D model, studies have been conducted for a number of river floodplain reaches (see, for example, Bates *et al.*, 1994, 1995, 1996). In addition to enabling a comparison of model predictions to field data it has also been possible to test the impact on model output of variations in mesh resolution (Section 7.2.1) and numerical solution technique (Section 7.2.2), as well as to comment on current practical application complexity and scale (Section 7.2.3). As indicated above, such testing is essential for a complete analysis of model behaviour which can enable conclusions to be drawn concerning model capabilities and limitations and to suggest areas for future development.

7.2.1 Mesh resolution effects

The impact of varying mesh resolution for applications of TELEMAC-2D to actual river channel/floodplain flow problems has been investigated for two UK river reaches: the River Culm in Devon and the River Stour in Dorset. For each site two mesh discretizations were constructed and a flood event simulated using identical flow and parameter conditions. This allowed models to be compared in terms of their ability to predict downstream discharge and inundation extent over the reach.

An 11 km reach of the River Culm was selected for this study between the river flow gauging stations at Woodmill (upstream) and Rewe (downstream). This river reach comprises a gravel-bed channel, approximately 10 m wide, meandering across a well-developed floodplain up to 450 m in width. This reach was selected due to the regularity with which substantial inundation events occur, typically on six occasions per year, and the complexity of the floodplain topography, which includes mill races, an embankment and a channel bifurcation. Data on flow conditions were provided by continuous stage recorders at Woodmill and Rewe. Rated sections also existed at these sites allowing discharge to be estimated from flow data. Boundary conditions for the numerical model consisted of an imposed flow rate as the upstream boundary condition, while at the downstream end of the mesh a water surface elevation condition was imposed. This combination of boundary conditions gave a well-posed mathematical problem according to the theory of Characteristics and, moreover, allowed the observed downstream discharge to be used as an independent data set with which to undertake model validation. The two finite element meshes constructed for this reach consisted of 2040 and 9600 linear triangular elements respectively (see Figures 5.71a and 7.16). In order to specify the catchment topography an elevation was assigned to each computational node in the discretization. In this case elevation data for each mesh were interpolated from a single topographic parameterization based on data from UK Ordnance Survey 1 : 2500 series maps enhanced with six surveyed cross-sections.

For the River Stour a similar scale reach, 12 km in length between the river gauging stations at Hamoon (upstream) and Blandford Forum (downstream), was selected. At the upstream gauging site, stage and discharge information were available. However, at the Blandford Forum station no rated section existed despite the presence of stage recording equipment. It was therefore necessary to construct a rating curve in order to obtain discharge data for model validation. This was achieved by surveying a section across the channel and floodplain and then using a standard modelling package, HYMO3 (Baird and Anderson, 1992), to generate a rating relationship. HYMO3 takes

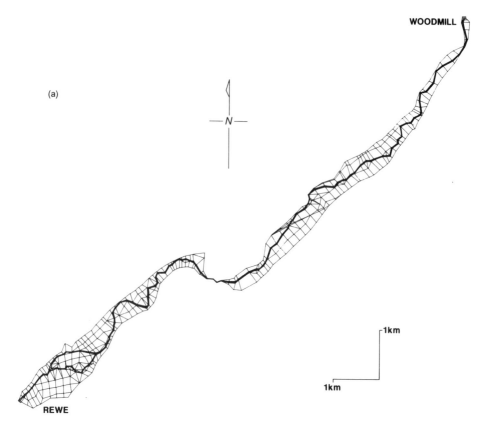

Figure 7.1 Finite element discretizations developed for an 11 km reach of the River Culm, Devon, UK. The two domains shown consist of (a) 2040 elements and (b) 9600 elements respectively

the specified cross-section and calculates the flow at a series of elevations using the Manning equation with momentum exchange between main channel and floodplain flows accounted for via a simple analytical formulae based on lines of zero shear (Knight and Hamed, 1984). Use of such a procedure obviously introduces a number of additional errors and uncertainties into the analysis of model results; however, such a level of data provision is typical of a number of river gauging stations in the UK. The modelling approach therefore raises questions regarding current field data capture procedures and the extent to which the development of high space–time resolution modelling approaches is becoming data limited. For the River Stour the two computational domains consisted of 4396 and 11265 elements respectively (see Figures 7.2a and 7.2b), thereby achieving less distortion and a smoother transition between areas of high and low element density than the equivalent River Culm discretization. Topography was again assigned from a single discretization, however, elevations were obtained from an experimental 10×10 m Digital Terrain Model (DTM) supplied by the UK Ordnance Survey and approximately 25 surveyed channel cross-sections made available by the UK National Rivers Authority (see Bates *et al.*, 1996). This provided a

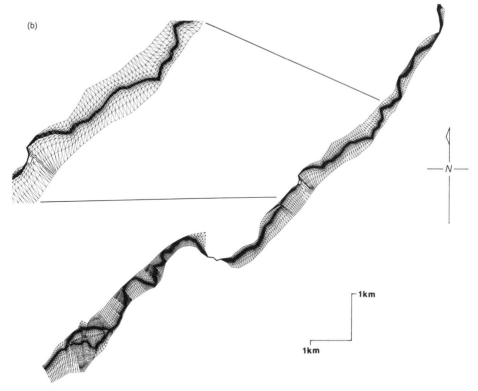

Figure 7.1 *(Continued)*

more accurate topographic parameterization which incorporated a greater degree of spatial variability (see Figures 7.3a and 7.3b) than the River Culm data set.

For each river reach a single medium-sized flood event was simulated at each mesh resolution giving a total of four simulations in all. An identical TELEMAC-2D set-up was used for each simulation. This employed a fractional step method (Marchuk, 1975) where advection terms are solved initially, separate from propagation, diffusion and source terms which are solved together in a second step. For the advection step several schemes may be chosen with the Method of Characteristics chosen here for the momentum equation. To ensure mass conservation and oscillation-free solutions with unrefined meshes the Streamline Upwind Petrov Galerkin (SUPG) method was applied for the advection of the water depth, h, in the continuity equation. According to this technique, standard Galerkin weighting functions are modified by adding a streamline upwind perturbation, a method which overcomes problems of artificially diffuse solutions produced by ordinary upwinding methods. The second step (propagation) made use of an implicit time discretization and solved the resulting system with a conjugate gradient type method.

To begin a dynamic flood computation it was necessary to first develop a stable steady-state flow at bankfull discharge for each finite element mesh. This could then be used as an initial condition for the flood wave simulation. This was achieved by

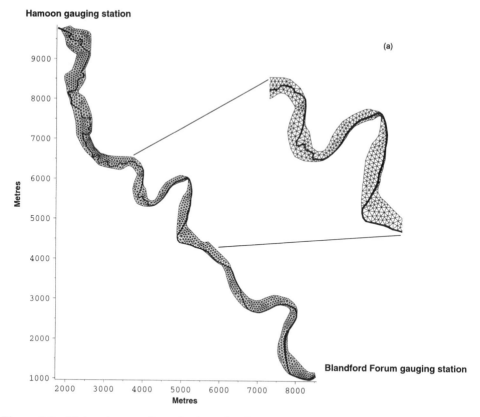

Figure 7.2   Finite element discretizations developed for a 12 km reach of the River Stour, Dorset, UK. The two domains shown consist of (a) 4396 elements and (b) 11 265 elements respectively

commencing with a simple hydraulic condition (no flow velocity) and boundary conditions representing the bankfull state. Velocities were then allowed to develop until all spurious oscillations had propagated out of the system and the outflow discharge had ceased to show variation. The discrepancy between imposed inflow and downstream outflow under steady-state conditions is approximately equivalent to the mass conservation error. This was computed and found to be within ±2% at all times. For the River Culm a 1 in 1 year recurrence interval flood event which took place over 15 hours on 30 January 1990 was simulated with the above model set-up. A model time-step of 2 s was chosen for this event as this gave the most efficient numerical simulation while maintaining a low Courant number (cf. Bates *et al.*, 1995). The Courant number is a measure of stability for explicit numerical approximation procedures and is calculated by:

$$C_r = u \frac{\Delta t}{\Delta x} \tag{7.1}$$

where C_r = Courant number; t = time; x = mesh size and u = flow velocity.

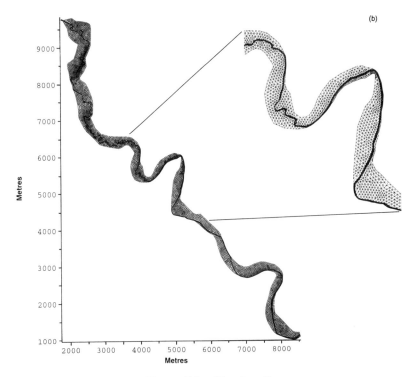

Figure 7.2 (*Continued*)

Instability occurs for such a scheme when the Courant number exceeds 1. TELEMAC-2D is a fully implicit scheme and is theoretically not subject to such constraints. In reality, however, as the Courant number has a sound physical basis it is often a good surrogate measure of the quality of the numerical solution and should therefore be taken into account when developing model applications. Given an event of 15 hours duration, 27 000 time-steps were thus required for each computation. For the River Stour a similar-sized event was chosen which took place over 62 hours on 20 December 1993. Here the above discretization criteria resulted in a simulation of 55 800 time-steps each with a duration of 4 s. A single set of boundary friction parameters was developed for each reach. These were determined for each low resolution model by a trial and error calibration which minimized the phase error between observed and predicted flow peaks. This calibration was then transferred directly to the relevant high resolution mesh. Separate boundary friction parameters were applied in main channel and floodplain areas with respective values 0.025 and 0.083 selected for the River Culm and 0.017 and 0.035 for the River Stour. The differences in the calibrations reflect variations in slope and flood wave travel times between the two reaches. The Culm has an average slope over the reach of 0.002 and a lag between peak discharge at the upstream and downstream gauges for the 30 January 1990 flood of approximately 9 hours. This compares with a much shallower average slope of 0.0008 on the River Stour and a flood wave travel time through the reach of 5 hours for the 20 December 1993 event. These data imply a much higher frictional energy loss over the River Culm reach and this is reflected in the selected calibration.

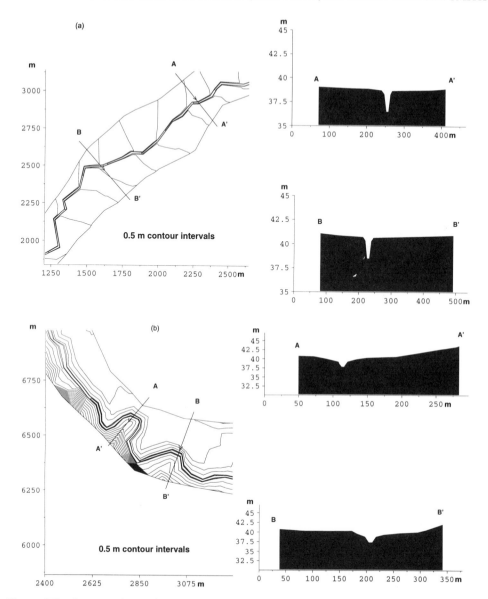

Figure 7.3 A comparison of the typical topographic surface complexity used to parameterize (a) the low resolution River Culm finite element model and (b) the low resolution River Stour finite element model

Figures 7.4a and 7.4b compare predicted downstream discharge from each model simulation with the relevant gauging station observations. It can be seen that all the simulations reported in this chapter show a good correspondence with the observed data. Some minor variations do occur. For example, in both cases the high resolution mesh consistently predicts a higher peak discharge than the low resolution version and the response could be considered to be more dynamic. In the case of the River Culm the initial portion of the simulation is not well approximated by the high resolution mesh possibly indicating some problems obtaining a true steady-state computation at this mesh resolution for this reach. However, after the hydrograph rise is reached, predictions from both high and low resolution River Culm meshes are broadly similar. In the case of the River Stour both models overestimate discharge on the rising and falling limb of the hydrograph. This may perhaps be due to a systematic error in the developed stage discharge rating equation for this site; however, more storm simulations would be required to confirm this. We can conclude from this study that even a low resolution mesh discretization for the TELEMAC-2D model can provide a reasonable approximation to the bulk flow through reach-scale floodplain systems. At this scale the effect of detailed local flow processes are averaged out and it becomes relatively easy to calibrate the model to obtain a satisfactory result in terms of predicted downstream discharge. However, studies with catchment hydrology models (Binley *et al.*, 1991; Fawcett *et al.*, 1995) have shown that significantly different model internal

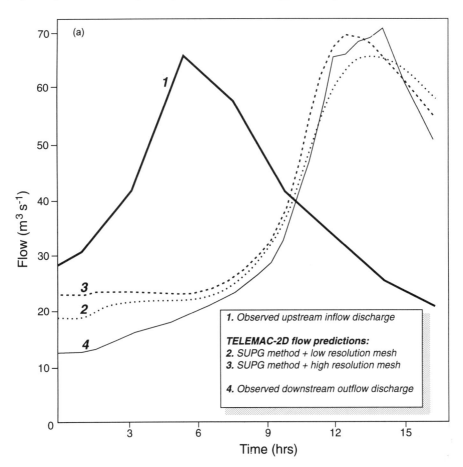

Figure 7.4 A comparison of downstream discharge predicted by the Streamline Upwind Galerkin method at high and low mesh discretization resolutions for a medium-sized flood event on (a) the River Culm and (b) the River Stour study reaches

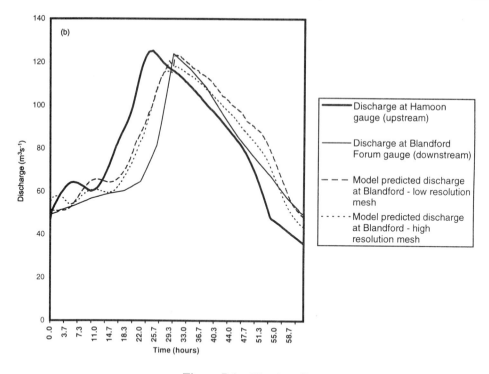

Figure 7.4 (*Continued*)

processes can produce identical bulk flow predictions at the domain outlet. Despite the fact that hydraulic models simulate a smaller number of processes at higher spatial and dimensional resolutions than typical catchment hydrology schemes, this remains a major problem for model validation. This is particularly so given the lack of reliable and comprehensive data sets for internal hydraulic model variables such as water depth, flow velocity and inundation extent. For this reason the impact of mesh resolution on predicted floodplain inundation extent was also examined. For the River Culm flood event an estimate of maximum flood inundation extent was available (Simm, 1993). This was obtained by amalgamating a number of different data sources including air and ground photographs taken during the flood event and post-event mapping of trash lines on the floodplain. This gave an estimated value for maximum inundation of 68% of the total floodplain area. Given the uncertainties involved in this procedure it was assumed that this value would be subject to errors of ±20% and would occur at a point equidistant in time between peak discharge at the upstream and downstream gauges ±1.5 hours. Despite these errors this is a unique record which could not be repeated for the River Stour reach where data availability is more typical of lowland river/floodplain systems. Ideally, a sequence of synoptic flood inundation extent maps would be required for a comprehensive model validation but at present no such data set exists. Floodplain inundation extent through time is given for the two river reaches in Figures 7.5a and 7.5b. In the case of the River Culm the high resolution mesh predicts a greater range of inundation states than the low resolution mesh, albeit with similar timing. All simulations conducted in the River Culm study are within or close to the upper portion of the "observed" inundation extent error band. For the River Stour model response to increasing resolution was exactly opposite to that noted for the Culm. Here the high resolution mesh produces a response in terms of floodplain inundation extent that is far more damped, with peak inundation being some 12% lower than that for the low resolution mesh in spite of having a higher peak discharge. Although no corresponding "observed" inundation extent data set was available for the River Stour, there appeared to be clear advantages to the use of a more highly refined topographic parameterization. In particular, the River Stour models appeared able to simulate significantly more exposed floodplain states, with simulations commencing with water contained in the channel only and, after substantial inundation, returning to this position by the end of the model run. This would seem to be a far more realistic representation of floodplain inundation phenomena. Inundation extent patterns during the flood event are also provided for the low resolution mesh from each reach in Figures 7.6a and 7.6b. These confirm the above statement showing that the River Stour simulation gives a more

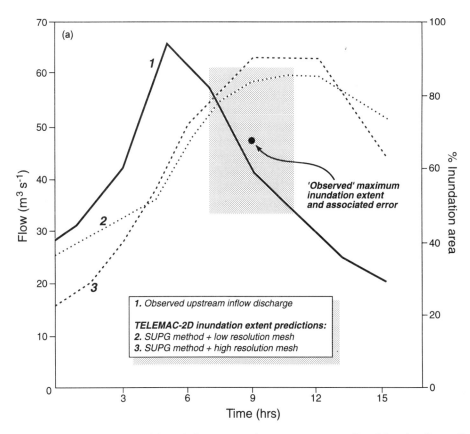

Figure 7.5 A comparison of inundation extent time sequences predicted by the Streamline Upwind Petrov Galerkin method at high and low mesh discretization resolutions for a medium-sized flood event on (a) the River Culm and (b) the River Stour study reaches. Note that for the River Culm simulation an inundation extent observation was available. This has been assumed to have a measurement error of ±20% in magnitude and ±1.5 hours in time

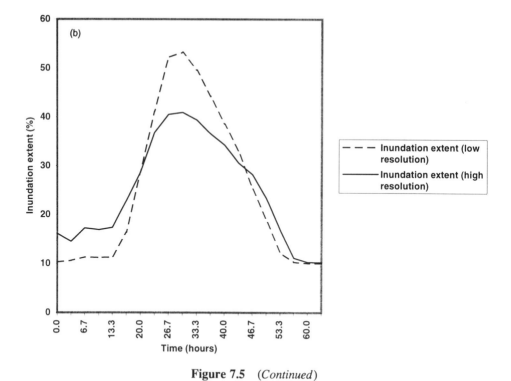

Figure 7.5 (*Continued*)

highly differentiated pattern of predicted inundation extent in both space and time that is more consistent with known process behaviour.

Mesh resolution can thus be seen to exert a major influence over the results obtained from a particular model application. A low resolution mesh can provide a reasonable approximation to the bulk flow characteristics of a reach and is a useful tool for model calibration and development studies. However, in terms of simulating spatially heterogeneous flow fields choice of mesh resolution becomes increasingly important. Mesh resolution affects model results in three main ways. Firstly, an insufficient density of computational points relative to gradients of computed variables can lead to an inaccurate numerical solution that in turn leads to a coarse representation of the domain under consideration. Secondly, mesh resolution can act as a filter on topographic information (see Figure 7.7) in situations where the nodal density is less than that of the terrain information database used to parameterize the model (cf. Band and Moore, 1995; Bates *et al.*, 1996). Lastly, there is an important interaction between process scale and mesh resolution that has, to date, not been adequately investigated. The process representation of a particular model is effectively limited to a particular range of scales beyond which the underlying assumptions implicit in the modelling procedure may not hold. For example, if we consider two dimensional fluid and sediment transport modelling, at a fine spatial scale the flow may exhibit significant vertical accelerations that are not represented in depth-averaged calculations. In such situations the scheme used to account for turbulent velocity fluctuations may also

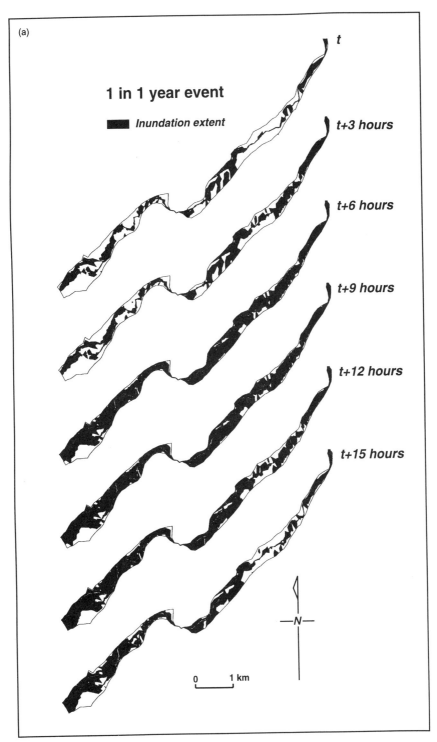

(a)

1 in 1 year event

■ *Inundation extent*

t

t+3 hours

t+6 hours

t+9 hours

t+12 hours

t+15 hours

—N—

0   1 km

Figure 7.6 Inundation extent time series predicted by the TELEMAC-2D model for a medium-sized flood event on (a) the River Culm low resolution finite element mesh and (b) the River Stour low resolution finite element mesh using a Streamline Upwind Petrov Galerkin finite element technique

(b)

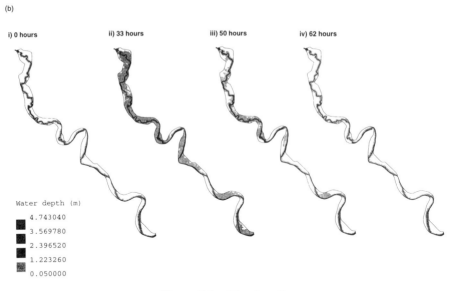

Figure 7.6   (*Continued*)

become invalid if properties of the turbulence field at this scale are not adequately represented. Much further research into mesh resolution effects is therefore reqiored to fully explore the above issues and to comment further on appropriate mesh resolutions for particular applications.

7.2.2   Comparison of numerical solution algorithms

A further area in which we wish to ascertain model robustness and attempt to determine potential model improvements, is the development and testing of numerical algorithms for solution of the model controlling equations. This forms a key component of research into flood inundation prediction (see, for example, Molinaro and Natale, 1994) and has been attempted for the River Culm study reach described in Section 7.2.1. Here three alternative numerical solution algorithms in two separate two-dimensional finite element models were tested for the flood event shown in Figure 7.4a. These were an SUPG solver in the TELEMAC-2D code detailed above, an alternative hybrid numerical solution scheme available in TELEMAC-2D and a standard Galerkin weighted residual solver implemented in the RMA-2 two-dimensional finite element model (King and Norton, 1978). Using the TELEMAC-2D hybrid method the SUPG scheme employed for the advection of the water depth, h, in the continuity equation is replaced by an algorithm consisting of a combination of the Method of Characteristics and weighted differences. The RMA-2 code solves an identical set of controlling equations to TELEMAC-2D and similarly uses the Boussinesq eddy-viscosity concept to represent turbulent flows (see Bates *et al.*, 1995) but differs in the approach chosen to solve the controlling equations. Here an implementation of the Galerkin weighted residual technique for a continuum of triangular or quadrilateral elements is used. The discretization is linear for water depth and quadratic for flow velocity giving six or

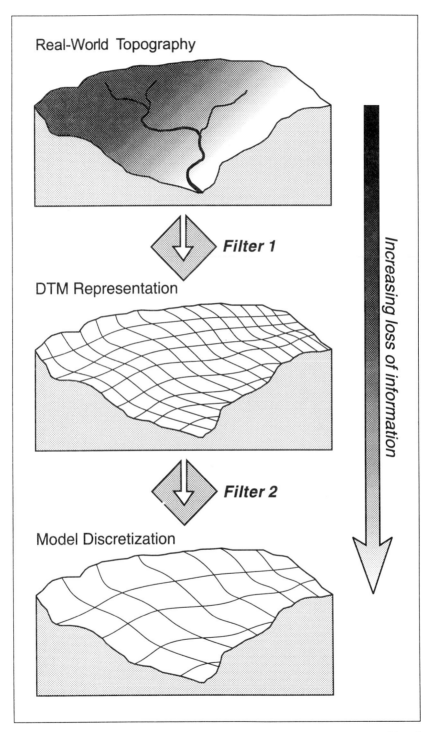

Figure 7.7   Schematic diagram showing the loss of topographic information resulting from the filtering effect of DTM and model discretization construction

eight nodes per triangular or quadrilateral element respectively. Due to the extreme non-linearity of the controlling equations, the numerical integration for the Galerkin procedure is performed iteratively using a Newton–Raphson type solver. Unlike TELEMAC-2D, which employs an element-by-element method, RMA-2 fully assembles the matrices of the linear system at each iteration. As these matrices are both large and sparse this can add significantly to the computational time and, as a consequence, the time-step used for all RMA-2 simulations reported here was 1800 s. Although this results in Courant numbers for the RMA-2 simulations in excess of 500, stability was maintained as the scheme is also fully implicit.

Simulations were attempted with each solver at both high and low mesh resolutions, giving a total of six model runs in total. For the RMA-2 simulations a number of triangular elements were amalgamated into quadrilaterals to improve accuracy and reduce storage requirements. Nodal coordinates were not changed in any way and the impact on model results of this change to the discretization was therefore assumed to be

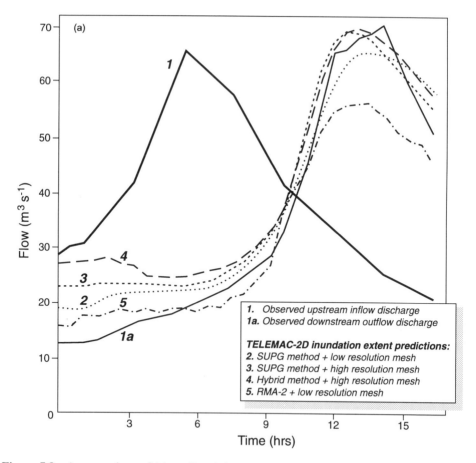

Figure 7.8 A comparison of (a) predicted downstream discharge and (b) inundation extent for the River Culm study reach obtained using a variety of numerical techniques and mesh resolutions

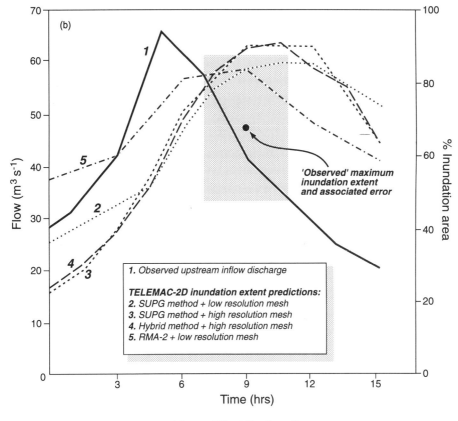

Figure 7.8 (*Continued*)

minimal. Of the calculations attempted, two simulations failed to produce any result. In the case of the RMA-2 simulation on the high resolution mesh, computer matrix storage requirements were too great for the computation to be attempted on a high specification workstation even when an accepted frontwidth reduction algorithm (King, 1970) was used. Solution of high resolution fluid flow problems would thus appear to be best performed with an element-by-element code where matrices are stored in their elementary form. TELEMAC-2D simulations employing the hybrid numerical scheme on the low resolution mesh displayed a number of problems including poor mass conservation properties, irregular velocity vectors and negative depths. The method was therefore rejected as an appropriate technique for use with such highly distorted meshings. Results from the remaining simulations are therefore shown for predicted downstream discharge and inundation extent in Figures 7.8a and 7.8b respectively.

Little difference in simulated discharge or inundation extent is shown between TELEMAC-2D calculations made with the SUPG and hybrid numerical solution schemes. The SUPG scheme appears to attain a more convincing initial steady state, but once the rising limb of the hydrograph is reached the hybrid numerical technique produces nearly identical discharge predictions. In terms of computation time, the SUPG method is more efficient (see Table 7.1), giving an approximate 28% reduction

Table 7.1 Computation times and efficiencies for simulations of a 1 in 1 year recurrence interval flood event conducted on a HP9000/735 workstation using a variety of numerical techniques and mesh resolutions

Mesh/Solver	No. of nodes	Computation time	Efficiency (Time per 1000 nodes per time-step)
Low resolution/RMA-2	1200	265 minutes	7.361 minutes
Low resolution/SUPG	1200	320 minutes	0.593 s
High resolution/SUPG	5600	2188 minutes	0.868 s
High resolution/hybrid	5600	2930 minutes	1.162 s

in computational requirements for this flow problem over the alternative hybrid scheme. Over the entire simulation the SUPG scheme also gave an approximate 20% reduction in the relative error on mass calculated by the model. The Galerkin weighted residual technique employed by RMA-2 correctly predicts the timing but not the magnitude of the observed discharge peak. Fully implicit Galerkin techniques of the type employed by RMA-2 have been shown by Fourier analysis to have an undesirable damping effect on the solution by reducing the amplitude of simulated waves. It may be that this is a cause of the underprediction of the flood peak by RMA-2 whereby the damping effect would not allow the model to capture the full dynamic range of the hydrograph. Further testing would be needed to determine this conclusively. Despite this, all simulations show a realistic initial decline in discharge magnitude. RMA-2 and TELEMAC-2D similarly estimate maximum inundation extent but differ in both the timing and range predicted. This may partially be a consequence of not controlling for roughness parameterization between the two schemes. In addition, both models commence the simulation with a partially wet floodplain; this implies either that the flow field along the reach prior to the flood event was not a steady state or that the channel topographic resolution was not sufficient to precisely identify bankful discharge in particular locations. Previous studies with TELEMAC-2D (Hervouet and Janin, 1994) and the River Stour simulations reported in Section 7.2.1 have shown that simulations can be developed with a completely dry floodplain and continued until the flood has receded. With RMA-2, however, model instability was noted for floodplain states that were predominately dry, particularly under conditions of flood wave recession.

Research into the impact of the numerical solution on model results has been shown to be an essential step in the further development of models of river channel/floodplain flow. A significant finding has been that research into numerical algorithms must take into account a possibly complex interaction with mesh resolution. Moreover, the study presented has shown that although the use of different numerical solutions may result in only a small improvement in the model predictions, for example between the SUPG and hybrid numerical solution schemes, improvements to the computational efficiency or mass conservation properties may be significant.

7.2.3   Potential application complexity and scale

The above studies have demonstrated the applicability of two dimensional finite element hydraulic models to river channel/floodplain problems over reaches of approximately 10–12 km and using up to 12 000 finite elements. Longer reaches have been simulated, albeit using a smaller number of elements. An example of such a study is the application by Gee *et al.* (1990) of the RMA-2 model to a 24-km reach of the River Fulda in Germany. Given likely developments in available computer power, considerable potential exists to increase the dimensionality, complexity and scale of model applications This does, however, raise a number of research issues which should be explored prior to such developments.

In terms of moving from two to three-dimensional simulations a major unresolved problem is the development of an ability to model the covering and uncovering of low lying floodplain areas. Hervouet and Van Haren (Chapter 6, this volume) present a three-dimensional code based on the TELEMAC-2D scheme. Here a two dimensional grid is specified that follows the bottom topography. Between this and the water surface the model linearly interpolates a user-specified number of prismatic finite elements. This scheme is an accurate representation of the physical system and is effective for steady-state computations or situations where the whole domain is inundated. For dynamic simulations involving floodplains a problem occurs as the water depth, h, at a particular node tends to zero. When $h = 0$, the linearly interpolated prismatic elements are destroyed and solution instability results. The solution here may ultimately be a full three-dimensional solution with a deformable finite element mesh. This does, however, remain a substantial research problem.

Increasing model complexity and scale merely generates a practical problem of data provision rather than a theoretical constraint. Figures 7.9a and 7.9b show two examples of complex long-reach finite element meshes generated for use with the TELEMAC-2D model for the River Severn, UK, and the Missouri River, USA. In the case of the River Severn, the finite element domain consists of a 40-km reach around the town of Shrewsbury between the gauging stations at Montford Bridge and Buildwas and is comprised of 12 000 elements. At present the Missouri River mesh is restricted to an in-channel area covering approximately 60 km of river downstream of Gavins Point Dam, Nebraska, and also consists of 12 000 elements. Both finite element models represent an initial attempt to model such problems and computational power now exists to enable mesh refinement through the addition of further elements. Studies suggest that the current potential maximum number of elements that can be simulated on a powerful workstation is perhaps 30 000, thus giving considerable scope for improvement (see, for example, Hervouet and Janin, 1994; Hervouet and Van Haren, Chapter 6, this volume). In effect the constraints on model application have shifted, from being driven by available computational resources, to having sufficient data to parameterize such models at a level commensurate with their resolution. We have seen with the River Culm finite element model that use of a high resolution mesh gives only a marginal improvement in model capability due to lack of an appropriate data set, particularly for topographic information. With increasing resolution such impacts can only become more significant.

Application of hydraulic models to actual flood flow problems can thus be seen to generate a number of issues requiring future research. These include the development of analysis techniques which can rigorously control for mesh resolution or topography impacts, compilation of a comprehensive model validation data set, the development of

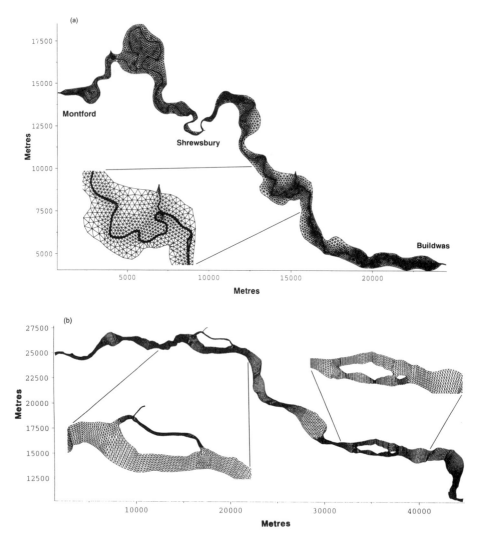

Figure 7.9 (a) TELEMAC-2D mesh consisting of approximately 13 000 finite elements constructed for a 40 km channel and floodplain reach of the River Severn, Shropshire, UK. (b) TELEMAC-2D mesh consisting of approximately 13 000 finite elements constructed for a 60 km channel-only reach of the Missouri River between Gavins Point Dam and Yankton river gauging station

numerical algorithms to simulate dynamic flood inundation problems in three dimensions, and model process developments (see Younis, Chapter 9, this volume). In this chapter we explore two such areas. Firstly, we attempt to conduct a number of standard sensitivity analysis procedures while controlling for mesh and topographic resolution (Section 7.3). Secondly, we attempt to further develop two-dimensional finite element hydraulic models for application to reach-scale river channel/floodplain problems by relaxing a number of previously impermeable boundary conditions in such schemes (Section 7.4).

7.3 ANALYSIS OF FLOOD INUNDATION MODELS USING NUMERICAL EXPERIMENTS

Standard methods for conducting numerical experiments with distributed models typically use a sensitivity analysis or uncertainty estimation methodology to investigate model response to variety of possible conditions. Analysis is conducted for real world applications as a means of determining robust calibration procedures (Romanowicz *et al.*, Chapter 10, this volume), investigating possible errors in the model structure (Binley *et al.*, 1991; Lane *et al.*, 1994) or model response to small variations in initial or boundary conditions (Bates and Anderson, 1996). Recognition of the impact of mesh resolution on model predictions (Bates *et al.*, 1995) and its complex relationship with model topographic parameterization (Bates *et al.*, 1996) have led to a need to conduct numerical experiments on flood inundation models in an environment where we can control for such effects. This can only be achieved using hydraulic models for hypothetical reaches where the modeller has precise control in these two areas. Such tests cannot replace application sensitivity analysis, but rather should run in parallel to enable maximum information concerning model behaviour to be gained and specific questions to be answered in greater detail. In this section we attempt two types of analysis using controlled reaches. Firstly, we apply a standard two-parameter sensitivity analysis using Monte Carlo simulation techniques to subsample the parameter space (Section 7.3.1). Secondly, we attempt a theoretical investigation of mesh resolution effects while controlling rigorously for topographic information content (Section 7.3.2).

7.3.1 Monte Carlo simulation methods

As the influence of a specific parameter can depend on the values of other parameters, the traditional single-parameter perturbation approach to sensitivity analysis may not fully capture all aspects of model behaviour. Ideally, therefore, more parameters should be varied to create a more accurate picture of model behaviour. Usually, however, only two parameters are varied simultaneously to allow clear presentation of the results, in the form of a response surface (Harlin and Kung, 1992), which, if the two parameters are the most influential, should represent almost the full range of model responses.

In conducting such an analysis the user has two basic decisions to make; which; parameters to vary and which output variables to inspect. The latter is a very important decision that can significantly alter the conclusions from the sensitivity analysis. The outflow hydrograph compared to an observed hydrograph is the classic measure for hydrologic models (Fawcett *et al.*, 1995) and is therefore shown here. For a distributed hydraulic model this is generally not sufficient as the main results are often internal values such as water depths over the domain or velocities. However, a simple measure of overall model performance must be applied to create a response surface. Such a measure which reflects the internal model performance is the maximum inundation extent, which is also examined in this study. Lane *et al.* (1994) illustrate the use of a distributed sensitivity analysis with single-parameter perturbations. Such distributed analysis could be adapted to two-parameter sensitivity to show a fuller range of model performance over the model domain.

To create a response surface, the parameter ranges must first be defined. Once this parameter space is specified parameter combinations must be chosen and the model run using these parameter sets. For a response surface fairly dense sampling must be employed to ensure all variation is shown. This could be done on a regular two-parameter perturbation scheme or by using a Monte Carlo methodology to randomly generate parameter combinations, a theme treated in much more detail in Chapter 10 (Romanowicz *et al.*, this volume). In this section the latter approach is taken as it is a well-tried technique that with some further refinement of the parameter distributions can be further applied to create uncertainty estimations (Binley *et al.*, 1991, Harlin and Kung, 1992; Romanowicz *et al.*, Chapter 10, this volume).

With TELEMAC-2D and other similar hydraulic models the two most dominant physical parameters have been shown to be the bed friction and the turbulent viscosity term, which determines turbulence effects (Bates, 1992). These are therefore the two parameters examined in this investigation. (Model structural parameters can also be

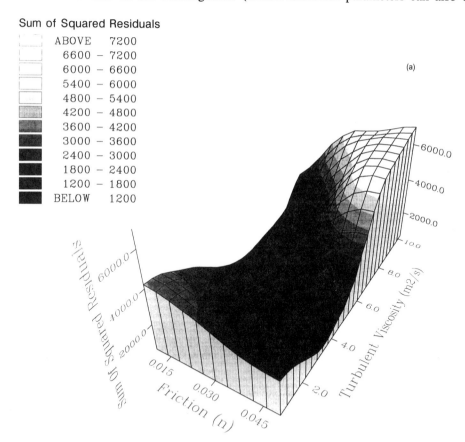

Sum of Squared Residuals

ABOVE	7200
6600 –	7200
6000 –	6600
5400 –	6000
4800 –	5400
4200 –	4800
3600 –	4200
3000 –	3600
2400 –	3000
1800 –	2400
1200 –	1800
BELOW	1200

(a)

Figure 7.10 Response surfaces from a Monte Carlo sensitivity analysis carried out on a small hypothetical reach. The two variable parameters were bed friction (shown as the channel component) and turbulent viscosity. Results were interpreted in terms of: (a) the sum of squared residuals (Green and Stephenson, 1986); (b) the model efficiency (Green and Stephenson, 1986). The efficiency approaches 1.0 as the compared hydrographs converge

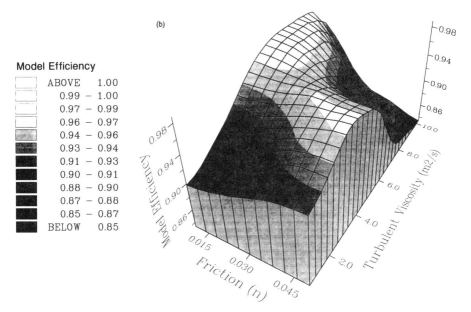

Figure 7.10   (*Continued*)

investigated in this manner. For example, although not considered in this investigation we will see in Section 7.3.2 that spatial resolution may be an important parameter to investigate using a Monte Carlo framework.)

This two-parameter sensitivity analysis was carried out on a model of a 1-km long hypothetical reach. This was both to control for mesh resolution and topography and allow a rigorous assessment of the technique prior to applying it to a real river reach. The ranges for the two parameters were deliberately taken to be fairly broad in order to asses the model behaviour under a wider range of conditions than would typically be used. The friction was taken in the range of Manning's n from 0.01 to 0.05 for the channel and a floodplain Manning coefficient of $3m + 0.01$ where m is the channel value, a range encompassing a large variety of channel conditions (Chow, 1959). The turbulent viscosity was taken in the range of 1.0 to 10.0 m^2 s^{-1}, this range being one that on initial investigations encompassed a wide range of results while maintaining model stability and a realistic output.

The response surfaces shown in Figures 7.10a and 7.10b are created using two different methods to compare the individual run hydrograph to a mean hydrograph. This is not intended to give an indication of the model's accuracy, which would require a measured hydrograph, but to look at the variation in the outflow hydrograph created by changing the parameter values. The two methods used are the sum of squared residuals, a commonly used comparative tool, and model efficiency, a simple dimensionless measure which increases towards unity as the fit of the hydrograph progressively improves (both measures are detailed in Green and Stephenson, 1986). From the sum of squared residuals response surface (Figure 7.10a), the parameter sets most closely comparable to the mean hydrograph are those with the lowest values. These are found in a broad, diagonal band across the full range of friction values and over about half the range of turbulent viscosity values. The model efficiency response surface (Figure 7.10b) shows a very similar pattern, this time the most closely comparable runs have an efficiency rating approaching 1.0. The most dramatic loss of performance comes as the turbulent viscosity and friction increase; in effect this creates a delayed and slightly attenuated outflow hydrograph which is susceptible to large errors at such values. In this case both parameters act to retard the hydrograph. In the opposite corner, with low friction and flow turbulent viscosity, both parameters act to facilitate a rapid passage of the hydrograph through the reach but the response change is much less marked. In the corners where the two parameters are operating against each other, the hydrograph change is, as would be expected, of smaller magnitude. These results show that simply by using a hydrograph comparison as a measure of performance, it is possible to produce results of equal quality across a very broad range of parameter values, some bordering the non-physical. Such a result shows there is a lot of parameter set equivalence which is a major concern when calibrating the model. Moreover, it confirms the conclusion drawn in Section 7.2.1 that it is relatively simple to obtain a calibration that produces a satisfactory simulation of the downstream reach hydrograph.

Another measure of model performance examined over the same parameter ranges and model reach is the peak percentage inundation extent (Figure 7.11). This shows clearly that for these simulations the inundation extent is principally governed by the turbulent viscosity and by the friction as a secondary factor, particularly when the turbulent viscosity is low. The results at extreme parameter combinations are not unexpected, low inundation when friction and turbulent viscosity are low and high inundation when

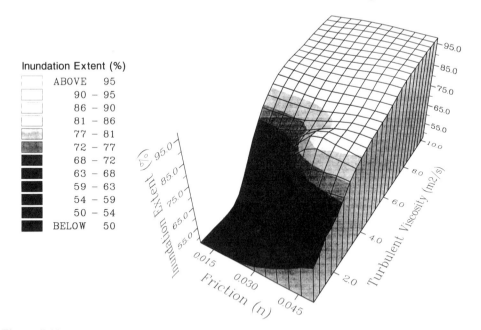

Figure 7.11   Peak inundation extent response surface from a Monte Carlo sensitivity analysis carried out on a small hypothetical reach. The two variable parameters were bed friction (shown as the channel component) and turbulent viscosity. The peak inundation values for each simulation were interpolated to form a response surface

friction and turbulent viscosity are high. With high turbulent viscosity and low friction the inundation extent is still very high showing velocity diffiusivity is clearly the dominant parameter in this region. At low turbulent viscosity and high friction the inundation extent is at a medium value showing both parameter are of influence in this area.

This result contrasts with the work of Bates *et al.* (1992) who found the bed friction was the single dominant factor controlling the model outputs. The most probable cause for this is the scale of the two studies. Bates *et al.* (1992) based their study on a the 11-km River Culm reach discussed in Section 7.2 utilizing the low resolution finite element mesh shown in Figure 7.1a. In this application computational constraints necessitated the use of large elements which are likely to show sensitivity primarily to bulk flow parameters such as friction. The simulation reported here, being conducted using much smaller scale and more finely resolved elements, results in the impact of turbulence effects becoming much more pronounced especially around the floodplain/channel interface at higher mesh densities. In examination of the hydrograph response surfaces and transposing the region of best fit onto the inundation extent response surface shows that while maintaining an equally acceptable hydrograph the maximum inundation extent can vary between 80 and 93% of the total floodplain area. Such a difference shows very clearly that a complex hydraulic model cannot be fully calibrated on a hydrograph comparison alone.

Thus a two-parameter sensitivity analysis can give an extra dimension to the study of the model behaviour, particularly as one can control for mesh resolution and topography impacts. It is still rather simplistic, only looking at two factors from a larger possible range. Model structural and physical parameters can both be treated in this manner, highlighting the relative importance of each and any interactions existing between them. It must, however, be remembered that there is a tremendous degree of complexity in such a distributed model. Many of the arguments of Beven (1989) regarding problems with distributed hydrologic models are equally applicable to distributed hydraulic models in terms of parameterization, calibration and validation. No "optimum" parameter set is possible for such a complex system and even finding an optimum parameter set for a single event is almost impossible. Further research in this area should investigate such considerations for a model of a real reach and an observed hydrograph. These methods, once improved in terms of specifying more accurate parameter distributions and perhaps using more parameters if necessary, can ultimately be utilized in the formulation of uncertainty estimations.

This simple test case has highlighted three important results. Firstly, that the influence of one parameter can depend on the value of another, as seen in the inundation extent response surface, a fact that would not be seen using a single-parameter sensitivity analysis. Secondly, a complex hydraulic model cannot be calibrated fully using only outflow hydrograph comparison techniques. Lastly, we have once again demonstrated the influence of mesh resolution on model results. In the following Section (7.3.2) we attempt a more detailed exploration of such effects for a series of finite element discretizations of varying mesh density but identical topography.

7.3.2 Investigation of mesh resolution effects

The question of determining optimum mesh resolution in the application of a model such as TELEMAC-2D to a fluvial environment and the effect of changing spatial resolution on model output, whether considering bulk flow characteristics or inundation extent, is one that requires considerable analysis. Earlier in the chapter comparative results between a low resolution and high resolution application on the Rivers Culm and Stour were discussed, indicating how the relationship of spatial density of computational points to gradients of system variables may condition the accuracy of the numerical solution obtained. Also discussed was the fact that a greater spatial density of computational points reduces the filter on domain characteristics, such as friction or topography, allowing a more variable spatial representation of the domain to be considered. Lack of data and computational demands can place limits on the extent to which such questions can be addressed for real applications. The solution adopted here is to run hypothetical tests in parallel to real applications enabling specific questions to be investigated. The results presented here continue the analysis of the effect of resolution on the model's output using a hypothetical test case. Here we initially study bulk flow characteristics and extend the analysis to consider at-a-point scalar flow rates at identical x, y coordinates for a variety of mesh resolutions.

Although the domain considered was purely synthetic, both the domain and flood events applied were scaled on realistic events that have been used in past analyses. The domain consisted of a 20-m wide 2-m deep sinuous channel flowing down the centre of a hypothetical reach 2000 m long by 800 m wide with a downstream gradient of 0.005. In total seven finite element meshes were constructed, labelled 1 to 7, with mesh 1 having the lowest resolution (see Table 7.2).

The actual difference in scale is best demonstrated in Figure 7.12, where in the particularly sensitive region of the channel/flood plain interface, there are on average 14 elements in mesh 7 compared to one in mesh 1. The same TELEMAC-2D set up as used in Section 7.2 was applied, except for an alteration in the boundary conditions; an imposed flow rate at the upstream end of the reach was retained while at the downstream end of the mesh all variables were allowed to vary freely. A synthetic downstream boundary condition could have been constructed but it was felt that this would unnecessarily constrain the model behaviour: the results obtained would primarily reflect the assumed boundary conditions. As no field measurements existed, the results obtained from mesh 4 were used as a control. The hydrograph was scaled from a previous analysis taking into account the dimensions of the domain and consisted of a flood event lasting 20 hours in real time with a peak discharge of 40 m^{-3} s^{-1}, and a total volume of approximately 1 728 000 m^3.

Table 7.2 Characteristics of the finite element discretizations used in numerical experiments on mesh resolution dependencies for two-dimensional finite element simulations of river flood inundation

Mesh	Nodes	% of nodes in channel	Elements	Average element size (m^2)
1	888	36.25	1669	71.54
2	1199	35.45	2284	58.73
3	1982	40.26	3824	43.33
4	2858	38.80	5578	32.24
5	3746	33.45	7310	30.48
6	4652	36.86	9128	24.88
7	6064	31.53	11890	24.11

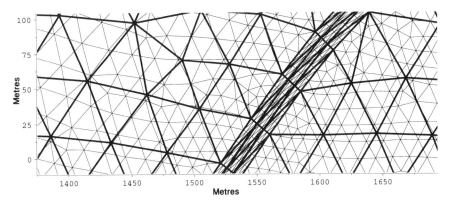

Figure 7.12 Variation in spatial resolution used in an investigation of mesh resolution effects on two-dimensional finite element model predictions for a hypothetical channel and floodplain domain

When bulk flow characteristics, such as the peak discharge, the volume of the outflow hydrograph and the mean flow are analysed, a pattern consistent with that identified for the Rivers Culm and Stour can be observed (see Figure 7.13). As the spatial resolution increases the peak flow increases from mesh 1 to 6, though decreases for mesh 7. There is an increase in the peak discharge from 23.6 $m^3 s^{-1}$ to 35.09 $m^3 s^{-1}$ (+48%) from mesh 1 to 6, and the peak discharge then decreases by 5.1% from mesh 6 to 7. A similar pattern with equivalent magnitudes is identified if other bulk flow variables are analysed. Although there appears to be wide variation between resolutions it is possible to put these results in perspective. From mesh 4 to mesh 7 there is minimal variation in peak discharge, with peak discharges increasing from 33.93 $m^3 s^{-1}$, to 34.08 $m^3 s^{-1}$, to 35.09 $m^3 s^{-1}$, then decreasing to 33.40 $m^3 s^{-1}$ for mesh 7, a range of only 5.1%. This indicates, as previously suggested, that there is a minimum mesh density which adequately computes a particular problem, and beyond which results may not significantly alter. However, increasing the resolution beyond the specified optimal resolution has no significant beneficial effect on our predictions of bulk flow characteristics. This is important when considering the computational cost of the simulation: for a 20-hour real-time simulation, the computation time is 26 hours on mesh 4 compared to 62.25 hours on mesh 7 when processed on a HP 9000/735 workstation.

The second area of interest is the within-domain results; the velocity measurements calculated at specific x, y coordinates within the domain. This type of analysis enables the distributed model to be used to its highest potential and is of particular importance if a second model, such as a sediment transport model, is to be initialized using hydrodynamic results. The results presented in Figure 7.14 concern the variation in scalar flow rate with spatial resolution and are taken at a single nodal point within the domain. This point has identical x, y coordinates in all seven meshes and is located within the channel bank region. The area of interest is located on the inside of the channel bend as this area is a particularly sensitive region when considering overbank flow in compound channels (cf. Knight and Shiono, Chapter 5, this volume).

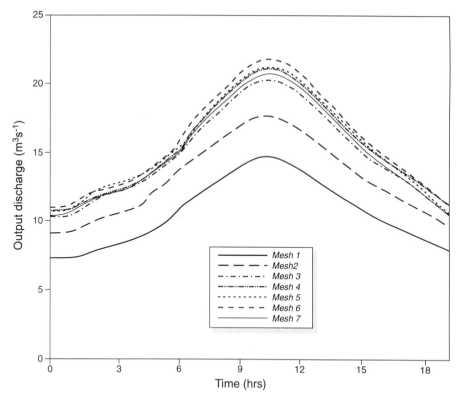

Figure 7.13 Downstream outflow hydrographs predicted at varying mesh resolutions by the TELEMAC-2D finite element model. Mesh resolutions 1–7 represent hypothetical finite element domains of dimensions 800×2000 m consisting of 1669 to 11 890 finite elements respectively (see Table 7.2). In these simulations mesh 4 (5578 elements) was arbitrarily taken as the control. Note the existence of a minimum element density necessary to obtain a robust solution

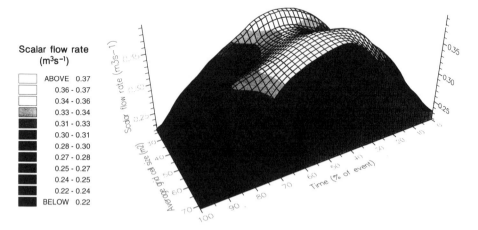

Figure 7.14 Variations in at-a-point scalar flow rate with changing mesh density during a flood event simulation using the TELEMAC-2D finite element model for the finite element discretizations detailed in Table 7.2

A reverse trend, but similar pattern, can be identified in the bulk flow characteristics if the at-a-point results are studied. The variation in scalar flow rate for the point shows that as the average element size decrease, the flow rate, throughout the flood simulation, decreases from mesh 1 to 6, then increases for the highest resolution mesh 7. At the time of peak discharge the greatest range in velocity is approximately 0.05 m s^{-1}. This is of significance if such models are used to identify spatial patterns for possible erosion or deposition zones. Again, there are cases where the spatial resolution is too low to obtain an acceptable finite element solution. Moreover, a similar pattern of minimal variation above mesh resolution 4 suggests the existence of an optimal mesh resolution. While it is not suggested that results for a single point within the domain apply equally elsewhere, Figure 7.14 succinctly demonstrates the potential of this analysis technique. It is likely that as we utilize a dynamic model solving a set of highly non-linear equations, further application of this analysis technique would reveal considerable complexity in the model response.

The above results have shown that there may be an optimum resolution for particular application purposes beyond which little improvement in predictive ability is obtained for this class and scale of model application. Although the increase in spatial resolution appears to improve the quality of the results, an optimum resolution for a particular scale needs to be identified. This relationship will obviously be complicated by spatial variability in domain characteristics such as topography, friction and turbulence. For applications where we are interested in simulating such spatial variability optimum mesh densities may be much higher than those used for bulk flow computations. This study has highlighted the fact that objective rules for domain discretization are currently lacking in hydraulic modelling despite a potentially significant impact of mesh density on model results, and the computational costs of model overspecification.

7.4 MODEL DEVELOPMENT

7.4.1 Background

Currently available hydraulic models of floodplain flow presume zero flux boundaries at all points except the upstream inflow and downstream outflow cross-sections. In this section we present a preliminary numerical approach for the simulation of water transfers in channel and near-channel areas during periods of river flood flow and floodplain inundation. Such regions represent one of the most hydrologically active parts of terrestrial catchments and encompass a variety of highly spatially variable processes with often steep gradients in process magnitude. In particular, floodplains represent a sink for fluvially transported sediment (Walling *et al.*, 1986, 1992, Chapter 12, this volume) and sediment-borne contaminants (Marron, 1989), as well being the origin of saturated soil conditions critical to flood generation and termed partial contributing areas (Anderson and Kneale, 1982). Floodplains can also act as regulators of large-scale nitrogen exchange between terrestrial and aquatic ecosystems during flood events by providing a buffer zone in which denitrification can take place (Brunet *et al.*, 1994;

Burt and Haycock, Chapter 14, this volume). The driving forces for the above processes are ultimately water transfers in channel (hydraulic) and near-channel (hillslope hydrology) areas. Currently the ability to fully model coupled hydraulic and hydrological processes during flood events does not exist, yet this is an essential prerequisite to a more complete understanding of many catchment processes.

Hydraulically, flood events in natural channels consist of a low amplitude flood wave passing through a river reach. At high flows this can lead to the extensive inundation of adjacent low-lying areas which act either as temporary storage locations or as an additional route for flow conveyance. Such flows across topographically complex terrain and incorporate a number of distinct mechanisms operating over a wide range of time and length scales (Bates and Anderson, 1993). These include turbulence (Rodi, 1980; Younis, Chapter 9, this volume), the development of shear layers (Sellin, 1964; Knight and Shiono, Chapter 5, this volume) and the transfer of water from channel to floodplain and its subsequent return (Ervine *et al.*, 1994; Sellin and Willetts, Chapter 8, this volume). A further set of highly spatially variable hydrological processes also operate in channel and near-channel environments during a flood event. Hillslopes adjacent to the floodplain contribute received precipitation to flood flows via overland and subsurface flow pathways. In terms of the subsurface component, flow is typically by Darcian water transfer through the soil matrix, although a number of other non-Darcian preferential flowpaths, such as soil macropores (Germann, 1990), have been observed in particular situations. A further interaction between channel flow and hillslope hydrology occurs via the development of soil saturation in areas adjacent to the channel. Such areas have been shown (Pearce *et al.*, 1986) to expand upslope into near-channel areas during flood events to create a "wedge" of saturated soil from which overland flow can more readily occur. The floodplain subsurface soil moisture store also receives water from flood flows and rainfall by surface infiltration which then drains under gravity to give a wetting front passing downwards through the soil profile. Lateral and downslope transfers of floodplain subsurface water may also occur, although these will be relatively slow processes due to low slope angles and possible discontinuities in the soil structure, such as infilled historic river channels (cf. Haycock and Burt, 1993). To give an indication of the impact of such hydrological sources and sinks on flood propagation we take the example of the floodplain subsurface moisture store. Assuming typical flow and infiltration values for a temperate lowland floodplain, the subsurface floodplain storage is approximately 6.4% of the total flow volume for a medium-sized flood event. (This calculation is based on a floodplain of 5 km^2 inundated for 18 hours, a flood volume of $0.5 \times 10^6 \text{ m}^3$, an infiltration rate of $0.1 \times 10^{-6} \text{ m s}^{-1}$, floodplain moisture at 90% saturation and saturation occurring at a water content of 0.3 m^3 per m^3 of soil.) Even at the assumed infiltration rate, available storage capacity is easily filled during the flood, resulting in a potentially significant impact on the total channel/floodplain flow. Furthermore, this calculation is likely to be an underestimate due to the assumption of a planar floodplain and no within-reach storage of flood waters. There is thus a need to couple hillslope (overland and subsurface flow) sources and floodplain flow sinks (flood inundation infiltration) to existing hydraulic models of floodplain flow (Bates and Anderson, 1993) in order to relax the zero flux boundaries currently established in hydraulic modelling.

7.4.2   Boundary condition relaxation

In order to implement zero flux boundary relaxation, linkages have been made from TELEMAC-2D to two separate hydrology components: a pseudo-three-dimensional finite difference hillslope hydrology model to represent slopes adjacent to the floodplain and a one-dimensional finite difference soil water model to represent infiltration into floodplain soils. Such a formulation limits flow beneath the floodplain to vertical movements only, thus ignoring any lateral displacement, and necessitates a number of assumptions to be made regarding water transfers at the floodplain perimeter/side slope boundary.

A pseudo-three-dimensional finite difference *hillslope hydrology model* has been developed based on a two-dimensional scheme in order to represent convergent and divergent flow processes in hillslopes adjacent to the floodplain. The basis of this model is Darcy's law. While there is no universally valid method for the *a priori* prediction of K based upon the soil suction, certain methods are available based on Poiseuille's law. In particular the Millington–Quirk method (1959), codified by Jackson (1972), has substantial acceptance (Hillel, 1980) and is thus used here.

Rainfall, infiltration, evaporation and detention storage processes at the soil surface are also represented in the model, with overland flow being initiated when infiltration and surface detention capacities have been exceeded. The modelling scheme also allows subdivision of the slope profile into three hydrologically distinct soil units and the whole equation system is solved for a slope cross-section discretized into a two-dimensional grid using a forward difference, explicit, block-centred finite difference scheme. Anderson and Lloyd (1991) describe this model in full. This basic model was further enhanced for the present study by assigning a variable width to each column of two-dimensional grid cells. This enabled slopes to be represented as a discretized "wedge" that may either expand or contract downslope, terminating at the floodplain perimeter (see Figure 7.15). In this way the impact of convex or concave hillslope morphologies can be spatially discretized to represent the topography and spatial variations in hillslope hydrology adjacent to the hydraulic model. To achieve linkage of the hydraulic and hillslope hydrological schemes, TELEMAC-2D has been modified to include a relaxed perimeter boundary condition through which fluxes may occur. In the original hydraulic model the flux at the perimeter boundary is given as:

$$\int_\Gamma \Psi . h . \vec{u} . \vec{n} . \mathrm{d}^\Gamma = 0 \qquad (7.2)$$

where Γ is the perimeter boundary, n is a unitary vector normal to the domain boundary pointing outwards, Ψ is a model basis function and h is the water depth.

To obtain a boundary flux we set this term to equal some real quantity at each perimeter node. This then becomes a source term in the hydraulic model continuity equation assembled for each nodal point and the resulting system is solved in the normal manner. The hillslope outflow surface and subsurface hydrographs are used to generate values of this spatially variable flux term at the hydraulic model perimeter boundary. As an initial step, both surface and subsurface hillslope outflow hydrographs were summed at each time-step and a value of the flux, given in $m^3 s^{-1}$ per metre of hillslope/floodplain boundary, added to the floodplain flow. To improve this preliminary coupling it was assumed that a boundary flux would only occur at a

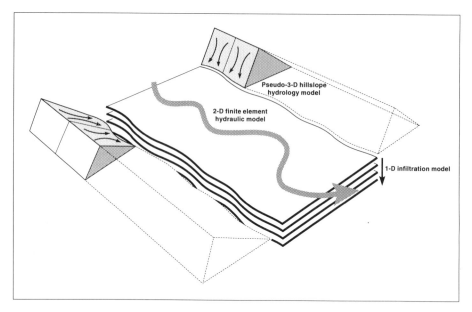

Figure 7.15 Schematic representation of a coupled hydraulic/hydrology model for simulation of the water transfer environment in channel and near-channel areas. Note that the hillslope wedges can be convex or concave in plan as required to represent hillslope topography and resultant soil water flowpaths (see text for full explanation)

particular perimeter node if the water depth at that point was greater than zero. At other times hillslope inflows were assumed to infiltrate at the base of the slope and move out of the system into groundwater stores. In order to avoid a sudden, and potentially numerically unstable, transition between the presence or absence of inflows, a scaling function was implemented that varied from 0 to 1 as the water depth at the floodplain perimeter varied from 0 to some small depth, σ. For simulations reported in this chapter a value for σ of 0.1 m was chosen. In this depth range, therefore, the inflow volume was taken to be the product of the actual inflow value and the scaling function. This representation has a certain physical basis and provides an initial mechanism for the relaxation of the zero flux boundaries, hitherto universally assumed for such two-dimensional hydraulic models.

A one-dimensional *floodplain/channel sink model* (Anderson and Howes, 1985) facilitates relaxation of the zero flux boundary, present at the interface between the flood flow and the subsurface soil moisture store in all current river floodplain hydraulic schemes. The finite difference model represents infiltration, evaporation and surface detention in an identical manner to the hillslope model above. Water movement in the saturated and unsaturated zones is restricted to one-dimensional vertical flow. This scheme is solved using a forward difference, explicit, block-centred finite difference scheme for a soil column beneath each computational node on the hydraulic model finite element mesh. Due to the ability of the hydraulic scheme to predict dry areas within the computational domain, the top boundary condition for the soil water model can be provided by either hydraulic or atmospheric inputs. If the node is dry, evaporation or rainfall drive the model and are used to update the soil moisture status. If the node is inundated, however, the infiltration rate is no longer supply limited and is set to its maximum. Calculation of the maximum infiltration rate for particular soil conditions gives the nodal vertical water flux from the hydraulic to the soil water model. This forms a sink term in the hydraulic model continuity equation and can also be used to update surface flow calculations at each time-step. Hitherto zero flux boundary conditions have thus been relaxed by the hillslope source and floodplain/ channel sink models described above.

7.4.3   Model testing and analysis

In order to begin investigating whether boundary condition relaxation can be achieved in a numerically stable manner, the scheme outlined in Section 7.4.2 has been tested on a hypothetical catchment using typical parameter values. The hydraulic model component for this test case is the River Culm low resolution mesh shown in Figure 7.1a with model boundary condition and parameterization data taken from 1 in 1 year recurrence interval event simulation shown in Figure 7.4a. To implement the hillslope hydrology scheme, a model of a single planar hillslope segment was simulated using data representative of catchment and storm conditions. Hillslope soils were assigned a hydraulic conductivity of 5.0×10^{-6} m s^{-1} with initial saturation in unsaturated cells at 70% of maximum. The computed downslope flux from this simulation was then applied as temporally varied but spatially uniform inflow along the hydraulic model perimeter. Hydrology inflows were omitted from inflow and outflow points to the reach to reduce the possibility of instability generation due to ill-posed boundary conditions. For the floodplain subsurface model component soils were assumed to be a silty loam with initial saturation at 95% of maximum and a hydraulic conductivity of 5.0×10^{-6} m s^{-1}.

Three simulations were then conducted comprising:

1. Simulation of channel flow with infiltration to the floodplain subsurface moisture store.
2. Simulation of channel flow with contributed inflows from hillslopes adjacent to the floodplain perimeter.
3. A fully coupled hydraulic/hydrology simulation linking both developed hydrology subcomponents to the channel flow model.

This latter scenario gives a complete representation of boundary condition relaxation in both the floodplain perimeter and floodplain surface contexts.

Results for the three simulations are given in Figure 7.16. It can be seen that while the infiltration scheme may not affect the timing of the predicted flood hydrograph it can have a significant impact on the flow volume and the magnitude of the flood peak, even when using conservative values for catchment hydrology parameters. In this case, addition of the infiltration model to the control simulation results in a 6.4% reduction in predicted peak flow and a predicted flux from channel flow to the floodplain subsurface soil moisture store of 149 600 m^3 of water (7.5% of the total control flood volume). This is consistent with the example calculation given in the introduction to this chapter. Addition of hillslope inflows to the control simulation also impacted on the peak flow, the flow volume and led to a slight retardation of the timing of the predicted

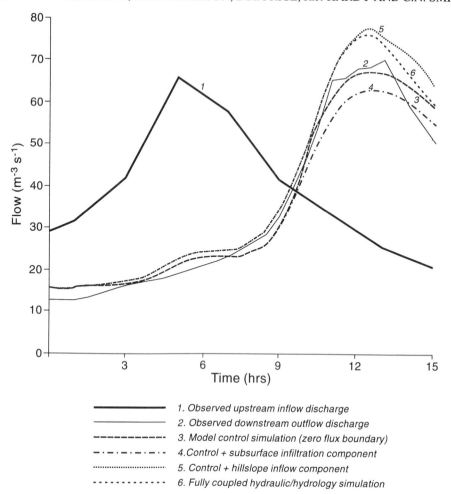

Figure 7.16   Evaluation of the impact of boundary condition relaxation and hydrology process inclusion on a two-dimensional finite element hydraulic model. Here the 1 in 1 year recurrence interval flood event showin in Figure 7.4a is simulated for the low resolution River Culm finite element domain shown in Figure 7.1a. Shown are conditions of (3) a model control simulation employing zero flux boundaries, (4) a model control simulation plus subsurface infiltration, (5) a model control simulation plus hillslope inflows and (6) a model control simulation plus infiltration and subsurface inflows

outflow hydrograph peak. Peak flow was increased by 15.3% and flow volume by 110 000 m^3 (5.5% of total control flood volume). The fully coupled simulation using all three components showed no difference from the hillslope inflow calculation on the rising limb of the hydrograph, but a significant difference in the peak and recession phases. Significant differences from the control simulation are also shown if we consider model predictions of flood inundation extent. Figure 7.17 shows the variation in inundation extent produced by simulations (1) and (2) above at a point approximately six hours into the flood event on the rising limb of the hydrograph. Both the infiltration component and the fully coupled model produce large changes in the predicted inundation field and thus other hydraulic and sediment transport variables. It is clear from this study that significant hydrological processes are subsumed within the hydraulic model calibration procedure. In the example presented, catchment hydrology processes and their linkage to the floodplain domain have been radically simplified. Neither has an attempt been made to calibrate either hydrology model component. Despite this, all three simulations reported in this chapter lie broadly within the error band of the observed downstream outflow data and within the error band assigned to observations of flood inundation extent.

To give an indication of the significance of the flow field changes (generated by the above boundary condition relaxation), the results of simulations conducted with the hydraulic component to test the impact of choice of numerical solver (Figure 7.8a) on model predictions can be compared to the data presented in Figure 7.16. For the high resolution River Culm finite element mesh the change to model predictions is marginal if the Streamline Upwind Petrov Galerkin solution for the advection of h in the continuity equation is replaced with an alternative hybrid scheme.

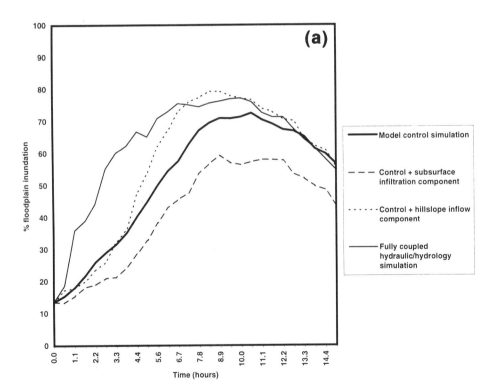

Figure 7.17   (a) Model predicted time sequences of inundation extent under varying boundary condition relaxation assumptions for the River Culm low resolution finite element mesh (Figure 7.1a) simulation of a 1 in 1 year recurrence interval flood event (Figure 7.4a). (b) and (c) Model predicted inundation extent for a section of the River Culm finite element mesh shown in Figure 7.1a for simulations (3), (4) and (6) in Figure 7.16. A model control simulation employing zero flux boundaries, a model control simulation plus subsurface infiltration and a model control simulation plus infiltration and subsurface inflows are shown

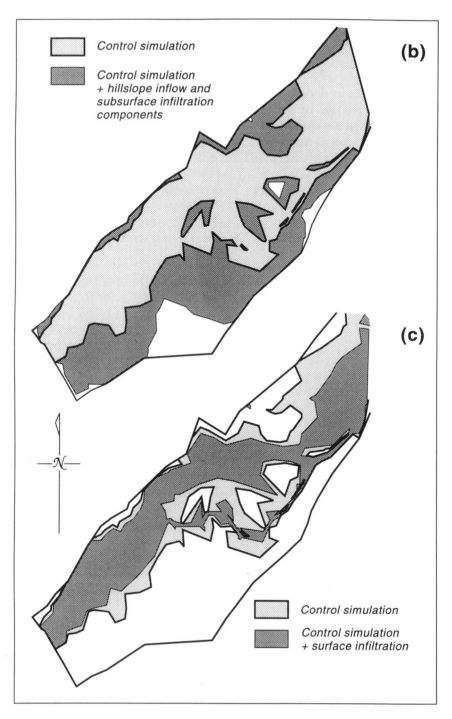

Figure 7.17   (*Continued*)

Refinement of numerical solution algorithms is an important research activity in hydraulic modelling (see, for example, Molinaro and Natale, 1994). This study indicates that boundary condition relaxation may be significantly more important in terms of the impact on model simulations and that greater emphasis should perhaps be accorded to such studies. This does, however, have a number of implications for the simulation of flood flow dynamics. Simulations indicate that to obtain a correct specification of flood flows and internal flow dynamics in reach-length hydraulic models, account must be taken of the flux across boundaries which have previously been considered impermeable While such process inclusion can apparently be significant, typical river discharge measurement techniques may not always be sufficiently accurate to identify actual flows or the net reach mass balance with which to satisfactorily validate such a high space/time resolution model. Given that all practical applications of hydraulic models involve calibration (due to the impossibility of knowing *a priori* exact initial conditions or model parameters), we must consider the extent to which correct process representation has been approximated by the calibration process. In addition, hydrology process inclusion in hydraulic models places greater emphasis on the need to improve data capture techniques for all aspects of model parameterization and validation.

7.5 DISCUSSION

Recent advances in fluid flow, river and floodplain hydraulics, outlined in Chapters 5 and 6 of this volume are capable of implementation at river reach scales of 10–50 km. This chapter has provided evidence of this implementation and also of the need for a detailed assessment of model performance and model assumption relaxation.

In what is generally a very data sparse environment, the role of models that are increasingly physically realistic is important in specifying future field monitoring programmes. Two aspects are seen as particularly significant in terms of research direction. First is the inclusion of uncertainty methods in the modelling structures to provide further insights into model parameterization needs, predictive outcomes and process dominance (this theme is further developed in Chapter 10 of this volume). Second is the nature of the linkages between the floodplain environment (hitherto treated as hydraulically isolated) and the hillslope and floodplain subsurface (hitherto modelled hydrologically without comprehensive regard for channel hydraulics).

It has been shown that the hitherto universally assumed zero flux boundary assumptions in such hydraulic models are not "neutral" assumptions. Relaxation of the boundaries (a physically realistic condition and of significance to hydrological and hydrochemical investigations) impacts on the numerical results in a manner which, while exemplified here, is deserving of further attention. In particular, it raises questions of developing full three-dimensional finite element solutions at the catchment scale and incorporating TELEMAC-2D style hydraulic codes in order that hillslope, floodplain and channel hydrological and hydraulic interactions are more fully described. This type of model in respect of hydrology simulations has been developed in small headwater catchments (see Binley and Beven, 1993) but has not yet been developed for application at the catchment scale with which we are here concerned.

From this discussion it is clear that the potential for coupled hydraulic/hydrology models for both flood prediction and catchment hydrology studies needs further examination. There appear to be considerable benefits to such an approach which result from additional process inclusion combined with a high resolution space/time representation. Such process complexity has hitherto been subsumed in the calibration procedure. Thus, there is now a powerful argument for complete investigation of these issues.

ACKNOWLEDGEMENTS

The research reported in this chapter has been made possible with the support of the UK Natural Environment Research Council (NERC) Grants GR3/8633 and GR3/09925, Electricité de France, the US Army Corps of Engineers and Hewlett Packard UK Ltd. The authors are grateful to the UK National Rivers Authority, the University of Exeter and the UK Ordnance Survey for the provision of data.

REFERENCES

Anderson, M.G. and Bates, P.D. (1994) Initial testing of a two dimensional finite element model for floodplain inundation. *Proceedings of the Royal Society of London, Series A*, **444**, 149–159.

Anderson, M.G. and Howes, S. (1985) Development and application of a combined soil water–slope stability model. *Quarterly Journal of Engineering Geology*, **18**, 225–236.

Anderson, M.G. and Kneale, P.E. (1982) "The influence of low angled topography on hillslope soil–water convergence and stream discharge. *Journal of Hydrology*, **57**, 65–80.

Anderson, M.G. and Lloyd, D.M. (1991) Using a combined slope hydrology–stability model to develop cut slope design charts. *Proceedings of the Institute of Civil Engineers*, **91**, 705–718.

Baird, L. and Anderson, M.G. (1992) Ungauged catchment modelling. I. Assessment of floodplain flow model enhancements. *Catena*, **19**, 17–31.

Baird, L., Gee, D.M. and Anderson, M.G. (1992) Ungauged catchment modelling. II. Utilization of hydraulic models for validation. *Catena*, **19**, 33–42.

Band, L.E. and Moore, I.D. (1995) Scale: landscape attributes and Geographical Information Systems. In: J.D. Kalma and M. Sivapalan (eds), *Scale Issues in Hydrological Modelling*, John Wiley and Sons, Chichester, 159–180.

Bates, P.D. (1992) Finite element modelling of floodplain inundation. Unpublished PhD thesis, University of Bristol, 220 pp.

Bates, P.D. and Anderson, M.G. (1993) A two dimensional finite element model for river flow inundation. *Proceedings of the Royal Society of London, Series A*, **440**, 481–491.

Bates, P.D. and Anderson, M.G. (1996) A preliminary investigation into the impact of initial conditions on flood inundation predictions using a time/space distributed sensitivity analysis. *Catena*, **26**, 115–134.

Bates, P.D., Anderson, M.G., Baird, L., Walling, D.E. and Simm, D.E. (1992) Modelling floodplain flows using a two dimensional finite element model. *Earth Surface Processes and Landforms*, **17**, 575–588.

Bates, P.D., Anderson, M.G. and Hervouet, J.-M. (1994) Computation of a flood event using a two dimensional finite element model and its comparison to field data. In: P. Molinaro and L. Natale (eds), *Modelling Flood Propagation over Initially Dry Areas*, American Society of Civil Engineers, New York, 243–256.

Bates, P.D., Anderson, M.G. and Hervouet, J.-M. (1995) An initial comparison of two

2-dimensional finite element codes for river flood simulation. *Proceedings of the Institute of Civil Engineers: Water, Maritime and Energy* (in press).

Bates, P.D., Anderson, M.G. and Horritt, M. (1996) Terrain information in geomorphological models: stability, resolution and sensitivity. In: S.N. Lane, K.S. Richards and J.H. Chandler (eds), *Landform Monitoring, Modelling and Analysis*, John Wiley and Sons, Chichester (in press).

Beven, K.J. (1989) Changing ideas in hydrology: the case of physically based distributed models. *Journal of Hydrology*, 157–172.

Binley, A.M. and Beven, K.J. (1993) Three dimensional modelling of hillslope hydrology. In: K.J. Beven and I.D. Moore (eds), *Terrain Analysis and Distributed Modelling in Hydrology*, John Wiley and Sons, Chichester, 107–119.

Binley, A.M., Beven, K.J., Calver, A. and Watts, L.G. (1991) Changing responses in hydrology: assessing the uncertainty in physically based model predictions. *Water Resources Research*, **27**(6), 1253, 1261.

Brookes, A.N. and Hughes, T.J.R. (1982) Streamline Upwind/Petrov Galerkin formulations for convection dominated flows with particular emphasis on the incompressible Navier–Stokes equations. *Computer Methods in Applied Mechanics and Engineering*, **32**, 199–259,

Brunet, R.C., Pinay, G., Gazelle, F. and Roques, L. (1994) Role of the floodplain and riparian zone in suspended sediment and nitrogen retention in the Ardour River, South-West France. *Regulated Rivers*, **9**, 55–63.

Chow, V.T. (1959) *Open Channel Hydraulics*. McGraw-Hill, New York, 680 pp.

Ervine, D.A., Sellin, R.J. and Willetts, B.B. (1994) Large flow structures in meandering compound channels. In: W.R. White and J. Watts (eds), *2nd International Conference on River Flood Hydraulics*, John Wiley & Sons, Chichester, 459–469.

Fawcett, K.R., Anderson, M.G., Bates P.D., Jordan, J.-P. and Bathurst, J.C. (1995) The importance of internal validation in the assessment of physically based distributed models. *Transactions of the Institute of British Geographers*, **20**, 248–265.

Gee, D.M., Anderson, M.G. and Baird, L. (1990) Large scale floodplain modelling. *Earth Surface Processes and Landforms*, **15**, 512–523.

Germann, P. (1990) Macropores and hydrologic hillslope processes. In: M.G. Anderson and T.P. Burt (eds), *Process Studies in Hillslope Hydrology*, John Wiley and Sons, Chichester, 327–363.

Green, I.R.A. and Stephenson, D. (1986) Criteria for comparison of single event models. *Hydrological Sciences Journal*, **31**, 395–411.

Harlin, J. and Kung, C.-S. (1992) Parameter uncertainty and simulation of design floods in Sweden. *Journal of Hydrology*, **137**, 209–230.

Haycock, N.E. and Burt, T.P. (1993) The role of floodplain soils in reducing the nitrate concentration of subsurface runoff: a case study in the Cotswolds, England. *Hydrological Processes*, **7**, 287–295.

Hervouet, J.-M. (1989) Comparison of experimental data and laser measurements with computational results of the TELEMAC-2D code (shallow water equations). In: C. Maksimovic and M. Radojkovic (eds), *Computational Modelling and Experimental Methods in Hydraulics (HYDROCOMP '89)*, Elsevier, Amsterdam, 237–242.

Hervouet, J.-M. (1993) Validating the simulation of dam breaks and floods. *Advances in Hydro-Science and Engineering Volume 1 Part A*, Washington, USA, 754–761.

Hervouet, J.-M. and Janin, J.-M. (1994) Finite element algorithms for modelling flood propagation. In: P. Molinaro and L. Natale (eds), *Modelling Flood Propagation over Initially Dry Areas*, American Society of Civil Engineers, New York, 102–113.

Hillel, D. (1980) *Fundamentals of Soil Physics*. Academic Press, New York, 413 pp.

Jackson, R.D. (1972) On the calculation of hydraulic conductivity. *Proceedings of the Soil Society of America*, **36**, 380–382.

King, I.P. (1970) An automatic reordering scheme for simultaneous equations derived from network systems. *International Journal of Numerical Methods in Engineering*, **2**, 523–533.

King, I.P. and Norton, W.R. (1978) Recent applications of RMA's finite element models for two dimensional hydrodynamics and water quality. In: C.A. Brebbia, W.G. Gray and G.F. Pinder (eds), *Proceedings of the Second International Conference on Finite Elements in Water Resources*, Pentech Press, London, 81–99.

Knight, D.W. and Hamed, M.E. (1984) Boundary shear in symmetrical compound channels. *Proceedings of the American Society of Civil Engineers, Journal of the Hydraulics Division*, **110**, 1412–1430.

Lane, S.N., Richards, K.S. and Chandler, J.H. (1994) Application of distributed sensitivity analysis to a model of turbulent open channel flow in a natural river channel. *Proceedings of the Royal Society of London Series A*, **447**, 49–63.

Marchuk, G.I. (1975) *Methods of Numerical Mathematics*. Springer-Verlag, New York, 316 pp.

Marron, D.C. (1989) The transport of mine tailings as suspended sediment in the Belle. *IAHS Publication*, **184**, 19–26.

Millington, R.J. and Quirk, J.P. (1959) Permeability of porous media. *Nature*, **183**, 387–388.

Molinaro, P. and Natale, L. (1994) *Modelling of Flood Propagation over Initially Dry Areas*. American Society of Civil Engineers, New York, 373 pp.

Pearce, A.J., Stewart, M.K. and Sklash, M.G. (1986). Storm runoff generation in humid headwater catchments. *Water Resources Research*, **22**, 794–804.

Rodi, W. (1980) *Turbulence Models and their Application in Hydraulics*. IAHR Publication, Delft, 104 pp.

Sellin, R.J., (1964) A laboratory investigation into the interaction between flow in the channel of a river and that over its floodplain. *La Houille Blanche*, **7**, 793–802.

Simm, D.J. (1993) *The deposition and storage of suspended sediment in contemporary floodplain systems: a case study of the River Culm, Devon*. Unpublished PhD thesis, University of Exeter.

Walling, D.E., Bradley, S.B. and Lambert, C.P. (1986) Conveyance loss of suspended sediment within a floodplain system. *IAHS Publication*, **159**, 119–132.

Walling, D.E., Quine, T.A. and He. Q. (1992) Investigating contemporary rates of floodplain sedimentation. In: G.E. Petts and P.A. Carling (eds), *Lowland Floodplain Rivers: A Geomorphological Perspective*, John Wiley and Sons, Chichester, 165–184.

8 Three-Dimensional Structures, Memory and Energy Dissipation in Meandering Compound Channel Flow

R.H.J. SELLIN* and **B.B. WILLETTS†**
* *Department of Civil Engineering, University of Bristol, UK*
† *Department of Engineering, University of Aberdeen, UK*

8.1 INTRODUCTION

Flow resistance in river channels is a matter of great complexity, even for inbank flows. It is closely linked to the energy dissipation rates and therefore to the internal flow structures in the channel. This degree of complexity results from the large number of sources of energy loss and the uncertainty in predicting channel flow resistance from our continued inability to identify and quantify each source separately. In this chapter we shall attempt to show that recent research has at least identified many of these energy loss processes even if empirical and sometimes quite arbitrary global methods are adopted to quantify them. The different sources of resistance include bed material grain drag (which requires a knowledge of the equivalent roughness size k_s), as well as bed-form drag caused by ripples, dunes and pool-riffle sequences as discussed by Engelund (1966) and Chang (1988).

River planform produces extra flow resistance (a type of large-scale from drag) usually associated with river bends, energy being lost in the helical motion and transverse shear as described by Rozovskii (1961) and Chang (1983). The presence of vegetation is another factor which is perhaps more difficult to quantify. Some guidance on the flow resistance due to grasses is provided by the US Soil Conservation Service (1954), and more recently by Kouwen and Li (1980).

When a river reaches the overbank flow regime, either in naturally occurring river floods or in artificially created two-stage channels (Purseglove, 1989; Sellin and Giles, 1989), the sources of energy dissipation and flow resistance are much more difficult to determine. The principal reason for this is extensive three-dimensional mixing of river and floodplain flows, especially in the case of meandering compound flows.

It is necessary at this point to look at the distinction between *natural floodplain systems* and *artificial multi-stage channels*. In attempting to distinguish between these two cases it should remembered that, despite their very different origins, there is usually

Floodplain Processes. Edited by Malcolm G. Anderson, Des E. Walling and Paul D. Bates.
© 1996 John Wiley & Sons Ltd.

more in common between them than there are differences. The principal differences lie in the characteristic values of the floodplain/main channel relative depth and also in the anticipated frequency of inundation of the floodplain. The relative depth ratio (defined as the ratio of the water depth over the floodplain to the water depth in the main channel) would usually be smaller for the natural system than for the man-made two-stage channel, while the frequency of inundation of the upper channel would be higher in the artificial system. These two factors both arise as a consequence of the fact that in the artificial channel the upper stage or berm channel is designed to be used fairly frequently and that room for its lateral extension is normally limited. A common design value for the return period for inundation of the upper channel would be 6 to 9 months.

In considering natural floodplain systems it is important to remember that while the floodplain may be in a reasonably natural state it is always possible that the river channel (the main channel) may have been subject to regulation or straightening on one or more occasions in the past. Until quite recently river engineers were subject to great pressures to remove meanders and to straighten channels to improve either inland navigation or flood prevention. Today a much higher value is placed on a natural planform and cross-section for rivers to preserve the balance in both geomorphological and habitat terms.

A more insidious pressure is still operating in floodplain regions due to the pressure on development land. There is clearly a strong case for floodproofing (a term used to describe the protection of low-lying land from flooding by the construction of bunds or floodwalls) existing residential and business areas. However, planning applications frequently request the extension of this protection to the present washlands or active floodplain areas. If this is allowed to proceed without proper control then the adequacy of the existing protection schemes becomes threatened by rising water levels in the increasingly restricted and inadequate flood channel. As an example of the undesirable effects of this kind of planning pressure the case of the River Roding flowing past the village of Abridge in Essex can be cited. Figure 8.1 shows the way in which the river floodplain has been "pinched" by development at Abridge, thus reducing the effective width of the river as it passes close to the north of the village.

It is unlikely that it will be possible to find a truly natural floodplain system for any river in a highly developed region like the southeast of England. Figure 8.2 shows the upper River Dee near Braemar in Scotland where the width of the valley in its upper reaches has allowed the natural development of a floodplain for this gravel river. If we are to provide protection for natural river and floodplain combinations and to reinstate others to a more satisfactory condition it is important that we understand the flow mechanisms that control these complex systems.

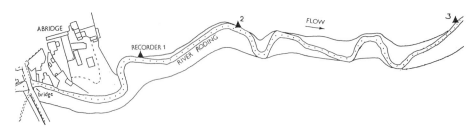

Figure 8.1 Plan of two-stage channel on River Roding downstream from Abridge, Essex, UK

Figure 8.2 The naturally developed floodplain of a gravel river: the River Dee above Braemar, UK

8.1.2   Artificial multi-stage channels

Artificial multi-stage channels can be classified by whether or not the natural floodplain is retained as the principal upper channel. The alternative is to excavate one or more berms below the natural floodplain level for that river reach. Thus the distinguishing feature of the first category is the limitation of the upper floodplain channel by the construction of artificial levees or floodbanks on one or both sides of the river. The second type does not need levees if the design discharge can be contained satisfactorily within the artificial two-stage channel, if not then it becomes a three-stage channel. Hence the adoption of the more general multi-stage channel term.

The berms in this type of channel are low and moist. Vegetation growing there will normally have uninterrupted access to soil moisture at its roots. Attempts to form grassland under these conditions have in the past encountered problems because gradual building up of the berm level by silting and the accumulation of organic debris occurs unless the area is grazed regularly or else cut mechanically, which is expensive. Apart from this potential loss of cross-section the unchecked growth of the type of plants which will thrive under these conditions leads to very high hydraulic roughness, and thus probably to a further reduction in discharge capacity. Both types of channel cross-section are illustrated in Figure 8.3.

When rigorous channel straightening is combined with equally straight levee construction then the result is far from natural in appearance and behaviour as can be seen in Figure 8.4. Although such geometrical precision may still be appropriate in severely limited urban situations, as in the situation shown in Figure 8.5, the trend today is to imitate natural planforms and cross-sections as far as is practical in the interests of both enhanced aesthetic values and diverse habitat creation. The River Roding improvement scheme is shown in outline plan in Figure 8.6. It falls in the second of the two categories defined above with no levees and therefore no backdrainage problems. The planform is interesting as it retains the original river for the lower channel while the berm limits are formed around this but with much reduced sinuosity. The extent to which the upper channel should mirror the sinuosity of the lower one is a matter of some interest at present as it has been found to exercise an important influence on the discharge capacity of the final compound channel. More will be said on this matter in the later sections of the present chapter.

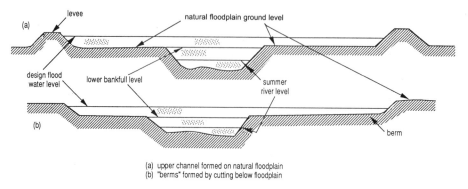

(a) upper channel formed on natural floodplain
(b) "berms" formed by cutting below floodplain

Figure 8.3   Two-stage channel cross-sections

Figure 8.4   River Kinzig, Baden-Wurtenberg, Germany

Figure 8.5   Two-stage channel on the north River Wey through Farnham, Surrey, UK

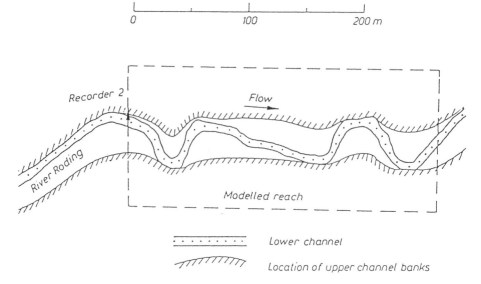

Figure 8.6   Details of two-stage channel study reach on the River Roding, Essex, UK

8.2 LABORATORY-BASED STUDIES

8.2.1 Early studies

The first published account of studies into compound meandering channels came from the US Army Corps of Engineers in Vicksburg (1956). Their aim was to investigate some of the parameters which affect flooding rivers, in particular the Mississippi. A number of large-scale model channels were constructed and tested in a flume 30.5 m long and 9.1 m wide. The dependence of the discharge capacity on the following parameters was investigated:

(i)   the sinuosity of the main channel and the relative radius of curvature of the bends,
(ii) the floodplain roughness, and
(iii) the relative area of the floodplain compared with that of the main channel.

The following conclusions were reached

(i)   an increase in sinuosity results in a decrease in discharge,
(ii) increasing the floodplain roughness reduces the discharge substantially, and
(iii) reducing the floodplain area reduces the overall discharge.

Toebes and Sooky (1967) presented results from a study carried out in a laboratory flume (7.3 m × 1.2 m). The tests were designed to provide information about the differences between straight, meandering and compound meandering channels. They confirmed the results from the Vicksburg report (1956) and made the following observations for compound meandering flow:

1. The interaction energy losses increase with decreasing mean velocity and show a maximum when varying the overbank flow depth.
2. Flow on the floodplain is found to affect the flow in the main channel. The maximum flow velocity in the cross-section is reduced, and the region of maximum velocity is shifted towards the receiving floodplain.
3. The helical secondary currents are different from and stronger than those observed in an inbank meandering channel. The currents for a compound meandering channel weaken and shrink laterally as the flow progresses round the bend. In the cross-over region (defined as the region containing the point of inflection in the meandering channel planform) there is a strong secondary current which rotates in the opposite direction to that observed under inbank flow conditions.
4. An imaginary horizontal interface, separating the main channel flow from the floodplain flow, should be used for analytical purposes.
5. Energy losses per unit length of meandering channel are up to 2.5 times greater than for a straight single channel at the same hydraulic radius and discharge.
6. Rivers will tend to straighten themselves in times of flood in order to maximize discharge for a given stage.

This work was extended by James and Brown (1977), who investigated the effect of sinuosity in a flume measuring 26.8 m × 1.5 m. From their results they concluded that:

1. With increasing sinuosity the resistance to flow increases and the velocity profiles become more distorted compared with the straight channel case.

2. Neither the Manning nor the Chezy equation give an accurate description of the stage–discharge relationships for meandering channels. They proposed a method of analysis based on a modified Manning equation and a correction factor.

Rajaratnam and Ahmadi (personal communication, 1985) investigated the nature of water flow in a compound meandering channel. They observed that, in the main channel, the region of maximum velocity is located at the inside of each bend. On the floodplain, regions of high velocity are found outside the meander belt away from interaction with the main channel. These observations have been summarized by Ervine and Jasem (1989) and by McKeogh and Kiely (1989) in terms of regions, as was first suggested by Toebes and Sooky. In the main channel, flow velocities are reduced due to:

(i)   friction on the bed and side walls,
(ii)   the creation and development of the secondary current,
(iii) turbulent shear at bankfull level due to the difference between the main channel and floodplain velocities, and
(iv)   enhanced eddy viscosity due to flow separation at the bends.

Within the meander belt region on the floodplain, McKeogh and Kiely identified three sources of energy loss:

(i)   bed friction
(ii)   expansion losses as the water from the upstream floodplain crosses the main channel, and
(iii) contraction losses as the flow returns to the downstream floodplain.

Outside the meander belt the local bed friction is thought to be the principal source of energy loss.

Having identified the mechanisms, McKeogh and Kiely attempted to quantify them using a modified version of the Colebrook–White equation combined with the Darcy–Weisbach equation in order to obtain a stage–discharge relationship. The resulting method was found to give a reasonable prediction of discharge for the higher floodplain depths.

The mechanisms responsible for the energy lost in the main channel have since been more clearly defined by Kiely (1990). Holden and James (1989) have also shown that the extent and therefore the importance of such mechanisms is a function of the main channel slope.

River floodplains frequently support large quantities of vegetation. Investigations with roughened floodplains have therefore been attempted by a number of groups: Toebes and Sooky (1967), James and Brown (1977) and Nalluri and Judy (1985). They all conclude, unsurprisingly, that this form of roughness significantly reduces discharge capacity. Nalluri and Judy conclude that the additional friction is affected by the shape, spacing and distribution of the roughness elements.

Work on one, two and three-dimensional model development – both conceptual and numerical – is still in its infancy due to:

(i)   the very small number of model facilities on which such work can be carried out,
(ii)   the limited knowledge of the mechanisms which control such flows,

(iii) the problem of defining cross-sections and frames of reference in both main
       channel and floodplain, and the way in which these change as the flow proceeds
       downstream, and
(iv)   the requirement of a prior detailed understanding of straight compound channel
       flow.

8.2.2 The SERC Flood Channel Facility

The UK Flood Channel Facility (FCF), funded originally by the Science and
Engineering Research Council (now the EPSRC), was constructed at HR Wallingford
in 1985–1986. It was intended to provide a large-scale indoor experimental facility for
research teams both from the universities and from industry to study the mechanics of
two-stage flow. The ultimate aim was to provide river engineers with a reliable design
manual for two-stage channels.

The FCF consists of a tank in which channel models are constructed. The usable
model area measures 56×10 m, making it very suitable both for wide floodplain
geometry and also for meandering channels. The channel gradient is fixed when the
model, which is given a cement mortar finish, is constructed. Water can be recirculated
with a maximum discharge of $1 \text{ m}^3 \text{s}^{-1}$. Comprehensive instrumentation is provided
together with the necessary access structures.

Figure 8.7 The Flood Channel Facility at HR Wallingford

The FCF, shown in Figure 8.7, was designed with three programmes in mind; the first involving straight compound channels is described in the previous chapter, the second for meandering two-stage channels is considered here, while the third, which includes sediment transport, but using the earlier geometry, started in 1994.

8.2.3 Meandering compound channels

The work described here was carried out in the FCF described above. Two sets of experiments were completed, each employing an inner or main channel consisting of a series of uniform meanders set in a flat floodplain having a "valley gradient" close to 0.001. In the first set the sinuosity was 1.37 and in the second 2.04. Details of the plan and cross-section geometry are shown in Figure 8.8. Preliminary investigation had shown that an inner channel with natural variation in cross-section through the bends gave significantly different results from one with a uniformly trapezoidal cross-section. Accordingly the main experimental programme used channels with cross-section variation following an idealized "natural" geometry obtained by inspection of 17 published sources of meander channel geometry in the field. This quasi-natural geometry was constructed within a uniform trapezoidal envelope as shown in Figure 8.9. An adjustable tailgate weir was used to shorten the drawdown curve at the downstream end of the flume and was reset for each flow rate.

The programme included systematic observations of stage, discharge, the velocity field and water surface levels. The horizontal velocity vector was determined (in magnitude and direction) at a large number of locations both in the channel and over the floodplain. This information was checked and supplemented by means of photographs of injected dye and surface floats.

These detailed measurements were made for each of three flow depths, 140 mm, 165 mm and 200 mm; the first of these represents inbank flow. These depths are measured from the deepest part of the channel cross-section which is 150 mm below the local floodplain.

Figure 8.10 compares the stage–discharge curves for four cases (two channel sinuosities and two floodplain roughnesses). It shows that in all cases a significant increase in depth above bankfull is required before any increase in flow capacity is achieved and, thereafter, a rapid increase of discharge with depth occurs. For a given discharge the depth is greater for the higher sinuosity, and at the deeper overbank flows floodplain roughness has a marked and adverse effect on depth for a given discharge. Thereafter, a rapid increase of discharge with depth occurs.

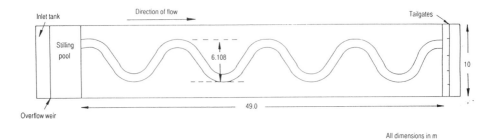

Figure 8.8 Plan of the 1.34 sinuosity two-stage meandering channel in the FCF

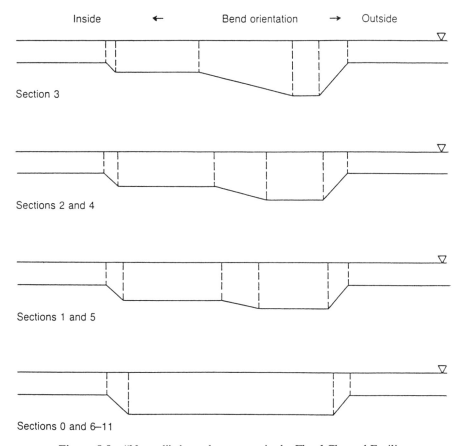

Figure 8.9   "Natural" channel geometry in the Flood Channel Facility

In summary, some specific features are worth noting. At a given discharge the depth of flow is greater for the higher sinuosity, while at greater depths the floodplain roughness has a marked (adverse) effect on discharge at any particular depth. At these greater depths the floodplain flow is the dominant component in the total discharge, and so the floodplain roughness which impedes this flow exerts an important influence on the behaviour of the system as a whole.

Before considering the flow structure and momentum exchange mechanisms in compound meandering channels it is worthwhile describing in outline the instruments and techniques adopted for use in the FCF to achieve this end. While the stage–discharge relationship, and its derivatives, reveal the global effect of this complex internal flow structure on the total discharge capacity (and by inference on its hydraulic resistance), it provides no direct information on the flow structures themselves. To obtain this information point velocity measurements are required and were obtained at selected cross-sections. Notional rectangular grids are constructed on each of the selected cross-sections to define the point measurements locations. Ideally the direction as well as the magnitude of each point velocity vector is needed although in practice it has proved difficult to obtain the required vertical component.

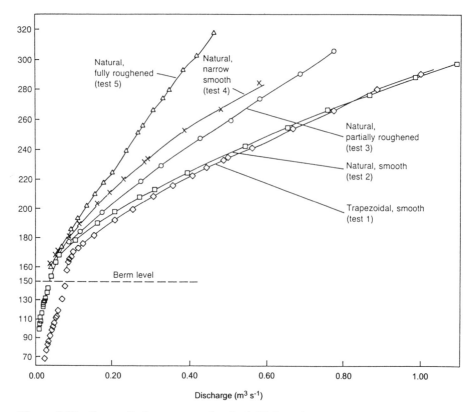

Figure 8.10   Stage–discharge curves for the 1.37 sinuosity compound channel in the FCF

The selective use of dye releases, recorded on video, has provided a valuable supplement to these velocity measurements. The dye traces of course follow any vertical currents present and these can often be inferred from the video record. Because the point velocity data include the transverse *horizontal* component it follows that lateral velocity profiles can be constructed for each cross-section investigated, as shown in Figure 8.11. Local flow direction in the x–y plane was first determined for each grid using a balanced vane attached by a vertical shaft to an angular displacement potentiometer. Then, using the stored direction data, a second traverse was made over the same grid points with a steerable miniature propeller current meter. Thus in order to enhance accuracy the propeller meter was rotated to align with the direction of the local horizontal velocity vector and thereby measure directly the magnitude of this vector. The importance of this procedure can be appreciated when it is realized that the local directions of flow on the floodplain and in the lower channel may differ by as much as 60° in a single vertical in the region of the channel cross-overs.

Examples of velocity distribution diagrams are shown as follows. Figure 8.12 shows the location of the measuring sections, numbered 0–11 for the 1.37 sinuosity (60° cross-over angle) channel. Measurements are made within a band extending 300 mm onto the floodplain on each side. It will be seen that sections 1–5 are located within the bend region while 6–11 are located within the straight channel reach at the cross-over.

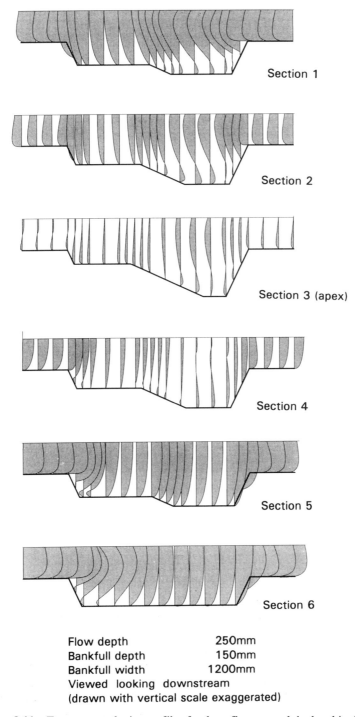

Section 1

Section 2

Section 3 (apex)

Section 4

Section 5

Section 6

Flow depth   250mm
Bankfull depth   150mm
Bankfull width   1200mm
Viewed looking downstream
(drawn with vertical scale exaggerated)

Figure 8.11   Transverse velocity profiles for deep flow around the bend in the FCF

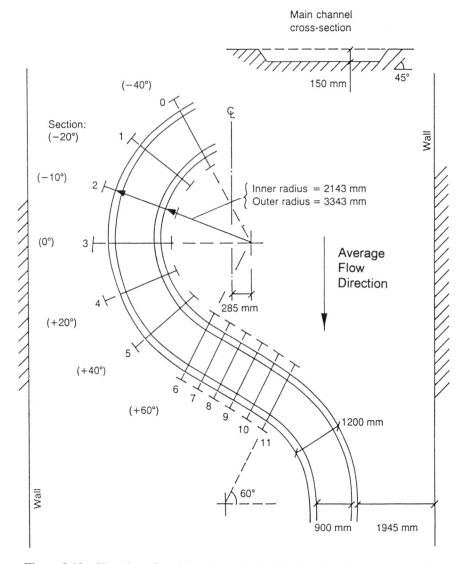

Figure 8.12   Plan view of model study reach showing location of measuring sections

Sections 0 and 6 are at the tangent points and section 11 also constitutes section 0 at the start of the next half-meander in this regular repetitive channel geometry.

   Longitudinal velocity distributions in the bend region, represented by "isovels", are shown in Figure 8.13 and in the cross-over channel in Figure 8.14 for the flow conditions indicated in the figure captions. These represent shallow overbank flow, as would obtain for a natural river floodplain, and the quasi-natural lower channel form as described above. It must be remembered that each of these cross-sections is perpendicular to the local lower channel centreline direction and the longitudinal velocity component is defined as being perpendicular to the plane of this measurement

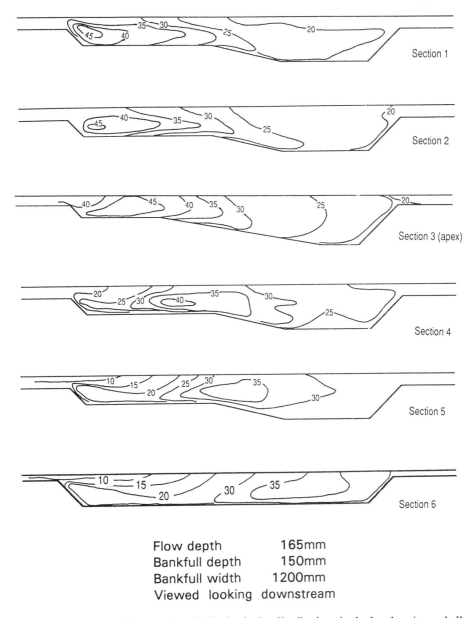

Section 1

Section 2

Section 3 (apex)

Section 4

Section 5

Section 6

Flow depth   165mm
Bankfull depth   150mm
Bankfull width   1200mm
Viewed looking downstream

Figure 8.13 FCF meander tests: longitudinal velocity distributions in the bend regions: shallow overbank flow

section. Figure 8.15 and 8.16 show the corresponding velocity distributions for deeper flow conditions over the floodplain, more appropriate to an artificial two-stage channel perhaps.

Transverse velocity component profiles are shown for the same sections and flow conditions in Figures 8.17 to 8.20. The transverse velocity profiles are drawn on the

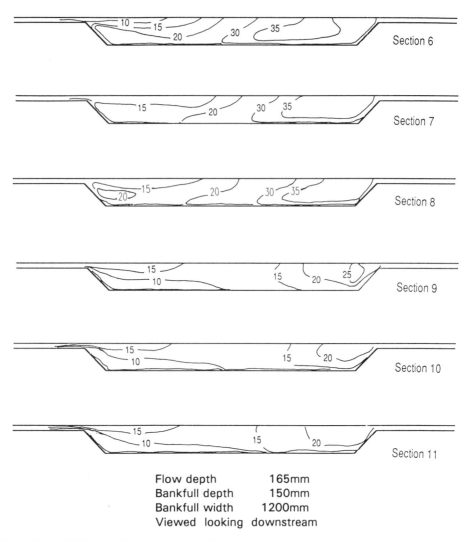

Figure 8.14 FCF meander tests: longitudinal velocity distributions in the cross-over regions: shallow overbank flow

verticals at which the point measurements were made. This method of data display is very sensitive to the presence of lateral currents and highlights quite small features associated with separated flow under the lower banks. Conditions at the meander bends (the apex regions) are probably less important than at the cross-over regions, but the interaction between floodplain and main channel flows is more obvious, as can be seen from Figures 8.17 and 8.19.

The interpretation of these data is helped by secondary current distribution diagrams. These are constructed from transverse velocity profiles referred to above and demonstrate the growth and decay of secondary circulation cells in the main channel. Figure 8.21 shows these cells at the meander bends, while Figure 8.22 shows them in

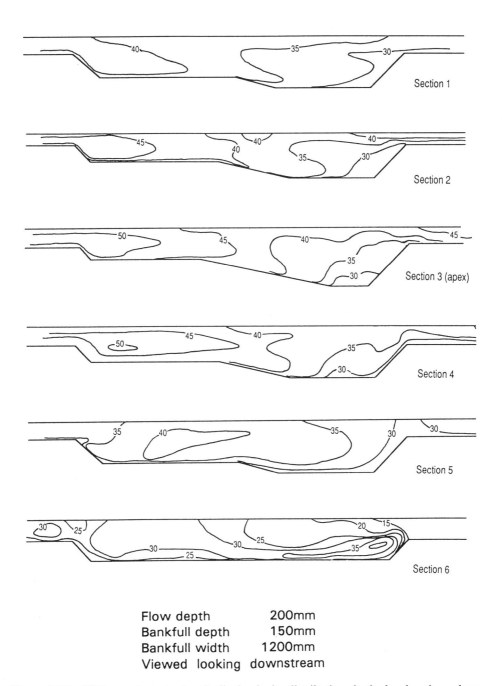

Flow depth   200mm
Bankfull depth   150mm
Bankfull width   1200mm
Viewed looking downstream

Figure 8.15 FCF meander tests: longitudinal velocity distributions in the bend regions: deeper overbank flow

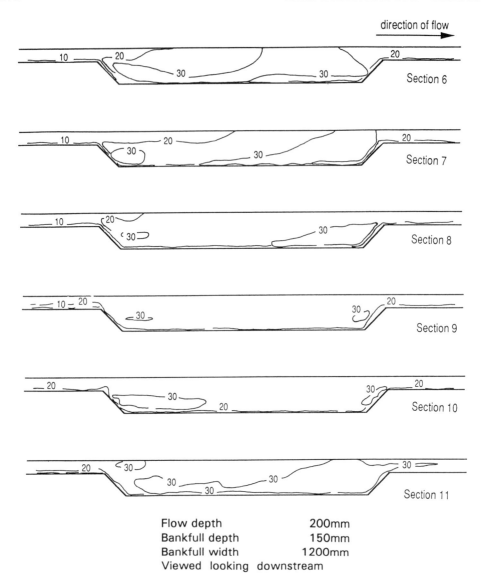

Flow depth   200mm
Bankfull depth   150mm
Bankfull width   1200mm
Viewed looking downstream

Figure 8.16 FCF meander tests: longitudinal velocity distributions in the cross-over regions: deeper overbank flow

the cross-over region, both for shallow floodplain flow. Figures 8.23 and 8.24 display the corresponding flow structure for deeper flow over the floodplain.

Considering first the flow conditions at the bends with shallow floodplain flow, an inspection of Figure 8.17 shows how the surface cross-flow current, well established at section 1, switches direction, relative to the channel thalweg, as the sections move downstream and around the bend. The rotational sense of the principal secondary flow cell (Figure 8.21) is reversed at around section 4 under the influence of the changing

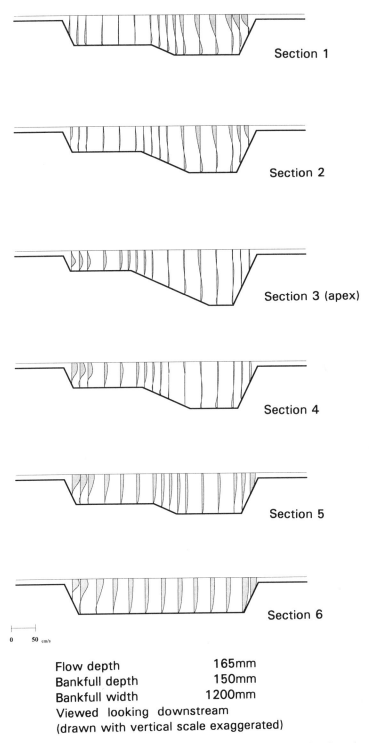

0	50	cm/s

Flow depth   165mm
Bankfull depth   150mm
Bankfull width   1200mm
Viewed looking downstream
(drawn with vertical scale exaggerated)

Figure 8.17 FCF meander tests: transverse velocity profiles in the bend regions: shallow overbank flow

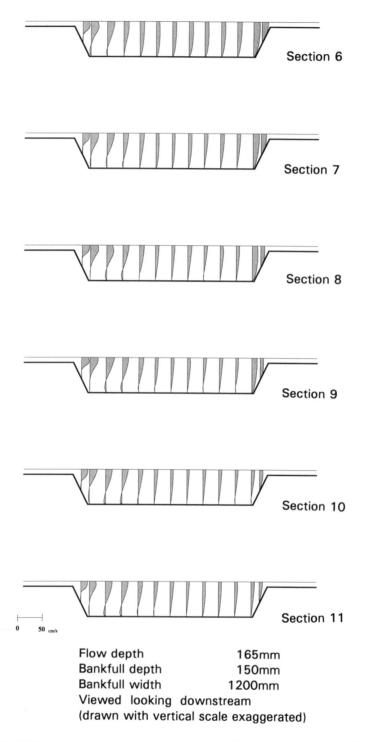

Section 6

Section 7

Section 8

Section 9

Section 10

Section 11

0   50 cm/s

Flow depth   165mm
Bankfull depth   150mm
Bankfull width   1200mm
Viewed looking downstream
(drawn with vertical scale exaggerated)

Figure 8.18   FCF meander tests: transverse velocity profiles in the cross-over regions: shallow overbank flow

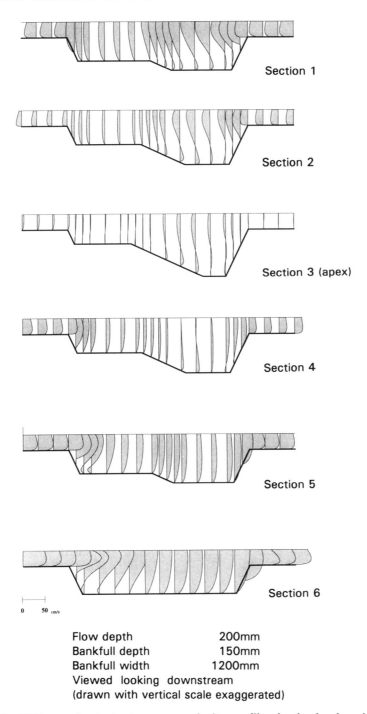

Flow depth   200mm
Bankfull depth   150mm
Bankfull width   1200mm
Viewed looking downstream
(drawn with vertical scale exaggerated)

Figure 8.19 FCF meander tests: transverse velocity profiles in the bend regions: deeper overbank flow

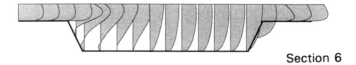

Section 6

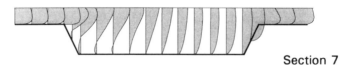

Section 7

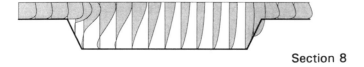

Section 8

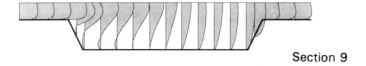

Section 9

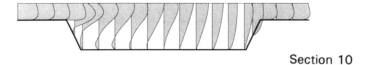

Section 10

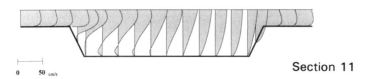

Section 11

0   50 cm/s

Flow depth   200mm
Bankfull depth   150mm
Bankfull width   1200mm
Viewed looking downstream
(drawn with vertical scale exaggerated)

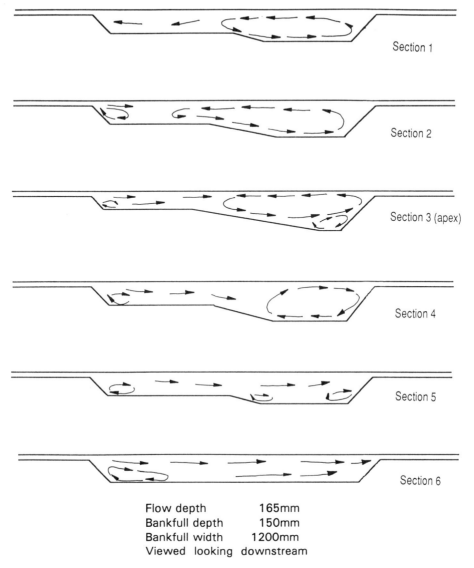

Flow depth   165mm
Bankfull depth   150mm
Bankfull width   1200mm
Viewed looking downstream

Figure 8.21 FCF meander tests: secondary circulation cells in the bend region: shallow overbank flow

cross-currents. The transverse velocity profiles demonstrate very well how this comes about.

Looking now at the deeper overbank flow at the bends, Figure 8.19 shows the transverse velocity profiles and Figure 8.23 the secondary flow cell structure. It is clear that the surface currents, aligned approximately with the "valley" direction are now

Figure 8.20 (*opposite*) FCF meander tests: transverse velocity profiles in the cross-over regions: deeper overbank flow

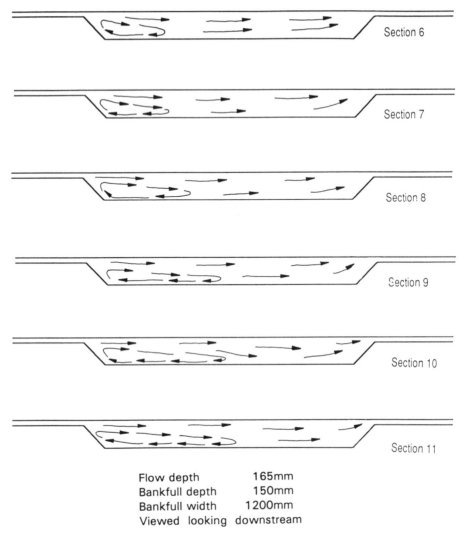

Flow depth   165mm
Bankfull depth   150mm
Bankfull width   1200mm
Viewed looking downstream

Figure 8.22   FCF meander tests: secondary circulation cells in the cross-over region: shallow overbank flow

dominant and the secondary flow cells correspondingly weaker. By section 5 these are almost completely suppressed and there is a strong cross-current present at all depths in the flow.

At the cross-overs, and with shallow floodplain flow, Figure 8.18 shows the transverse velocity profiles and Figure 8.22 the secondary flow cells. The flow structure in this straight channel reach seems very simple. The flow in the lower channel is dominant with a weak but steady cross-current maintained by the floodplain flow. A weak secondary flow cell develops steadily under the "upstream" bank.

For the deeper floodplain flow, Figures 8.20 and 8.24, flow conditions are similar to those observed for the shallow floodplain condition. The one significant difference is

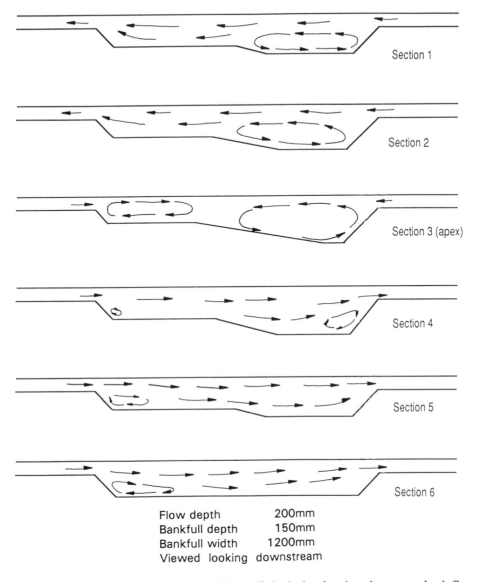

Section 1

Section 2

Section 3 (apex)

Section 4

Section 5

Section 6

Flow depth   200mm
Bankfull depth   150mm
Bankfull width   1200mm
Viewed looking downstream

8.23 FCF meander tests: secondary circulation cells in the bend region: deeper overbank flow

that the floodplain-generated surface currents appear much stronger this time as would be expected. The secondary flow cell grows in a similar manner and location.

The velocity vectors can be resolved in the longitudinal or downstream direction (with reference to the lower channel). Figures 8.13 to 8.16 show this information in isovel or velocity contour form.

In the bend region Figure 8.13, which relates to shallow overbank flow, shows the maximum longitudinal velocity zone to stay close to the inside bank of the lower channel until section 4 is reached. From this point onwards it weakens and moves

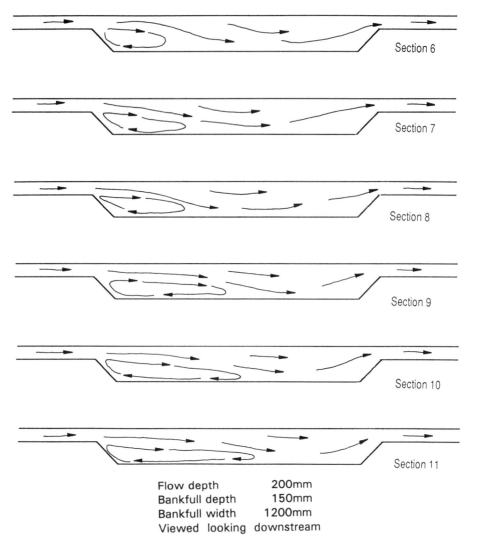

Flow depth   200mm
Bankfull depth   150mm
Bankfull width   1200mm
Viewed looking downstream

Figure 8.24 FCF meander tests: secondary circulation cells in the cross-over region: deeper overbank flow

progressively across the channel until by the time section 6 is reached it is firmly established close to the right hand bank.

With inbank flow, the position of the zone of maximum longitudinal velocity is observed to be near the outer bank at and beyond the bend apex in laboratory studies. Field observations are less consistent on this point, because of the variability of plan geometry in the field – see Thorne *et al.* (1985). However, the high longitudinal velocity zone for inbank flow is never found as near the inner bank as is shown in figure 8.13, sections 1, 2 and 3. The reason for the difference between inbank and overbank longitudinal velocity distribution lies with the reversed direction of the major secondary circulation cell, seen in Figure 8.17. In overbank flow, the sense of this cell has the effect of sweeping towards the inner bank water flowing near the surface of the stream at high velocity. This traps, close to that bank, the strong current created there by the oblong floodplain flow in the cross-over region, and already noted to have been established there by the time that section 6 had been reached. For a fuller description of this process see Ervine *et al.* (1994).

Figure 8.15 demonstrates the same general velocity pattern for the case of deeper overbank flow. However, the much stronger floodplain flow associated with this relative water depth appears to complicate the longitudinal velocity distribution compared with that shown in Figure 8.13. The filament of maximum velocity still remains firmly located close to the inside channel bend at the apex (section 3) and it is rather slower to move across to the other side of the channel. However, it can be seen that a very concentrated high velocity zone has appeared tucked under the outside bank at section 6. This is related to the observed strong flow out of the inner channel and onto the floodplain in this region, as demonstrated by the transverse velocity profiles for section 6 for this flow depth in Figure 8.20. The combination of these two strong and spatially superimposed streams, flowing in different directions, suggests a zone of high shear and resulting high turbulence in this area. This is confirmed by the measured high values of boundary shear stress on the floodplain here. Further details of the shear stress distribution on the floodplain are to be found in Knight *et al.* (1992).

In the cross-over reaches there is a markedly skewed distribution of the longitudinal velocity. At the 165 mm depth, Figure 8.14, all the highest velocity fluid is ejected from the lower channel region on to the floodplain and this process is complete by the time section 9 has been reached. Thereafter, the longitudinal velocity component builds up again as the next bend is approached.

In summary, some specific features are worth noting. For a given discharge the flow depth increases with channel sinuosity. At greater depths the floodplain roughness has a marked (adverse) effect on discharge at any particular depth. At the greater depths studied the floodplain flow is the dominant component in the total discharge, and so the floodplain roughness which impedes this flow exerts an important influence on the behaviour of the system as a whole.

The information contained on the velocity and circulation diagrams considered in this section so far can be used to build up an overall picture of the principal flow mechanisms involved. Figure 8.25 shows such a synthesis for the 1.37 sinuosity channel. Comparing this with Figures 8.21 and 8.23 shows that the rotational cell proposed for the bend region is present at both depths studied. It is growing very weak by section 4 and has vanished by section 5. It is interesting to notice that, at 165 mm depth, a small rotational cell has started to form against the inside bank at section 2 accompanied by a reversal of flow at the surface over this bank before the apex of the bend (section 3) has been reached. At 200 mm depth a similar structure tries to form at section 3 but is suppressed by the strong flow coming off the floodplain at section 4 and does not form properly until section 5 is reached.

If we now examine the flow structures in the straight cross-over reaches we find high (transverse) velocities leaving the lower channel for the downstream floodplain between sections 6 and 9 at both flow depths and then loosing strength as the next bend approaches. On the other side of the channel a powerful rotational cell builds in the lee of the upstream bank at both flow depths. The "plunge" area shown in this figure was

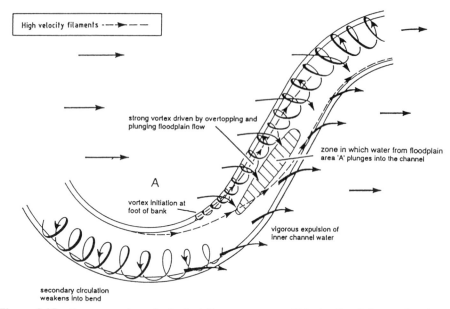

High velocity filaments -- -▶- -- --

strong vortex driven by overtopping and
plunging floodplain flow

zone in which water from floodplain
area 'A' plunges into the channel

A

vortex initiation at
foot of bank

vigorous expulsion of
inner channel water

secondary circulation
weakens into bend

Figure 8.25 Representation of principal flow structures within the flooded meander channel

identified from dye traces but its existence is also supported by the concentration of the high (longitudinal) velocity region towards the bed of the channel between sections 6 and 8, as shown in Figures 8.14 and 8.16.

8.3   FIELD-BASED RESEARCH PROGRAMMES – ARTIFICIAL COMPOUND CHANNELS

8.3.1   Quasi-straight channels – River Maine, Co Antrim, Northern Ireland

When creating an artificial two-stage channel, the designer can choose either to make the floodplain boundary geometry provide an envelope for the inner channel, but have geometry otherwise independent of the inner channel (see section on River Roding, below), or alternatively to make the floodplain boundary follow the curves of the sinuous inner channel. The latter choice leads to floodplains of roughly uniform width and cross-sections, taken normal to the course of the inner channel, which are uniform. This makes for simple construction and low cost, in the absence of constraints imposed by the terrain or by prior development in the vicinity of the river. It also leads to overbank flows which have greater similarities with inbank flows (more closely co-linear) than are found when floodplain geometry is only loosely related to channel plan geometry and floodplain width is irregular. For this reason, it is sensible to deal with the two cases separately. We do so here and have adopted the term "quasi-straight compound channel" for systems in which the floodplain boundary is parallel to the inner channel bank on each side of the channel. The term is appropriate because flow in this sort of system is predominately in the channel direction, as it is in a straight compound channel, with the consequence that the two types of system share important characteristics.

The difficulties of making observations when the river banks are overtopped are more manageable in this kind of engineered system than they are when the floodplain is irregular. An "improved" reach of the River Maine in Ulster which falls into this category has been closely studied by Myers with a number of collaborators (Myers, 1990; Martin and Myers, 1991; Lyness and Myers, 1994a,b). The reach they studied was 1 km long, was essentially straight but contained a single bend of roughly 45°. The channel top width was approximately 14 m and the floodplains varied in width only between 6.5 and 8.2 m, so that the cross-section of two floodplains plus the channel was between 27.3 and 30.4 m. The inner channel was between 0.9 and 1.0 m deep and, it is important to note, the floodplains sloped inwards towards the inner channel. At one particular cross-section, the presence of a bridge made it possible to measure flow velocities at different points in the cross-section in some detail, even when the river was in flood.

Data collected at this cross-section showed marked similarity in velocity distribution to the better known use of straight channels in straight, parallel floodplains. In particular, channel velocities were retarded at the channel/floodplain interface by interaction between the channel and floodplain flows. This was attributed to the effect of the shear layer between the sluggish flow on the floodplains, associated with the shallow flow depth there, and the vigorous, parallel flow in the inner channel, as is conventional for the straight channel case. The sinuosity of the reach was assigned no influence. While it would be necessary to make observations of the uniformity of conditions at other cross-sections to confirm explicitly the validity of this interpretation, the flow distribution in the cross-section and the stage–discharge relationship conform well with what would be expected of a straight channel of the same cross-section. Significantly, quite good stage–discharge prediction was achieved on the basis of dividing the cross-section into three by means of vertical (zero stress) surfaces placed above the inner channel banks, with modest under-prediction of depth at small overbank depths, as would be found for the equivalent straight system.

8.3.2 Quasi-natural channels

There had been growing public concern that man-made river changes decrease the richness of wildlife habitats and the quality of the landscape. This must include many schemes carried out prior to 1980 both in the UK and in other developed countries. Figure 8.4 gives a good example of such an artificial two-stage channel now considered unsatisfactory on this count. A potential conflict of interest exists in the design of river schemes as often the most efficient hydraulic solution may not give the most acceptable environmental solution, especially if the geomorphological implications of the changes are poorly grasped. As a result of this rediscovered appreciation of the role of rivers in our landscape, the use of straight, even uniform cross-section, channels is being replaced wherever possible by either singly or doubly meandering compound channels.

River Roding, UK

Compound or two-stage river channels are frequently adopted in the design of flood-alleviation schemes. In order to achieve wildlife conservation and habitat enhancement objectives the natural river planforms are commonly retained. One such scheme was carried out on the River Roding at Abridge in Essex (Figures 8.1 and 8.6) during the 1980s. An extensive post-implementation study was carried out over the ensuing five-year period to examine the hydraulic behaviour, the habitat development and the maintenance problems associated with this type of river design (Sellin *et al.*, 1990).

There are two types of compound channel design:

1. Retention of the natural floodplain as the upper channel while limiting the extent of inundation by the construction of levees or floodwalls – see Figure 8.26.
2. The excavation of a wholly artificial upper channel, or berm, below the natural floodplain level – see Figure 8.27.

Berms in type (b) channels are low and moist. Vegetation established there will normally have uninterrupted access to soil moisture and strong growth will ensue. Maintenance will therefore be required unless satisfactory arrangements for grazing can be made. The Roding channel was of this type.

The main reason for carrying out the Roding study was that it soon became obvious after the completion of the remodelled channel that its discharge capacity was well below the design value (some 30%). Water-level recorders were installed along the length of the study reach, as shown in Figure 8.1, and discharge values were interpolated from neighbouring flow gauging structure records. It was soon found that the resulting stage–discharge relationships were strongly dependent on the state of the berm vegetation. Major input parameters were therefore the season and the maintenance policy in force over the period. Figure 8.28 shows two experimentally determined stage–discharge curves, one, P1, for a winter season in which no berm cutting or clearing had taken place during the previous growing season, and the other, P2, for a winter in which the berms had been cleared during the October at the start of the period. Figure 8.29 shows the ratio between these two curves, a 50% difference being reached at the design flood water level. It must be remembered that the P2 condition did still not achieve the design discharge value for the channel.

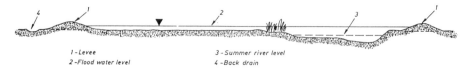

1 - Levee   3 - Summer river level
2 - Flood water level   4 - Back drain

Figure 8.26 Cross-section of two-stage channel: natural floodplain bounded by levees

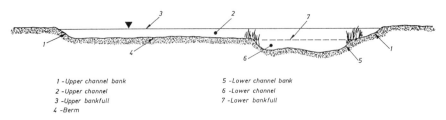

1 - Upper channel bank   5 - Lower channel bank
2 - Upper channel   6 - Lower channel
3 - Upper bankfull   7 - Lower bankfull
4 - Berm

Figure 8.27 Cross-section of two-stage channel: upper channel formed by cutting down into the natural floodplain

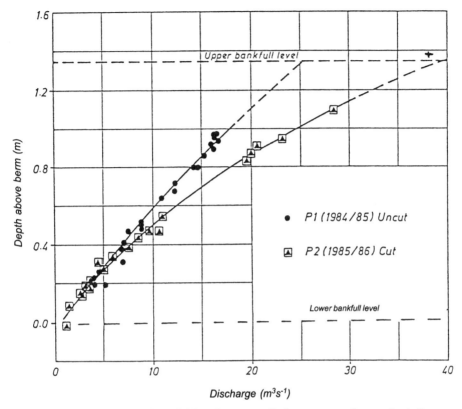

Figure 8.28 River Roding field study: stage–discharge curves for overbank flow

Data collected during the summer season were more variable. The highest discharges occurred during August as the result of localized thunderstorms. Figure 8.30 shows values of the Manning roughness coefficient n calculated from the stage–discharge points obtained from the depth and discharge measurements on the river. The curves P1 and P2 represent the winter behaviour of this channel.

A hydraulic model (horizontal scale 1 : 50, vertical scale 1 : 16) was constructed of this reach and used to explore the effects of selective cutting on the berm. Figure 8.31 shows the scheme adopted for these tests. Selected berm areas were designated 1–5 as indicated in this diagram. First the retention of a marginal strip (1, 2) along either bank of the inner channel was explored for two reasons: it is the most difficult to cut, and this area is frequently colonized by well-developed reed and sedge growths which provide a valuable riverside habitat. The next test involved the partial removal of this strip (from zone 2), thus reducing flow resistance in critical areas associated with strong flows onto or off the berm. This suggested the concept of an unimpeded high-level flood channel which was further developed by the retention of vegetation in slack water areas (4). Zone 3 therefore became the principal flood discharge channel and vegetation cutting was given a higher priority here. The berm upper banks (5) were steep and more difficult to cut. They could also be left untended over longer periods without serious consequences.

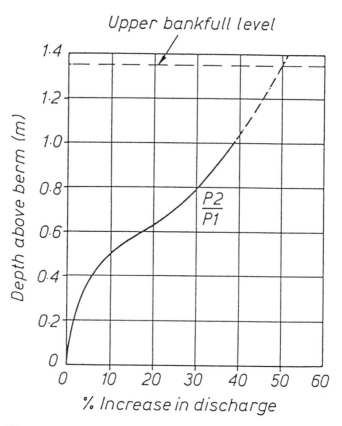

Figure 8.29 River Roding field study: effect of cutting vegetation on berm expressed as percentage increase in discharge: P2 (cut berm) discharge compared with P1 (uncut berm)

Stage–discharge curves obtained as a result of these model tests are shown in Figure 8.32. The code used is as follows:

M3 – complete cut in all five zones;
M4 – marginal strip retained (1 and 2);
M5 – zone 3 only cut;

M1 – no berm cutting for two seasons, corresponding with condition P1 from the prototype river tests.
    Conclusions drawn from the River Roding study included the following:

(i)   Channel elements should be sized to match seasonal needs; the lower channel to hold normal summer flows with some freeboard, while the whole channel cross-section must be designed to take the designated flood discharge. If this policy is followed the lower channel will not be over-sized and should provide a good habitat for aquatic invertebrates and fish species especially if natural channel sinuosity is adopted.
(ii)   A full berm cut may not be required each time, as outlined above.

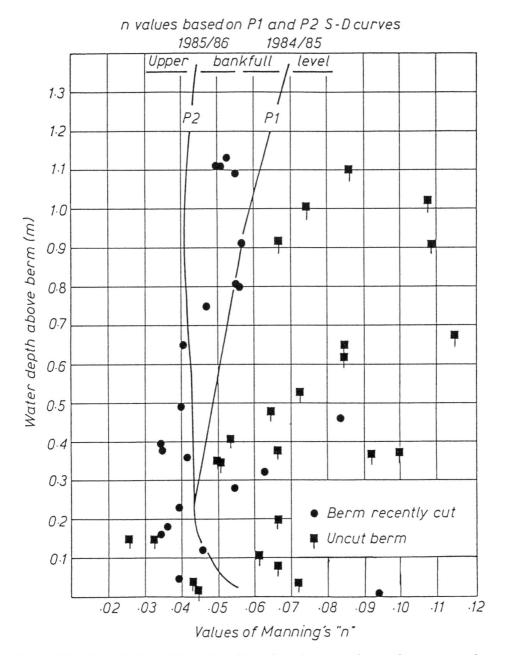

Figure 8.30 River Roding field study: effect of cutting vegetation on berm expressed as experimentally determined Manning's *n* values: isolated points represent summer conditions, lines represent winter behaviour

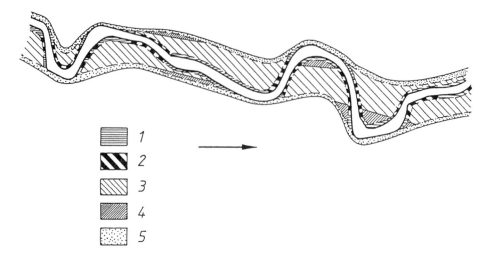

Figure 8.31 River Roding model study: definition of berm sub-zones used in selective cutting tests: 1 – marginal strip, 2 – marginal strip (areas of strong cross-flow), 3 – main berm, 4 – low velocity berm areas, 5 – upper channel banks

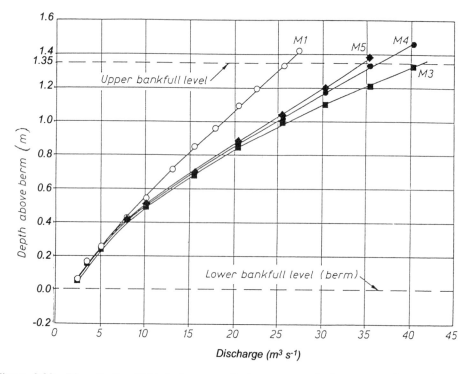

Figure 8.32 River Roding field study: stage–discharge curves for berm vegetation cutting tests (see text for test details)

(iii) If a low-level berm is used in a design, annual vegetation growth will be large and provision must be made for regular maintenance on the berm.

(iv) To minimize maintenance problems berm areas should be designed to give easy access for suitable equipment. Steep upper banks are not necessary and a continuous access strip should be provided along this bank top on either side of the river.

(v) Sinuous two-stage channel designs can be adopted to provide a valuable wildlife habitat in the river corridor. Reed and sedge stands should be left uncut during the breeding season to provide suitable habitats for marsh bird species. The resulting additional resistance must be allowed for in the hydraulic sizing of the channel elements.

River Blackwater, Hampshire, UK

The construction of a major trunk road along the River Blackwater valley in Hampshire resulted in the need to relocate a length of that river. This offered the UK National Rivers Authority the opportunity to design the new channel and undertake a multidisciplinary study of an environmentally acceptable channel design with a view to developing a set of guidelines for more general use.

Because the collection of field data is a chancy business in that the occurrence of high discharges in a river during a certain study period cannot be guaranteed, the decision was taken to construct and test hydraulic models of the test reach in order to extend the parameter range under controlled conditions. Two models were built to undistorted scales of 1/5 and 1/25. Thus it was possible at the same time to explore the effect of scale on the validity of the model results obtained. The large 1/5 scale model had a length of 56 m and was constructed in the National Flood Channel Facility at HR Wallingford and tested in 1993. The small-scale model was constructed at the University of Bristol in 1994 for a more exhaustive test programme involving a larger list of parameters. The reconstruction of the prototype river channel and the installation of the instrumentation were completed in 1994.

The new channel design was based on discharge values obtained from a detailed numerical model of the upstream catchment (ONDA), to take account of the effect of the new road construction on the river hydrographs.

Instrumentation installed in the river study reach consisted of five recording water-level gauges and an electromagnetic flow gauge installed just upstream of the test reach shown in outline in Figure 8.33. It is expected that field data will be collected during the period 1995–1997.

Objectives for the field measurement programme include the following:

1. The determination of friction factor values for different channel and berm vegetation states.

2. To carry out a detailed morphological study of the channel and berm as this develops.

3. To undertake an environmental study. This will be carried out by NRA (now EA) Thames Region, and will include the colonization of the new channel and berm, using different degrees of new planting and the introduction of aquatic invertebrates. A pool and riffle system will be included in the channel bed design.

4. A study of river colonization by fish species.

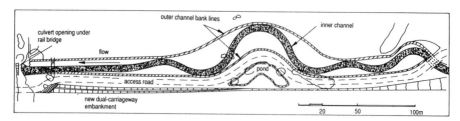

Figure 8.33   Plan of two-stage channel test reach on River Blackwater, Hampshire, UK

Objectives for the FCF-based and small-scale model programmes include:

1. The determination of the effect of berm cross-fall on the hydraulic capacity of the whole channel.
2. To make a detailed examination of flow conditions in the flood channel at or near design flood conditions.
3. To investigate scale effects in laboratory river models.
4. To investigate the effect of using a distorted vertical scale in small laboratory river models.

8.4   DISCHARGE PREDICTION MODELS

One possible method of estimating channel conveyance in meandering compound flows is to estimate the energy dissipation due to each source including bed friction, planform/bend losses, expansion/contraction losses and other flow interaction losses. This type of hydromechanics approach has been attempted by Ervine and Ellis (1987) with limited success. A similar approach has been attempted by James and Wark (1992), combining bed friction losses with lumped estimates of interaction losses between different zones within the meandering compound cross-section. A similar approach has been adopted by Greenhill and Sellin (1993).

In the meantime an alternative approach has been developed that does not quantify individual sources of energy loss, but instead provides a global correction factor to a theoretical estimate of the discharge based on a simple bed-friction and straight channel only model. The correction factors are therefore related to the range of parameters believed to affect compound meandering flow and discussed above. Discharge estimates may also be influenced by the Reynolds number of the flow, and therefore careful regard must be paid to scale effects where scale model data is used for this purpose.

8.4.1   Single channel scheme

Figure 8.34 identifies two distinct types of meandering channel/floodplain system. In Section 8.3 we called them quasi-straight and quasi-natural systems. From what has already been said, it should be clear that methods of predicting depth from discharge, or the converse, should take careful account of this distinction. Indeed,

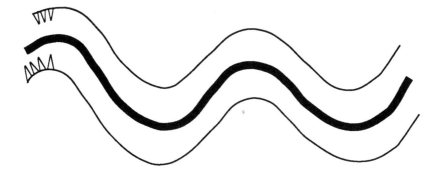

Quasi-straight compound channel system

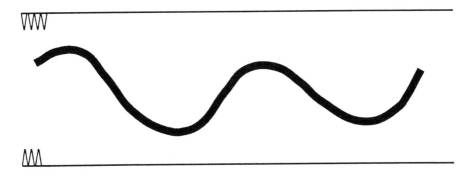

Sinuous channel in non-compliant floodplain

Figure 8.34   Schematic channel layouts

the first step in any calculation should be to examine the geometry of the system and identify its type. Only then can a proper decision be made about the appropriate prediction scheme.

Whereas experiments performed to investigate flow resistance in compound channels usually employ steady, uniform flow, the practical circumstances in which the derived knowledge is applied invariably concern unsteady flows (flood hydrographs). When interest is confined to one reach of only one stream, it is sometimes satisfactory to make calculations for the peak flow treated as if it were steady, but more generally it is necessary to route unsteady flood waves through tributary and main channels in a catchment. This involves the use of equations like 8.1i, ii and iii below, of which 8.1 iii introduces the experimental evidence about flow resistance in one of a small number of well-known formulations.

$$\frac{\partial A}{\partial t} + \frac{\partial Q}{\partial x} = q \tag{8.1i}$$

$$\frac{\partial Q}{\partial t} + \frac{\partial}{\partial x}\left(\frac{Q^2}{A}\right) + gA\left(\frac{\partial h}{\partial x} + S_f\right) = 0 \tag{8.1ii}$$

$$Q = \frac{A}{n} \cdot R^{2/3} S_f^{1/2} \tag{8.1iii}$$

Here x is the displacement in the flow direction, h and A are the flow depth and cross-sectional area, Q is the flow rate, q is the rate of lateral inflow, S_f is the hydraulic gradient, R is the hydraulic radius, t is time, and g is gravitational acceleration. n is the Manning resistance coefficient, one of several ways of expressing flow resistance. See also the guideline design report: NRA (1994).

As we have discussed, the flow resistance in the case of overbank flow arises not only from boundary shear stress in the inner channel and on the floodplains, but also from energy dissipation associated with the interaction between the channel and floodplain flows. This derives from vorticity in the vertical shear layers between these flows when they flow sensibly parallel to one another, as in straight and quasi-straight systems. When the floodplain boundaries do not follow with the inner channel plan geometry, these vertical shear layers do not build up and the important sources of energy loss now include the turbulence associated with vigorous mixing of the channel and floodplain waters where flow directions do not coincide, and strong secondary currents are set up in the channel cross-section. In either case, flow rate can be calculated for the whole cross-section, making allowance as is appropriate for the several dissipation processes, or for each of a number (usually three or four) of zones into which the cross-section is divided for this purpose, the zone sub-flows then being aggregated to give a system flow rate. In this section we are concerned with the first of these approaches.

In Figure 8.35 the upper sketch shows the cross-section of a uniform compound channel, the lower one a system in which the inner channel meanders across the floodplain within the limits indicated by the vertical lines. In both, the wetted perimeter increases drastically when the flow first goes overbank, with very little change in the cross-sectional area of flow. Therefore a resistance coefficient like n in equation 8.1iii, which is formulated in terms of R, the ratio of area wetted perimeter can be expected to behave oddly at small overbank depths. This is indeed true, and the changes in the value of the resistance coefficient must be incorporated in calculations to achieve reasonable accuracy, whether n or an alternative coefficient is employed. In the present state of knowledge, this has to be done on the basis of experimental evidence. For compound channels of the upper type, such evidence is plentiful and has been organized with design calculation in mind (Ackers, 1991). While it is now unconventional to do so, it would be possible, for systems of this type, to calculate discharge using equation 8.1iii with area, wetted perimeter and thus hydraulic radius evaluated for the complex cross-section as a whole, provided that due account were taken of the behaviour of the resistance coefficient when flow first goes overbank (Myers and Brennan, 1990). For those of the lower type, which involve many more

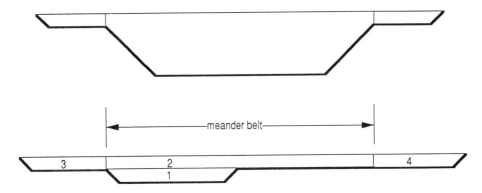

Figure 8.35 System cross-sections. The upper one has a uniform inner channel symmetrically placed in a uniform floodplain; the lower one has a channel which meanders within the limits indicated by the vertical lines in a floodplain that can be considered straight

variables and much more complex flows, the resistance behaviour is less comprehensively understood and methods of expressing it in calculations are more tentative. In these systems it is rarely wise to attempt to base discharge calculations on the lumped resistance of channel and floodplains. (James and Brown (1977) devised a method of this kind, but were uneasy about its obvious dependence on very limited data for meandering channels.) An exception to this rule occurs when depth on the floodplain is very large and therefore most of the flow conveyed is above bankfull level. In this case, particularly when the inner channel sinuosity is high, the system may be considered as a single channel (the floodplain) in which the inscribed inner channel provides hydraulic roughness. The answer obtained will be inaccurate but, because of the large discharge carried by the floodplain, the inaccuracy may be unimportant. Yen and Yen (1983) produced a single-channel method which took no account of the contribution to system discharge of the inner channel flow, but treated the inner channel as a source of flow resistance only. The inescapable error incurred by doing this clearly diminishes in significance when depth on the floodplain increases sufficiently for floodplain flow to dominate the system discharge. However, it is significant that publications by James and Wark (1992), and by Ervine *et al.* (1993), designed to help with calculations for such systems, recommend a method based on the conveyance of sub-zones in the cross-section with subsequent correction for the interactions between the sub-zone flows. Calculations of this type will be discussed in a later section.

Single channel schemes of calculation are severely limited in usefulness by two intractable difficulties. On one hand the provision of a database sufficient to define the influence on discharge of the bewildering array of relevant variables would require an impossibly long and costly experimental programme. On the other hand the physics of interaction between different zones of the flow, which would have to be properly understood to enable a sound theoretical method to be developed for the calculation of the compound flow, involves features for which the numerical modelling techniques available present extreme difficulty and are too complex to resolve without them.

8.4.2   Divided channel configurations

The way in which system cross-sections may be divided to assist in producing reliable predictions about discharge is illustrated in Figure 8.35. The flow in each zone into which the section is divided can be estimated on the assumption that it is restrained by the portion of the solid boundary in contact with the water in that zone; then a correction must be applied to the system discharge obtained by the aggregation of the zonal sub-flows, to account for interaction at the interfaces between them. Although several studies have evaluated motional shear stress transferred at the interfaces for particular circumstances, it is usual, in making the preliminary calculation of zonal sub-flows, to ignore the interaction at that stage. The calculation of the sub-flows then shares the simplicity of routine calculation of discharge for the straightforward case of inbank flow, and also its difficulty, namely the attribution of a value for the resistance coefficient based on limited knowledge of the topography, texture and vegetation growth at the boundaries.

Modification of the aggregated flow thus obtained to account for the particular complications of two-stage (or multi-stage) flow is vital because the resistance associated with these special features is greater than that arising from boundary shear alone. Two distinct approaches are possible, one based on estimates of the energy dissipation rate associated with the interactions between the zonal flows, and the other making direct use of experimental findings about the influence on discharge of the several parameters known to affect it. These are listed by Ervine et al. (1993) as:

1.   Sinuosity of the main channel.
2.   Relative roughness of the floodplain boundary compared with the main channel.
3.   Aspect ratio of the main channel.
4.   Meander belt width relative to the total floodway width.
5.   Relative depth on the floodplain compared with the main channel.
6.   The main channel cross-sectional shape, including the side slope of its banks.
7.   Floodplain topography, and in particular lateral slope toward the main channel.

All of these influences have been investigated at laboratory scale, and all but the sixth at the large scale afforded by the UK Flood Channel Facility. However, the limited range of variation of the parameters, and the artificiality of some features of the laboratory channels, restricts the confidence with which these data can be directly applied to design calculations. Since all the calculation methods that have been advocated have important dependency on empiricism, these shortcomings of the data sources are serious, and emphasize the urgent need for further research.

In following the recent development of methods of introducing zonal interaction into the calculation of system discharge, it is helpful to highlight three studies, each of which uses the available experimental evidence differently. This selection is based on their relationship to one another rather than on any supposed superiority to other recent contributions on the subject. All three use the scheme of subdivision of the system shown in Figure 8.35.

Ervine and Ellis (1987) based their calculation of discharge on semi-empirical formulae for energy loss in the flow features they considered important in compound meandering channels, but avoided resort to data from multi-stage channel experiments wherever possible. Thus, for example, the energy loss at the flow expansion which was postulated to occur where floodplain flow within the meander belt width (zone 2) encounters the inner channel was calculated by the application of the force–momentum principle in two dimensions, and the contraction loss where it re-encounters the floodplain by means of loss coefficients presented by Rouse (1950). The inner channel flow (zone 1) was assumed to be restrained by boundary friction and by energy lost to secondary currents, and not by interaction with other elements of the cross-section. It was recognized that, in the peculiar circumstances of compound flow with a meandering inner channel, secondary currents grow and decay in a manner unlike that found in the case of inbank flow and, for that reason, the result of Chang (1983) for energy loss to secondary currents was halved for out-of-bank calculations.

This piecemeal application of energy loss calculations developed for other circumstances was relatively successful, giving discharge predictions that compare well, for most data, with those of other methods (e.g. James and Wark, 1992).

James and Wark themselves (1992), having the advantage of a large and new set of careful observations made on the SERC Flood Channel Facility at large scale, set out to produce a physically based deterministic model incorporating the loss mechanisms which had been experimentally observed. They were unable to carry this approach to a successful conclusion because of the weakness of our present understanding .of the interaction between the flows in zones 1 and 2. Accordingly, they arrived at methods which retained physical arguments where they could be sustained, but depended heavily on experimental evidence of the influence of the boundary geometry and flow depth on the flow resistance. Such evidence used in the development of the procedures was drawn from two sources (experimental evidence in the SERC Flood Channel Facility and at Aberdeen University), and data from other investigations were then used to test the procedures.

Several variants of the basic method were produced and tested. The best of them were sound for circumstances rather similar to those in which the two underlying data sets were produced. However, when tested against all the data available to the authors, some evidence was found of frailty, particularly in cases involving rough floodplains. Comparative tests of the variants emphasized the importance of the empirically derived input.

Ervine *et al.* (1993) adopted a device of Ackers (1991) to systematize a purely empirical approach to the calculation of compound channel discharge. This uses a non-dimensional discharge coefficient, F^*, the ratio of measured discharge to discharge calculated by the multi-zone method. F^* was found to vary with six of the seven influences listed above (the seventh was not investigated) when applied to experimental results and it was implied that the observed pattern of variation of F^* could be applied in field situations, with due caution. The sparsity of the data available to define this variation indicates vividly the need for considerable programmes of further observation before not only this, but all other methods of discharge prediction for compound channels, can be thought of as reliable.

The performance of all these methods, and the reservations with which the authors present them, make it clear that their reliability is limited by the inadequate understanding that we presently have of the mechanisms of energy loss in multi-stage systems in which the inner channel meanders and the floodplain boundaries do not comply with the inner channel plan geometry. These methods generate confidence only for cases falling within or close to the range represented by the database from which they are constructed. Because the same database, broadly, has been used in the construction of all the recent numerical models, there is a tendency for them to share the same areas of strength and weakness. In particular, they remain generally untested for full-scale application to natural systems. Recent advances, derived from a concentration of research effort in the problem, have reduced the risks inherent in design calculations with a weak physical base, but much remains to be done yet.

REFERENCES

Ackers, P. (1991) Hydraulic design of straight compound channels. Report SR 281, HR Wallingford.

Chang, H.H. (1983) Energy expenditure in curved open channels. *J. Hyd. Engng, ASCE*, **109**(HY7), 1012–1022.

Engelund, F, (1966) Hydraulic resistance of alluvial streams. *J. Hyd. Engng, ASCE*, **92**(HY2), 315–326.

Ervine, D.A. and Ellis, J. (1987) Experimental and computational aspects of overbank flood-plain flow. *Trans. Roy. Soc. Edin., Series A*, **78**, 315–475.

Ervine, D.A. and Jasem, H.K. (1989) Flood mechanisms in meandering channels with flood-plain flow. *Proc. 23rd IAHR Congress, Ottawa, Canada*, pp. B449–B456.

Ervine, D.A., Willetts, B.B., Sellin, R.H.J. and Lorena, M. (1993) Factors affecting conveyance in meandering compound flows. *J. Hyd. Engng, ASCE*, **119**(12), 1383–1399.

Ervine, D.A., Sellin, R.H.J. and Willetts, B.B. (1994) Large flow structures in meandering compound channels. In: *Proc. 2nd Int. Conf. on River Flood hydraulics*, W.R. White, and J. Watts (eds), John Wiley, pp. 459–469.

Greenhill, R.K. and Sellin, R.H.J. (1993) Development of a simple method to predict discharges in compound meandering channels. *Proc. Inst. Civ. Engrs., Water, Maritime and Energy J.*, **101**, 37–44.

Holden, A.P. and James, C.S. (1989) Boundary shear distribution on flood plains. *J. Hydraulic Research, IAHR*, **27**(1), 78–89.

James, C.S. and Wark, J.B. (1992) Conveyance estimation for meandering channels. Report SR 329, HR Wallingford.

James, M. and Brown, B.J. (1977) Geometric parameters that influence flood-plain flow. Corps of Engineers, US Army, Waterways Experiment Station, Vicksburg.

Kiely, G.K. (1990) Overbank flow in meandering compound channels – the important mechanisms. In: *Proc. Int. Conf. on River Flood Hydraulics*, W.R. White (ed.), John Wiley, 207–217.

Knight, D.W., Yuan, Y.M. and Fares, Y.R. (1992) Boundary shear in meandering river channels. *Proc. Int. Symp. on Hydraulic Research in Nature and Laboratory*, Yangtze River Scientific Research Institute, Wuhan, China, Nov., Vol. 2, pp. 102–106.

Kouwen, N. and Li, R.M. (1980) Biomechanics of vegetative channel linings *J. Hyd. Div., ASCE*, **106**(HY6), 1085–1103.

Lyness, J. and Myers, W.R.C. (1994a) Comparisons between measured and numerically modelled unsteady flows in a compound channel using different representations of friction slope. *2nd Int. Conf. on River Flood Hydraulics*, W.R. White and J. Watts (eds), John Wiley, 383–391.

Lyness, J. and Myers, W.R.C. (1994b) Velocity coefficients for overbank flows in a compact compound channel and their effects on the use of one dimensional flow models. In: *Proc. 2nd Int. Conf. on Hydraulic Modelling*, A.J. Saul (ed.) Mech. Eng. Publications, 379–398.

Martin, L.A. and Myers, W.R.C. (1991) Measurement of overbank flow in a compound river channel. *Proc. Inst. Civ. Engrs, Part 2*, **91**, 645–657.

McKeogh, E.J. and Kiely, G.K. (1989) Experimental study of the mechanisms of flood flow in meandering channels. *Proc. 23rd IAHR Congress, Ottawa, Canada*, pp. B491–B498.

Myers, W.R.C. (1990) Physical modelling of a compound river channel. In: *Proc. Int. Conf. on River Flood Hydraulics*, W.R. White (ed.), John Wiley, 381–390.

Myers, W.R.C. and Brennan, E. (1990) Flow resistance in compound channels. *J. Hydraulic Research*, **28**(2) 141–155.

Nalluri, C. and Judy, N.D. (1985) Interaction between main channel and flood-plain flow. *Proc. 21st IAHR Congress, Melbourne, Australia*, pp. 378–382.

NRA (1994). Design of straight and meandering compound channels – interim guidelines on hand calculation methodology. NRA R&D Report No. 13, National Rivers Authority.

Purseglove, J. (1989) *Taming the Flood, a History and Natural History of Rivers and Wetlands.* " Oxford University Press with Channel 4 Television.

Rouse, H. (ed.) (1950) *Engineering Hydraulics*, John Wiley.

Rozovskii, I.L. (1961) *Flow of Water in Bends of Open Channels*. Acad. of Sci. of the Ukraine (translated from the Russian by the Israel Programme for Scientific Translations).

Sellin, R.H.J. and Giles, A. (1989) Flow mechanisms in spilling meander channels. *Proc. 23rd IAHR Congress, Ottawa, Canada*, pp. B499–B506.

Sellin, R.H.J., Giles, A. and van Beesten, D.P. (1990) Post implementation study of a two-stage channel in the R Roding, Essex. *J. IWEM*, **4**, 119–130.

Thorne, C.R., Zevenbergen, L.W., Pitlick J.C. Rais, S. Bradley, J.B. and Julien, P.Y. (1985) Direct measurements of secondary currents in a meandering sand-bed river. *Nature*, **315**, 746–747.

Toebes, G.H. and Sooky, A.A. (1967) Hydraulics of meandering rivers with flood-plains. Proc. ASCE, J. of Waterways and Harbours Division, **93**(2), 213–226.

US Soil Conservation Service (1954) Handbook of channel design for soil and water conservation. SCS-TP-61, Washington, DC, 34/pp.

US Army Corps of Engineers (1956) Hydraulic capacity of meandering channels in straight floodways. Waterways Experiment Station, Vicksburg, Tech. Mem. 2–429.

Yen, B.C. and Yen, C.L. (1983) Flood flow over meandering channels. *Proc. Conf. Rivers '83*, ASCE, pp. 554–561.

9 Progress in Turbulence Modelling for Open-Channel Flows

B.A. YOUNIS

Department of Civil Engineering, City University, London, UK

9.1 INTRODUCTION

The goal of mathematical simulations of open-channel flows is to predict, to acceptable engineering accuracy, the dependence of flow parameters such as the mean velocity, the boundary stresses and the turbulence field on geometric parameters such as the channel slope, sinuosity, cross-sectional shape and the surface roughness. There is, *in principle*, no difficulty in attaining this goal: the equations that describe the instantaneous fluid motions are both exact and well known, algorithms for solving these equations numerically for complex boundaries are now commonplace and the computer resources, that have for so long formed an effective deterrent to serious simulations, are increasingly cheap and abundant. The difficulty in the *practical* simulation of turbulent flows, of course, from the impossibility of conducting simulations with mesh- and time-step sizes that are sufficiently small to capture the details of the instantaneous motion. Time-averaging the instantaneous equations removes the need to resolve the small-scale, high-frequency motions but introduces unknown correlations, the turbulent (or Reynolds) stresses, which have now to be approximated in terms of known or knowable quantities. This is then the need for a *model of turbulence* and its primary requirement; to represent the effects of the most complicated form of fluid motion with a number of equations that are sufficiently few and well-posed to be usable in routine engineering calculations. In this chapter, we consider some of the difficulties associated with the development and verification of appropriate turbulence models for open-channel flows and report on some recent progress in this field.

Nearly all computational methods in current use in engineering utilize a turbulence model which is based on Boussinesq's assumption that the Reynolds stresses $(\overline{u_i u_j})$ are linearly proportional to the local mean rate of strain. thus:

$$-\overline{u_i u_j} = \nu_t \left(\frac{\partial U_i}{\partial x_j} + \frac{\partial U_j}{\partial x_i} \right) - \frac{2}{3} \delta_{ij} k \qquad (9.1)$$

In the above, ν_t is the eddy or turbulent viscosity, which has to be determined from algebraic relations or from the solution of differential transport equations.

Floodplain Processes. Edited by Malcolm G. Anderson, Des E. Walling and Paul D. Bates.
© 1996 John Wiley & Sons Ltd.

Boussinesq's assumption, while adequate in many turbulent flows, does not provide the basis for a genuinely *predictive* model for open-channel flows. That this is the case is due to one or more of the following reasons:

1. The assumption of a *linear* stress–strain relation produces incorrect levels of turbulence anisotropy. Nowhere is this more serious than in man-made and natural channels of non-circular cross-section where the turbulence anisotropy is the primary mechanism for the generation of turbulence-driven secondary motions in planes normal to the main flow direction (Prandtl, 1926). These motions are generally quite small (being of the order of 4% of the maximum streamwise velocity) but are nevertheless quite influential in determining the overall flow behaviour especially, for example, in transferring momentum between a main channel and its floodplains (see Figure 9.1 for geometry and coordinates). Models based on Equation 9.1 yield no secondary flow whatsoever.

2. The absence of a mechanism to reflect the presence of a free surface. There is evidence from experiments (Komori *et al.*, 1982) and from Direct Numerical Simulations (Handler *et al.*, 1993) to suggest that the free surface modifies the fluctuating pressure field in its vicinity in such a way as to transfer energy from the turbulent velocity fluctuations normal to the free surface into the components parallel to it. Such inter-component transfer is not accounted for in Equation 9.1.

3. Insensitivity to the effects of longitudinal streamline curvature. Turbulent shear layers are known to be very sensitive to streamline curvature in the plane of the mean shear (Bradshaw, 1973) where the extra rates of strain introduced by the curvature enhance the turbulence activity on the concave (de-stabilized) side but suppress it on the convex (stabilized) side leading, in conditions of strong curvature, to the collapse of the turbulent shear stress in regions of finite shear (So and Mellor, 1973). Turbulence models based on Equation 9.1 fail to reproduce such effects unless modified in some *ad hoc* way.

4. Neglect of "history" effects (by which is meant the transport of turbulence quantities by mean flow convection and by turbulent fluctuations). This adversely affects the models' performance in two important situations. Firstly, in

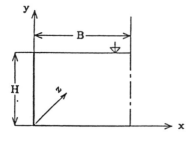

(a) Rectangular channel

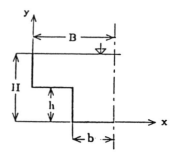

(b) Compound channel

Figure 9.1   Channel flow coordinates and notation (a) Rectangular channel (b) Compound channel

strongly three-dimensional flows, as would occur for example in a main channel with skewed floodplains: the eddy viscosity is no longer isotropic (ν_t is universally taken as a scalar quantity) and the stress and mean rate of strain directions would no longer be aligned. Secondly, in rapidly evolving flows produced, for example, by sudden change in boundary conditions (slope, surface roughness, etc.). Here, the mean flow and the turbulence fields are no longer in equilibrium and the turbulent stresses would no longer be related to the local strain field in a simple way.

It thus seems necessary in the search for a reliable and generally applicable turbulence model for open-channel flows to abandon Boussinesq's stress–strain relationship in favour of alternative means for representing the unknown Reynolds stresses. In this chapter, we consider two such alternatives: a *non-linear* stress–strain relationship used in conjunction with a two-equation model and a complete Reynolds stress transport closure in which the unknown stresses are obtained, directly, from the solution of their own differential transport equations. The mathematical basis of these models will be described in the next section followed, in Section 9.3, by the presentation of sample results for a variety of flows in ducts and channels. Closing remarks are provided in Section 9.4.

9.2 TURBULENCE MODELS CONSIDERED

9.2.1 A non-linear eddy-viscosity model (NKE)

Several alternatives to Boussinesq have been reported in the literature. Lumley (1970), for example, proposed a generalized constitutive relationship for $\overline{u_i u_j}$ which he arrived at by identifying the appropriate scaling parameters that enter in the relation and then expressing it as a finite tensor polynomial. Pope (1975) pointed out that Lumley had made "illicit use" of the alternating tensor density and that rendered the results invalid. Pope proposed an alternative form which is restricted to two-dimensional flows (he remarked that the three-dimensional form was "so intractable as to be of no value"). Speziale (1987) proposed a model which is *quadratic* in the mean rate of strain. This was arrived at by postulating a particular form of the stress tensor and then specifying it by the application of a number of fundamental constraints such as invariance, realizability and material frame indifference in the limit of two-dimensional turbulence. This proposal may be expressed as:

$$-\overline{u_i u_j} = 2\nu_t S_{ij} - \frac{2}{3}\rho k \delta_{ij}$$

$$+ C_D L (S_{im} S_{mj} - \frac{1}{3} S_{mn} S_{mn} \delta_{ij})$$

$$+ C_E L (\overset{\circ}{S}_{ij} - \frac{1}{3} \overset{\circ}{S}_{mm} \delta_{ij}) \tag{9.2}$$

where S_{ij} is the mean rate of strain tensor:

$$S_{ij} = \frac{1}{2}\left(\frac{\partial U_i}{\partial x_j} + \frac{\partial U_j}{\partial x_i}\right) \tag{9.3}$$

and $\overset{\circ}{S}_{ij}$ is the Oldroyd derivative, defined as:

$$\overset{\circ}{S}_{ij} = \frac{\partial S_{ij}}{\partial t} + U_m \frac{\partial S_{ij}}{\partial x_m} - \frac{\partial U_i}{\partial x_m}S_{mj} - \frac{\partial U_j}{\partial x_m}S_{mi} \tag{9.4}$$

In Equation 9.2, ν_t, the eddy viscosity, is obtained from:

$$\nu_t = C_\mu \frac{k^2}{\varepsilon} \tag{9.5}$$

and L is the turbulence length scale defined as:

$$L = 4C_\mu^2 \frac{k^3}{\varepsilon^2} \tag{9.6}$$

Note that the first line of Speziale's model is in fact Boussinesq's relation (Equation 9.1) which expresses the turbulent stresses as a linear function of S_{ij}. Line 2 reveals the quadratic nature of the model: when expanded, this line will produce terms that contain velocity gradients multiplied by themselves or by other velocity gradients. Line 3, which contains the Oldroyd derivative, introduces the dependence of $\overline{u_i u_j}$ on the "history" of the mean rate of strain. Note the introduction now of terms which contain gradients of the velocity gradients and yet more velocity gradients multiplying other gradients. Some appreciation of the complexity involved may be gained from inspection of the relation for the normal stress component $(-\overline{u^2})$ which, for a fully developed three-dimensional flow, may be written as:

$$-\overline{u^2} = 2\nu_t \frac{\partial U}{\partial x} - \frac{2}{3}k$$

$$+ \frac{LC_D}{4}\left\{4\left(\frac{\partial U}{\partial x}\right)^2 + \left(\frac{\partial W}{\partial x}\right)^2 + \left(\frac{\partial U}{\partial y} + \frac{\partial V}{\partial x}\right)^2\right.$$

$$-\frac{4}{3}\left[\left(\frac{\partial U}{\partial x}\right)^2 + \left(\frac{\partial U}{\partial y}\right)^2 + \frac{1}{2}\left(\left(\frac{\partial W}{\partial x}\right)^2 + \left(\frac{\partial W}{\partial y}\right)^2 + \left(\frac{\partial U}{\partial y} + \frac{\partial V}{\partial x}\right)^2\right)\right]\right\}$$

$$-LC_E\left\{2\left(\frac{\partial U}{\partial x}\right)^2 + \left(\frac{\partial U}{\partial y}\right)^2 - \left(\frac{\partial U}{\partial y}\frac{\partial V}{\partial x}\right) + U\frac{\partial^2 U}{\partial x^2} - V\frac{\partial^2 U}{\partial y \partial x}\right.$$

$$-\frac{2}{3}\left[\left(\frac{\partial U}{\partial x}\right)^2 + \left(\frac{\partial U}{\partial y}\right)^2 + \frac{1}{2}\left(\left(\frac{\partial W}{\partial x}\right)^2 + \left(\frac{\partial W}{\partial y}\right)^2 + \left(\frac{\partial U}{\partial y} + \frac{\partial V}{\partial x}\right)^2\right)\right]\right\} \tag{9.7}$$

Clearly, this model introduces many more terms than Boussinesq and this, at the very least, will increase the computational effort involved in evaluating the Reynolds stresses. Moreover, the task of obtaining a converged numerical solution to the mean flow and turbulence equations becomes somewhat more demanding due to the difficulties involved in evaluating some of the velocity gradient terms in the near-wall region. This model is undoubtedly complicated but so, also, is the phenomenon being modelled.

The turbulence kinetic energy k and its dissipation rate, ε, which appear in the definition of the eddy viscosity and the length scale, are obtained from the solution of the standard differential transport equations:

$$U_j \frac{\partial k}{\partial x_j} = \frac{\partial}{\partial x_j}\left(\frac{\nu_t}{\sigma_\varepsilon}\frac{\partial k}{\partial x_j}\right) + P_k - \varepsilon \tag{9.8}$$

$$U_j \frac{\partial \varepsilon}{\partial x_j} = \frac{\partial}{\partial x_j}\left(\frac{\nu_t}{\sigma_\varepsilon}\frac{\partial \varepsilon}{\partial x_j}\right) + C_{\varepsilon_1}\frac{\varepsilon}{k}P_k - C_{\varepsilon_2}\frac{\varepsilon^2}{k} \tag{9.9}$$

The form of the k and ε equations used in the non-linear eddy-viscosity model is therefore identical to that of the more familiar linear model, the sole difference being the precise relation used to link the Reynolds stresses to the velocity field. A number of empirical coefficients are involved, and those are assigned the values recommended in Speziale (1987).

9.2.2 A Reynolds stress transport model (RSM)

In models of this type, the concept of an explicit stress–strain relationship is abandoned altogether and, instead, the Reynolds stresses are obtained from the solution of differential transport equations of the form:

$$
\underbrace{U_k \frac{\partial \overline{u_i u_j}}{\partial x_k}}_{\text{convection}} = -\underbrace{\left(\overline{u_i u_k}\frac{\partial U_j}{\partial x_k} + \overline{u_j u_k}\frac{\partial U_i}{\partial x_k}\right)}_{\text{production}}
$$

$$
-\underbrace{\frac{\partial}{\partial x_k}\left[\overline{u_i u_j u_k} + \frac{1}{\rho}\overline{(p'u_j}\delta_{ik}) - \nu\frac{\partial \overline{u_i u_j}}{\partial x_k}\right]}_{\text{diffusion}}
$$

$$
-\underbrace{2\nu\left(\frac{\partial u_i}{\partial x_k}\frac{\partial u_j}{\partial x_k}\right)}_{\text{dissipation}} + \underbrace{\frac{p'}{\rho}\left(\frac{\partial u_i}{\partial x_j} + \frac{\partial u_j}{\partial x_i}\right)}_{\text{redistribution}} \tag{9.10}
$$

Equation 9.10 contains several unknown correlations which need to be modelled. Of the three components of the diffusion term, viscous diffusion is retained in its original form, pressure diffusion is neglected as nothing is known about it, and the triple correlations (which represent the transport of $u_i u_j$ by the turbulent fluctuations) are normally modelled using Daly and Harlow's (1970) gradient-transport hypothesis:

$$-\overline{u_i u_j u_k} = C_s\frac{k}{\varepsilon}\overline{u_k u_l}\frac{\partial \overline{u_i u_j}}{\partial x_l} \tag{9.11}$$

The dissipation of $\overline{u_i u_j}$ by viscous action is assumed to be isotropic at high turbulence Reynolds numbers (Rotta, 1951) to obtain:

$$2\nu \overline{\left(\frac{\partial u_i}{\partial x_k}\frac{\partial u_j}{\partial x_k}\right)} = \frac{2}{3}\delta_{ij}\varepsilon \tag{9.12}$$

The dissipation rate (ε) is obtained from the solution of an equation similar to that used in eddy-viscosity models, viz.:

$$U_j \frac{\partial \varepsilon}{\partial x_j} = \frac{\partial}{\partial x_k}\left(C_\varepsilon \frac{k}{\varepsilon}\overline{u_k u_l}\frac{\partial \varepsilon}{\partial x_l}\right) + \frac{\varepsilon}{k}(C_{\varepsilon_1}P_k - C_{\varepsilon_2}\varepsilon) \tag{9.13}$$

where P_k is the rate of production of turbulence kinetic energy. Equation 9.13 is identical to that used in conjunction with the non-linear $k-\varepsilon$ model, except for the turbulent diffusion term which is modelled along the lines suggested by Equation 9.11.

The last term in Equation 9.10 is responsible for the redistribution of the turbulence kinetic energy among the three fluctuating velocity components and is customarily taken to consist of three separate contributions:

$$\overline{\frac{p'}{\rho}\left(\frac{\partial u_i}{\partial x_j} + \frac{\partial u_j}{\partial x_i}\right)} = \Phi_{ij,1} + \Phi_{ij,2} + \Phi_{ij,w} \tag{9.14}$$

The separate elements represent, respectively, purely turbulent interactions, interactions between the mean strain field and fluctuating velocities and, finally, corrections needed to account for the damping effects of a solid wall on the fluctuating pressure field in its vicinity.

$\Phi_{ij,1}$ is modelled after Rotta's (1951) proposal which makes it proportional to the turbulence anisotropy:

$$\Phi_{ij,1} = -C_1\varepsilon\left(\frac{\overline{u_i u_j}}{k} - \frac{2}{3}\delta_{ij}\right) \tag{9.15}$$

Launder et al. (1975) proposed to model $\Phi_{ij,2}$ as:

$$\Phi_{ij,2} = -\frac{C_2 + 8}{11}\left(P_{ij} - \frac{2}{3}\delta_{ij}P_k\right) - \frac{30C_2 - 2}{55}k\left(\frac{\partial U_i}{\partial x_j} + \frac{\partial U_j}{\partial x_i}\right)$$

$$-\frac{8C_2 - 2}{11}\left(D_{ij} - \frac{2}{3}\delta_{ij}P_k\right) \tag{9.16}$$

where P_{ij} is the rate of production of $\overline{u_i u_j}$ defined in Equation 9.10, and

$$D_{ij} = -\left(\overline{u_i u_k}\frac{\partial U_k}{\partial x_j} + \overline{u_j u_k}\frac{\partial U_k}{\partial x_i}\right)$$

Recently, Speziale *et al.* (1991) proposed the following model for the sum of $\Phi_{ij,1}$ and $\Phi_{ij,2}$:

$$\Phi_{ij} = -(C_1\varepsilon + C_1^* P_k)b_{ij} + C_2\varepsilon\left(b_{ik}b_{ij} - \frac{1}{3} b_{mn}b_{mn}\delta_{ij}\right)$$

$$+[C_3 - C_3^*(b_{mn}b_{mn})^{1/2}]kS_{ij} + C_4k\left(b_{ik}S_{jk} + b_{jk}S_{ik} - \frac{2}{3} b_{mn}S_{mn}\delta_{ij}\right)$$

$$+C_5k(b_{ik}W_{jk} + b_{jk}W_{ik}) \tag{9.17}$$

where b_{ij} and W_{ij} are, respectively, the Reynolds stress anisotropy and the mean vorticity tensors.

This model has become quite popular recently as it proved superior to the Launder *et al.* formulation in a number of aspects (Basara and Younis, 1995) and some results obtained with it will be presented later in this chapter.

$\Phi_{ij,w}$, the term needed to modify the redistribution process in the presence of a solid wall, consists of separate corrections to the rapid and the return-to-isotropy parts, thus:

$$\Phi_{ijw} = \Phi'_{ij,1} + \Phi'_{ij,2} \tag{9.18}$$

Shir (1973) and Gibson and Launder (1978) proposed the following models for $\Phi'_{ij,1}$ and $\Phi'_{ij,2}$ respectively:

$$\Phi'_{ij,1} = C'_1 \frac{\varepsilon}{k} \left(\overline{u_k u_m} n_k n_m \delta_{ij} - \frac{3}{2} \overline{u_k u_i} n_k n_j - \frac{3}{2} \overline{u_k u_j} n_k n_i\right) f\left(\frac{L}{n_i r_i}\right) \tag{9.19}$$

$$\Phi'_{ij,2} = C'_2 \left(\Phi_{km,2} n_k n_m \delta_{ij} - \frac{3}{2} \Phi_{ki,2} n_k n_j - \frac{3}{2} \Phi_{kj,2} n_k n_i\right) f\left(\frac{L}{n_i r_i}\right) \tag{9.20}$$

This combination of $\Phi_{ij,2}$ and $\Phi_{ij,w}$ was previously used by Younis (1982) for predicting a number of two-dimensional boundary layer flows where it was found to perform quite well. In that study, the coefficients C'_1 and C'_2 were set equal to 0.5 and 0.1, respectively, so that the combination of Equations 9.15, 9.16 and 9.18 gives the appropriate relative stress levels in the uniform-stress layer of a plane-wall boundary layer in local equilibrium.

The function f in Equations 9.19 and 9.20 is defined as the ratio of a typical turbulent eddy size (L) to the normal distance to a wall (n); its value reflects the relative importance of the wall in damping the fluctuating pressure field in its vicinity.

The effects of a free surface on the turbulence structure in its vicinity are not completely understood, though there is some evidence to suggest that its effects are not dissimilar to those associated with a completely rigid boundary. In particular, the observed reduction in eddy viscosity with approach to the free surface, together with the redistribution of energy from the component normal to the surface into the components parallel to it, imply that there is need to modify the pressure-strain term to account for this effect. A convenient approach to adopt here is to utilize the wall-damping model

(Equations 9.19 and 9.20) in conjunction with a free-surface damping function f_f. Naot and Rodi (1982) defined this function as:

$$f_f = \left[\frac{L}{\left\langle \frac{1}{y_f^2} \right\rangle^{-1/2} + C_f L} \right]^2 \tag{9.21}$$

C_f is assigned the value of 0.16 and the turbulence length scale here is obtained from:

$$L = \frac{C_\mu^{3/4}}{\kappa} \frac{k^{3/2}}{\varepsilon} \tag{9.22}$$

The complete model contains a number of coefficients whose values are determined by reference to data from simple, homogeneous, shear flows. The values used in this work can be regarded as fairly standard (see, for example, Younis, 1982; Demuren and Rodi, 1984). For a complete list see Cokljat and Younis (1995).

9.3   EXAMPLE APPLICATIONS

9.3.1   Rectangular ducts

Comparisons are made with the data of Leutheusser (1963) obtained in a fully developed turbulent flow in a smooth-walled rectangular duct for $Re = 5.6 \times 10^4$ and $AR = 3$. The computations were obtained with a grid of 42×22 covering one quarter of the solution domain. Figure 9.2 compares the results of the models of the previous section with the measured contours of streamwise velocity (normalized by $\overline{W_\tau}$) and, also, with the Algebraic Stress Model (ASM) predictions of Nakayama *et al.* (1983). The plots show that the main velocity profiles are distorted by the presence of the turbulence-driven secondary motions which act to transport high-momentum fluid along the corner bisectors and lower momentum fluid along the walls. All models reproduce these effects to varying degrees of success with the non-linear model yielding results that are essentially identical to the full second-order closure. The predicted and measured normal stress differences responsible for driving these secondary motions are compared in Figure 9.3 with Hoagland's (1960) measurements in a duct of aspect ratio of 3 at $Re = 2. \times 10^4$. The trends are fairly well reproduced but the qualitative agreement is rather poor with the non-linear model underestimating the degree of anisotropy present in the flow. This is not altogether surprising since normal stress differences are partially due to transport effects and these are absent from the non-linear stress–strain relationships.

9.3.2   Compound ducts

For the more difficult case of compound, asymmetric duct, comparisons are made with the data of Knight and Lai (1985a,b) at $Re = 8.23 \times 10^4$ with a side-to-main ducts width ratio of 1.93 and a depth ratio of 0.51. Figure 9.4 shows that the effects of the secondary

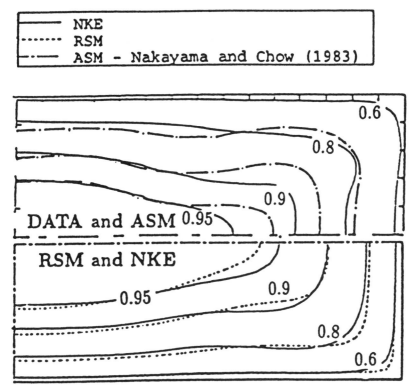

Figure 9.2   Mean velocity contours W/W_{max}. Data of Leutheusser (1963)

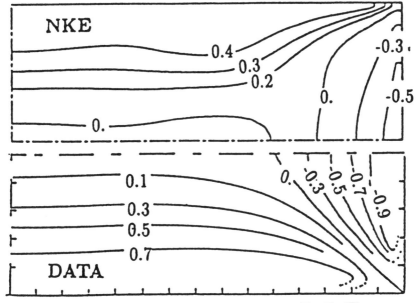

Figure 9.3   Normal stress anisotropy contours $(\overline{u^2} - \overline{v^2})/W_\tau^2$

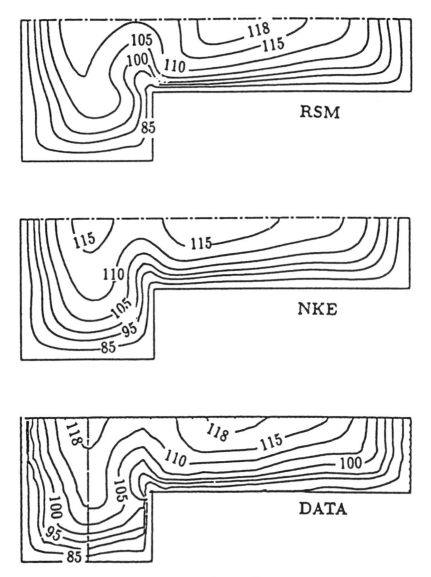

Figure 9.4   Mean velocity contours $W/W_b \times 10^2$. Data of Knight and Lai (1985a)

motions have been to distort the main velocity contours quite severely everywhere, especially in the vicinity of the inner corner where these motions are strongest. There are now two quite separate regions of maximum main flow, one in the main duct and the other in the side duct; the locations of both are fairly well predicted by the present models. Figure 9.5 compares the predicted and measured boundary shear stresses. Also shown there are standard k–ε model predictions (where no secondary motions are produced), which demonstrate the influence exerted by these motions on the near-wall flow behaviour.

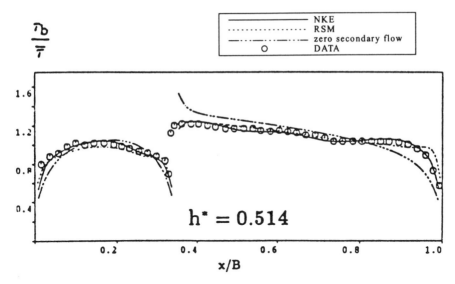

Figure 9.5   Bed shear stress in compound asymmetric duct. Data of Knight and Lai (1985)

9.3.3   Trapezoidal ducts

Comparisons are made with the data sets of Rodet (1960) for $Re = 2.4 \times 10^5$ and $AR = 0.82$ and of Prinos *et al.* (1988) for $Re = 2.25 \times 10^5$ and $AR = 3.5$. The results for the Rodet flow, obtained on a grid of 34×42, are presented in Figure 9.6. The non-linear model reveals the existence of a complex system of secondary-flow cells leading to main flow distortion that compares quite well with the measurements. The turbulence anisotropy is somewhat better predicted than before, which is surprising in view of the additional complexity wrought by asymmetry. The effects of changing the aspect ratio of the trapezoidal duct are clearly evident in Figure 9.7 where predictions obtained on a grid of 61×40 are compared with the data of Prinos *et al.* Agreement for the main flow contours is rather indifferent: the data show that the maximum velocity occurs well away from the plane of symmetry (suggesting very strong secondary flow along this plane) and this improbable feature is not reproduced by the model. Near the corners, the model's results are similar to those obtained for Rodet's flow and for the 3 : 1 rectangular duct, but the data do not show a consistent trend. The turbulence kinetic energy is surprisingly well predicted, particularly with respect to the position and magnitude of the maximum kinetic energy level. The model also seems to reproduce the trends in the measured boundary shear stresses (Figure 9.8), though quantitative differences are apparent.

9.3.4   Rectangular channels

Detailed measurements of fully developed turbulent flows in open rectangular channels have been reported by Tominaga *et al.* (1989) for channels of aspect ratios of 2.01, 3.94 and 8.0. The calculated and measured secondary velocity vectors are compared in Figures 9.9a–c for the three aspect ratios. Superimposed on those results are the contours of secondary flow stream function, non-dimensionalized by (W_{max}). Both predictions and measurements show that the secondary flow is organized in two distinct regimes which, in the terminology of Tominaga *et al.* amount to a large "free-surface vortex" and a smaller "bottom vortex" confined in the main to the corner region of the channel. As the width of the channel is increased relative to its depth, the free-surface vortex is subjected to preferential elongation leading eventually to its reorganization into two separate contra-rotating vortices. This behaviour is broadly reflected in both the calculations and the measurements, though some quantitative differences between the two are apparent. The effects of the secondary flow on the primary mean velocity can be seen from Figure 9.10a–c where the predicted and measured contours of this quantity are presented. The free-surface vortex acts to transport slow-moving fluid from the vicinity of the side walls into the central portion of the channel. This distorts the primary velocity field quite significantly, leading in some cases to the displacement of the position of maximum velocity to below the free surface. For $B/H = 2.01$, the maximum velocity is predicted to lie about $0.33^{*}H$ below

(a)

(b)

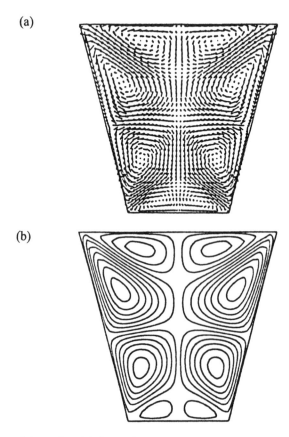

Figure 9.6 NKE model results: (a) Secondary velocity, (b) secondary velocity streamlines, (c) mean velocity contours W/W_{max}, (d) turbulence anisotropy $(\overline{u^2} - \overline{v^2})/\overline{W^2}$

(c)

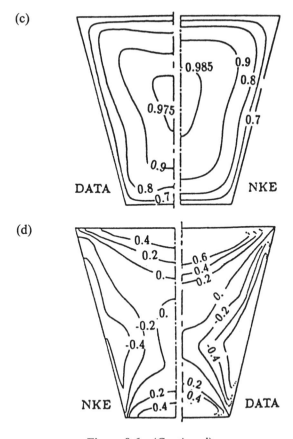

Figure 9.6   (*Continued*)

the free surface, while the measurements show this position to be at "about $0.2H–0.3H$". For higher values of B/H, the agreement is somewhat less satisfactory than before, which, is surprising in view of the diminishing importance of the secondary flows with increasing aspect ratio.

Contours of the turbulence anisotropy $(\overline{u^2} - \overline{v^2})$ for $B/H = 2.01$ are reported and those are compared with the Reynolds stress model predictions in Figure 9.11. Of particular interest here is the shaded line which is the loci of points where $\overline{u^2}$ and $\overline{v^2}$ are equal. In a closed duct, the flow is symmetrical around the corner bisector and this line would therefore lie diagonally along it. The presence of the free surface imparts an asymmetry to the flow leading to the observed bias towards the side wall. The position of this dividing line is very well matched in the predictions and the measurements, particularly in the central parts of the flow. Where the contour lines are very close, gradients of the normal stress anisotropy are high and hence strong secondary velocities are generated there. Agreement, overall, is fair though some overprediction is observed near the free surface, which suggests that the model for the free-surface damping has exaggerated the reduction in the vertical fluctuating velocity component there.

(a)

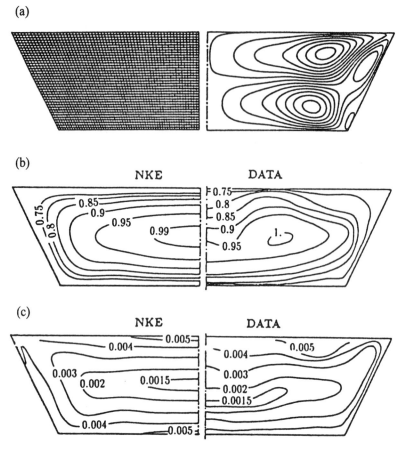

(b)

Figure 9.7 NKE model results: (a) Grid 61×40 and secondary flow streamlines, (b) main velocity contours W/W_{max}, (c) turbulence kinetic energy k/W^2_{max}. Data of Prinos *et al.* (1988)

Figure 9.12 compares the predicted and measured bed shear-stress distributions for the three aspect ratios measured. The distortion to the primary velocity profiles wrought by the secondary flow produces a localized region of wall-stress deficit which is seen to move towards the side wall with increasing aspect ratio. This trend is well represented by the calculations though some quantitative differences remain, most notably for the intermediate aspect ratio. The shear-stress profiles were made non-dimensional with their averaged values and it should be noted that. in the calculations, those values satisfied the momentum integral balance in each case.

The measurements of Knight *et al.* (1984) in open channels at several aspect ratios provide an opportunity for a further assessment of the model's near-wall behaviour. Figure 9.13 compares the predicted and measured average wall and bed shear stresses as a function of channel aspect ratio. The agreement here is very close except, perhaps, for the largest aspect ratio measured ($B/H = 6$) where the measured values on the two sides are somewhat different, which indicates some departure of the flow from exact symmetry.

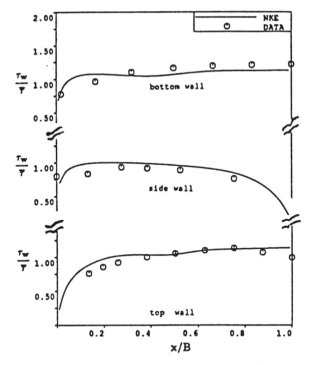

Figure 9.8   Wall shear stresses in trapezoidal duct. Data of Prinos *et al.* (1988)

9.3.5   Compound channels

Figure 9.14 compares the predicted and measured (Tominaga and Ezaki, 1988) mean velocity in a compound channel of depth ratio $h^* = 0.5$. Strong secondary currents are generated along the diagonal of the main-channel/floodplain junction and these are deflected by the free surface to form a down-wash along the channel's axis. The effects of these motions on the distribution of the main streamwise velocity are two-fold: the contours of the latter are severely distorted along the same diagonal line, while slow-moving fluid transported from the corner region causes the position of maximum velocity to occur well below the free surface. This is obtained in both the measurements and the predictions with the latter showing the details of the floodplain flow more clearly. In specifying the wall-boundary conditions in the calculations it was assumed that, immediately next to the wall, the streamwise component of mean velocity follows the standard logarithmic law of the wall distribution with a Karman constant of 0.4187 and an additive constant of 5.45. A check on the validity of this assumption is made by comparing the predicted and measured distributions of the bed shear stresses. This is done in Figure 9.14 where these quantities are presented, non-dimensionalized by the averaged boundary shear stress. The correlation between the two is quite satisfactory especially for the observed waviness over the floodplain bed.

The influence on the streamwise velocity is seen in Figure 9.15: the distorted contours near the junction and the displacement of the position of maximum velocity to

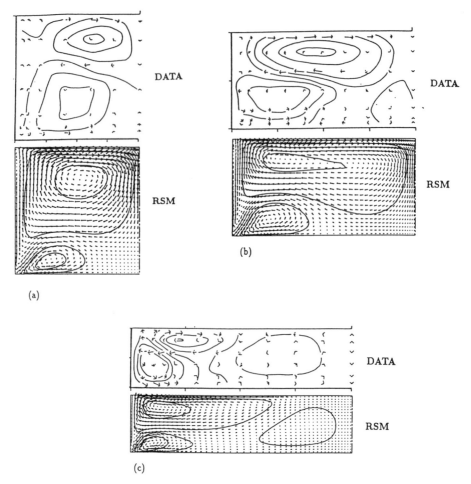

Figure 9.9 Predicted and measured secondary flow in open channels of aspect ratios 2.01 (a), 3.94 (b) and 8.0 (c). Data of Tominaga *et al.* (1989)

below the free surface at the plane of symmetry are direct consequences of the patterns of secondary flow there.

The parameters that are of most practical interest, namely the variation of total discharge with the depth ratio and also, the proportions of this quantity carried in the main channel and the floodplains are compared in Figure 9.16. The measurements are those of Knight and Demetriou (1983), obtained with a propeller current meter for a range of h^*. Two width ratios were considered, viz. $B/b = 2$ and 3. The predicted total discharge agrees well with the measured variation of this quantity with h^* and with the width ratio. Moreover, the model also predicts fairly accurately the proportions of the total discharge obtained in the main channel and in the floodplains, which suggests that the effects on momentum transport of vortices with vertical axes – which are probably present in the real flow but not captured here – are negligible.

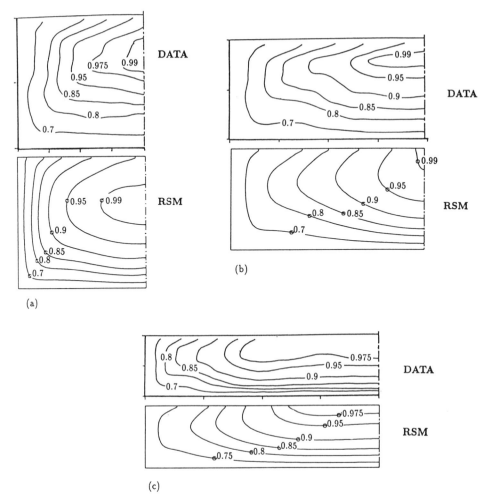

Figure 9.10 Contours of main velocity in open channel. Data of Tominaga *et al.* (1989) for aspect ratios 2.01 (a), 3.94 (b) and 8.0 (c)

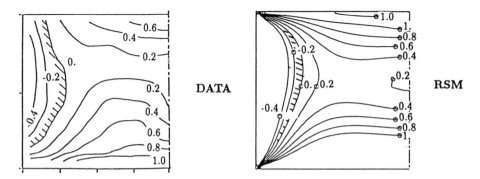

Figure 9.11 Predicted and measured turbulence anisotropy for channel with $AR = 2.01$. Data of Tominaga *et al.* (1989)

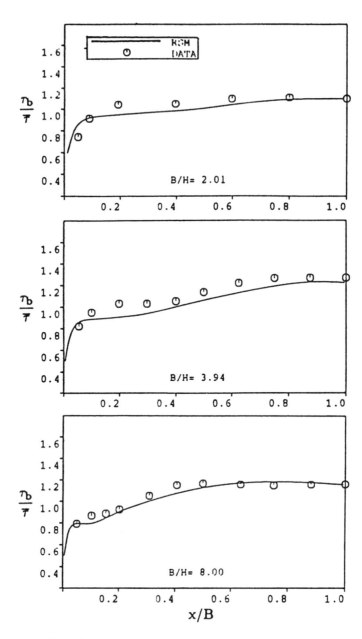

Figure 9.12 Variation of bed shear-stress distribution with channel aspect ratio. Data of Tominaga *et al.* (1989)

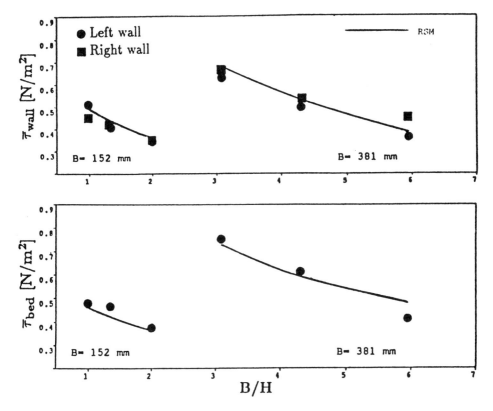

Figure 9.13 Variation of average wall (top) and bed shear stress with channel aspect ratio. Data of Knight *et al.* (1984)

9.3.6 Asymmetric compound channels

The present predictions with a Reynolds stress model are compared in Figures 9.17a and b with the data of Tominaga and Nezu (1991) for $h^* = 0.5$. The data show that two vortices are now present in the main channel and that both reach the free surface and meet at a point approximately 1.6 times the depth from the side wall. This position is predicted almost exactly by the model which also captures the main features of the streamwise velocity contours. In particular, the effects of the secondary currents in causing the main flow to bulge away from the floodplain/main channel junction is well predicted and so is the displacement in the position of maximum streamwise velocity to below the free surface. This latter feature of the flow was absent from the recent calculations of Naot *et al.* (1993) who used the Algebraic Stress Model (ASM) of Naot and Rodi (1982), which suggests that the levels of turbulence anisotropy were somewhat underpredicted by that model.

The predicted bed shear-stress distribution for the main channel and the floodplain, non-dimensionalized by the averaged shear stress, is compared with the measurements in Figure 9.18. Quantitatively, the correlation between predictions and data is astonishingly bad bearing in mind the reasonable correspondence observed for the

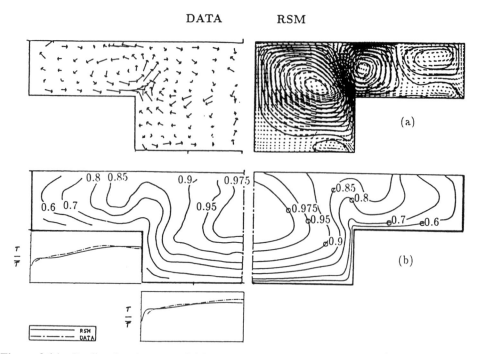

Figure 9.14 Predicted and measured (a) secondary and (b) streamwise (W/W_{max}) velocities in compound channel $h^* = 0.5$

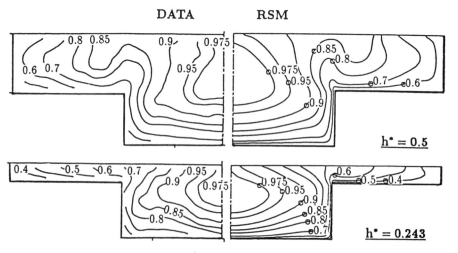

Figure 9.15 Contours of main flow (W/W_{max}) in compound channels. Data of Tominaga *et al.* (1988)

streamwise velocities. It should be mentioned here that the channel slope was not provided in Tominaga and Nezu (1991) but had to be determined in the computations by trial and error to match the quoted values of the bulk velocity. The possibility of an error of interpretation cannot therefore be discounted bearing in mind that the averaged

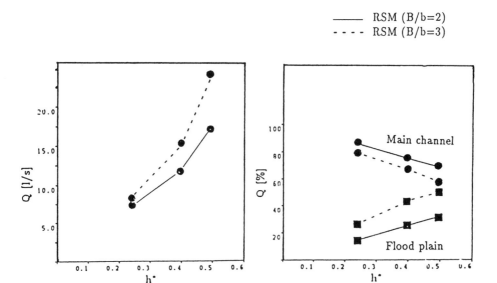

Figure 9.16 Variation of discharge with h^* and B/b

boundary shear stress value satisfied overall momentum balance exactly in the computations but only within 6% in the experiments. Qualitatively, the agreement is somewhat better in that the expected waviness in the bed shear-stress distribution is clearly present in the data and the predictions and the occurrence of the maximum boundary shear stress at $x/(b_1 + b_r)$ of about 0.6 is also obtained in both.

The turbulence anisotropy which is responsible for driving the secondary flow is shown in Figure 9.19. As expected, this quantity is greatest in the neighbourhood of the bed and the free surface where the damping effect of each will have caused an enhancement of $\overline{u^2}$ at the expense of $\overline{v^2}$, the vertical fluctuations. The measured line of isotropic turbulence is only very crudely reproduced but the experimental errors involved in evaluating the differences of two measured quantities cannot be discounted.

The two main components of the Reynolds shear stresses, $\overline{uw}$ and $\overline{vw}$, are compared in Figure 9.20 and 9.21, respectively. These quantities, like the turbulence anisotropy, were made non-dimensional by the averaged wall friction velocity, a quantity which itself is subject to some uncertainty. The correspondence between the predictions and the measurements is broadly in line with what is normally obtained for such parameters. It is interesting to note that the signs of these quantities correlate very closely with the signs of the appropriate rates of strain over large portions of the flow but there are regions obtained in the predictions where the two are of opposite sign. These regions occur in between the loci of the points where the shear stresses and the rates of strain are zero which it is obtained in the predictions, do not generally coincide. Similar behaviour is found in a variety of asymmetric flows (e.g. wall jets and differentially roughened channels) and can only be reproduced by using models that fully take into account the transport effects on the turbulence field.

(a)

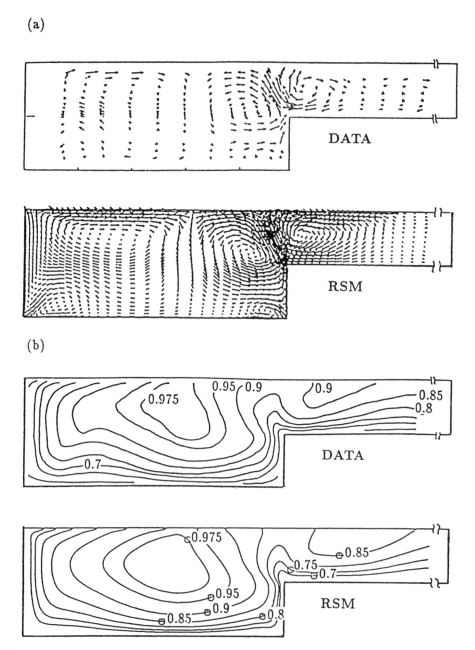

(b)

Figure 9.17   Asymmetric compound channel. Secondary (a) and streamwise (W/W_{max}) (b) velocities for $h^* = 0.5$

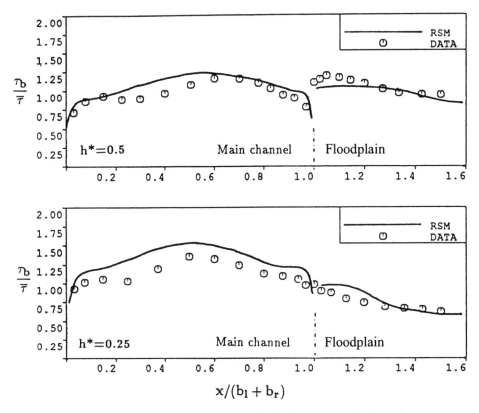

Figure 9.18   Bed shear-stress distribution in asymmetric channel

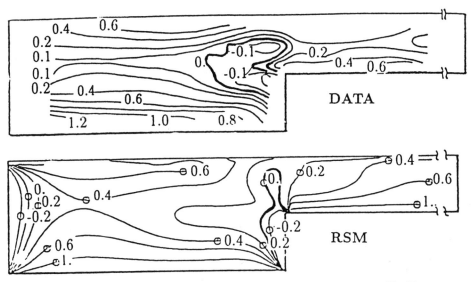

Figure 9.19   Asymmetric compound channel. Turbulence anisotropy $(\overline{u^2} - \overline{v^2})/W_\tau^2$

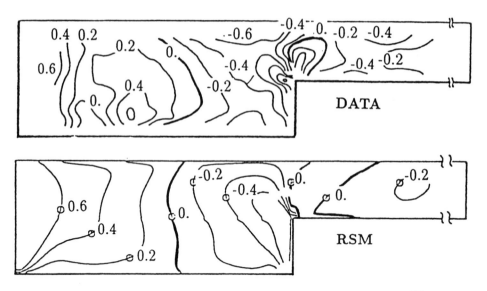

Figure 9.20   Asymmetric compound channel. Shear-stress component $-\overline{uw}/W_\tau^2$

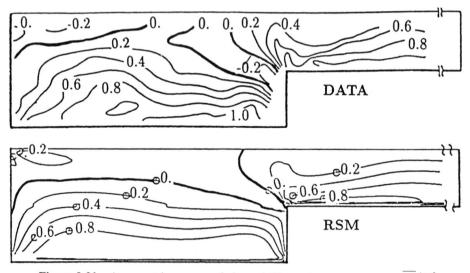

Figure 9.21   Asymmetric compound channel. Shear-stress component $-\overline{vw}/W_\tau^2$

9.3.7   Compound channels with roughness

Predictions are next reported for the experiments of Knight and Hamed (1984) in which the floodplain beds of a symmetric channel were roughened through the application of strip roughness elements. In the computations, the effects of roughness are introduced via the wall boundary conditions for the streamwise velocity. The wall shear stresses, used as momentum flux conditions, are deduced from the rough-wall correlation used in Knight and Hamed (1984), viz.:

$$\frac{\tilde{W}}{W_\tau} = 6.06 \log_{10}\left(\frac{H - h}{\chi}\right) \tag{9.23}$$

where $\tilde{W}$ is the depth-averaged streamwise velocity and x is a roughness parameter whose value (which depends on the size, spacing and shape of the roughness elements used) is here taken as 3.4924, which was found appropriate by Knight and Macdonald (1979) to square-sectioned elements with a spacing-to-height ratio of 10.

The predicted and measured contours of streamwise velocity are presented in Figure 9.22. The severe distortion seen in the calculated contours along the diagonal of the floodplain/main-channel junction is reminiscent of that observed in the smooth-walled case and indicates a strong secondary flow current in the same direction. This current, when eventually deflected by the free surface, is responsible for the downward shift in the position of velocity maximum at the centreline (somewhat exaggerated in the predictions) and for the creation of a local velocity maximum in the floodplain, also below the free surface. The comparative effects of floodplain roughness on the overall conveyance of the channel is most clearly seen in Figure 9.23 which compares the predicted and measured spanwise variation of the depth-averaged streamwise velocity. As expected, the flow rate is lower in the floodplain, and correspondingly higher in the main channel, relative to the smooth-walled case.

9.3.8 Trapezoidal channels

In the experiments of Tominaga *et al.* (1988), the floodplains were rectangular in cross-section but the main-channel walls were inclined at 45 °. The depth ratio was 0.5. The slope was not provided and had to be estimated by trial and error to match the quoted Reynolds number. The predicted and measured secondary and primary velocities are compared in Figures 9.24a and 9.24b respectively. The predicted secondary motion does not differ substantially from that obtained in the rectangular case: a strong jet directed along the junction diagonal leading to the formation of contra-rotating vortices on either side of it. The measurements are rather too sparse for clear patterns to be discernible but, somewhat surprisingly, they appear to show that the flow along the inclined wall is in fact away from the junction rather than towards it. The consequences

DATA   RSM

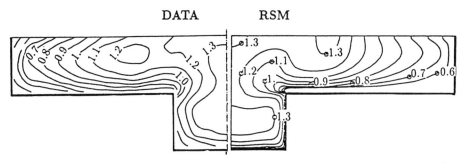

Figure 9.22 Compound channel with roughened floodplains. Streamwise velocity (W/W_{max}) for $h^* = 0.5$

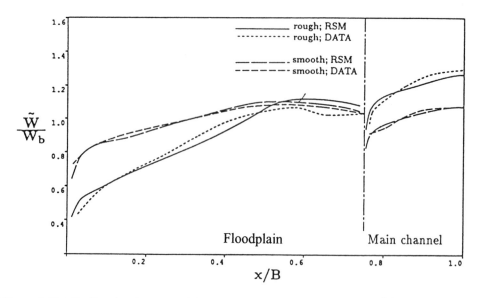

Figure 9.23 Predicted and measured depth-averaged streamwise velocity ($\tilde{W}/W_b$) in symmetric compound channel with roughened floodplains

of such apparent discrepancy are not serious enough to affect the close correspondence between predicted and measured contours of streamwise velocity seen in Figure 9.24b.

A systematic study of the effects of the depth ratio on the flow in compound channels in which both sections are trapezoidal was reported by Yuen and Knight (1990). The walls were again inclined by $45°$, and B/b set equal to 3.0. Three depth ratios were considered: viz. $h^* = 0.5$, 0.45 and 0.35. The predicted secondary flow patterns (not shown) are fairly uneventful and are very similar to those found in the rectangular case. Their consequences on the primary velocity distributions are clear from Figure 9.25: local velocity maxima occur in the floodplains and the contours around the mid-channel assume their now familiar distorted shapes. The maximum relative difference between the predicted and measured discharge rates for the three depth ratios is of the order of 2%.

9.3.9 180° turn-around channel

Monson *et al.* (1989) report measurements in a 180° turn-around duct formed from a rectangular channel with an aspect ratio of 10. The flow is two-dimensional along the mid-plane and results are reported for the two values of the Reynolds number (based on duct width and bulk velocity) of 10^5 and 10^6. The computational domain and a sample grid are shown in Figure 9.26. Two grids were used: one formed of 208×42 meshes in the streamwise and cross-stream directions respectively, and another of 208×84 meshes (shown). The meshes were non-uniformly distributed in either direction, being concentrated near the walls and at inlet and exit from the bend. All the results reported below were obtained with the second finer grid and we have established by comparing the results of the two grids that the reported solutions are grid-independent (see Basara *et al.*, 1995, for details). Simulations were carried out for both values of Reynolds

DATA   RSM

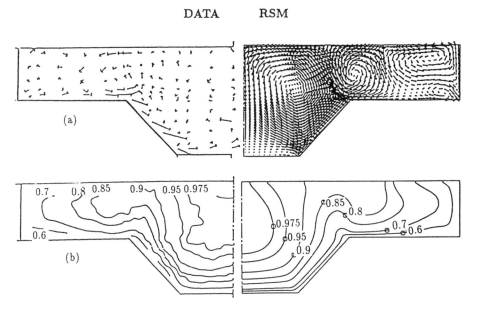

Figure 9.24 Compound trapezoidal channel. Predicted and measured secondary (a) and streamwise (W/W_{max}) (b) velocities for $h^* = 0.5$

number, but only the results for the higher value (where larger separation was observed) are reported here.

Figure 9.26 also shows the mean flow streamlines predicted by both models. It is immediately clear that the $k-\varepsilon$ model predicts a separated-flow zone which is far shorter and thinner than that captured by the Reynolds stress closure. The point where the flow actually detaches from the inner wall is also predicted to occur much later by the two-equation model. In this connection it should be noted that our predictions for the $Re = 10^5$ with the $k-\varepsilon$ model failed to predict separation altogether, in sharp contrast with the Reynolds stress model which correctly obtained this feature. That the two models should behave so differently is directly attributed to their response to the effects of streamline curvature. Near the convex wall, the radial momentum increases with distance from the wall and hence the sense of curvature is stabilizing (Bradshaw, 1973), leading to the suppression of turbulence activity relative to the straight wall. A reduction in the turbulence levels implies that the boundary layer is less able to withstand the adverse pressure gradients encountered on entry to the duct with the results that it detaches from the surface to form the separated flow region. This is demonstrated fairly clearly by the Reynolds stress model. Figure 9.27a,b compares the predicted and measured wall skin-friction coefficient C_f on both the inner and outer walls. On the inner wall C_f actually rises on entry to the bend $(s/H = 4)$ in response to a favourable pressure gradient, but is then seen to reduce sharply as the effects of curvature take hold. Shortly after exit from the bend, C_f passes through zero and reverses sign as would be expected within the reversed-flow zone. In contrast to this, the $k-\varepsilon$ model, after initially responding to the favourable pressure gradient, predicts almost no response to the effects of curvature till well downstream of the start of the

RSM   DATA

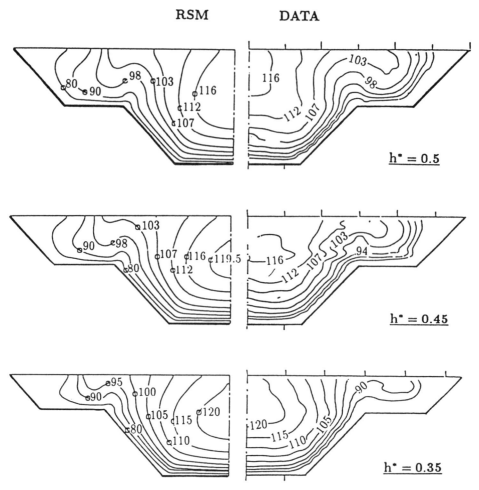

Figure 9.25 Compound trapezoidal channel. Effects of depth ratio on streamwise velocity $(W/W_b \times 10^2)$

bend. It eventually drops to zero, but recovers fairly quickly after that. The behaviour of this quantity on the stabilized outer wall, is less remarkable, though interestingly, the $k-\varepsilon$ model seems to predict the downstream recovery fairly well.

Figures 9.28a–c compare the predicted and measured axial mean-velocity profiles at three locations within the bend, viz at $\theta = 0°$, $90°$ and $180°$. The cross-stream coordinate is normalized by the duct's width: its zero value is located on the inner wall. The effects of the favourable pressure gradient in accelerating the flow adjacent to the inner wall are clearly well predicted with both models at $\theta = 0°$. By $\theta = 180°$, and after the effects of streamline curvature have been felt, the differences between the two model results are quite marked: the Reynolds stress model obtains the expected negative velocity values close to the surface and captures the remainder of the profile shape fairly well. Some departure from the measurements is observed near the point where the maximum velocity occurs but this is dwarfed by the discrepancy observed in the $k-\varepsilon$ results.

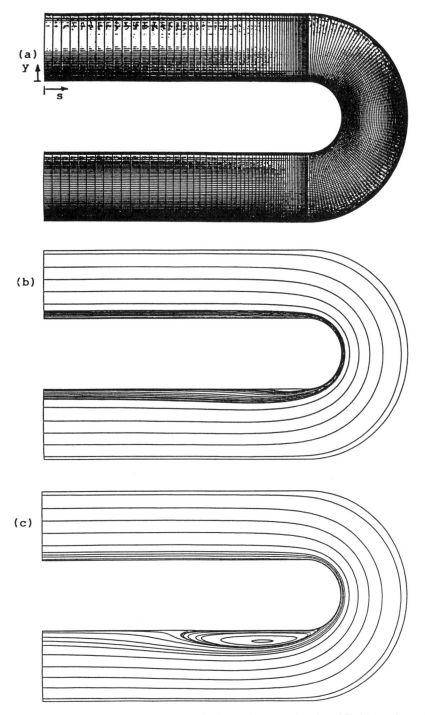

Figure 9.26 Two-dimensional turn-around duct. Computational grid (a) and predicted streamlines with k-ε (b) and Reynolds stress (c) models

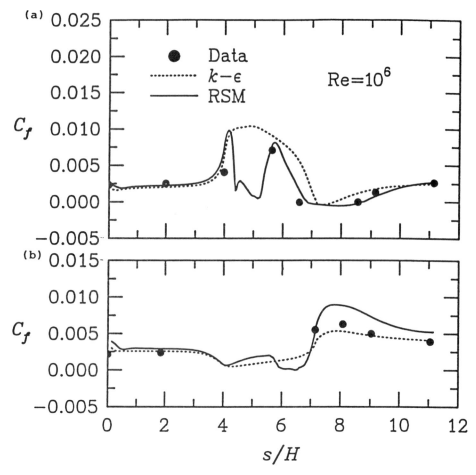

Figure 9.27 Two-dimensional turn-around duct. Predicted and measured skin-friction coefficient on inner (a) and outer (b) walls

9.4 CLOSING REMARKS

The chapter focused on the presentation of results for a variety of open-channel flows obtained with turbulence models more sophisticated than those normally used in engineering practice. The motivation has been to show that the complex physical processes present in these flows may be captured, fairly accurately, without the use of *ad hoc* modifications. The level of closure found necessary to achieve this result was the complete Reynolds stress transport modelling approach, which involves the solution of seven differential transport equations for the turbulence variables. The genuine *prediction* of open-channel flows to acceptable engineering accuracy is thus an attainable goal, albeit with considerable computer resources.

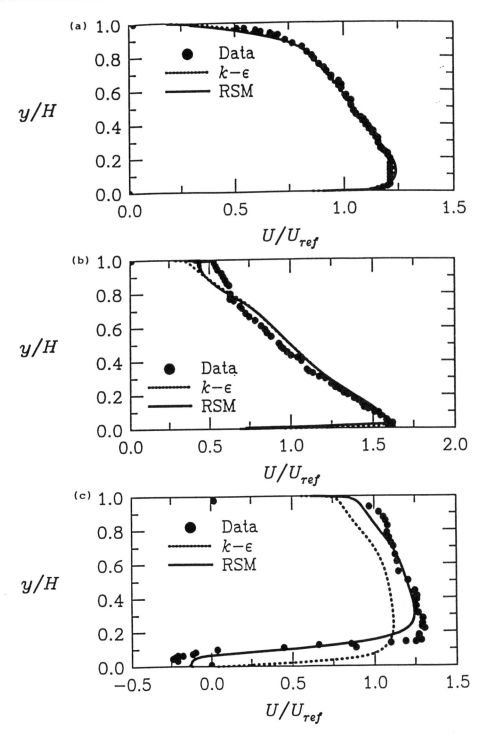

Figure 9.28   Mean-velocity profiles at: (a) 0°, (b) 90° and (c) 180°. Two-dimensional flow

9.5 NOTATIONS

Latin symbols

AR	Aspect ratio of the channel
B	Half width of channel
b	Half width of main part of channel
C_1, C_1', C_2, C_2'	Reynolds stress model coefficients
$C_{\varepsilon 1}, C_{\varepsilon 2}'$	Coefficients of ε equation
C_s, C_ε	Diffusion-model coefficients
C_D, C_E	Non-linear model coefficients
C_f	Skin-friction coefficients
f_f	Free-surface damping function
f_x, f_y	Wall damping functions
g	Gravitational acceleration
H	Depth of open channel
h	Height of floodplain
h^*	Ratio between depths of main-channel and floodplain
k	Turbulence kinetic energy
L	Length scale of turbulence
n_i	Unit vector normal to the surface
P_k	Production of turbulent kinetic energy
p, p'	Time-averaged and fluctuating values of pressure
s	Distance along centre line
S_{ij}	Mean rate of strain tensor
$\overset{\circ}{S}_{ij}$	Oldroyd derivative
U, V, W	Time-averaged velocity components in x, y and z directions
u, v, w	Fluctuating velocity components in x, y and z directions
$\overline{u_i u_j}$	Reynolds stress tensor
W_{ij}	Mean vorticity tensor
W_τ	Friction velocity
$\overline{u^2}, \overline{v^2}, \overline{w^2}$	Normal-stress components in x, v and z directions
$\overline{uv}, \overline{uw}, \overline{vw}$	Shear-stress components
x, y, z	Spanwise, normal and streamwise coordinate directions
y_f	Distance from free surface

Greek symbols

δ_{ij}	Kronecker delta
ε	Dissipation of turbulent kinetic energy
θ	Slope of the channel bed
κ	von Karman constant
ν	Molecular kinematic viscosity
ρ	Density
σ_ε	Coefficient of turbulent diffusion of ε
$\bar{\tau}$	Average shear stress

Subscrips

b   Bed
c   Channel
f   Free surface
t   Turbulent
w   Wall

REFERENCES

Basara, B. and Younis, B.A. (1995) Assessment of the SSG pressure–strain model in two-dimensional turbulent separated flows. *Applied Scientific Research*, **55**, 39–61.
Basara, B., Cokljat, D. and Younis, B.A. (1995) Assessment of eddy-viscosity and Reynolds-stress transport closures in two- and three-dimensional turn-around ducts. *ASME, FED*, vol. 217, Separated and Complex flows, 249.
Bradshaw, P. (1973) Effect of streamline curvature on turbulent flow. *AGARDograph 169*.
Cokljat, D. and Younis, B.A. (1995) Second-order closure study of open channel flows. *J. Hydraul. Eng. ASCE*, **121**, 94–107.
Daly, B.J. and Harlow, F.H. (1970) Transport equations in turbulence. *Phys. Fluids*, **13**, 2634.
Demuren, A.O. and Rodi, W. (1984) Calculation of turbulence driven secondary motion in non-circular ducts, *J. Fluid Mech.*, **140**, 189.
Gibson, M.M. and Launder, B.E. (1978) Ground effects on pressure fluctuations in the atmospheric boundary layer. *J. Fluid Mech.*, **86**, 491.
Handler, R.A., Swean, T.F., Leighton, R.I. and Swearingen, J.D. (1993) Length scales and the energy balance for turbulence near a free surface. *AIAA Journal*, **31**, 1998.
Hoagland, L.C. (1960) Fully developed turbulent flow in straight rectangular ducts. PhD thesis, Dept of Mech. Eng., MIT.
Irwin, H.P.A.H. (1973) Measurements in a self-preserving plane wall jet in a positive pressure gradient, *J. Fluid Mech.*, **61**, 33.
Knight, D.W. and Demetriou, J.D. (1983) Flood plain and main channel flow interaction. *J. of Hydraul. Eng., ASCE*, **109**, 1073.
Knight. D.W. and Hamed. M.E. (1984) Boundary shear in symmetrical compound channels. *J. Hydraul. Eng., ASCE*, **110**, 1412.
Knight, D.W. and Lai, C.J. (1985a) Turbulent flow in compound channels and ducts. *Int. Symp. on Refined Flow Modelling and Turbulence Measurements*, Iowa, USA.
Knight, D.W. and Lai, C.J. (1985b) Compound duct flow data. Summary of Experimental Results, Dept. of Civil Eng., University of Birmingham, UK.
Knight, D.W. and Macdonald, J.A. (1979) Hydraulic resistance of artificial strip roughness. *J. Hydraul. Div., ASCE*, **105** (HY6), 675.
Knight, D.W., Demetriou, J.D. and Hamed, M.E. (1984) Boundary shear in smooth rectangular channels. *J. Hydraul. Eng.*, *ASCE*, **110**, 405.
Komori, S., Ueda. H., Ogino, F. and Mizushina, T. (1982) Turbulence structure and transport mechanism at the free surface in an open channel flow. *Int. J. Heat Mass Transfer*, **25**, 513.
Krishnappan, B.G. and Lau, Y.L. (1986) Turbulence modelling of flood plain flows. *J. Hydraul. Eng., ASCE*, **112**, 251.
Launder, B.E., Reece, G.J. and Rodi, W. (1975) Progress in the development of a Reynolds stress turbulence closure. *J. Fluid Mech.*, **68**, 537.
Leutheusser, H.J. (1963) Turbulent flow in rectangular ducts. *J. Hydraul. Div., ASCE*, **89** (HY3), 1.
Lumley, J.L. (1970) Toward a turbulent constitutive relation. *J. Fluid Mech.*, **41**, 413.
Monson, D.J., Seegmiller, H.L. and McConnaughey, P.K. (1989) Comparison of LDV measurements and Navier–Stokes solutions in a two-dimensional 180-degree turn-around duct. *AIAA Paper* 89–0275, Reno-Nevada.

Nakayama, A., Chow, W.L. and Sharma, D. (1983) Calculation of fully developed turbulent flows in ducts of arbitrary cross-section. *J. Fluid Mech.*, **128**, 199.

Naot. D. and Rodi, W. (19.92) Calculation of secondary currents in channel flow. *J. Hydraul. Div.*, *ASCE*, **108**(HY8), 948.

Naot, D., Nezu, I. and Nakagawa, H. (1993) Hydrodynamic behavior of compound rectangular open channels. *J. Hydraul. Eng.*, *ASCE*, **119**, 390.

Nezu, I. and Rodi, W. (1985) Experimental study on secondary currents in open channel flow. *Proc. 21st Cong. of IAHR, Melbourne. Australia*, 378.

Nezu, I., Nakagawa, H. and Tominaga, A. (1985) Secondary currents in a straight channel flow and the relation to its aspect ratio. In: *Turbulent Shear Flows*, Springer-Verlag, **4**, 246.

Patankar, S.V. (1980) *Numerical Heat Transfer and Fluid Flow*. McGraw-Hill Book Co., Inc., New York, N.Y.

Patankar, S.V. and Spalding, D.B. (1972) A calculation procedure for heat, mass and momentum transfer in three-dimensional parabolic flows. *Int. J. Heat Mass Transfer*, **15**, 1787.

Pope, S.B. (1975) A more general effective-viscosity hypothesis. *J. Fluid Mech.*, **72**, 331.

Prandtl, L. (1926) Uber die ausgebildete Turbulenz. *Verh. 2nd Int. Kong. fur Tech. Mech. Zurich* (Trans. NACA Tech. Memo. no. 435) 62.

Prinos, P., Tavoularis, S. and Townsend. R. (1988) Turbulence measurements in smooth and rough-walled trapezoidal ducts. *J. Hydraul. Eng.*, *ASCE*, **114**, 43.

Reece, G.J. (1977) A generalized Reynolds-stress model of turbulence. PhD thesis, Imperial College, University of London.

Rodet, E. (1960) Etude de l'ecoulement d'un fluide dans un tunnel prismatique de section trapezoidalei. *Publ. Sci. et Tech. du Min. de l'Air*, no. 369.

Rodi, W. (1976) A new algebraic relation for calculating the Reynolds stresses. *ZAMM*, **56**, 219.

Rotta, J.C. (1951) Statistische theorie nichthomogener turbulenz. *Z. Phys.*, **129**, 547.

Shir, C.C. (1973) A preliminary study of atmospheric turbulent flow in the idealized planetary boundary layer. *J. Atmos. Sci.*, **30**, 1327.

So, R.M.C. and Mellor, G.L. (1973) Experiments on convex curvature effects in turbulent boundary layers. *J. Fluid Mech.*, **60**, 43.

Speziale, C.G. (1987) On nonlinear k−l and k−ε models of turbulence. *J. Fluid Mech.*, **178**, 459.

Speziale, C.G., Sarkar, S. and Gatski, T.B. (1991) Modeling the pressure−strain correlation of turbulence: an invariant dynamical systems approach. *J. Fluid Mech.*, **227**, 245.

Tominaga, A. and Ezaki, K. (1988) Hydraulic characteristics of compound channel flow. *6th Congress Asian and Pacific Regional Division, IAHR*, 20−22 July, Kyoto, Japan.

Tominaga, A. and Nezu, I. (1991) Turbulent structure in compound open-channel flows. *J. Hydraul. Eng.*, *ASCE*, **117**, 21.

Tominaga. A., Ezaki, K. and Nezu, I. (1988) Turbulent structure in compound channel flows with rectangular and trapezoidal main channel. *Proc. Third Int. Symp. on Refined Flow Modeling and Turbulence Measurements*, 26−28 July, Tokyo, Japan.

Tominaga, A., Nezu, I., Ezaki, K. and Nakagawa, H. (1989) Three-dimensional turbulent structure in straight open channel flows. *J. Hydraul. Research*, **27**, 149.

Younis, B.A. (1982) Boundary layer calculations with Reynolds stress turbulence models. Mechanical Engineering Department Report FS/82/25, Imperial College, London.

Younis, B.A. (1984) On modeling the effects of streamline curvature on turbulent shear flows. PhD thesis, Imperial College, University of London.

Yuen, K.W.H. and Knight. D.W. (1990) Critical flow in a two stage channel. *Proc. Int. Conf. on River Flood Hydraulics*, W.R. White (ed.), John Wiley, Paper G4, 267.

10   Bayesian Calibration of Flood Inundation Models

R. ROMANOWICZ, K.J. BEVEN and J. TAWN

Institute of Environmental and Biological Sciences, University of Lancaster, UK

Far better an approximate answer to the right question,
which is often vague,
than an exact answer to the wrong question,
which can always be made precise.

(John W. Tukey, 1962)

10.1   FORECAST EQUIFINALITY AND UNCERTAINTY IN MODELLING FLOOD INUNDATION

The problem of modelling the river flow in a channel with floodplains is very important from the point of view of both the security and economics of a lowland community. This has been highlighted recently by the widespread flooding throughout Europe in January and February 1995. The prediction of flood inundation has been the subject of research for many years and still has not been completely resolved. This is partly because, although the hydraulics of well-defined systems are well understood, flood inundation is a distributed problem that does not have well defined boundary conditions (either in geometry, "roughness coefficients" or upstream and lateral inputs), nor well defined parameterizations of the energy losses in such complex transient flows. This has started to be remedied in studies such as the large-scale flood channel facility funded by the UK Science and Engineering Research Council at Hydraulics Research, Wallingford (Knight, 1989), work that has led to new formulations of energy losses at the boundary between the main channel and the floodplain flows (see also Knight *et al.*, 1989).

However, the distributed nature of the inundation problem requires a distributed solution. In the past there have been one dimensional (downstream) solutions of the St Venant equations and two-dimensional solutions (in plan) of the depth-integrated Navier–Stokes equations used in flood inundation prediction (see Anderson and Bates, 1994, and Bates *et al.* Chapter 7, this volume). Three-dimensional solutions are also now being used in hydraulics and other areas of fluid dynamics (Cokljat and Younis

Floodplain Processes. Edited by Malcolm G. Anderson, Des E. Walling and Paul D. Bates.
© 1996 John Wiley & Sons Ltd.

1995), but the higher the dimension of the solution the more parameters will be required, including eddy viscosity and roughness parameterizations that cannot in general be estimated *a priori*, particularly for these complex flows where depths and velocity gradients may change rapidly in the flow domain.

Thus, there are three important aspects of predicting flood inundation that require further research. One is a proper formulation of the appropriate equations to represent the hydraulics of the flow, including improved understanding of where approximations might be allowable; the second is the development of efficient numerical algorithms that can cope with the wetting and drying of the flow domain and the changing spatial pattern of velocity and shear throughout a flood event; and the third is the calibration of the distributed parameters of the model. In the research reported here, we consider the flood inundation problem, focusing on the problem of parameter calibration and the uncertainties of model predictions resulting from uncertainty in the values of calibrated model parameters. We use a relatively simple quasi-two-dimensional model formulation, rather than a more complex numerical algorithm to explore the hypothesis that accuracy in predicting flood inundation is more a problem of representing the geometry of the floodplain and the roughness parameterization than of using a more sophisticated closure scheme for the Navier–Stokes equations.

The problem of calibration of distributed inundation models belongs to the class of inverse problems but, in common with many distributed modelling problems, is severely overparameterized in the sense that the number of parameters that must be specified cannot be supported by the calibration data available (Beven, 1989). Ideally, using forward modelling, we want to find the parameters of that model which give simulated outputs (water levels or flows) that are close to the observed data on water levels or flows. Thus, forward modelling consists of predicting the error-free data, Y_t, that would correspond to the given model parameter set θ.

$$Y_t = G(\theta, r_t) \qquad (10.1)$$

where $G(.)$ is the non-linear operator working from the parameter space into the model space and r_t denotes the input data.

The predicted values will not be identical to the true observed values because of (i) errors in the observations, including the boundary conditions, and (ii) errors in the model structure. Both sources of error can be represented by statistical models and these will be discussed further below.

The inverse problem can then be formulated as follows: find the parameter set θ, which minimises

$$J(\theta) = |\, Y_t - Z_t \,| \qquad (10.2)$$

where Z_t denotes the observed data at time t and $|\,.\,|$ denotes some assumed measure of performance.

In order to evaluate quantitatively the goodness-of-fit of model generated data Y_t, a whole range of functions for $J(\theta)$ can be considered. We call these measures of goodness-of-fit, or objective functions. The choice of objective function is determined by the goal of our modelling. There are many objective functions that could be used, including the sum of square errors, the sum of absolute errors, or the minimum of the maximum difference. For distributed dynamic models, the objective function may also take account of the space–time correlation of the errors. If the interest is in θ then the expected distribution of the modelling errors will be highly influential in determining the form of the objective function.

Given that hydraulic models are highly non-linear, with thresholds for wetting and drying, together with the cumulative nature of our goodness-of-fit function over some simulation period, the large number of distributed parameters required results in an ill-posed calibration problem. Thus, we can expect to obtain the same or very similar values of the objective function for different parameter sets (non-uniqueness of the solution of the inverse problem), or different inflow series (for different flood events in this case) can give a different relative goodness of fit for the same parameter sets. The idea that, for a given model structure, there is some optimum parameter set that can be used to simulate the system is not tenable in this context. There is rather an equifinality of parameter sets, where equally acceptable parameter sets might be found in very different parts of the parameter space. This then leads to uncertainty in the calibration problem (parameter uncertainty) and related uncertainty in the model predictions (Beven, 1993).

As a result, we should look for a different approach to that typical of the control engineers' approach to model calibration. The Generalized Likelihood Uncertainty Estimation (GLUE) methodology, introduced by Beven and Binley (1992), is one such an attempt. It transforms the problem of searching for an optimum parameter set into a search for the sets of parameter values, which would give reliable simulations for a wide range of inflow series. This approach, developed within a Bayesian framework, also enables confidence limits of the predicted model outputs (water levels or the outflows) to be obtained. This feature seems to be advantageous from the point of view of hydraulic modelling, where assessing the reliability of the predictions is very important.

We shall assume that the information about the parameters, model inputs and model outputs can be represented in the form of a distribution. The inverse problem given by equation 10.2 will be presented in the Bayesian form (Box and Tiao, 1992), as given by:

$$f(\theta \mid \mathbf{z}) = \frac{f(\theta)M(\mathbf{z} \mid \theta)}{f(\mathbf{z})} \tag{10.3}$$

where $\mathbf{z}$ is the vector of observations, $f(\theta \mid \mathbf{z})$ is the posterior distribution (probability density) of parameters given the data, $f(\theta)$ is the prior probability density of the parameters, $f(\mathbf{z})$ is the scaling factor and $M(\mathbf{z} \mid \theta)$ represents the theoretical information on the relationship between $\mathbf{z}$ and θ, obtained from the forward modelling (equation 10.1) (the likelihood measure for $\mathbf{z}$ given the parameter set θ). In other words, we transform the inverse problem as stated by equation 10.2 into the study of the posterior probability distribution of parameters, given the prior information on parameters and the information on the physical structure of the modelled process. In this formulation we do not restrict the solution of the calibration problem to finding the parameters which minimize any particular objective function. From the posterior we can obtain any sort of information on model parameters we need: mean, median, error bars (see Tarantola, 1987). There is no problem of non-uniqueness, as we are interested in the evaluation of the admissible or "behavioural" sets of parameters, rather than one "optimum" set. The non-existence of the solution in this sense would mean that there is no admissible parameter sets in the range of parameters we were searching through.

10.2   FORECASTING FLOOD INUNDATION WITHIN MATLAB

In this study the problem of calibration of the flood inundation model has been combined with the visualization of the resulting predictions of the flow of water on the floodplains. The software package MATLAB/SIMULINK is used to allow the simultaneous modelling and visualization of the results. The choice of model formulation was based on the need for both simplicity, so that it would be possible to run many different parameter sets, and the need to incorporate in the model the important non-linearities arising from the real, complicated geometry of the floodplains. A formulation similar to the quasi-two-dimensional model of Cunge (1976) was chosen. This model uses resistance terms in the form of Manning–Strickler laws, and neglects the inertial terms in the flow equations. It assumes that the flow builds up slowly on the floodplains and hence the resistance terms prevail in the flow equation. The model requires the specification of initial conditions for the storage and boundary conditions at each end of the modelled reach. This model can be easily written in the SIMULINK iconographic language, in which each element of the flood channel is represented as a storage element.

In the formulation used here each reach of the flood channel is represented by one channel and two floodplain elements, one on each bank. Each element is represented by a non-linear interconnected storage reservoir (cell) as in Cunge (1976). It is assumed that there is unique relation between the storage of each cell and the height of the water surface, as well as the cross-section/height and wetted perimeter/height relations at the boundaries of each cell. These functions are derived from the geometry of the river and the floodplains, taking account of the valley slope in each element, and have the form of look-up tables.

The basic continuity equation of the model for one cell is given as follows:

$$\frac{dV_i}{dt} = Q_{in}(h, t) - Q_{out}(h, t) + Q_l(h, t) + Q_r(h, t) \qquad (10.4)$$

where V_i denotes the storage of the ith cell (channel or floodplain reach), h denotes the height of water level in the cell, Q_{out} is the downvalley outflow from that cell, Q_{in} the inflow from upstream, and Q_l and Q_r represent the left and right side flow respectively. The relation between the flow and the height of the water surface in the reaches is given by:

$$Q_{ik} = KA(\bar{h})^{5/3} P(\bar{h})^{-2/3} S_{ik}^{1/2} \qquad (10.5)$$

where P is the wetted perimeter, A is the cross-sectional area, K is the roughness (Strickler) coefficient [$m^{1/3} s^{-1}$] (K_c for channels, K_f for the floodplain elements), and $S_{ik} = (h_i - h_k)/\Delta x$ is the slope of the water surface between cells i and k.

For the side flow between channel and floodplain elements, we use either the relation for a free overflow weir (where there is a flood bank at the edge of the channel) or for submerged flow conditions. In the current study it was necessary to use a linear interpolation of elevations between the known cross-sections. This means that the river banks are essentially constant in slope in each reach, and for the particular case of a fully submerged flow above some threshold elevation h_{i0}, with a constant depth of flow h_{ik} along the length of the cell, the left side flow can be written as:

$$Q_1 = K_1(\bar{h}_{ik} - h_{i0})^{5/3} l_{ik} S_{ik}^{1/2} \tag{10.6}$$

where K denotes the effective roughness coefficient for the lateral flow, and l_{ik} denotes the length of the cross-section between elements i and k. The same relationship is used for the right hand side lateral flow. In the general case, these side flows can be represented in the same look-up table form as in equation 10.5, using all the available data on the irregular geometry of the boundary between channel and floodplain, including any threshold for a weir-type flow.

The roughness coefficients K will be used as the calibration parameters in the calibration stage of the problem. For each reach of the channel at least five roughness coefficients need to be specified, for downstream flow in the channel, for flow on each of the floodplain elements, and for the exchanges between the channel and floodplain elements (which might change with the transition from weir flow to submerged conditions).

The inflow data correspond to the sum of inflows on the floodplain and the channel and, therefore, we have to find the inflow partitioning dependent on the velocity of water. This was achieved using the relation for the flow (equation 10.5), and assuming a uniform flow parallel to the bed slope, for each floodplain and the channel at the boundary. The sum of the downstream element flows must then correspond to the inflow $Q(t)$. From this relation we can find the corresponding water levels and inflow partitioning, $Q_{fpl}(h,t)$, $Q_c(h,t)$ and $Q_{fpr}(h,t)$:

$$Q(t) = Q_{fpl}(h,t) + Q_c(h,t) + Q_{fpr}(h,t) \tag{10.7}$$

which gives the effective height h at the inflow boundary corresponding to the total inflow $Q(t)$.

As explained in Section 10.4, the calibration of the model is performed using the "observations" of water levels taken from the simulation results of the two-dimensional

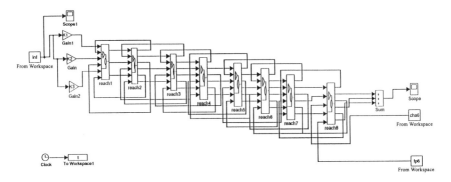

Figure 10.1 SIMULINK representation of the flood inundation model for eight subreaches

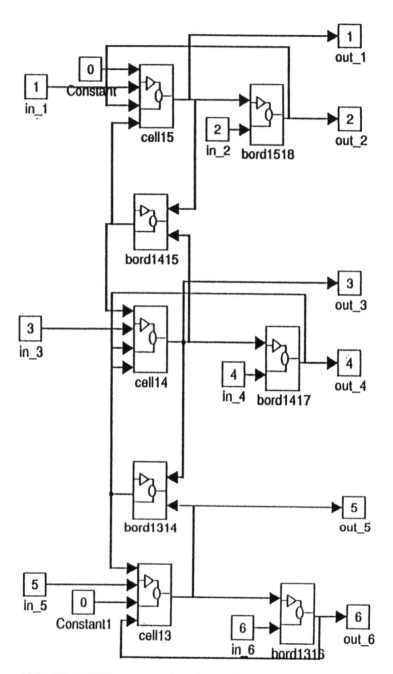

Figure 10.2   SIMULINK representation of one subreach of the flood inundation model

finite element model developed by Anderson and Bates (1994), and Bates *et al.* (1992), (see also Chapter 7, this volume). To be consistent for the downstream boundary condition, the "observed" discharge from the finite element model was used to find the corresponding water levels, under the same assumption of uniform flow parallel to the river slope. Lateral inflows at the boundaries have been ignored in the current study. The reader should note that the model used is equivalent to a coarse mesh approximation to the reduced parabolic depth-averaged flow equations (Cunge, 1976). Information can propagate upstream in the hydraulic model as the flow between cells depends on the corresponding gradient of the water surface allowing for the prediction of backwater effects.

This form of the model is easily implemented in the SIMULINK framework, where we can use up to six different implicit finite difference schemes to obtain the solution of the flood routing problem. A SIMULINK presentation of the model is given in Figure 10.1. Each cell corresponds to a storage in the channel reach or on the floodplains. They are modelled using continuous integrators. The SIMULINK representation of one reach containing three cells (two floodplains and the channel) and two flow evaluation units is shown in Figure 10.2. Each flow evaluation unit is calculating the flow value, corresponding to the solution of the heights, according to the prescribed area and wetted perimeter look-up tables and flow formulae.

Figure 10.3 presents the SIMULINK scheme of one cell. All the look-up tables are prepared in the MATLAB program, which also sets all the necessary initial conditions and parameters and flow partitioning. SIMULINK allows for the graphical tracking of the time changes of all the variables as required, by simply adding a graphical element to the structure. That means that we can follow the time changes of the storage, water levels and flows in the channel and on the floodplains as the simulation proceeds.

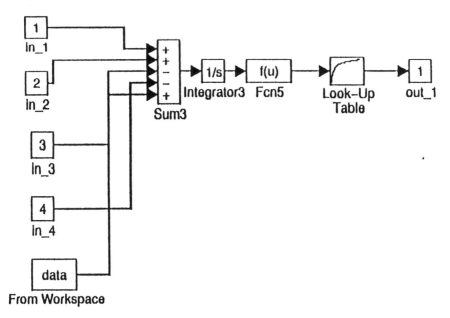

Figure 10.3 SIMULINK representation of the storage cell of the flood inundation model

However, spatial visualization of the water depths in the channel and on the floodplains needs additional MATLAB transformations and will be discussed later.

For the purpose of the uncertainty analysis we have made many simulation runs for different parameter sets using a Monte Carlo technique. This has been carried out by a FORTRAN implementation of the model on the Lancaster University PARAMID parallel processing system, consisting of multiple Intel i860 processors linked by transputers. In this case an implicit method of time integration has been used, with a check on the size of time-steps to maintain stability. To minimize the influence of specifying a distributed initial condition, we calculate a steady-state solution for each new set of parameters, given the specified initial discharge through the reach.

10.3   GENERALIZED LIKELIHOOD UNCERTAINTY ESTIMATION

10.3.1   Bayesian uncertainty estimation

Following the formulation of the inverse problem given by equation 10.3, we can treat the calibration problem in a Bayesian uncertainty estimation framework introducing the probability measures as the descriptors of the state of information. The Bayesian approach is based on the assumption that all the uncertainties can be represented and evaluated as probabilities (Box and Tiao, 1992). Thus we assume that there is a probability distribution of prediction errors, dependent on the model structure and some parameters θ, which have a prior probability distribution $f(\theta)$. The prior probability distribution should be based on the modeller's experience. In the current study the prior is taken to be a uniform probability density function (pdf) over a range of each parameter specified by the modeller. In the case when the estimated parameter values have finite limits and hence may lay in the middle of the parameter range, the uniform distribution on priors is well justified (Woodbury and Ulrych, 1993). Another way to define the priors is the use of retrospective calibration, mentioned in Draper (1995) and introduced by Dawid (1984).

The posterior density function defined by equation 10.3 can be used in the calculation of the confidence limits for the predictions of the model. As equation 10.3 is defined over the whole parameter space, any parameter interactions will be implicitly reflected in the calculated posterior distributions. This feature of the approach is especially advantageous in the case of a distributed model, where the parameters are very interdependent. The marginal distributions for individual parameters or groups of parameters can be calculated by an integration of the posterior distribution over the rest of the parameters as necessary. Equation 10.3 can be applied sequentially as new data become available and the existing posterior distribution, based on $(N-1)$ calibration periods, is used as the prior for the new data in the Nth calibration period. Thus

$$f(\theta \mid \mathbf{z}_1, ..., \mathbf{z}_N) \propto f(\theta \mid \mathbf{z}_1, ..., \mathbf{z}_{N1})L(\theta \mid \mathbf{z}_N) \tag{10.8}$$

where $L(\theta \mid \mathbf{z}_N)$ is the information about θ from the Nth calibration period.

In the case of the flood inundation model the new incoming data are the observations of the water levels on the subsequent cross sections, as the simulations proceed downstream. As they correspond to different points in space, and in our formulation, different state variables, this updating of the posterior distributions will have a different meaning than in the case of the calibration of lumped or quasi-distributed model (e.g. the applications to the rainfall–runoff model TOPMODEL reported in Beven, 1993; Romanowicz *et al.*, 1994; Freer *et al.*, 1996).

The interdependence of a new data set and parameters of the model is reflected, as previously, in the forward model equations, which provide the simulated data values. As already mentioned in Section 10.1, simulated and observed values differ. The best way to describe the uncertainties resulting from the forward modelling is through the analysis of the associated error model.

10.3.2 Formulation of the error model

It is assumed that the errors between the observed and predicted variables (water levels) have the additive form:

$$Z_t = Y_t(r_t, \theta) + \delta_t, \ t = 1, \dots, T \tag{10.9}$$

where $\mathbf{Z}_t$ denotes the observation vector, $\mathbf{Y}_t$ denotes the simulated levels, $\boldsymbol{\delta}_t$ is the vector of model errors, $\mathbf{r}_t$ denotes the input data and T is the number of discrete time periods.

The major statistical assumption which must be made is the choice of a structure of the statistical model for $\boldsymbol{\delta}_t$. In order to take into account the correlated errors and a potential imbalance between input and output we shall model $\boldsymbol{\delta}_t$ by the Gaussian first order autoregressive model AR(1) with non-zero mean:

$$\boldsymbol{\delta}_t - \boldsymbol{\mu} = \sum_{i=1}^{p} \mathbf{A}_i \left(\boldsymbol{\delta}_{t-i} - \boldsymbol{\mu} \right) + \boldsymbol{\varepsilon}_t \tag{10.10}$$

where the vector mean, $\boldsymbol{\mu}$, of the vector $\boldsymbol{\delta}_t$ process is constant in time, the $\mathbf{A}_i$ is the matrix of auto-regressive parameters for the error series and the $\boldsymbol{\varepsilon}_t$ is assumed to be a normal white noise vector error with covariance matrix Σ.

Hence, the $\boldsymbol{\delta}_t$ will also be normally distributed and the likelihood function of the predicted flows can be expressed as the likelihood of the error variate $\boldsymbol{\delta}_t$ with parameters $\boldsymbol{\phi} = (\boldsymbol{\mu}, \Sigma, \mathbf{A})$, depending on the hydraulic model parameters θ. In what follows it will be assumed that the error series can be described adequately by a first-order auto-regressive process ($p = 1$). This selection of p is purely illustrative and in practice a higher order process may be found to be a better description of the errors. Moreover, the Gaussian model itself may be inadequate. Following the discussion given by Draper (1995), such a choice of a single model structure does not allow the model structural uncertainty to be taken into account. Rather than a single model we could take various families of feasible model structures and introduce the uncertainty connected with each structure into the Bayes formula (equation 10.3) when evaluating the predictive uncertainty of parameters within each structure. The limit is not methodological, but rather the practical one of the computer time available. Hence, given the intrinsically large uncertainty in θ this additional complication is considered unnecessary here.

10.3.3   The likelihood function and the posterior distribution

For the sake of simplicity of presentation the one-dimensional case of the error model (equation 10.10) will be described (one observation site only), which can be easily generalized for the multidimensional case. For the one-dimensional case, let $A = \alpha$, $\Sigma = \sigma$. The error ε_t is assumed to be white Gaussian and hence, its likelihood function will consist of the T multiplications of the conditional probability function for each observation of the error ε_t given previous data. After substitution of the variables from equations 10.9 and 10.10 we can write the joint posterior distribution of parameters of the hydrological and statistical models as follows (for more detailed discussion see Romanowicz *et al.*, 1994):

$$f(\theta, \phi \,|\, z) = C \times f(\theta, \phi) \times (2\pi\sigma^2)^{-T/2}(1 - \alpha^2)^{1/2} \exp\left[-\frac{1}{2\sigma^2}\,\Psi(\mu, \alpha, \theta)\right] \quad (10.11)$$

where z denotes the observations (water levels in the channel and on the floodplains), C is a normalizing constant and

$$\Psi(\mu, \alpha, \theta) = (1 - \alpha^2)(\delta_1^* - \mu)^2 + \sum_{t=2}^{T} [\delta_t^* - \mu - \alpha(\delta_{t-1}^* - \mu)]^2 \quad (10.12)$$

while $f(\theta, \phi)$ denotes the prior distribution of the parameters of the hydraulic model and the error model. It can be calculated from:

$$f(\theta, \phi) = f(\theta) \times f(\phi \,|\, \theta) \quad (10.13)$$

In the general case the distribution of the error model parameters may be expected to depend on the parameters θ, but adopting a position of ignorance here we take $f(\theta, \phi) = f(\theta)f(\phi)$.

It is possible to obtain some prior information about the range of the values of the roughness parameters K. Thus according to the discussion given in Section 10.3.2 we can use the uniform distribution over that range. Hydraulic modellers will have limited prior knowledge about ϕ. However, we can evaluate those values approximately from Monte Carlo simulations of water levels in the modelled reach.

In spite of the fact that we are mostly interested in the posterior probability space described by equation 10.11, it can be very illustrative to write the log-likelihood function for this error model. It takes the form:

$$l(\theta, \phi) = -\frac{T}{2}\ln\sigma^2 + \frac{1}{2}\ln(1 - \alpha^2) - \frac{\Psi(\mu, \alpha, \theta)}{2\sigma^2} \quad (10.14)$$

The maximum likelihood estimates of the parameters of the hydrological model and the noise process can be found in the usual way by setting the differentials of equation 10.14 to zero and solving for the resulting optimal parameter values. Similarly standard likelihood techniques (Cox and Hinkley, 1974) can be used to develop confidence intervals for θ and ϕ. However, within the Bayesian framework used in this flood inundation application, the confidence limits for the predicted water levels can be simply obtained from the posterior distribution of the parameter sets, without the need to assume that an optimum parameter set exists.

It can be seen that the function given in equation 10.14 corresponds to the standard sum of squares goodness-of-fit criterion, transformed by the error model. As the multiplication of the exponent functions is equivalent to the summation of their exponents, the updating of the posteriors is equivalent to adding the new errors to the sum of the old ones. Hence, after all the information on the model parameters is included, our resulting posterior distribution will give the solution of the entire calibration problem. Another important conclusion is that by using a non-Gaussian error model a different objective function would be obtained. Looking at this problem from the calibration point of view, we can try to find among the generalized Gaussian distributions (which combine, among others, absolute value error, square error and minimax criterion, see Box and Tiao, 1992; Tarantola, 1987), a distribution that would best describe the error model and give the greatest sensitivity to the model parameters. In general, we should chose among the distributions which have exponential form, which assures that our global posterior is equivalent to the sum of local posteriors and, hence, the global solution is consistent with the local ones.

10.3.4 Evaluation of predictive uncertainty

The posterior distribution $f(\theta, \phi \mid z)$ of the parameters can be used in estimating the predictive uncertainty of the model under the assumption that the distribution of the error series will be the same in prediction as in calibration. The distribution of predictions then depends upon the (non-linear) hydraulic model and the statistical error model. For the structure of the statistical model assumed here, the cumulative distribution of the error term at any time, given a particular set of hydraulic and statistical model parameters θ and ϕ, will be given by

$$P(\delta_t < \delta \mid \theta, \phi) = \Phi\left(\frac{\delta - \mu}{\sigma/(1 - \alpha^2)^{1/2}}\right) \qquad t = 1, \ldots, T \qquad (10.15)$$

where Φ is a standard normal distribution function. The predictive distribution of water levels Z_t' conditioned on the calibration data z will be then given by

$$P(Z_t < y \mid z) = \int_\theta \int_\phi \Phi\left(\frac{y - \mu - y_t}{\sigma/(1 - \alpha^2)^{1/2}}\right) f(\theta, \phi \mid z) \, d\phi \, d\theta \qquad (10.16)$$

where y_t is the hydrological model output at time t based on using hydrological model parameters θ. In discrete form, as used here,

$$P(Z_t < y \mid z) = \sum_\theta \sum_\phi \Phi\left(\frac{y - \mu - y_t}{\sigma/(1 - \alpha^2)^{1/2}}\right) f(\theta, \phi \mid z) \qquad (10.17)$$

We then seek for chosen confidence limits, the lower and upper percentage points $[z_L, z_u]$ of the distribution of Z_t for example such that

$$P(Z_t < z_L \mid z) = 0.1$$
$$P(Z_t < z_U \mid z) = 0.9$$

for a 80% confidence limit at time t. Clearly here equation 10.17 is the predictive distribution of water levels at time t and $[z_L, z_u]$ a predictive 80% confidence interval for water levels. As the time length of our observations is not big (only 37 time-steps), the evaluation of confidence limits does not involve intensive computations in MATLAB, once the Monte Carlo simulations results are available.

An alternative approach, which is more consistent with the methods of obtaining confidence intervals for flows used in the GLUE procedure (Beven and Binley, 1992) is to estimate confidence intervals of the predictive distribution without consideration of the structure of the errors. This gives:

$$P(Y_t < y \,|\, \mathbf{z}) = \int_{\theta} P(Y_t < y \,|\, \theta, \mathbf{z}) f(\theta \,|\, \mathbf{z}) d\theta \qquad (10.18)$$

where $f(\theta \,|\, z)$ is the marginal posterior of θ. However, as Y_t is a deterministic function of the θ parameters equation 10.18 reduces to:

$$P_r(Y_t < y \,|\, \mathbf{z}) = \int_{B} f(\theta \,|\, \mathbf{z}) d\theta \qquad (10.19)$$

where B is the set of θ such that Y_t is less than y, given both θ and z. Clearly in this case computational problems are simplified at the expense of obtaining confidence intervals for the hydrological model predictions rather than resultant water levels.

10.4   MODEL CALIBRATION

10.4.1   The data description

The model has been applied to a reach of the River Culm, in Devon, about 12 km long. The elevation data available are in the form of a vector of irregular (x, y, z) values, giving the elevations of points situated on cross-sections of the channel and the floodplains. The whole river reach is represented as 95 cross-sections (Figure 10.4). A single event has been studied for this model calibration. Time series of inflow and outflow discharges for the whole reach are available for 40 half-hour time periods.

In one part of the river, the flow divides into two channels. For this study, this reach was modelled in the same way as the rest of the river, with spillage of water towards one of the floodplains, the look-up tables reflecting the proper cross-sectional areas, wetted perimeters and storage values for a given elevation as defined by the available elevation data.

It was mentioned in Section 10.2 that for the calibration study reported here we shall use the "observations" of water levels in the channel and on one floodplain in six different cross-sections along the river (12 observation sites all together) taken from the simulation results of the two-dimensional finite element model RMA-2 developed in Bristol (Bates *et al.*, 1992). This strategy of comparing with model results has been taken to allow the investigation of the effects of multiple observation sites on the predicted uncertainty.

10.4.2   Prior probabilities of parameters

Monte Carlo simulations are made for 12 different roughness parameters: one for both floodplains and one for the channel in each of the six subreaches between the

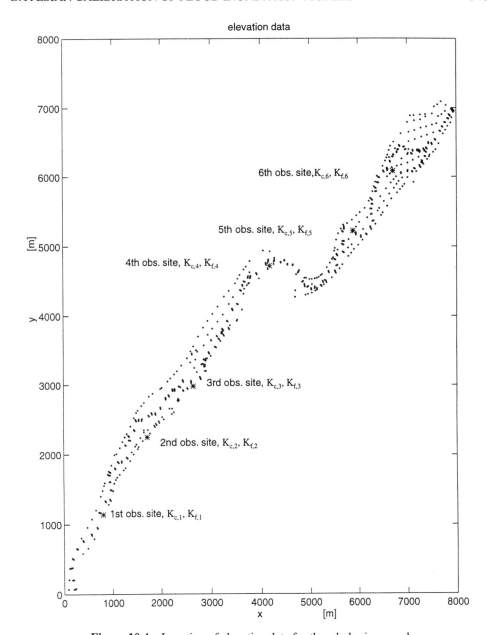

Figure 10.4   Location of elevation data for the whole river reach

measurement points. Initial sensitivity analysis suggested that the results were not strongly sensitive to the roughness parameters for the side exchanges in equation 10.5 so these values were fixed.

This choice of the calibration parameters was dictated by the requirement of minimizing the number of parameters to be calibrated (allowing one parameter for each of the 12 sites where measurements are available). As was mentioned in Section 10.3.1, a uniform prior distribution for each parameter is well justified in the case when no other information is available. Let $K_{c,i}$ and $K_{f,i}$ denote the roughness coefficients in the ith reach between measurement sites of the channel and the floodplain respectively. After the initial sensitivity analysis of the model performance using wide parameter ranges, it was decided that the priors for the roughness coefficients both for the floodplains and the channel will take the range of values as given in Table 10.1.

The choice of the parameter range was influenced by three factors: (i) the physical meaning of parameters, (ii) the computation time and (iii) numerical stability restrictions. According to the first factor, it would be expected that the floodplain roughness coefficients should take bigger values than the channel. It was decided to leave this differentiation to the model itself and the regions for both roughness coefficients were taken to be the same. As the number of Monte Carlo simulations was restricted to 3000 only, the parameter range had to be kept as small as possible while including the expected best values. This procedure can lead to the necessity of resampling of the parameter space as new data become available, as discussed by Beven and Binley (1992). The choice of the limits for the parameter ranges was based on the results of a preliminary exploration of the shape of the likelihood response surface. The limits used gave the maximum sampling density given the limited number of Monte Carlo samples possible around the expected likelihood peak for each parameter.

10.4.3 Simulation results

The sensitivity analysis of the model to parameters was performed using Monte Carlo simulations over the range of the parameter values and subsequent analysis of the resulting likelihood response surfaces evaluated for each measurement site. The response surfaces were obtained on the basis of equation 10.11 with $\Sigma = I$ (identity matrix), $\mu = 0$ and $A = 0$. Hence they describe the probability density functions (pdf) of the error model (equation 10.10) restricted to the special case of $N(0, 1)$ white Gaussian noise. As was noted in Section 10.3.3, those probability density functions are equivalent in shape to the standard least square goodness of fit response surfaces. The results of the simulations showed that the model performance is sensitive only to six channel parameters and essentially is not sensitive to the roughness coefficients on the floodplains, except for the last one, $K_{f,6}$. This result is in agreement with the observation made by Cunge et $al.$ (1976), that the floodplains act primarily as additional storage, while the main body of the wave is transported by the main channel. The notation for $K_{c,i}$ and $K_{f,i}$ (the roughness coefficients in the ith reach between measurement sites of the channel and the floodplain respectively) is kept as in Section 10.4.2. Then $K_{c,1}$ has influence on the differences between simulated and observed

Table 10.1 Ranges for the parameter values used in simulations (all values in $m^{1/3}\ s^{-1}$)

$K_{f,1}$	$K_{c,1}$	$K_{f,2}$	$K_{c,2}$	$K_{f,3}$	$K_{c,3}$	$K_{f,4}$	$K_{c,4}$	$K_{f,5}$	$K_{c,5}$	$K_{f,6}$	$K_{c,6}$
15	15	4	4	5	5	5	5	1	1	1	1
25	25	25	25	30	30	50	50	25	25	25	25

water levels in the channel in the first cross-section only; $K_{c,2}$ influences the likelihood measure for the first cross-section and to a smaller extent, the second cross-section; $K_{c,3}$ influences both likelihood measures for the second and the third cross-sections; $K_{c,4}$ influences the likelihood measure only in the third cross-section; $K_{c,5}$ influences the likelihood measure in the fourth cross-section and (to a lesser extent) also in the fifth cross-section; $K_{c,6}$ influences the likelihood measure for both the sixth and the fifth cross-sections.

From the computations it was found that $K_{f,6}$ also influences the likelihood measure for the sixth reach. This is caused by the fact that the sixth reach consists of the two parallel channels which divide the main stream of water into two parts.

Those results do not mean that the exchange of water between the floodplain and the channel does not influence water flow in the model nor that the floodplains act as mere storage areas. From these results we can conclude only that the corresponding roughness coefficients did not influence the measure of model performance given the data available. Those parameters controlling the channel to floodplain exchange also constitute an important part of the model description. In particular, too small lateral exchange coefficients were resulting in significant differences between the channel and floodplain water levels, while too big values were causing instability of the solution. However, the information contained in the available observations was not sufficient to infer what values the exchange coefficients should take with any precision. As for the floodplain roughness coefficients, the only place where the information contained in the observations was enough to infer its range was the reach where the channel divides and there is a large enough flow on the floodplain element (observation site 6, floodplain roughness coefficient $K_{f,6}$).

10.4.4   Statement of the calibration problem

The general deterministic goodness-of-fit function (the inverse problem, equation 10.2) for the entire model, taking into account the insensitivity of the results to certain parameters, will have the form:

$$J(\theta) = \sum_{t=1}^{T} \sum_{i=1}^{12} g_i(\varepsilon_{i,t}) \tag{10.20}$$

where ε denotes the difference between the simulated and observed water levels in all 12 observation sites; $\theta = [K_{c,1}, K_{c,2}, K_{c,3}, K_{c,4}, K_{c,5}, K_{f,6}, K_{c,6}]$ is the vector of roughness coefficients affecting the simulated water levels and g_i denotes the goodness-of-fit function for the ith measurement site (equivalent to measure of performance in equation 10.2).

When using the concept of joint information as in equation 10.3, equation 10.20 is equivalent to the search for the joint posterior distribution over the parameter space. Application of the error model (Equation 10.10) to the general calibration problem, for the parameter vector θ involves introducing the 12-dimensional vector of mean values μ, together with the $[12 \times 12]$ matrix of correlation coefficients α and the $[12 \times 12]$ covariance matrix Σ. Assuming a Gaussian error model we would get the formulation for the posterior distributions as given by equation 10.11. Taking into account the number of additional parameters that we have to introduce, the solution of that problem seems to be too troublesome.

The analysis of the individual responses of the model for each observation site gives an indication of the spatial correlation between the parameters. The spatial interactions between parameters of the model will be primarily limited to the nearest neighbouring sites, which should allow the simplification of the calibration problem. According to the results of the simulations, we can approximately subdivide the deterministic goodness-of-fit function (equation 10.20) as follows:

$$J(\theta) = J_1(K_{c,1}, K_{c,2}) + J_2(K_{c,2}, K_{c,3}) + J_3(K_{c,3}, K_{c,4}) + J_4(K_{c,5})$$
$$+ J_5(K_{c,5}, K_{c,6}) + J_6(K_{f,6}, K_{c,6}) \quad (10.21)$$

In this subdivision each goodness of fit function on the right hand side of equation 10.21 corresponds to a different observation site (channel and the floodplain taken together).

Moving back into the Bayesian framework, we shall try to solve the equivalent to equation 10.20 Bayesian problem using the idea of recursive (Bayesian) estimation in order to break it into smaller dimension subproblems, as was done in the deterministic case (equation 10.21). Assuming the independence of the errors and Bayes' theorem (equation 10.8), the joint posterior probability for the entire problem, has the form:

$$f(\theta \mid z_1, \ldots, z_6) = f_1(\theta \mid z_1) f_2(\theta \mid z_2) f_3(\theta \mid z_3) f_4(\theta \mid z_4) f_5(\theta \mid z_5) f_6(\theta \mid z_6) / f_0^5(\theta)$$
$$(10.22)$$

where $f_i(.)$ denotes the posterior distribution for the joint channel and floodplain observation errors at each observation site and $f_0(\theta)$ denotes the uniform prior distribution on parameters.

Following the initial Monte Carlo simulations and associated sensitivity analysis we can make the following approximations:

$$f_1(\theta \mid z_1) = f_1(K_{c,1}, K_{c,2} \mid z_1), \ f_2(\theta \mid z_2) = f_2(K_{c,2}, K_{c,3} \mid z_2),$$
$$f_3(\theta \mid z_3) = f_3(K_{c,3}, K_{c,4} \mid z_3), \ f_4(\theta \mid z_4) = f_2(K_{c,5} \mid z_4),$$
$$f_5(\theta \mid z_5) = f_5(K_{c,5}, K_{c,6} \mid z_5), \ f_6(\theta \mid z_6) = f_6(K_{f,6}, K_{c,6} \mid z_6)$$

where the parameters $[K_{c,1}, K_{c,2}, K_{c,3}, K_{c,4}, K_{c,5}, K_{f,6}, K_{c,6}]$ belong to the part of the parameter vector θ to which the distributions are sensitive.

Taking the logarithm of the posterior of equation 10.22 changes the product into summation and divides the calibration problem into the subproblems, i.e.

$$\log(f(\theta \mid z_1, \ldots, z_6)) = \log(f_1(K_{c,1}, K_{c,2} \mid z_1)) + \log(f_2(K_{c,2}, K_{c,3} \mid z_2))$$
$$+ \log(f_3(K_{c,3}, K_{c,4} \mid z_3)) + \log(f_4(K_{c,5} \mid z_4)) + \log(f_5(K_{c,5}, K_{c,6} \mid z_5))$$
$$+ \log(f_6(K_{f,6}, K_{c,6} \mid z_6)) + -5\log(f_0(\theta)) \quad (10.23)$$

Equation 10.23 is equivalent to equation 10.21 and leads to the following subdivision of the calibration problem:

1.   The posterior distribution for $K_{c,1}$ is given by

$$f(K_{c,1} \mid z_1) = \int_{K_{c,2}} f_1(K_{c,1}, K_{c,2} \mid z_1) \, dK_{c,2}. \quad (10.24a)$$

2. The posterior distribution for $K_{c,2}$ is given by

$$f(K_{c,2}|\mathbf{z}_1,\mathbf{z}_2) = \int_{K_{c,3}}\int_{K_{c,1}} f_1(K_{c,1},K_{c,2}|\mathbf{z}_1)f_1(K_{c,2},K_{c,3}|\mathbf{z}_2)\,\mathrm{d}K_{c,1}\,\mathrm{d}K_{c,3}. \qquad (10.24\mathrm{b})$$

3. The posterior distribution for $K_{c,3}$ is given by

$$f(K_{c,3}|\mathbf{z}_2,\mathbf{z}_3) = \int_{K_{c,2}}\int_{K_{c,4}} f_2(K_{c,2},K_{c,3}|\mathbf{z}_2)f_3(K_{c,3},K_{c,4}|\mathbf{z}_3)\,\mathrm{d}K_{c,4}\,\mathrm{d}K_{c,2}. \qquad (10.24\mathrm{c})$$

4. The posterior distribution for $K_{c,4}$ is given by

$$f(K_{c,4}|\mathbf{z}_3) = \int_{K_{c,3}} f_3(K_{c,3},K_{c,4}|\mathbf{z}_3)\,\mathrm{d}K_{c,3}. \qquad (10.24\mathrm{d})$$

5. The posterior distribution for $K_{c,5}$ is given by

$$f(K_{c,5}|\mathbf{z}_4,\mathbf{z}_5) = \int_{K_{c,6}} f_5(K_{c,5},K_{c,6}|\mathbf{z}_5)f_4(K_{c,5}|\mathbf{z}_4)\,\mathrm{d}K_{c,6}. \qquad (10.24\mathrm{e})$$

6. The posterior distribution for $K_{f,6}$ is given by

$$f(K_{c,6}|\mathbf{z}_6) = \int_{K_{c,6}} f_6(K_{f,6},K_{c,6}|\mathbf{z}_6)\,\mathrm{d}K_{c,6}. \qquad (10.24\mathrm{f})$$

7. The posterior distribution for $K_{c,6}$ is given by

$$f(K_{c,6}|\mathbf{z}_5,\mathbf{z}_6) = \int_{K_{c,6}}\int_{K_{c,5}} f_5(K_{f,5},K_{c,6}|\mathbf{z}_5)f_6(K_{f,6},K_{c,6}|\mathbf{z}_6)\,\mathrm{d}K_{c,5}\,\mathrm{d}K_{c,6}. \qquad (10.24\mathrm{g})$$

In the case when we do not want to assume the independence of the errors for the cases 2, 3 and 4 we would have a joint posterior $f_{2,3,4}(K_{c,2},K_{c,3},K_{c,4})$ and for the cases 5, 6 and 7 – the joint posterior $f_{5,6,7}(K_{c,5},K_{f,6},K_{c,6})$. In each case the dependence structure is derived from the model for the dependent errors.

Our solution is conditioned only on those observations which have been used for the calibration. The solution of the subproblems (Equation 10.24) is still consistent with the general solution (equation 10.22), under the conditions that we did not omit information which was important and that the errors are independent where assumed to be so. This last assumption can be justified by the fact that all the interactions between the model parameters are taken into account by the model structure and are reflected in the parameter response surfaces obtained.

In the computations made so far we have assumed independence of the errors between the reaches, i.e. we follow steps 1 to 7 of the described calibration procedure. For each error ε_i we build the posterior probability distributions (equation 10.11), assuming the error structure given by equation 10.10. As an illustration of the dependence of the different model performance measures on the parameters Figure 10.5 shows the results for the channel and floodplain predictions for the second observation site and Figure 10.6 presents similar results obtained for the third observation site. In those figures each point represents a run of the model with randomly chosen parameter values. Figures 10.5 and 10.6, A, B, C, D show the likelihood response surfaces projected onto one parameter axis and calculated from equation 10.11 with fixed noise parameters ($\Sigma = \mathbf{I}$, $\mu = 0$ and $\mathbf{A} = 0$). As mentioned before, these response surfaces, together with the eight others for the rest of the observation sites, were used in the

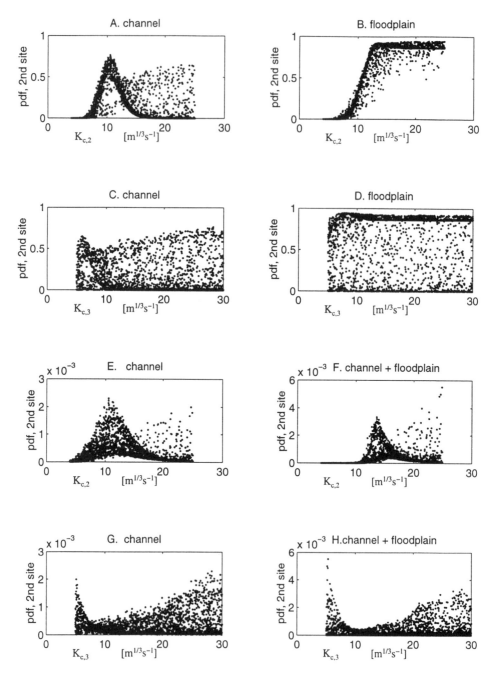

Figure 10.5 The dependence of the second observation site results. Each point represents a run of the model with randomly chosen parameter values: A. posterior pdf with $\Sigma = I$, $\mu = 0$ and $A = 0$ on $K_{c,2}$, for the channel; B. posterior pdf with $\Sigma = I$, $\mu = 0$ and $A = 0$ on $K_{c,2}$, for the floodplain; C. posterior pdf with $\Sigma = I$, $\mu = 0$ and $A = 0$ on $K_{c,3}$, for the channel; D. posterior pdf with $\Sigma = I$, $\mu = 0$ and $A = 0$ on $K_{,3}$, for the floodplain; E. posterior pdf on $K_{c,2}$, for the channel; F. joint posterior pdf on $K_{c,2}$, for channel and the floodplain; G. posterior pdf on $K_{c,3}$, for the channel; H. joint posterior pdf on $K_{c,3}$, for the channel and the floodplain

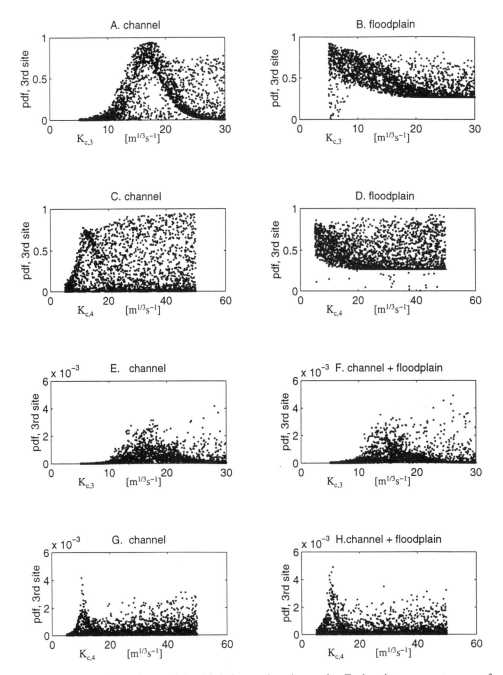

Figure 10.6 The dependence of the third observation site results. Each point represents a run of the model with randomly chosen parameter values: A. posterior pdf with $\Sigma = I$, $\mu = 0$ and $A = 0$ on $K_{c,3}$, for the channel; B. posterior pdf with $\Sigma = I$, $\mu = 0$ and $A = 0$ on $K_{c,3}$, for the floodplain; C. posterior pdf with $\Sigma = I$, $\mu = 0$ and $A = 0$ on $K_{c,4}$, for the channel; D. posterior pdf with $\Sigma = I$, $\mu = 0$ and $A = 0$ on $K_{c,4}$, for the floodplain; E. posterior pdf on $K_{c,3}$, for the channel; F. joint posterior pdf on $K_{c,3}$, for channel and the floodplain; G. posterior pdf on $K_{c,4}$, for the channel; H. joint posterior pdf on $K_{c,4}$, for channel and the floodplain

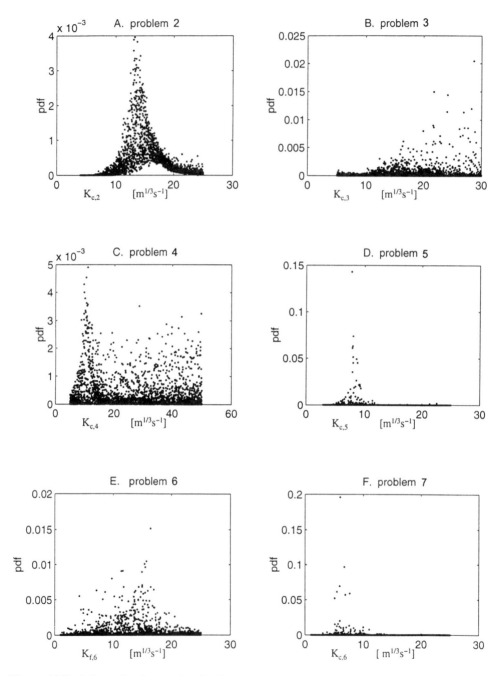

Figure 10.7   Joint updated posterior distributions. Each point represents a run of the model with randomly chosen parameter values: A. Problem 2 (equation 10.24b); B. Problem 3, (equation 10.24c); C. Problem 4 (equation 10.24d); D. Problem 5 (equation 10.24e); E. Problem 6 (equation 10.24f); F Problem 7 (equation 10.24g)

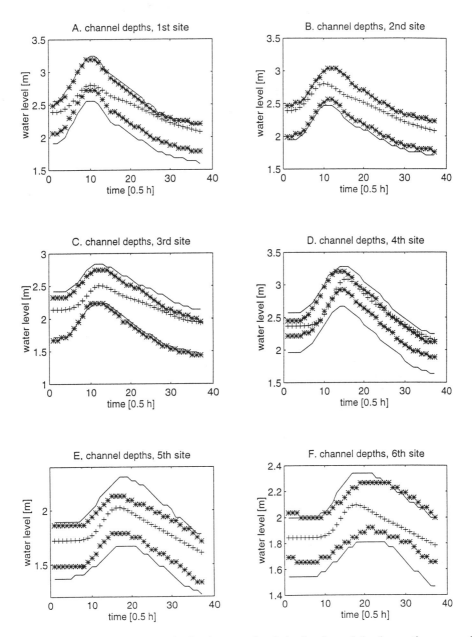

Figure 10.8 80% confidence limits for the water levels in the channel (+observations, — pdf without updating, * joint updated pdf): A. 1st observation site, joint updated pdf conditioned on sites 1 and 2; B. 2nd observation site, joint updated pdf conditioned on sites 1 and 2; C. 3rd observation site, joint updated pdf conditioned on sites 1, 2 and 3); D. 4th observation site, joint updated pdf conditioned on sites 4 and 5; E. 5th observation site, joint updated pdf conditioned on sites 5 and 6; F. 6th observation site, joint updated pdf conditioned on sites 5 and 6

sensitivity analysis of the whole calibration problem. Marginal posterior distributions for the depth predictions (i.e. summed over the error model parameter distributions) are shown on Figures 10.5 and 10.6 E and G. Marginal joint posterior probability distributions for the predicted depths shown on Figures 10.5 and 10.6 F and H were obtained using Bayesian updating of the posteriors for the channel with the prior taken as the posterior distribution after conditioning using the measured floodplain depth data. Usually, with more information used, the admissible region for the parameters is found to narrow (as seen on Figures 10.5 E, F and G, H), but for the third observation site the information contained in the floodplain observations is not adding anything to the channel results. As a result, the joint posteriors for the channel and the floodplain (Figure 10.6 F and H) have a wider admissible parameter range than the channel posteriors without updating (Figures 10.6 E and G). From the presented figures it can also be seen that the same parameter (e.g. $K_{c,3}$) may have different "optimal" values for different observation sites (Figure 10.5 G and Figure 10.6 E), which illustrates the necessity for the use of the proposed calibration procedure, stated in the form of the subproblems 1 to 7 for the different reaches (equation 10.24). The joint updated posterior distributions for the subproblems 2–7 are given in Figure 10.7.

We used formula 10.17 to obtain the confidence limits on our predictions of water levels along the channel at six observation sites on the channel and the floodplain. The resulting 80% confidence limits for the channel water levels are given in Figure 10.8. Figure 10.9 shows the 80% confidence limits for the predicted floodplain depths. The observed water levels are shown together with the confidence limits obtained without updating and the confidence limits based on the updated probabilities.

When analysing the confidence limits for the floodplains it occurred that the errors for small water levels (at the beginning and the end of the event) are too big, which was causing unacceptable results. This was due to the fact that the early version of the RMA-2 finite element model from where our "observed" water levels were taken, maintains a depth of water in partially wet elements at the beginning of the event to help avoid stability problems. Taking only the time periods corresponding to the four biggest water levels on the floodplains in the calculation of the likelihoods and posterior probability eliminated this problem. However, the fourth observation site, where the water levels are raised due to the geometry of the floodplain terrain, showed better results when all the time periods were taken into account. This procedure of the elimination of any information known to be biased or in error is consistent with the proposed methodology, which allows the user to choose the information used in the updating. It is also a good illustration of the way the confidence limits may serve as the tool to evaluate the model performance.

When new data become available we can use equation 10.8 to update the posteriors to include any new information on the model. We can also use the updated posteriors to find new confidence limits for the water levels, or any other variables in the model.

10.4.5 Spatial visualization of inundation predictions

Confidence limits for the water levels on the floodplains can be used to derive a map of the probability of inundation along the river reach using the different probability values for the water level in each river reach. Hence, at each time step, each water level on the

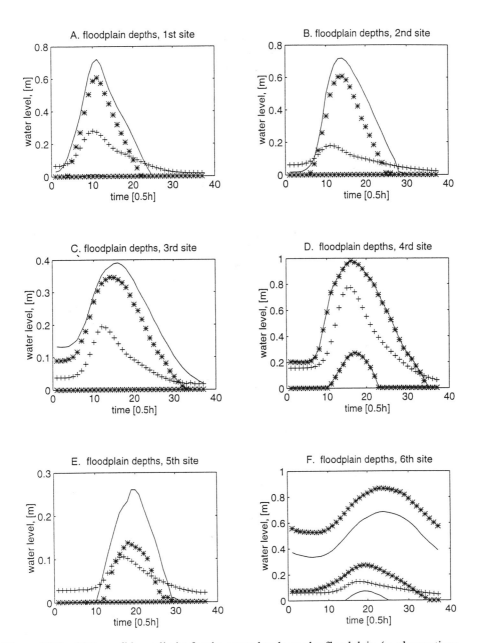

Figure 10.9 80% confidence limits for the water levels on the floodplain (+ observations, — probabilities without updating, * joint updated pdf): A. 1st observation site, joint updated pdf conditioned on sites 1 and 2; B. 2nd observation site, joint updated pdf conditioned on sites 1 and 2; C. 3rd observation site, joint updated pdf conditioned on sites 1, 2 and 3; D. 4th observation site, joint updated pdf conditioned on sites 4 and 5; E. 5th observation site, joint updated pdf conditioned on sites 5 and 6; F. 6th observation site, joint updated pdf conditioned on sites 5 and 6

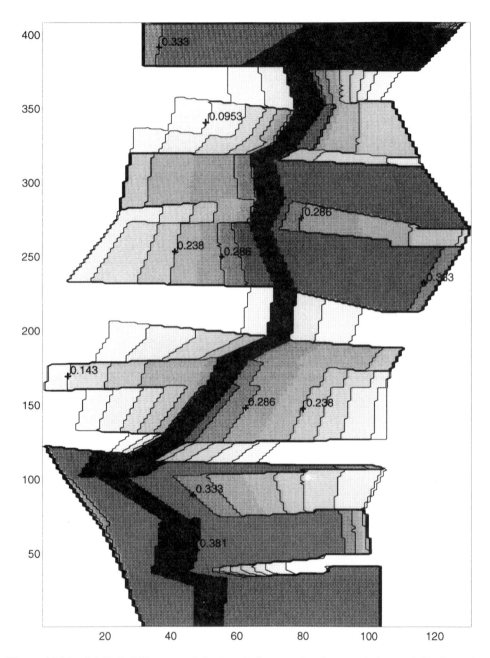

Figure 10.10   (a) Probability map of the inundations on the river reach for $t = 6.5$ h from the beginning of the event. Figures are predicted probabilities of inundations for different contours. (b) Probability map of the inundations on the river reach for $t = 10$ h from the beginning of the event. Figures are predicted probabilities of inundations for different contours

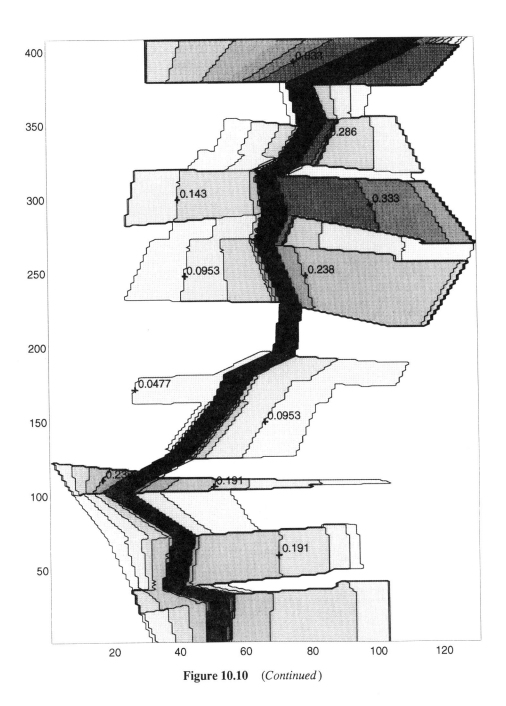

Figure 10.10   (*Continued*)

floodplains has associated with it a probability of being exceeded. Displaying the corresponding probability lines on the map of the terrain allows the risk of inundation to be visualized. The probability maps will be different for each discrete time period. In this way it is possible to visualize the propagation of the peak of the flood wave along the river reach. An example of mapping the probability of inundation is given in Figures 10.10a and 10.10b for two time-steps ($t = 6.5$ h and $t = 10$ h). Due to the array dimensions of the maps, only 18 of the 95 cross-sectional subreaches are shown, containing the second and the third observation sites. The confidence limits for all these 18 cross-sections were evaluated using the joint posterior probabilities for the floodplains and the channel from the second and the third observation sites. As was mentioned in the previous section, only the four highest water levels on the floodplains were used for the evaluation of those posteriors.

The effect of the different storage/water-level functions in the subreaches, derived by linear interpolation between the surveyed cross-sections at the reach boundaries, is clearly seen in these figures. The probability maps on the floodplains would be also more realistic if the model had more detailed description of the floodplains (i.e. three to four cells instead of one on each floodplain subreach). At the present there is the discrepancy between the resolution of the elevation data and the model used. A more detailed terrain map of the floodplain and a greater number of the floodplain cells would result in more irregular inundation predictions using this model.

10.5   CONCLUSIONS

This chapter has demonstrated the use of the Generalized Likelihood Uncertainty Estimation methodology to obtain first estimates of the uncertainty associated with inundation predictions of a simplified flood routing model. In this approach the standard search for an optimum parameter set is transformed into an examination of the posterior parameter distribution over the feasible parameter space. From the posterior distribution all the necessary information about parameters, predictions and uncertainties can be obtained. The choice of a statistical model for the errors within the GLUE framework can be viewed as being equivalent to the choice of a goodness-of-fit criterion for standard methods of calibration. However, the choice can be objectively assessed by examining the suitability of the error model used. It has also been demonstrated how the Bayesian nature of the approach allows the likelihood distributions and uncertainty estimates for the model predictions can be updated as new data become available.

The next stage in this work will be to implement the procedures into a real-time forecasting system in a way that will allow the revision of the pattern of risk of flooding as an event proceeds. This was a primary consideration in the choice of a simplified flood routing model in the current work. In such a system, the prior estimates of the parameter sets available before an event will be updated according to how well each model performs in forecasting mode during the event. At gauged sites it will be possible to evaluate forecasts ahead of time using the type of transfer function real-time forecasting of downstream levels used by Lees *et al.* (1994). The aim is then to provide updated risk of inundation forecasts in real time, overlain onto a map of the floodplain within a GIS system.

ACKNOWLEDGEMENTS

This work has been supported by NERC grant GR3/8950. Paul Bates and Richard Hardy from Bristol University are thanked for the data which allowed the calibration of the model.

REFERENCES

Anderson, M.G. and Bates, P.D. (1994) Initial testing of a two dimensional finite element model for floodplain inundation. *Proceedings of the Royal Society Series A*, **444**, 149–159.

Bates, B.C. and Townley, L.R. (1988) Nonlinear discrete flood event models. 1. Bayesian estimation of parameters. *J. Hydrology*, **99**, 61–76.

Bates, P.D., Baird, L., Anderson, M.G., Walling, D.E. and Simm, D. (1992) Modelling floodplain using a 2-dimensional finite element model. *Earth Surface Processes and Landforms*, **17**, 577–588.

Beven K.J. (1989) Changing ideas in hydrology: the case of physically-based models. *J. Hydrol.*, **105**, 157–172.

Beven, K.J. (1993) Prophecy, reality and uncertainty in distributed hydrological modelling. *Adv. Wat. Resour.*, **16**, 41–51.

Beven K.J. and Binley, A. (1992) The future of distributed models: model calibration and uncertainty prediction. *Hydrol. Process.*, **6**, 279–298.

Binley, A.M. and Beven, K.J. (1991) Physically based modelling of catchment hydrology: a likelihood approach to reducing predictive uncertainty. In: *Computer Modelling in the Environmental Sciences*, D.G. Farmer and M.J. Rycroft (eds), Clarendon, Oxford, 75–88.

Binley, A.M., Beven, K.J., Calver, A. and Watts, L. (1991) Changing responses in hydrology: assessing the uncertainty in physically-based predictions. *Wat. Resour. Res.*, **27**, 1253–1262.

Box, G.E.P. and Cox, D.R. (1964) An analysis of transformations (with discussion). *J. Roy. Statist. Soc., B*, **26**, 211–252.

Box, G.E.P. and Tiao, G.C. (1992) *Bayesian Inference In Statistical Analysis*, Wiley, Chichester.

Cokljat, D. and Younis, B.A. (1995) Second order closure study of open channel flows. *J. Hydraul. Eng. ASCE*, **121**, 94–107.

Cox, D.R. and Hinkley, D.V. (1974) *Theoretical Statistics*. Chapman & Hall, London.

Cunge, J.A., Holly, F.M. Jr and Verwey, A. (1976) *Practical Aspects of Computational River Hydraulics*, Pitman, London.

Dawid A.P. (1984) Statistical theory: the prequential approach. *J. Roy. Statist. Soc. A*, **147**, 278–292.

Draper D. (1995) Assessment and propagation of model uncertainty, *J. Roy. Statist. Soc. B*, **37**, 45–98.

Freer, J., Bevan, K.J. and Ambrose, B. (1996) Bayesion estimation of uncertainty in runoff prediction and the value of data: an application of the GLUE approach, *Water Resour. Res.* (in press)

Gelfand, A.E. and Smith, A.F.M. (1990) Sampling-based approaches to calculating marginal densities. *J. Amer. Statist. Assoc.*, **85**, 398–409.

Knight D.W. (1989) Hydraulics of flood channels. In: *Floods: Hydrological, Sedimentological and Geomorphological Implications*, K.J. Beven and P.A. Carling (eds), Wiley, Chichester, 83–105.

Knight D.W., Shiono, K. and Pirt, J. (1989) Prediction of depth, mean velocity and discharge in natural rivers with overbank flow. In: *Proc. Int. Conf. on Hydraulic and Environmental modelling of Coastal, Estuary and River Waters*, R.A. Falconer, P. Goodwin and R.G.S. Mathew (eds), Gower Technical Press, Aldershot, 419–428.

Lees, M., Young, P.C., Ferguson, S., Beven, K.J. and Burns, J. (1994) An adaptive flood warning scheme for the River Nith at Dumfries. In: *River Flood Hydraulics*, W.R. White and J. Watts (eds), Wiley, Chichester, 65–75.

Romanowicz, R., Beven, K. and Tawn, J. (1994) Evaluation of predictive uncertainty in nonlinear hydrological models using a Bayesian approach. In: *Statistics for the Environment 2, Water Related Issues*, V. Barnett and K.F. Turkman (eds), Wiley, Chichester, 297–315.

Tarantola, A. (1987) *Inverse Problems Theory, Methods for Data Fitting and Model Parameter Estimation*. Elsevier, Netherlands.

Tarantola, A. and Valette, B. (1982) Inverse problems = quest for information. *J. Geophysics*, **50**, 159–170.

Tukey, J.W. (1962) The future of data analysis. *Ann. Math. Stat.*, **33**, 1–67.

Woodbury, A.D. and Ulrych, T.J. (1993) Minimum relative entropy: forward probabilistic modelling. *Water Resour. Res.*, **29**, 2847–2860.

11 Modelling Sediment Transport and Water Quality Processes on Tidal Floodplains

R.A. FALCONER and Y. CHEN
Department of Civil and Environmental Engineering, University of Bradford, UK

11.1 INTRODUCTION

Tidal floodplains are complex ecosystems, which include tidal wetlands and mangrove swamps along intertidal shores. They are often dominated by complex channel networks, areas of marsh, fen or peatlands, or estuarine trees in mangrove swamps. They draw many of their physical, chemical and biological characteristics from the influences of coastal waters, inflowing fresh water and upland forests. They serve as meeting points between land and sea, and occupy specific ecological niches of considerable importance: to conservation groups (e.g. for their flora and fauna), to industrialists (e.g. for their sheltered access to coastal waters for shipping) and to local communities (e.g. for recreation and for efficient and high dispersion of effluent waste). They are often an important source of food for coastal aquaculture, whereby they act as natural filters for suspended material and pollutants, and offer effective flood protection for low-lying areas. In India alone the area of tidal floodplains, in the form of mangrove swamps, has been estimated at 356 000 ha, with the local communities being dependent upon fishing the adjacent estuarine and coastal waters for food and their livelihoods (Pardeshi, 1991).

In recent years there has been an increasing world-wide demand to develop tidal floodplains by converting part or all of the wetland to industrial and/or housing estates, or for agriculture – including rice fields, oil-palm plantations, and aquaculture ponds. For example, in Malaysia about 1% of the tidal floodplains have been lost to alternative uses every year for the past 30 years. Various studies on the gut content of fish and prawns have confirmed that the mangrove detritus forms a sizeable portion (up to 22%) of the food of these organisms (Ong *et al.*, 1991). Also, the statistical relationship between the areal extent of mangroves and prawn catch has been established for various mangrove systems. However, the quantitative link between the outwelling of nutrients from the mangrove systems and the effects of such outwelling on adjacent fisheries are not known.

If a tidal floodplain is to be enhanced, restored or studied (e.g. for environmental

Floodplain Processes. Edited by Malcolm G. Anderson, Des E. Walling and Paul D. Bates.
© 1996 John Wiley & Sons Ltd.

impact assessment purposes), it is important to identify the significant hydrodynamic, sediment flux and water quality processes occurring at the site. The changes to these physical and bio-chemical processes as a result of a project management plan, etc., must be predictable and compatible with the aims of the project. This chapter is devoted to the establishment of the governing differential equations describing the fluid flow, sediment transport and water quality processes occurring on these strategically and ecologically important regions and to the numerical modelling of these complex processes in a two-dimensional horizontal reference plane. Finally, examples are cited of the application of a typical two-dimensional numerical (or computational) model for predicting the flow, sediment transport and water quality processes along two large tidal floodplains in the UK, namely, Poole Harbour in Dorset and the Humber Estuary along the northern coast of England.

11.2   GOVERNING HYDRODYNAMIC EQUATIONS

11.2.1   Three-dimensional equations

The numerical models used by engineers, scientists and planners to predict the flow structure and solute transport processes on tidal floodplains are based on first solving the governing hydrodynamic equations on a rotating earth. For three-dimensional numerical models, the fluid is generally assumed to be isothermal and the vertical acceleration is assumed to be small compared to that due to gravity, i.e. giving a hydrostatic pressure distribution. Using the notation illustrated in Figure 11.1 the velocity and water elevation fields can be obtained by solving the Reynolds and continuity equations written in the following general form (Falconer, 1993):

$$\underbrace{\frac{\partial u}{\partial t}}_{1} + \underbrace{\frac{\partial u^2}{\partial x} + \frac{\partial uv}{\partial y} + \frac{\partial uw}{\partial z}}_{2} =$$

$$\underbrace{X}_{3} - \underbrace{\frac{1}{\rho}\frac{\partial P}{\partial x}}_{4} + \underbrace{v\left[\frac{\partial^2 u}{\partial x^2} + \frac{\partial^2 u}{\partial y^2} + \frac{\partial^2 u}{\partial z^2}\right]}_{5} - \underbrace{\left[\frac{\partial \overline{u'u'}}{\partial x} + \frac{\partial \overline{u'v'}}{\partial y} + \frac{\partial \overline{u'w'}}{\partial z}\right]}_{6} \quad (11.1)$$

$$\frac{\partial v}{\partial t} + \frac{\partial vu}{\partial x} + \frac{\partial v^2}{\partial y} + \frac{\partial vw}{\partial z} =$$

$$Y - \frac{1}{\rho}\frac{\partial P}{\partial y} + v\left[\frac{\partial^2 v}{\partial x^2} + \frac{\partial^2 v}{\partial y^2} + \frac{\partial^2 v}{\partial z^2}\right] - \left[\frac{\partial \overline{u'v'}}{\partial x} + \frac{\partial \overline{v'v'}}{\partial y} + \frac{\partial \overline{v'w'}}{\partial z}\right] \quad (11.2)$$

$$Z = \frac{1}{\rho}\frac{\partial P}{\partial z} \quad (11.3)$$

$$\frac{\partial u}{\partial x} + \frac{\partial v}{\partial y} + \frac{\partial w}{\partial z} = 0 \quad (11.4)$$

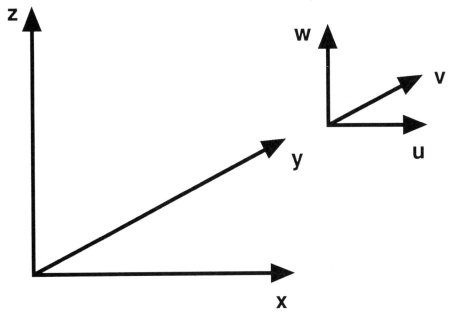

Figure 11.1 Cartesian coordinate system

where u, v, w = velocity components in x, y, z coordinate directions, t = time, X, Y, Z = body forces in x, y, z directions, ρ = fluid density, P = fluid pressure, v = kinematic viscosity ($= 10^{-6} \text{ m}^2 \text{s}^{-1}$ at STP), $\rho u' u'$, $\rho u' v'$, $\rho u' w'$, etc., = Reynolds (or apparent) stresses in x direction on x, y, z plane, etc.

The numbered terms in the momentum equation (11.1) refer to: local acceleration (term 1), advective acceleration (term 2), body force (term 3), pressure gradient (term 4), laminar shear stresses (term 5) and turbulent shear stresses (term 6).

If the effects of the earth's rotation are neglected and assuming that z is the vertical coordinate, as shown in Figure 11.1, then the body forces are given as:

$$\left. \begin{aligned} X &= 0 \\ Y &= 0 \\ Z &= -g \end{aligned} \right\} \tag{11.5}$$

However, in modelling large tidal floodplains, the effects of the earth's rotation cannot be ignored and the coriolis acceleration (Dronkers, 1964) has to be included, giving:

$$\left. \begin{aligned} X &= 2v\omega \sin \varphi \\ Y &= -2u\omega \sin \varphi \end{aligned} \right\} \tag{11.6}$$

where ω = speed of earth's rotation ($= 7.3 \times 10^5$ radians per second) and φ = earth's latitude at site of interest. The main effects of the Coriolis acceleration along tidal floodplains are to set up transverse water surface slopes across the floodplain and to bend streamlines, including deep channel jets, etc.

The pressure term (P) can be expressed in terms of the water surface elevation by integrating equation (11.3) with respect to z and using the notation in Figure 11.2. Thus:

$$P(z) = \rho g (\zeta - z) + P_a \tag{11.7}$$

where $\zeta =$ water surface elevation above (positive) datum and $P_a =$ atmospheric pressure, with the corresponding derivatives in equations (11.1) and (11.2) being expressed as:

$$\left. \begin{aligned} \frac{\partial P}{\partial x} &= \rho g \, \frac{\partial \zeta}{\partial x} + \frac{\partial P_a}{\partial x} \\[2mm] \frac{\partial P}{\partial y} &= \rho g \, \frac{\partial \zeta}{\partial y} + \frac{\partial P_a}{\partial y} \end{aligned} \right\} \tag{11.8}$$

with spatial differences in atmospheric pressure generally being negligible over tidal floodplains. The water surface elevation term obtained by substituting equation (11.8) in equations (11.1) and (11.2) can now be eliminated by integrating the continuity equation (11.4) over the depth and combining with the kinematic free surface condition to give:

$$\frac{\partial \zeta}{\partial t} = - \frac{\partial}{\partial x} \int_{-h}^{\zeta} u \, dz - \frac{\partial}{\partial y} \int_{-h}^{\zeta} v \, dz \tag{11.9}$$

The only unknown terms remaining in equations (11.1) and (11.2), before being able to solve numerically equations (11.1) to (11.4) to give the water elevation field and the three-dimensional velocity components, are the Reynolds stress terms. In solving for these stresses, Boussinesq (Goldstein, 1938) proposed that they could be represented in

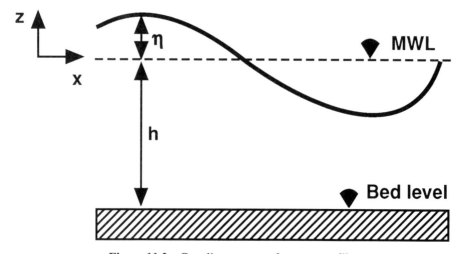

Figure 11.2   Coordinate system for wave profile

a diffusive manner, whereby for the x-direction [i.e. equation (11.1)] we get:

$$-\overline{u'u'} = \varepsilon\left[\frac{\partial u}{\partial x} + \frac{\partial u}{\partial x}\right]$$

$$\left.-\overline{u'v'} = \varepsilon\left[\frac{\partial u}{\partial y} + \frac{\partial v}{\partial x}\right]\right\} \qquad (11.10)$$

$$-\overline{u'w'} = \varepsilon\left[\frac{\partial u}{\partial z} + \frac{\partial w}{\partial x}\right]$$

where ε = kinematic eddy viscosity. In general, for tidal floodplains $\varepsilon \gg v$ and the laminar viscosity term [i.e. term 5 in equation (11.1)] is usually neglected in equations (11.1) and (11.2).

To determine the Reynolds stress formulations given in equation (11.10) many different and varyingly complex expressions exist for determining the eddy viscosity ε. These can be summarized as follows:

1. The simplest approach is to assume a constant value for ε based on field data. However, the range of values for ε is large, with typical values being between 10^2 m^{2}s^{-1} and 10^5 m^{2}s^{-1} (Bengtsson, 1978; Fischer et al., 1979).
2. The representation most widely used in tidal flows is to apply a zero-equation turbulence model similar to that prescribed by Prandtl's mixing length hypothesis (Goldstein, 1938) where:

$$\varepsilon = l^2 J \qquad (11.1)$$

where l = some characteristic mixing length and:

$$J = \sqrt{2\left(\frac{\partial u}{\partial x}\right)^2 + 2\left(\frac{\partial v}{\partial y}\right)^2 + 2\left(\frac{\partial w}{\partial z}\right)^2 + \left(\frac{\partial w}{\partial y} + \frac{\partial v}{\partial z}\right)^2 + \left(\frac{\partial u}{\partial z} + \frac{\partial w}{\partial x}\right)^2 + \left(\frac{\partial v}{\partial x} + \frac{\partial u}{\partial y}\right)^2}$$

$$(11.12)$$

with this expression for J generally being simplified due to the dominance of one or more velocity gradients. In determining the mixing length l, Prandtl suggested $l = \kappa z$ near the bed or a wall, where κ = von Karman's constant (=0.41) and z = coordinate perpendicular to the bed or wall, with $z = 0$ at the bed. Other expressions can be used for l, including, for example, von Karman's representation, given for the x-direction as:

$$l = \kappa\left[\frac{\partial u}{\partial z} \middle/ \frac{\partial^2 u}{\partial z^2}\right] \qquad (11.13)$$

3. For more complex turbulence models, transport equations exist for such parameters as the turbulent kinetic energy and the energy dissipation, giving the κ–ε turbulence model, or the Reynolds stresses can be solved directly, using the algebraic stress model. Further details of these more advanced models are given in Rodi (1984) and Falconer and Li (1994).

By substituting equations (11.12), (11.11), (11.10), (11.9), (11.8) and (11.6), or similar, into equations (11.1) to (11.4), it is now possible to solve numerically for u, v, w, and ζ across any tidal floodplain for a time varying flow field. However, in studying and modelling floodplain processes, the depth of flow is generally small and the flow can often be assumed to be well-mixed vertically. This allows the flow to be approximated by a two-dimensional horizontal flow formulation, which significantly reduces the computational cost of solving for the hydrodynamic processes and thereby allows for a much longer computational simulation of sediment transport and water quality processes across the floodplain.

11.2.2 Depth-integrated equations

In solving for the depth-integrated velocity field, equations (11.1) to (11.4) are first integrated over the depth and combined with the no-slip bed boundary condition, given as:

$$u_{-h} = v_{-h} = w_{-h} = 0 \tag{11.14}$$

where the subscript $-h$ denotes conditions on the bed, and the kinematic free surface condition, which assumes that a fluid particle remains on the surface (Vallentine, 1969), is given by:

$$\left. \frac{d\zeta}{dt} \right|_{\zeta} = \left[\frac{\partial \zeta}{\partial t} + u \frac{\partial \zeta}{\partial x} + v \frac{\partial \zeta}{\partial y} \right]_{\zeta} = w_{\zeta} \tag{11.15}$$

where the subscript ζ denotes conditions on the free surface. Carrying out this integration (Falconer, 1993) leads to the corresponding momentum and continuity equations:

$$\frac{\partial UH}{\partial t} + \beta \left[\frac{\partial U^2 H}{\partial x} + \frac{\partial UVH}{\partial y} \right] = fVH + gH \frac{\partial \zeta}{\partial x} + \frac{\tau_{xw}}{\rho} - \frac{\tau_{xb}}{\rho}$$

$$+ 2 \frac{\partial}{\partial x} \left[\bar{\varepsilon} H \frac{\partial U}{\partial x} \right] + \frac{\partial}{\partial y} \left[\bar{\varepsilon} H \left(\frac{\partial U}{\partial y} + \frac{\partial V}{\partial x} \right) \right] \tag{11.16}$$

$$\frac{\partial VH}{\partial t} + \beta \left[\frac{\partial UVH}{\partial x} + \frac{\partial V^2 H}{\partial y} \right] = -fUH + gH \frac{\partial \zeta}{\partial y} + \frac{\tau_{yw}}{\rho} - \frac{\tau_{yb}}{\rho}$$

$$+ \frac{\partial}{\partial x} \left[\bar{\varepsilon} H \left(\frac{\partial U}{\partial y} + \frac{\partial V}{\partial x} \right) \right] + 2 \frac{\partial}{\partial y} \left[\bar{\varepsilon} H \frac{\partial V}{\partial y} \right] \tag{11.17}$$

$$\frac{\partial \zeta}{\partial t} + \frac{\partial UH}{\partial x} + \frac{\partial VH}{\partial y} = 0 \tag{11.18}$$

where U, V = depth-averaged velocities in x, y directions, H = total depth of flow ($= h + \zeta$), β = momentum correction factor for non-uniform vertical velocity profile,

τ_{xw}, τ_{yw} = surface wind shear stress components in x, y directions, τ_{xb}, τ_{yb} = bed shear stress components in x, y directions and $\bar{\varepsilon}$ = depth-averaged eddy viscosity.

In practical model studies of flow on tidal floodplains, and in the absence of extensive field data on the vertical velocity profiles, the momentum correction factor is either set to unity or a vertical velocity profile is assumed. For an assumed logarithmic vertical velocity profile, as derived from Prandtl's mixing length hypothesis (Goldstein, 1938), it can readily be shown that the value for β is given by:

$$\beta = \left[1 + \frac{g}{C^2 \kappa^2} \right] \qquad (11.19)$$

where C = dé Chezy bed roughness coefficient. Alternatively, an assumed seventh-power law velocity profile leads to a constant value for β of 1.016.

For the surface wind shear stress components a quadratic friction law is assumed, based on a balance of the horizontal forces for a steady uniform flow, giving:

$$\left. \begin{array}{l} \tau_{xw} = C_s \rho_a W_x W_s \\ \tau_{yw} = C_s \rho_a W_y W_s \end{array} \right\} \qquad (11.20)$$

where C_s = air–water resistance coefficient, ρ_a = air density, W_x, W_y = wind velocity components in x, y directions and W_s = wind speed ($= \sqrt{W_x^2 + W_y^2}$). Various constant values and formulae have been proposed for the resistance coefficient C_s with one of the most widely used representations being the piecewise formulation proposed by Wu (1969), giving:

$$\left. \begin{array}{ll} C_s = 1.25 \times 10^{-3} W_s^{-0.2} & \text{for} \quad W_s \leqslant 1 \text{ m s}^{-1} \\ C_s = 0.5 \times 10^{-3} W_s^{-0.5} & \text{for} \quad 1 < W_s \leqslant 15 \text{ m s}^{-1} \\ C_s = 2.6 \times 10^{-3} & \text{for} \quad W_s > 15 \text{ m s}^{-1} \end{array} \right\} \qquad (11.21)$$

For most tidal floodplain studies the bed stress is also represented in the form of a quadratic friction law, as given by:

$$\left. \begin{array}{l} \tau_{xb} = \rho g U V_s C^{-2} \\ \tau_{yb} = \rho g V V_s C^{-2} \end{array} \right\} \qquad (11.22)$$

where V_s = depth-averaged fluid speed ($= \sqrt{U^2 + V^2}$).

To determine the Chezy value either a constant value can be specified directly for C, giving typically $30 \text{ m}^{1/2} \text{ s}^{-1} < C < 100 \text{ m}^{1/2} \text{ s}^{-1}$, or C can be evaluated from the Manning equation, given as:

$$C = \frac{1}{n} H^{1/6} \qquad (11.23)$$

where n = Manning roughness coefficient, with typical values for n being in the range 0.015 to 0.04 over most tidal floodplains. Alternatively, the Colebrook–White equation can be used (Henderson, 1966) giving:

$$C = \sqrt{\frac{8g}{f}} = -\sqrt{32g}\, \log_{10}\left[\frac{\kappa_s}{14.84H} + \frac{1.255C}{RE\sqrt{2g}}\right] \qquad (11.24)$$

where f = Darcy–Weisbach resistance coefficient, k_s = Nikuradse equivalent sand grain roughness and Re = Reynolds number for open-channel flow, given as $4\,V_s H/v$. In the numerical model DIVAST, outlined later in this chapter, the Colebrook–White equation has been used to calculate C, since the value of the roughness coefficient can be more closely related to bed features such as ripples and dunes. Furthermore, unlike the Chezy and Manning representations, this formulation includes the region of transitional turbulent flow (or non-fully developed turbulent rough flow). Recent numerical simulation refinements reported by Falconer and Owens (1987) and Falconer and Chen (1991) have shown that this improved and more comprehensive representation of the bed friction coefficient can be particularly important when modelling flooding and drying over tidal floodplains, where Reynolds number effects may be significant due to shallow depths and low velocities.

Finally, for the turbulent diffusion terms in equations (11.16) and (11.17), these terms can first be simplified by neglecting the gradients of $\bar{\varepsilon}H$ and the divergence of the mean flow (Kuipers and Vreugdenhil, 1973), giving for the x-direction:

$$2\frac{\partial}{\partial x}\left[\bar{\varepsilon}H\,\frac{\partial U}{\partial x}\right] + \frac{\partial}{\partial y}\left[\bar{\varepsilon}H\left(\frac{\partial U}{\partial y} + \frac{\partial V}{\partial x}\right)\right] = 2\bar{\varepsilon}H\,\frac{\partial^2 U}{\partial x^2} + \bar{\varepsilon}H\,\frac{\partial}{\partial y}\left[\frac{\partial U}{\partial y} + \frac{\partial V}{\partial x}\right]$$

$$+\, 2\frac{\partial U}{\partial x}\frac{\partial \bar{\varepsilon}H}{\partial x} + \left[\frac{\partial U}{\partial y} + \frac{\partial V}{\partial x}\right]\frac{\partial \bar{\varepsilon}H}{\partial y} \cong \bar{\varepsilon}H\left[\frac{\partial^2 U}{\partial x^2} + \frac{\partial^2 U}{\partial y^2}\right] \qquad (11.25)$$

and similarly for the y-direction:

$$2\frac{\partial}{\partial y}\left[\bar{\varepsilon}H\,\frac{\partial V}{\partial y}\right] + \frac{\partial}{\partial x}\left[\bar{\varepsilon}H\left(\frac{\partial U}{\partial y} + \frac{\partial V}{\partial x}\right)\right] \cong \bar{\varepsilon}H\left[\frac{\partial^2 V}{\partial x^2} + \frac{\partial^2 V}{\partial y^2}\right] \qquad (11.26)$$

In determining the value of $\bar{\varepsilon}$, this can either be estimated from field data or, assuming that bed-generated turbulence dominates over free shear layer turbulence, a logarithmic velocity profile can be assumed giving (Elder, 1959):

$$\bar{\varepsilon} = 0.167\, \kappa U_* H \qquad (11.27)$$

where U_* = shear velocity $(=\sqrt{g}\,V_s/C)$. However, field data by Fischer (1973) for turbulent diffusion in rivers has shown that the value for $\bar{\varepsilon}$ is generally greater than that given by equation (11.27) and is more accurately represented by:

$$\bar{\varepsilon} = 0.15 U_* H \qquad (11.28)$$

For most practical tidal floodplain simulation studies, even this value is low compared to measured data recorded in well-mixed estuaries (Fischer et al., 1979), with values for $\bar{\varepsilon}/(U_* H)$ ranging from 0.42 to 1.61.

11.3   GOVERNING SOLUTE TRANSPORT EQUATION

11.3.1   Three-dimensional equation

In modelling numerically the fluxes of sediments or water quality constituents within tidal floodplains, the conservation of mass equation can be written in general terms for any substance introduced into a fluid medium. For a three-dimensional turbulent flow field, the corresponding general solute transport or advective-diffusion equation for a solute φ can be written as (Harleman, 1966):

$$\underbrace{\frac{\partial \varphi}{\partial t}}_{1} + \underbrace{\frac{\partial u\varphi}{\partial x} + \frac{\partial v\varphi}{\partial y} + \frac{\partial w\varphi}{\partial z}}_{2} + \underbrace{\frac{\partial}{\partial x}(\overline{u'\varphi'}) + \frac{\partial}{\partial y}(\overline{v'\varphi'}) + \frac{\partial}{\partial z}(\overline{w'\varphi'})}_{3} = \underbrace{\varphi_o + \varphi_d + \varphi_k}_{4}$$

$$(11.29)$$

where φ = time-averaged solute concentration, u, v, w = time averaged velocity components in the x, y, z directions, φ = turbulent fluctuating solute concentration, u', v', w' = turbulent fluctuating velocity components in the x, y, z directions, φ_o = source or sink solute input (e.g. an outfall), φ_d = first-order decay or growth rate of solute and φ_k = total kinetic transformation rate for solute. The individual terms in the advective-diffusion equation (11.29) refer to:- local effects (term 1), transport by advection (term 2), turbulence effects (term 3), and source (or sink), decay (or growth) and kinetic transformation effects (term 4).

The cross-product terms $\overline{u'\varphi'}$, etc., represent the mass flux of solute due to the turbulent fluctuations and, by analogy with Fick's law of diffusion, it can be assumed that this flux is proportional to the gradient of the mean concentration and is in the direction of decreasing concentration (Harleman, 1966). Hence, the terms can be written as:

$$\left. \begin{aligned} \overline{u'\varphi'} &= -D_{tx}\frac{\partial \varphi}{\partial x} \\[2mm] \overline{v'\varphi'} &= -D_{ty}\frac{\partial \varphi}{\partial y} \\[2mm] \overline{w'\varphi'} &= -D_{tz}\frac{\partial \varphi}{\partial z} \end{aligned} \right\}$$

$$(11.30)$$

where D_{tx}, D_{ty}, D_{tz} = turbulent diffusion coefficients in x, y, z directions. For open-channel flow it is common to assume isotropic turbulence and to set the horizontal turbulent diffusion terms to the depth mean diffusion coefficient as given by Fischer (1973) wherein:

$$D_{tx} = D_{ty} = 0.15 U_* H$$

$$(11.31)$$

Likewise, for the vertical diffusion coefficient, in the absence of stratification it is common to assume a linear shear stress distribution and a logarithmic velocity distribution giving (Vieira, 1993):

$$D_{tz} = U_* \kappa z \left(1 - \frac{z}{H}\right) \tag{11.32}$$

where z = elevation above the bed. However, as for the velocity predictions on tidal floodplains, the flow is generally assumed to be well-mixed vertically due to the shallowness of the water column and hence more often than not the sediment transport and water quality processes are modelled using a two-dimensional depth-integrated approach.

11.3.2 Depth-integrated equation

In solving for the depth integrated solute distribution, equation (11.30) is first substituted into equation (11.29), with the resulting equation then being integrated over the depth. Use of the bed and kinematic free surface conditions, in that there can be no solute flux across the bed or free surface, and depth integration leads to the following general governing solute transport equation (Falconer, 1991):

$$\frac{\partial \phi H}{\partial t} + \frac{\partial \phi UH}{\partial x} + \frac{\partial \phi VH}{\partial y} - \frac{\partial}{\partial x}\left[HD_{xx}\frac{\partial \phi}{\partial x} + HD_{xy}\frac{\partial \phi}{\partial y}\right]$$

$$- \frac{\partial}{\partial y}\left[HD_{yx}\frac{\partial \phi}{\partial x} + HD_{yy}\frac{\partial \phi}{\partial y}\right] - H[\phi_o + \phi_d + \phi_k] = 0 \quad (11.33)$$

where ϕ = depth-averaged solute concentration, ϕ_o, ϕ_d, ϕ_k = corresponding depth-averaged concentrations for φ_o, φ_d, φ_k respectively, and D_{xx}, D_{xy}, D_{yx}, D_{yy} = depth-averaged longitudinal dispersion and turbulent diffusion coefficients in x, y directions respectively.

For the dispersion-diffusion terms, these coefficients can be shown to be of the following form in two-dimensions (Preston, 1985):

$$\left. \begin{aligned} D_{xx} &= \frac{(D_l U^2 + D_t V^2)H\sqrt{g}}{V_s C} + D_w \\ D_{yy} &= \frac{(D_l V^2 + D_t U^2)H\sqrt{g}}{V_s C} + D_w \\ D_{xy} &= D_{yx} = \frac{(D_l - D_t)UVH\sqrt{g}}{V_s C} + D_w \end{aligned} \right\} \tag{11.34}$$

where D_l = longitudinal depth-averaged dispersion constant, D_t = depth averaged turbulent diffusion constant and D_w = wind-induced dispersion-diffusion coefficient. For values of D_l and D_t these can be set to minimum values assuming a logarithmic velocity distribution, wherein $D_l = 5.93$ (Elder, 1959) and $D_t = 0.15$ (Fischer, 1976). However, in practical studies these values tend to be rather low (Fischer et al. 1979), with measured values of D_l and D_t ranging from 8.6 to 7500 and 0.42 to 1.61 respectively. In the absence of extensive dye dispersion field data, typical initial values used in the computer model simulations outlined herein are $D_l = 13.0$ and $D_t = 1.2$, with these values being adjusted by calibration when limited spatial solute concentration distribution data are available.

11.4 SEDIMENT TRANSPORT PROCESSES AND EQUATIONS

11.4.1 Introduction

In modelling sediment transport processes over tidal floodplains, the sediment is generally classified as being either cohesive (mud) or non-cohesive (sand and silt) in nature, and with these two types of sediment being described by different formulations. In general, sediment is described as being cohesive if the particle (or grain) diameter is less than about 60 μm (0.06 mm), with the particles having cohesive properties due to electrostatic forces comparable with gravity forces active between the particles (van Rijn, 1993). Furthermore, the total load for non-cohesive sediment transport is subdivided into two different modes of transport: bed load and suspended load. The bed load is defined as that part of the total load where the sediment is almost continuously in contact with the bed, being carried by rolling, sliding or hopping, whereas the suspended load is that part of the total load which is maintained in suspension for considerable periods of time by the turbulence of the flow (van Rijn, 1993). These two components of total load transport are also represented by two different formulations, since the transport mechanisms are different.

In formulating the fluxes of suspended and bed load sediment transport, numerous theories and empirical equations have been postulated based on both field and laboratory measured data. However, in recent model developments the formulations outlined by van Rijn (1984a, b) have been widely accepted for both suspended and bed load transport and are the formulations outlined herein and included in the model studies reported later.

11.4.2 Suspended load transport

Most sediment transport formulations for prescribing suspended sediment flux predictions in a two-dimensional numerical model are based on the numerical solution of the depth-integrated advective-diffusion equation (11.33), rewritten in the following form:

$$\frac{\partial SH}{\partial t} + \left[\frac{\partial SUH}{\partial x} + \frac{\partial SVH}{\partial y}\right] + \frac{\partial}{\partial x}\left[HD_{xx}\frac{\partial S}{\partial x} + HD_{xy}\frac{\partial S}{\partial y}\right] - \frac{\partial}{\partial y}\left[HD_{yx}\frac{\partial S}{\partial x} + HD_{yy}\frac{\partial S}{\partial y}\right] = E$$

(11.35)

where S = depth-averaged suspended sediment concentration and E = net erosion or deposition per unit area of the bed. The depth-averaged net erosion or deposition rate can be expressed in the form (Owens, 1987):

$$E = \gamma W_f (S_e - S) \tag{11.36}$$

where S_e = depth-averaged equilibrium concentration, determined from an appropriate sediment transport formula (e.g. the van Rijn formula), W_f = particle settling velocity and γ = a profile factor given by the ratio of the bed concentration S_a (i.e. the concentration at an elevation "a" above the bed), to the depth-averaged concentration S. Equilibrium conditions are said to occur in the flow when the sediment flux vertically upwards from the bed due to turbulence is in equilibrium with the net sediment flux downwards due to the fall velocity (or gravity).

For the depth-averaged equilibrium concentration many formulations have been proposed, with the van Rijn (1984b) formulation now being one of the most widely used in computational models. For this more complex and rigorous formulation, the depth-integrated sediment flux per unit width is first calculated according to the following steps:

1. Compute the characteristic particle diameter D_*:

$$D_* = D_{50} \left[\frac{(S_s - 1)}{v^2} \right]^{1/3} \tag{11.37}$$

   where D_{50} = sediment diameter for which 50% of the bed material is finer, S_s = specific density ($= \rho_s/\rho$, where ρ_s = sediment density and ρ = fluid density), v = kinematic fluid viscosity.

2. Compute the critical bed shear velocity U_{*cr} from Shield's diagram (Raudkivi, 1990):

$$D_* \leqslant 4, \quad \theta_{cr} = 0.24(D_*)^{-1}$$

$$4 < D_* \leqslant 10, \quad \theta_{cr} = 0.14(D_*)^{-0.64}$$

$$10 < D_* \leqslant 20, \quad \theta_{cr} = 0.04(D_*)^{-0.10}$$

$$20 < D_* \leqslant 150, \quad \theta_{cr} = 0.013(D_*)^{0.29}$$

$$D_* > 150, \quad \theta_{cr} = 0.055$$

   where:

$$\theta_{cr} = \frac{(U_{*cr})^2}{(S_s - 1)gD_{50}}$$

   or:

$$U_{*cr} = \sqrt{\theta_{cr}(S_s - 1)gD_{50}} \tag{11.38}$$

3. Compute the Chezy coefficient related to the grain size C':

$$C' = 18 \log_{10} \left[\frac{4R_b}{D_{90}} \right] \tag{11.39}$$

   where R_b = hydraulic radius related to the bed and D_{90} = sediment diameter for which 90% of the bed material is finer.

4. Compute the effective bed-shear velocity U'_*:

$$U'_* = \frac{\sqrt{g}V_s}{C'}$$ (11.40)

5. Compute the transport stage parameter T:

$$T = \frac{(U'_*)^2 - (U_{*cr})^2}{(U_{*cr})^2}$$ (11.41)

6. Compute the reference level "a" for the elevation of the boundary transition from bed load to suspended load transport:

$$a = 0.5\Delta$$ (11.42)

where

$$\Delta = 0.11H\left[\frac{D_{50}}{H}\right]^{0.3}(1 - e^{-0.5T})(25 - T)$$

7. Compute the reference concentration S_a at elevation a:

$$S_a = \frac{0.015D_{50}T^{1.5}}{aD_*^{0.3}}$$ (11.43)

8. Compute the characteristic particle size of the suspended sediment D_s:

$$D_s = D_{50}[1 + 0.011(\sigma_s - 1)(T - 25)]$$ (11.44)

where σ_s = geometric standard deviation of bed material, defined as:

$$\sigma_s = \frac{1}{2}[D_{84}/D_{50} + D_{50}/D_{16}]$$

9. Compute the fall velocity W_f:

$$W_f = \frac{(S_s - 1)gD_s^2}{18\upsilon}$$ (11.45)

10. Compute the dimensionless parameter ψ:

$$\psi = 2.5\left[\frac{W_f}{U_*}\right]^{0.8}\left[\frac{S_a}{S_o}\right]^{0.4}$$ (11.46)

where S_o = maximum bed concentration ($=0.65$)

11. Compute the suspension parameters Z and Z':

$$Z = \frac{W_f}{\beta\kappa u_*} \quad \text{and} \quad Z' = Z + \psi$$ (11.47)

12. Compute the factor F, defined as:

$$F = \frac{\left[\dfrac{a}{H}\right]^{Z'} - \left[\dfrac{a}{H}\right]^{1.2}}{\left[1 - \dfrac{a}{H}\right]^{Z'} (1.2 - Z')}$$

(11.48)

13. Finally, compute the depth integrated suspended load per unit width q_s:

$$q_s = FV_s HS_a$$

(11.49)

Having calculated the depth-integrated suspended load flux per unit width, the equilibrium depth-averaged sediment concentration can now be calculated from the following relationship:

$$S_e = \frac{q_s}{q}$$

(11.50)

where q = depth-integrated fluid speed ($=V_s H$).

Thus, using equations (11.37) to (11.50) to calculate S_e and knowing the fall velocity from equation (11.45), the depth-averaged suspended load flux can be evaluated from equation (11.35), following evaluation of the water elevation and velocity fields across the domain.

11.4.3 Bed load transport

For the bed load transport, van Rijn's (1984a) formulation can again be used, giving:

$$q_b = \frac{0.053 T^{2.1} (\Delta g)^{0.5} D_{50}^{1.5}}{D_*^{0.3}}$$

(11.51)

where D_* and T are as given in equations (11.37) and (11.41) respectively. Summing the depth-integrated suspended load flux (i.e. obtained from the product of S – calculated from equation (11.35) – and q from the depth-integrated velocity field) and the depth-integrated bed load flux [i.e. obtained from equation (11.51)], gives the total sediment flux across the computational domain for non-cohesive sediment transport.

11.4.4 Cohesive sediment transport

In modelling cohesive sediment transport the governing depth-integrated advective-diffusion equation (11.35) is used, but with the net erosion term E (on the right hand side of equation (11.35)) being rewritten in the following form:

$$E = \frac{dm}{dt}$$

(11.52)

where m = mass of sediment per unit area. This replacement term given in the above equation is used to define the erosion or deposition according to the following criteria:

$$\text{Deposition:} \quad \frac{dm}{dt} = W_f S \left[\frac{\tau_b}{\tau_d} - 1 \right] \quad \text{when } \tau_b \leqslant \tau_d$$

$$\text{Erosion:} \quad \frac{dm}{dt} = M \left[\frac{\tau_b}{\tau_e} - 1 \right] \quad \text{when } \tau_b \geqslant \tau_e$$

$$\text{Neither:} \quad \frac{dm}{dt} = 0 \quad \text{when } \tau_d < \tau_b < \tau_e$$

where τ_b = bed shear stress, τ_d = critical shear stress for deposition, τ_e = critical shear stress for erosion, and M = empirical erosion constant.

Most of the parameters included in the above formulations for equation (11.52) very much depend upon the sediment characteristics locally and in model studies for floodplain flows the values used must be chosen with extreme care. Typical values of the critical shear stresses for erosion and deposition are given in van Rijn (1993) for a range of different mud types. For the empirical erosion coefficient (M), reported values are typically in the range of (0.00001–0.0005) for soft natural mud.

The settling velocity can be calculated using Stoke's law, according to the average size of flocs rather than the individual particle size. The density also needs to reflect the effect of flocculation. On the one hand, the settling velocity will be increased since the size of a floc is greater than that of a particle. On the other hand, the settling velocity will be reduced due to hindered settling when the concentration is greater than about $10 \, \text{mg} \, l^{-1}$. These effects are incorporated in the following relationship for the settling velocity:

$$W_{fm} = W_f (1 - \gamma S_v)^m \tag{11.53}$$

where W_{fm} = hindered settling velocity, γ = coefficient, m = coefficient (in the range 3 to 5) and S_v = volume concentration. The coefficient m may be set to zero in order to disable the effect of hindered settling.

These formulations for both non-cohesive and cohesive sediment transport have been included in the authors' model DIVAST (Depth Integrated Velocities And Solute Transport), and will be used in the examples of sediment transport predictions based on the use of this model and cited later in the chapter.

11.5   WATER QUALITY PROCESSES AND EQUATIONS

11.5.1   Introduction

In modelling water quality processes over tidal floodplains, a large range of constituents are often modelled, including:

1. Physical indicators of water quality:
   (i)   Suspended solids – including inorganics, e.g. sand, and organics, e.g. sewage solids.

(ii)   Turbidity – representing water cloudiness due to colloidal matter in suspension.
(iii) Temperature – important in governing the flow conditions and the chemical and biological processes for many water quality constituents.
(iv)   Other physical indicators – such as colour, conductivity and radioactivity.
2. Chemical indicators of water quality:
   (i)   Dissolved oxygen – which is important for water quality since it is essential for supporting most forms of aquatic life.
   (ii)   Biochemical oxygen demand – which is a measure of the consumption of oxygen by bacteria during the aerobic degradation of organic matter.
   (iii) Nitrogen – which is an essential nutrient for biological growth and a major constituent of domestic effluents. Nitrogen may be present in a variety of chemical forms, including: organic nitrogen, ammoniacal nitrogen, nitrite and nitrate nitrogen.
   (iv)   Phosphorus – which is also an essential nutrient for biological growth and which appears exclusively as phosphate in aquatic environments. Phosphate may be present in several forms, including: orthophosphate, condensed phosphates (i.e. pyro-, meta- and poly-phosphates) and organically based phosphates.
   (v)   Chlorides – commonly occurring in the form of salinity in estuaries and an indication of sewage pollution in rivers.
   (vi)   Metals – particularly those which are easily dissolvable in water and are toxic, e.g. arsenic, cadmium, chromium, lead and mercury. These metals occur mainly as a result of industrial discharges, mine water and agriculture and, although only usually present in low quantities, can accumulate in the food chain and cause toxicity problems.
3. Biological indicators of water quality:
   (i)   Pathogens – including the faecal coliform group, with the principal bacteria being *Escherichia coli*.
   (ii)   Other biological organisms – such as algae.

Some of the main water quality constituents frequently modelled in tidal floodplain flows, including mangrove swamps and wetlands, will be discussed below. The formulations outlined herein and included in the DIVAST model are based on the US EPA formulations included in the QUAL2E model (Brown and Barnwell, 1987). For convenience of writing the advective-diffusion equation for the various water quality constituents, the dispersion-diffusion terms will be neglected and equation (11.33) will be simplified to give:

$$\frac{D\phi H}{Dt} = H[\phi_o + \phi_d + \phi_k] \qquad (11.54)$$

where D/Dt = total derivative.

11.5.2 Salinity (S)

Salinity can be regarded as a conservative tracer and is included in models in the form of equation (11.54), but with all terms on the right hand side of the equation being set to zero. Salinity is an important parameter to model in studying tidal floodplains since it affects the flora and fauna, and aquaculture, and can influence the behaviour of other water quality constituents, such as saturated dissolved oxygen level and heavy metals.

11.5.3 Total and faecal coliforms (E)

Total and faecal coliforms are included in models as indicators of pathogen contamination in floodplain waters. Both coliform types are generally expressed as first-order decay functions, taking account of die-off rates, and are modelled in equation (11.54) in the following form:

$$\frac{DEH}{Dt} = -K_5 EH \qquad (11.55)$$

where E = concentration of coliforms (counts per 100 ml), and K_5 = coliform die-off rate (in the range 0.05 to 4.0 day^{-1}).

11..5.4 Biochemical oxygen demand (L)

The concentration of BOD is generally expressed in the form of a first order reaction to describe the deoxygenation of ultimate carbonaceous BOD. The BOD function as expressed in most models takes into account additional BOD removal due to sedimentation, flocculation and scour, thereby giving:

$$\frac{DLH}{Dt} = -K_1 LH - K_3 LH \qquad (11.56)$$

where L = concentration of ultimate carbonaceous BOD (mg l^{-1}), k_1 = deoxygenation rate coefficient (in the range 0.02 to 3.4 day^{-1}), and K_3 = rate of loss of carbonaceous BOD due to settling, etc. (in the range -0.36 to 0.36 day^{-1}).

11.5.5 Phytoplanktonic algae—chlorophyll a(A)

Algae are the dominant component of the primary producers in many systems, particularly in lakes and estuaries. Algal dynamics is also closely linked with nutrient dynamics and dissolved oxygen levels in water quality models. Poor turbidity, taste and odour problems are often caused by algal blooms. The commonly used approach, which has been adopted in the present DIVAST model, is to aggregate all algae into a single constituent, i.e. chlorophyll a, and this approach is described herein.

The general governing processes for chlorophyll a can be expressed as:

$$\frac{DAH}{Dt} = (\mu - R - G)AH \qquad (11.57)$$

where A = concentration of chlorophyll a (mg l^{-1}), μ = gross growth rate (day^{-1}), R = total phytoplankton loss rate (day^{-1}), $R = r + m + e + s$, in which r = respiration rate, m = non-predatory mortality or decomposition rate, e = excretion rate, s = loss rate due to grazing, and G = settling rate for algae (day^{-1}). The gross growth rate is further limited by the availability of light and nutrients and is expressed as:

$$\mu = \mu_{max} f_1(L) \left[\frac{f_2(N) + f_3(P)}{2} \right]$$ (11.58)

where μ_{max} = local maximum growth rate (day^{-1} temperature dependent), and $f_1(L)$, $f_2(N)$ and $f_3(P)$ = growth limiting functions for light, nitrogen and phosphorus respectively. The depth and time integrated half saturation algal–light relationship is used in the model and has the following form:

$$f_1(L) = \frac{T_p}{\lambda H} \ln \left(\frac{K_L + I_0}{K_L + I_0\, e^{-\lambda H}} \right)$$ (11.59)

where T_p = photoperiod (expressed as a fraction of the day), λ = light extinction coefficient (m^{-1}), K_L = light intensity at which growth rate is half of the maximum growth rate, and I_0 = average light intensity at the water surface during daylight hours. The nitrogen and phosphorus limiting functions are respectively expressed as follows:

$$f_2(N) = \frac{N_e}{K_N + N_e}$$ (11.60)

$$f_3(P) = \frac{P_2}{K_P + P_2}$$ (11.61)

where K_N = half-saturation constant for nitrogen (mg l^{-1}), N_e = effective local concentration of available dissolved inorganic nitrogen (ammonia nitrogen plus nitrate nitrogen), K_P = half-saturation constant for phosphorus (mg l^{-1}), and P_2 = local concentration of available dissolved inorganic phosphorous (mg l^{-1}).

11.5.6   Nitrogen cycle (N)

In natural floodplain waters there is a stepwise transformation from organic (N_4) to ammonia (N_1) to nitrite (N_2) and nitrate (N_3) nitrogen. This transformation is included in the DIVAST model as follows:

1.   Organic nitrogen (N_4)

$$\frac{DN_4 H}{Dt} = -\beta_3 N_4 H - \sigma_4 N_4 H + \alpha_N \rho_N RAH$$ (11.62)

where N_4 = concentration of organic nitrogen (mg l^{-1}), β_3 = rate constant for hydrolysis of organic nitrogen to ammonia nitrogen (in the range 0.02 to 0.4 day^{-1}), and σ_4 = rate coefficient for organic nitrogen settling (in the range 0.001 to 0.1 day^{-1}), α_N = ratio of organic nitrogen to chlorophyll a, and ρ_N = fraction of dead and respired phytoplankton recycled to organic nitrogen pool.

2. Ammonia nitrogen (N_1)

$$\frac{DN_1H}{Dt} = \beta_3 N_4 H - \beta_1 f_4 N_1 H + \alpha_N (1 - \rho_N) RAH - \beta_A \alpha_N \mu AH \qquad (11.63)$$

where N_1 = concentration of ammonia nitrogen (mg l^{-1}), β_1 = rate constant for the biological oxidation of ammonia nitrogen (in the range 0.1 to 1.0 day^{-1}), f_4 = oxygen limitation factor = $O_2/(K_0 + O_2)$, where K_0 = half-saturation constant (=0.5 to 2.0 mg O_2l^{-1}), and β_A = ammonia preference factor:

$$\beta_A = \frac{N_1}{K_N + N_3} \left[\frac{N_3}{K_N + N_1} + \frac{K_N}{N_1 + N_3} \right] \qquad (11.64)$$

where N_3 = concentration of nitrate nitrogen.

3. Nitrite nitrogen (N_2)

$$\frac{DN_2H}{Dt} = f_4 \beta_1 N_1 H - \beta_2 N_2 H \qquad (11.65)$$

where N_2 = concentration of nitrite nitrogen (mg l^{-1}), and β_2 = rate constant for the oxidation of nitrite nitrogen (in the range 0.2 to 2.0 day^{-1}).

4. Nitrate nitrogen (N_3)

$$\frac{DN_3H}{Dt} = \beta_2 N_2 H - \alpha_N \mu (1 - \beta_A) AH \qquad (11.66)$$

where N_3 = concentration of nitrate nitrogen (mg l^{-1}).

In the case where only the one-stage process is considered for nitrification (i.e. oxidation of ammonia to nitrate directly, rather than to nitrite first and then to nitrate), this process can be simulated by modelling the conservation of nitrate nitrogen directly, which can be written as:

$$\frac{DN_3H}{Dt} = f_4 \beta_1 N_1 H - \alpha_N \mu (1 - \beta_A) AH \qquad (11.67)$$

11.5.7 Phosphorus cycle (P)

The phosphorus cycle is similar to the nitrogen cycle. Phosphorus in its organic forms is produced by algal death, which converts to the dissolved inorganic state for primary production. Phosphorus, when discharged from sewage treatment works, is generally in the dissolved inorganic form and can readily be taken up by algae. Phosphorus is included in the existing DIVAST model in the following form:

1. Organic phosphorus (P_1)

$$\frac{DP_1H}{Dt} = \beta_4 P_1 H - \sigma_5 P_1 H + \alpha_p P_p RAH \qquad (11.68)$$

where P_1 = concentration of organic phosphorus (mg l^{-1}), β_4 = organic phosphorus decay rate (in the range of 0.01 to 0.7 day $^{-1}$), σ_5 = organic phosphorus settling rate (day $^{-1}$), α_p = ratio of phosphorous to chlorophyll a, and P_p = fraction of dead and respired phytoplankton recycled to the organic phosphorus pool.

2. Dissolved phosphorus (P_2)

$$\frac{DP_2H}{Dt} = +\beta_4 P_1 H - \sigma_6 P_2 H - \alpha_p \mu A H \qquad (11.69)$$

where P_2 = concentration of inorganic or dissolved phosphorus (mg l^{-1}), and σ_6 = dissolved phosphorus settling rate (day $^{-1}$).

11.5.8 Dissolved oxygen (O)

The balance of oxygen in floodplain waters depends upon the capacity of the waterbody to re-aerate itself. Apart from being dependent upon the advection and diffusion processes, this capacity is a function of major sources and sinks of oxygen, including: atmospheric re-aeration (source), biochemical oxidation of carbonaceous BOD and nitrogenous organic matter, sediment oxygen demand, algae production and consumption by respiration. Thus the dissolved oxygen formulation in the model described herein is given as:

$$\frac{DOH}{Dt} = K_2(O^* - O)H - K_1 LH - K_4 - \alpha_5 f_4 \beta_1 N_1 H - \sigma_6 \beta_2 N_2 H + \alpha_3 \mu A H - \alpha_4 r A H \quad (11.70)$$

where O = concentration of dissolved oxygen (mg l^{-1}), K_2 = re-aeration rate in accordance with Fick's law for diffusion, O^* = saturation concentration of dissolved oxygen at local temperature, K_4 = sediment oxygen uptake (in the range 1.5 to 9.8 mg O_2 m $^{-2}$ day $^{-1}$), α_5 = rate of oxygen uptake per unit of ammonia nitrogen oxidation (in the range 3.0 to 4.0 mg O_2 mg $^{-1}$N), α_6 = rate of oxygen uptake per unit of nitrite nitrogen oxidation (in the range 1.0 to 1.14 mg O_2 mg $^{-1}$N), α_3 = oxygen production per unit of chlorophyll a, and α_4 = oxygen uptake per unit of chlorophyll a (mg O_2 mg $^{-1}$ Chl a). For the re-aeration rate K_2, the value is generally assumed to vary spatially in accordance with the widely used O'Connor and Dobbins (1958) relationship, given as:

$$K_2 = \frac{\sqrt{D_m V_s}}{H^{1.5}} \qquad (11.71)$$

where K_2 = re-aeration rate (day $^{-1}$), V_s = local (or grid square) depth averaged velocity (ft s $^{-1}$), H = local (or grid square) depth (ft), and D_m = molecular diffusion (ft^2 day $^{-1}$), which is given as:

$$D_m = 1.91 \times 10^{-3}(1.037)^{T-20} \qquad (11.72)$$

where T = local temperature ($°$C).

For the saturation level of dissolved oxygen within the water column, this decreases with temperature and salinity, with the usual representation in models being that given by Bowie *et al.* (1985):

$$O^* = 14.6244 - 0.367134T + 0.0043972T^+ - 0.0966S$$
$$+ 0.00205ST + 0.0002739S^2 \quad (11.73)$$

where T = temperature ($^\circ$C) and S = salinity (ppt).

11.5.9 Temperature dependence

All of the rate constants quoted herein are temperature dependent, with the range of values given being applicable at the standard temperature of 20 °C. These constant values are generally first included in numerical models at the 20 °C value and are then corrected for temperature according to a Streeter–Phelps type formulation, given as:

$$K_T = K_{20}\theta^{(T - 20^\circ)} \quad (11.74)$$

where K_T = value of the rate constant for the local fluid temperature (day^{-1}), K_{20} = rate constant at standard temperature (i.e. at 20 °C), and θ = empirical constant for each reaction constant, with typical values for θ being given by Brown and Barnwell (1987).

11.6   MODELLING FLOODING AND DRYING PROCESSES

In modelling numerically the flow and solute transport processes on tidal floodplains, it is first necessary to model the complex hydrodynamic processes of flooding and drying. Falconer and Chen (1991) have developed a finite difference numerical model DIVAST, which is now widely used by industry and academia for modelling two-dimensional flow, sediment transport and water quality processes in tidal waters with flooding and drying. The model solves the basic two-dimensional depth-integrated equations outlined previously in this chapter, with the finite difference representation of the governing differential equations being expressed in an alternating direction implicit form and being centred by iteration. The discrete values of the variables are represented by a space staggered grid scheme, as shown in Figure 11.3, in which water elevations ζ and solute concentrations ϕ are prescribed at different grid point locations to the depth-integrated velocities, or discharges per unit width q_x and q_y, and the depths below datum. This scheme has the advantage in that for each variable operated upon in time there exists a centrally located spatial derivative for each of the other variables.

In particular, specific refinements have been made to the model to improve the numerical representation of flooding and drying. In the first of these refinements, the depths below datum are defined in the model grid at the mid-point along the side of each grid square, i.e. at the velocity points, rather than at the four corners of each grid square as for the traditional grid (Falconer and Owens, 1987). Using this representation, twice as much bathymetric data can be included for a similar grid configuration, which means that the bed topography can be represented more accurately – particularly for non-linear bed variations.

In modelling specifically the flooding and drying of shallow floodplain reaches, considerable numerical difficulties also frequently arise as a result of the discretization

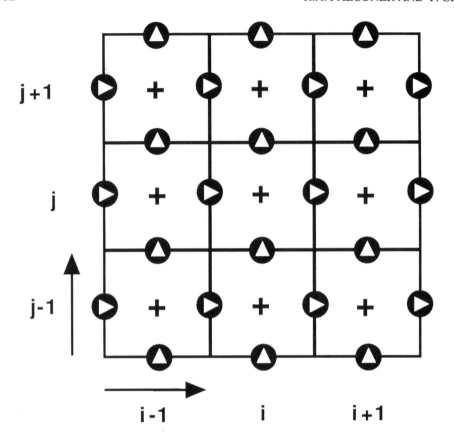

+ Water elevation above datum (η) and solute (ϕ)
▶ x - component discharge per unit width (q_x)
▲ y - component discharge per unit width (q_y)
● Depth below datum (h)

Figure 11.3   Finite difference grid structure

of these complex hydrodynamic processes which generally occur in nature in a smooth manner. Previous studies undertaken by Falconer (1984, 1986) on modelling the flooding and drying processes in Holes Bay and Poole Harbour, in Dorset, showed that modifications to a widely used scheme reported by Leendertse and Gritton (1971) for modelling flooding and drying gave numerically stable and, on the whole, encouraging predictions of the tidal velocities in both coastal basins. However, for the Poole Harbour study, the numerically predicted water elevations across the harbour gave greater phase lags than those measured in the field and the velocities immediately adjacent to a dry grid square were often higher than those measured. Although this scheme generally appeared to give satisfactory predictions for the tidal elevations in Poole Harbour, the same scheme exhibited pronounced oscillatory predictions of the water elevations and currents when applied to the Humber Estuary (Falconer and Owens, 1987). Hence extensive tests have been undertaken by the authors on various representations of the flooding and drying processes for a range of idealized floodplain bathymetries. These tests have highlighted a number of deficiencies in the previous schemes and a new scheme has been developed for modelling these complex hydrodynamic processes. Apart from proving to be more accurate and robust than the previous schemes reported in the literature, the new scheme has also compared favourably with widely used schemes, such as that outlined by Stelling *et al.* (1986), when predicted results are compared with field data for Poole Harbour.

The new scheme developed after extensive tests is described herein, with the drying tests outlined first:

1. At the end of every half time-step the total depth is calculated first at the centre of each side of every wet grid square. The depth at the centre of all sides for each wet grid square is then compared with the bed roughness k_s. If any side depth is less than k_s, then the corresponding depth and velocity component across that side are both equated to zero. Thus no flow or solute flux is permitted across the grid square side until flooded again.
2. If all four side depths for any grid square are less than k_s, then the grid square is effectively dry and the cell is removed from the computational domain. The water elevation and solute levels for the grid square are set to the corresponding values at the end of the previous half time step when the cell was last wet.
3. In addition to the above checks, the total depth is also calculated at the centre of every wet grid square, for each half-time step. If this depth is less than k_s, then the cell is again assumed to be dry and removed from the domain as before.
4. For the final drying check, all wet grid squares are considered to be potentially dry if the depth at the centre is greater than k_s but less than a predetermined value PRESET (typically 20 mm). If at least one water depth around the four sides of the grid square is greater than k_s, and the depth at the centre is less than PRESET, then this cell is assumed to be dry unless the flow direction from the adjacent wet grid square is towards the potentially dry grid square. If the flow direction is out of the potentially dry grid square then the cell is removed as before.

Following these drying checks, all previously dry grid squares are then considered for possible flooding. In these tests a currently dry grid square will be returned to the computational domain if all of the following conditions are satisfied:

1. Any of the four surrounding grid squares is wet.
2. The water elevation of the surrounding wet grid square(s) is higher than that of the dry grid square.
3. The total depth of the cross-section connecting the wet grid square(s) and the dry grid square is greater than k_s.
4. The depth at the centre of the dry grid square is greater than k_s.

In developing this new method for treating numerically the processes of flooding and drying, tests were concentrated on two idealized basins. These included: a one-dimensional estuary with a uniformly sloping beach, of slope 1 in 2760, length = 13 800 m and width = 4800 m, and an identical estuary but with the head extended horizontally for a further 6000 m, see Figure 11.4. For both tests a sinusoidal wave, of amplitude 2 m and period 32 000 s, was fed in at the entrance and tests were undertaken beginning at low tide, mean water level and high tide respectively. An

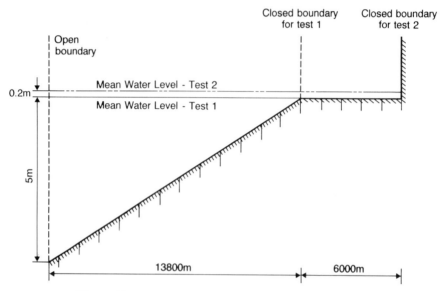

Figure 11.4   Schematic illustration of idealized test basins

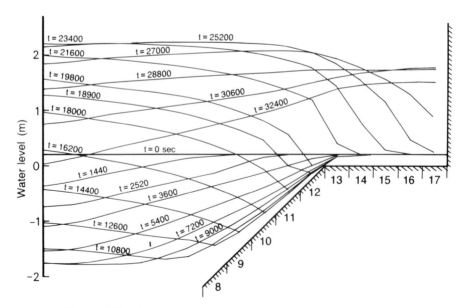

Figure 11.5   Flooding and drying sequence using the refined scheme

example simulation of the more severe second test is shown in Figure 11.5, with the new scheme showing a numerically stable and physically realistic drying and flooding of the basin. During ebb tide, the grid squares dry later than before as the horizontal plane dries first and a reduced maximum high water level occurs at the estuary on subsequent flooding. None of the other schemes considered was found to produce such satisfactory and stable predictions for this extremely severe test case (Falconer and Chen, 1991).

11.7 CASE STUDY APPLICATIONS

11.7.1 Introduction

The differential equations outlined in this chapter have been solved numerically using the model DIVAST for a number of idealized and site specific studies, with each having a particular research objective in mind. In two of these studies, namely Poole Harbour and the Humber Estuary, particular emphasis has been focused on modelling flow and solute transport processes on the tidal floodplains, primarily in connection with Environmental Impact Assessment (EIA) studies. Brief details of both of these studies are given below.

11.7.2 Poole Harbour, Dorset

Poole Harbour (Figure 11.6) is one of the largest natural harbours in Europe, with a perimeter of approximately 112 miles (180 km), and is connected to the English

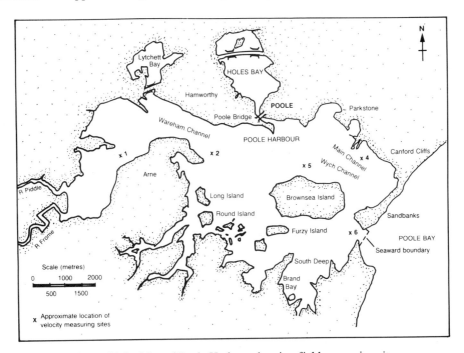

Figure 11.6 Map of Poole Harbour showing field measuring sites

Channel by a narrow entrance which is only approximately 300 m wide. The harbour is a Site of Special Scientific Interest (i.e. a SSSI). Extensive areas of salt marsh and phragmites reed exist within the basin, with these areas being of botanical and ornithological interest and supporting nationally rare populations of plants and birds. The estuary also provides valuable breeding and nursery areas for several species of fish, including bass, pollack and mullet. The harbour is generally shallow and the tidal range is relatively small, i.e. typically 2 m. The basin floods and dries extensively during the tidal cycle, with the plan surface area varying by over 50% during spring tides. At low tide a variety of wildlife abounds on the exposed shores, particularly in the undeveloped areas to the south and west of the harbour. In contrast, the town of Poole – with its quay and commercial port – forms part of an expanding and important trading centre for shipping links within Europe, and the ecological interest of the basin could be put at risk from severe pollution – particularly in view of the potential for poor flushing with the relatively narrow entrance. For further details on Poole Harbour see Falconer and Liu (1994).

With these considerations in mind, particular emphasis has been focused on modelling the numerical representation of the complex hydrodynamic processes of flooding and drying over the tidal floodplains and predicting the distribution of a range of water quality constituents, from agricultural runoff passing through the rivers Frome and Piddle and sewage effluents from the treatment works at Keysworth, Lytchett Minster and, in particular, Poole.

The region of Poole Basin was originally simulated in the DIVAST model using a mesh of 58×65 grid squares, with a constant grid spacing of 150 m. However, more recently the basin has been simulated using a 75 m grid and a mesh of 116×130 grid squares. At the centre of the side of each grid square a representative depth below datum was included from the Admiralty Chart No. 2611 and additional information provided by Poole Harbour Commissioners. In modelling the hydrodynamic features in Poole Harbour and Holes Bay, it was necessary to introduce time-varying water-level variations at the open seaward boundary as shown in Figure 11.6. For this purpose tidal data were obtained both for the entrance and Poole Bridge from Poole Harbour Commissioners. Other hydrodynamic inputs into the numerical model included the bed friction and the surface wind stress. For the bed resistance the Darcy friction formula was used, together with the Colebrook–White equation (11.24), with a bed roughness height (k_s) of 20 mm being assumed to represent the typical ripple features apparent on the bed.

Apart from the open and bed boundary conditions, initial conditions had to be included in the model for water elevations, velocities, water quality constituents and sediment fluxes for all potentially wet grid squares across the domain. The water elevations were initially set to high tide level everywhere and the initial velocities and sediment fluxes were set to zero. The initial water quality constituent levels were generally set to those base values expected seawards of the basin; e.g. the initial salinity level was set at 35 ppt everywhere. For the river discharges, a linear variation was assumed between the discharges recorded by the National Rivers Authority (Wessex Region) at the start and end of the simulation period and with a constant daily average throughflow being assumed for each of the sewage treatment works.

For the results of this study full details are given in Falconer and Chen (1991). In particular, the new flooding and drying scheme gave more accurate predictions of the water elevations at Poole Bridge, in comparison with measured field data, and the scheme showed no oscillatory predictions of the water elevation and velocity fields as for other schemes, such as Stelling *et al.*, (1986), and as shown in Figures 11.7 and 11.8 respectively.

Simulations were also undertaken to predict the nitrogen distributions across the tidal floodplains and, in particular, to establish the influence of Poole sewage treatment works on the nitrogen levels. The velocity and nitrogen concentration distributions for the existing inputs in Poole Harbour were compared with field measurements recorded at 12 sites within the basin. On the whole the agreement between both sets of results

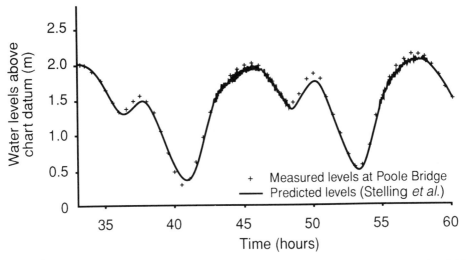

Figure 11.7   Predicted water elevations at Poole Bridge using method by Stelling *et al.* (1986)

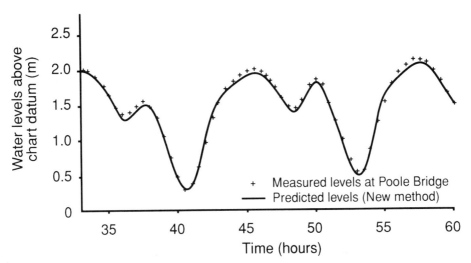

Figure 11.8   Predicted water elevations at Poole Bridge using method by Falconer and Chen (1991)

was very satisfactory, with the main practical results suggesting that the nitrogen levels in Poole Harbour would not be significantly reduced if nitrogen input from Poole sewage treatment works was to be removed. On the other hand, the nitrogen concentration in Holes Bay appeared to be reduced considerably if the nitrogen inputs from the sewage works were removed. Typical examples of the velocity field and

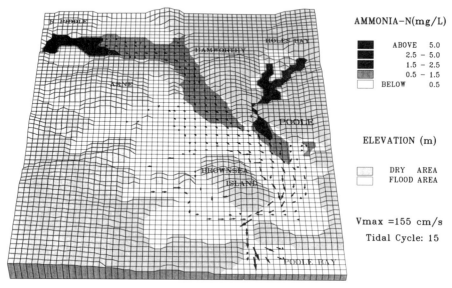

Figure 11.9 Predicted velocity and ammonia nitrogen distribution across Poole Harbour at low water level with input from Poole Sewage Works

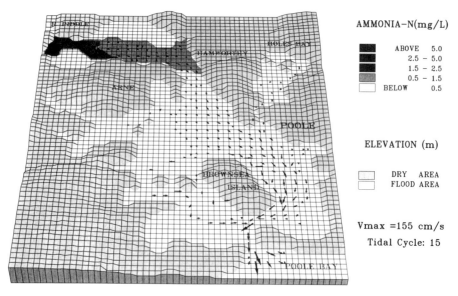

Figure 11.10 Predicted velocity and ammonia nitrogen distribution across Poole Harbour at low water level with no input from Poole Sewage Works

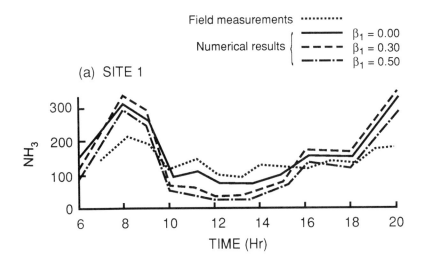

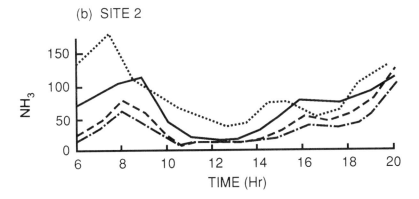

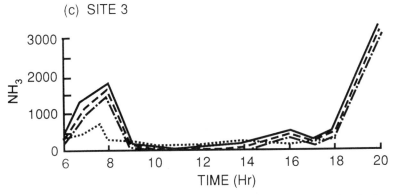

Figure 11.11 Comparison of predicted and measured ammonia nitrogen levels (in $\mu g \, l^{-1}$) at three sites across Poole Harbour for three decay rates

ammonia nitrogen concentration distributions are shown in Figures 11.9 and 11.10 for input and no input of nitrogen from Poole sewage works respectively.

As can be seen from the typical velocity field prediction for Poole Harbour, the effects of flooding and drying are smooth, with the velocity distributions across the floodplain being numerically stable and consistent for repetitive tides. An example of the level of agreement between the field and measured ammonia nitrogen levels across the floodplain is shown in Figure 11.11, with better agreement being obtained for most other sites in the main body of the flow field. Apart from ammonia nitrogen, a range of other water quality constituents – including sediment fluxes – were simulated for Poole Harbour, with the level of agreement between the predicted and measured results being generally better than that illustrated in Figure 11.11 (Falconer and Liu, 1994).

11.7.3   Humber Estuary, northeast England

The Humber Estuary (Figure 11.12) is a large well-mixed estuary, situated on the east coast of northern England, and providing an outlet to the North Sea for the rivers Trent and Ouse, and shipping access to a number of ports, including Hull, Immingham and Grimsby. The estuary is the largest in the UK and encompasses several Sites of Special Scientific Interest along the extensive floodplains on both banks of the estuary. There are 13 chemical and industrial works along the estuary, as well as domestic effluent discharges from several large towns adjacent to the floodplains (e.g. Hull and Cleethorpes). The management and monitoring of the water quality of the estuary is co-ordinated by the Humber Committee of the Anglian, Severn Trent and Yorkshire Regions of the Environment Agency, and with sediment flux data being acquired regularly by Associated British Ports (ABP).

The tidal currents outside the estuary are predominantly north–south, parallel to the Yorkshire and Lincolnshire coastlines. During flood tide the flow is southwards with a

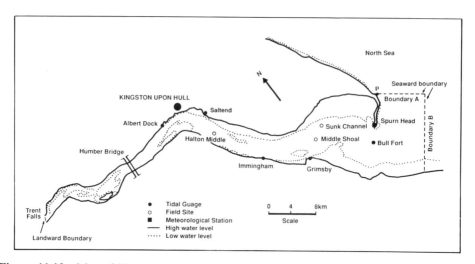

Figure 11.12   Map of Humber Estuary showing location of boundaries, grid outline and field measuring sites

band of water, approximately 8 km wide, separating the main alongshore current and that flowing into the estuary. This pattern is reversed during ebb tide, with the seaward boundary (shown in Figure 11.12) being parallel to, and southeast of, Spurn Head and acting effectively as a free streamline.

The region of the Humber Estuary was simulated in the numerical model DIVAST using a finite difference mesh of 118×56 grid squares and with an equal grid spacing of 500 m in both directions (Falconer and Owens, 1990). More recently the basin has also been simulated using a boundary fitting curvilinear grid (Lin and Falconer, 1995). The bathymetric data were included in the model based on data from the Admiralty Chart No. 109 and additional data provided by ABP.

In order to drive the hydrodynamic model, water elevation data recorded by ABP were used at both the seaward and landward boundaries, i.e. just beyond Spurn Head and at Trent Falls. Although it is generally preferable to avoid using water elevations at both boundaries for such a study, in this case no time-varying flow data were available at Trent Falls. Field measurements of water elevations, velocities and various water quality constituents were also available at several sites in between the two open boundaries. Data were available for different tidal ranges at each site, with a complete tidal cycle being monitored at quarter-hour intervals. For the various discharges along the estuary, mean daily rates were input at the outfall cells with linear interpolation being used to obtain intermediate values. For the water quality constituents from the rivers Wharfe, Aire, Don, Trent and Ouse, the boundary values used were obtained by summing the individual components as given by Mallowney (1982). Other inputs into the model included bed friction and surface wind stress. For the bed friction the Darcy–Weisbach resistance formula was again used, with a k_s value of 20 mm being assumed across the domain in the absence of field data and following calibration.

The model was run for several repetitive tides and for a range of different input scenarios. The resulting velocity field predictions were reproduced at low, mean water level flood, high and mean water level ebb tide relative to tidal conditions at the seaward boundary. Similarly, the water quality constituents and sediment fluxes were reproduced at high tide only.

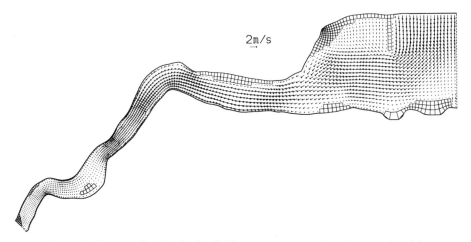

Figure 11.13   Predicted velocity field near mean water level for a spring tide

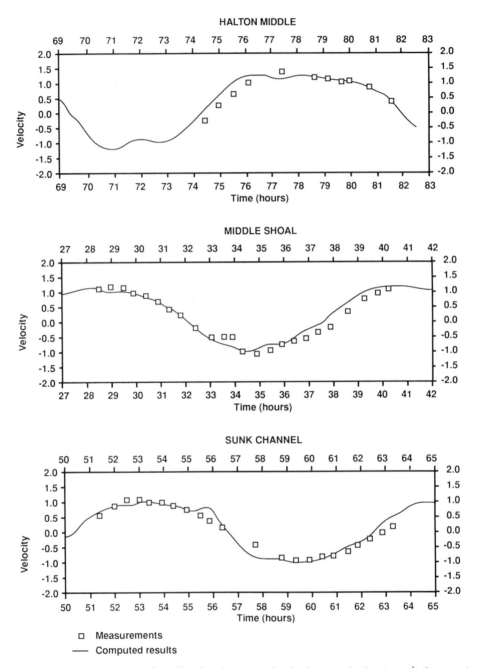

Figure 11.14 Comparison of predicted and measured velocity magnitudes (m s^{-1}) for a spring tide

The corresponding velocity field predictions confirmed that the velocity distribution was highly two-dimensional across the floodplains, particularly in the lee of Spurn Head and towards Trent Falls. At high tide a pronounced eddy was observed on the seaward side of the causeway. At mean water level flood tide a stagnation region was predicted and observed along the floodplain just northwest of Saltend, as the incoming tidal current opposed the flow from the river input at Trent Falls. For all stages of the tide strong currents, in excess of $1\ \mathrm{m\,s}^{-1}$, were observed in the region of Hull, confirming that the region is a site of high natural dispersion for effluent discharge. A typical predicted velocity distribution is shown in Figure 11.13 for a Cartesian grid, with comparisons between the predicted and measured velocities at Middle Shoal, Sunk Channel and Halton Middle (Figure 11.14) showing close agreement.

For the water quality constituents and sediment fluxes, simulations were made of salinity, BOD, organic nitrogen, oxidized nitrogen, dissolved oxygen, bed load and suspended load. The resulting numerically predicted distributions along the estuary and over the floodplains also agreed closely with the field measured results for all of the constituents considered, with typical comparisons for salinity, oxidized nitrogen and suspended sediments being shown in Figures 11.15, 11.16 and 11.17 respectively. In this study the model was then used to predict the environmental impact of siting a new sewage treatment works at Hull, with the results indicating that a new works would improve the coliform levels in the estuary, but only marginally improve other water quality constituent levels in the basin. The model has more recently been used to predict

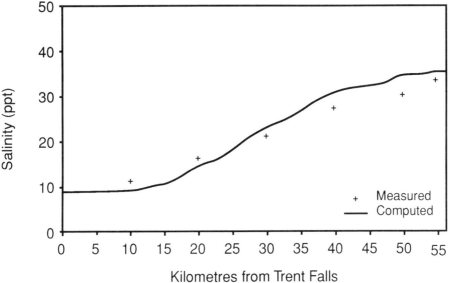

Figure 11.15 Comparison of predicted and measured salinity levels along the estuary

heavy metal concentrations along the estuary, and from a chemical effluent discharge at Saltend, with particular interest being focused on the metal concentrations along the wetlands, or floodplains, which constitute Sites of Special Scientific Interest. This study has indicated much lower levels of heavy metal concentrations than first expected, due to the high natural dispersion of the estuary. This project is still continuing, with recent research interest being focused on improving the representation of the bio-geochemical processes relating to the dynamic partitioning of the heavy metals into dilution and adsorption on to the sediments.

11.8 CONCLUSIONS

Details are given of the hydro-ecological importance of floodplains – including wetlands and mangrove swamps – and the increasing attempts being made by engineers and scientists to manage better these important waterways, with the aid of numerical hydrodynamic and solute transport predictive models. An outline derivation and explanation has been given of the three and two-dimensional forms of the governing hydrodynamic and solute transport equations, with the advective-diffusion equation being modified for cohesive and non-cohesive sediment fluxes and for a number of key water quality constituents, including physical, chemical and biological indicators.

Details have also been given of a refined numerical scheme for predicting flooding and drying of tidal floodplains, with the scheme having been tested against other documented

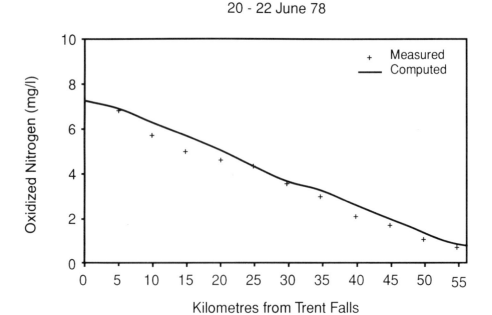

Figure 11.16   Comparison of predicted and measured oxidized nitrogen levels along the estuary

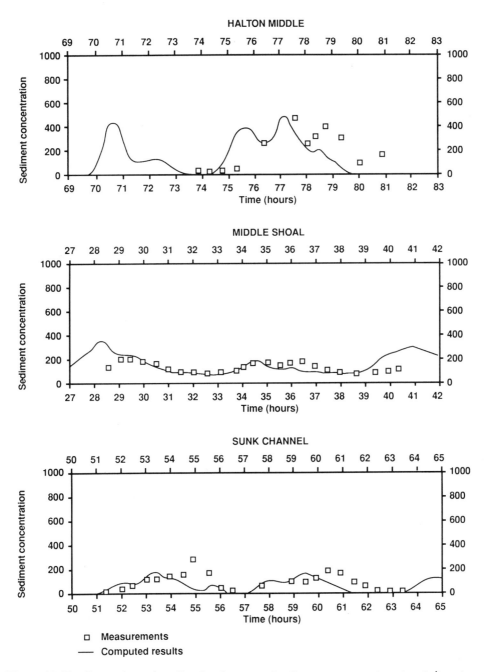

Figure 11.17 Comparison of predicted and measured sediment concentrations (mg l^{-1}) at three sites along the estuary

schemes for a severe idealized test case. The refined scheme gave numerically stable and realistic water elevation predictions in a complex tidal basin with extensive floodplains.

Finally, two model case studies have been outlined in the chapter, including hydrodynamic, sediment flux and water quality indicator predictions in Poole Harbour, Dorset, and the Humber Estuary, along the north east coast of England. Both basins have large areas of tidal floodplains, with the numerical model results agreeing favourably with independently measured field data in both studies. Much emphasis has been focused in these studies on the nutrient levels and sediment fluxes over the floodplains, particularly since these wetlands constitute key Sites of Special Scientific Interest. The model described herein has recently been extended to three dimensions and continuous improvements are being included in representing both the hydrodynamic and solute transport processes on floodplains.

ACKNOWLEDGEMENTS

The numerical model studies reported herein were funded by a number of organizations, including: Engineering and Physical Sciences Research Council, Natural Environment Research Council, European Union, IBM UK Ltd, Yorkshire Water plc, Wessex Water plc and NATO. The authors are also grateful to the following organizations for providing data: Environment Agency, ABP Research and Consultancy Ltd, HR Wallingford Ltd and Poole Harbour Commissioners.

The authors would also like to thank their colleague Dr Lin Binliang for his assistance with the Humber Estuary study.

REFERENCES

Bengtsson, L. (1978) Wind-induced circulation in lakes. *Nordic Hydrology*, **9**, 75–94.
Bowie, G.L. *et al.* (1985) *Rates, Constants and Kinetics Formulations in Water Quality Modelling. Environmental Research Laboratory*, US EPA, Athens, GA, Report No EPA/ 600/3-85/040, June, 475 pp.
Brown, L.C. and Barnwell, T.O. Jr (1987) *The Enhanced Stream Water Quality Models QUAL2E and QUAL2E-UNCAS: Documentation and User Manual.* Environmental Research Laboratory, US EPA, Athens, GA, Report No. EPA/600/3-87/007, May, 189 pp.
Dronkers, J.J. (1964) *Tidal Computations in Rivers and Coastal Waters.* North Holland Publishing Co., Amsterdam.
Elder, J.W. (1959) The dispersion of marked fluid in turbulent shear flow. *Journal of Fluid Mechanics*, **5**(4), 544–560.
Falconer, R.A. (1984) A mathematical model study of the flushing characteristics of a shallow tidal bay. *Proceedings of the Institution of Civil Engineers*, **77**(2), 311–332.
Falconer, R.A. (1986) A water quality simulation study of a natural harbour. *Journal of Waterway, Port, Coastal and Ocean Engineering*, ASCE **112**(1), 15–34.
Falconer, R.A. (1991) Review of modelling flow and pollutant transport processes in hydraulic basins. *Proceedings of First International Conference on Water Pollution: Modelling, Measuring and Prediction*, Southampton, UK, Computational Mechanics Publications, 3–23.
Falconer, R.A. (1993) An introduction to nearly horizontal flows. In: *Coastal, Estuarial and Harbour Engineers' Reference Book*, M.B. Abbott and W.A. Price (eds), E. & F.N. Spon Ltd, London, Chapter 2, 27–36.

Falconer, R.A. and Chen, Y. (1991) An improved representation of flooding and drying and wind stress effects in a two-dimensional tidal numerical model. *Proceedings of the Institution of Civil Engineers*, **91**(2), 659–678.

Falconer, R.A. and Li, G. (1994). Numerical modelling of tidal eddies in narrow entranced coastal basins using the $k-\varepsilon$ turbulence model. In: *Mixing and Transport in the Environment* K. Beven *et al.* (eds)), John Wiley & Sons Ltd, Chichester, Chapter 17, 325–350.

Falconer, R.A. and Liu, S.Q. (1994) Numerical modelling of hydrodynamic and water quality processes in an enclosed tidal wetland. In: *Wetland Management* R.A. Falconer and P. Goodwin (eds), Thomas Telford, London, Chapter 9, 119–129.

Falconer, R.A. and Owens, P.H. (1987) Numerical simulation of flooding and drying in a depth-averaged tidal flow model. *Proceedings of the Institution of Civil Engineers*, **38**(2), 161–180.

Falconer, R.A. and Owens, P.H. (1990) Numerical modelling of suspended sediment fluxes in estuarine waters. *Estuarine, Coastal and Shelf Science*, **31**, 745–762.

Fischer, H.B. (1973) Longitudinal dispersion and turbulent mixing in open channel flow. *Annual Review of Fluid Mechanics*, **5**, 59–78.

Fischer, H.B. (1976) Mixing and dispersion in estuaries. *Annual Review of Fluid Mechanics*, **8**, 107–133.

Fischer, H.B., List, E.J., Koh, R.C.Y., Imberger J. and Brooks, N.H. (1979) *Mixing in Inland and Coastal Waters*. Academic Press Inc., San Diego, 483 pp.

Goldstein, S. (1938) *Modern Developments in Fluid Dynamics*, Vol. 1. Oxford University Press, Oxford.

Harleman, D.R.F. (1966) Diffusion Processes in stratified flow. In: *Estuary and Coastline Hydrodynamics*, A.T. Ippen (ed.), McGraw Hill Book Co. Inc., New York, Chapter 12, 575–597.

Henderson, F.M. (1966) *Open Channel Flow*. Collier-Macmillan Publishers, London, 522 pp.

Kuipers, J. and Vreugdenhil, C.B. (1973) *Calculations of Two-Dimensional Horizontal Flows*. Delft Hydraulics Laboratory Technical Report, No. S163, Part 2, Delft Hydraulics, October, 44 pp.

Leendertse, J.J. and Gritton, E.C. (1971) *A Water Quality Simulation Model for Well-Mixed Estuaries and Coastal Seas: Vol. 2. Computation Procedures*. The Rand Corporation, Santa Monica, Report No. R-708-NYC, July, 53 pp.

Lin, B. and Falconer, R.A. (1995) Modelling sediment fluxes in estuarine waters using curvilinear co-ordinate grid system. *Estuarine, Coastal and Shelf Science*, **41**, 413–428.

Mallowney, B.M. (1982) *The Moving-Sediment Steady-State Model of Dissolved Oxygen in the Humber Estuary*. Report for Humber Estuary Working Party, WRc Environmental Protection, August, 30 pp.

O'Connor, D.J. and Dobbins, W.E. (1958) Mechanism of reaeration in natural streams. *Transactions of the ASCE*, **123**, 641–684.

Ong, J.E. *et al.* (1991) Characterisation of a Malaysian mangrove estuary. *Estuaries*, **14**(1), 38–48.

Owens, P.H. (1987) Mathematical modelling of sediment transport in estuaries. PhD thesis, University of Birmingham, UK, 220 pp.

Pardeshi, P.S. (1991) Mangroves in peril. *Sanctuary Asia*, **XI**(2), 40–46.

Preston, R.W. (1985) *The Representation of Dispersion in Two-Dimensional Shallow Water Flow*. Central Electricity Research Laboratories, Report No. TPRD/U278333/N84, May, 13 pp.

Raudkivi, A.J. (1990) *Loose Boundary Hydraulics*, 3rd edition. Pergamon Press plc, Oxford, 538 pp.

Rodi, W. (1984) *Turbulence Models and their Application in Hydraulics*, 2nd edition. International Association for Hydraulic Research, Delft, the Netherlands, 104 pp.

Stelling, G.S. *et al.* (1986) Practical aspects of accurate tidal computations. *Journal of Hydraulic Engineering*, ASCE, **112**(9), 802–817.

Vallentine, H.R. (1969) *Applied Hydrodynamics*, 2nd edition. Butterworth and Co. (Publishers) Ltd.

van Rijn, L.C. (1984a) Sediment transport part 1: bed load transport. *Journal of Hydraulic Engineering*, ASCE, **10**, 1431–1456.

van Rijn, L.C. (1984b) Sediment transport part 2: suspended load transport. *Journal of Hydraulic Engineering*, ASCE, **10**, 1613–1641.

van Rijn, L.C. (1993) *Principles of Sediment Transport in Rivers, Estuaries and Coastal Seas.* Aqua Publications, the Netherlands, 673 pp.

Vieira, J.R. (1993) Dispersive processes in two-dimensional models. In: *Coastal, Estuarial and Harbour Engineers' Reference Book*, M.B. Abbott and W.A. Price (eds), E. & F.N. Spon Ltd, London, Chapter 2, 179–190.

Wu, J. (1969) Wind stress and surface roughness at air–sea interface. *Journal of Geophysical Research*, **74**, 444–455.

12 Floodplains as Suspended Sediment Sinks

D.E. WALLING*, Q. HE* and A.P. NICHOLAS†
*Department of Geography, University of Exeter, UK
†School of Geography, University of Leeds, UK

12.1 THE CONTEXT

Geomorphological studies of floodplains have traditionally focused on the formation and evolution of floodplain landforms (cf. Lewin, 1978; Nanson and Croke, 1992). As such, most attention has been directed to river channel processes and more particularly the interactions between channel migration and floodplain construction/destruction (e.g. Wolman and Leopold, 1957; Brackenridge, 1988; Howard, 1992). Less attention has, in general, been given to the fine overbank deposits which mantle large areas of most floodplains, although such deposits have been widely recognized as an important element of floodplain landforms. This emphasis on channel processes and the mobilization, transport and deposition of coarse sediment in response to lateral migration is, in part, a reflection of the fact that for most "active" rivers, channel migration provides the most graphic evidence of floodplain evolution and change, and many studies have understandably focused on areas where such activity is well-developed and readily documented. The slow rates of accretion commonly associated with overbank deposition on the floodplain surface are less readily documented and provide less tangible evidence of landform evolution. Furthermore, Wolman and Leopold (1957) have suggested that overbank deposits account for only approximately 10–20% of the total alluvial deposits associated with floodplain systems. It should, however, be recognized that for many lowland river floodplains, particularly those where channelization and river training works limit or prevent channel migration and the reworking of floodplain deposits, the overbank deposition of fine sediment is likely to represent the dominant component of floodplain development. Thus Ritter *et al.* (1973) have estimated that overbank deposits dominate the alluvial deposits associated with the floodplain of the Delaware River, USA.

Recent interest in catchment-wide assessments of erosion and sediment delivery and the establishment of suspended sediment budgets has also highlighted the role of lowland floodplains, and more particularly their fine overbank deposits, as important sinks or stores for suspended sediment transported through the channel system. It is well

Floodplain Processes. Edited by Malcolm G. Anderson, Des E. Walling and Paul D. Bates.
© 1996 John Wiley & Sons Ltd.

known that the majority of the suspended sediment yield from a drainage basin will be transported during flood events, and inundation of the floodplain at such times can provide the opportunity for deposition of a significant proportion of the suspended sediment load and therefore for a significant conveyance loss. Thus, while they may, perhaps, be seen as being of limited relevance in studies of floodplain evolution. the low rates of accretion associated with overbank deposition of fines can, nevertheless, represent substantial quantities of sediment when viewed in terms of the large surface area involved. Lambert and Walling (1987), for example, measured the suspended sediment load entering and leaving a 11 km reach of the lower River Culm in Devon, UK, where the floodplain was regularly inundated during flood events, and estimated that approximately 28% of the suspended sediment load was deposited on the floodplain within this reach. Middelkoop and Asselmann (1994) were also able to combine an assessment of the mass of fine sediment deposited on a ca. 100 km reach of the floodplain of the River Waal in the Netherlands between the German border and Gorkum during a major (40-year) flood that occurred between December 1993 and January 1994, with information on the suspended sediment load transported by the river, to estimate that the overbank deposition accounted for ca. 19% of the total suspended sediment load transported into the reach during the event. Using a different approach, based on an assessment of the fate of Chernobyl-derived radiocaesium moving through the river system, Walling and Quine (1993) estimated that 23% of the total suspended sediment transported through the main channel system of the River Severn, UK, during the period 1986 to 1989, was deposited on the floodplain.

There are many other examples of rivers where the suspended sediment load declines downstream and where overbank deposition on the floodplain undoubtedly accounts for a substantial proportion of this reduction. The importance of river floodplains as suspended sediment sinks can increase further when land-use change causes accelerated erosion and releases large volumes of sediment into a river system which is adjusted to much lower sediment loads. The classic work of Trimble (1976, 1983) on the sediment budget of the 200 km^2 catchment of Coon Creek in the Driftless Area of Wisconsin, USA, has, for example, demonstrated that more than 50% of the sediment mobilized from the slopes of this basin during the period of accelerated soil loss from agricultural land, which extended from 1850 to 1938, was deposited on the floodplains of the lower and middle valley and its tributaries. Coon Creek is itself a small tributary of the Mississippi River and Kesel et al. (1992) have constructed a tentative sediment budget for the entire Lower Mississippi valley for the period 1880–1911 during which the river channel was in an essentially natural state. This sediment budget indicates that ca. 24% of the sediment input to the Lower Mississippi during this period was deposited as overbank deposits on the river floodplain.

The longer-term significance of these transmission losses associated with the overbank deposition of fine sediment will clearly depend on the rate of channel migration within the floodplain system, since the deposited sediment may be subsequently remobilized by lateral erosion. Meade (1994), for example, cites the case of the floodplain of the middle reaches of the Amazon River between the mouths of the Purus and Negro rivers, where overbank sediment deposits are as much as 20 m thick. Measurements undertaken by Mertes (1990) indicate that overbank deposition on the 3800 km^2 of inundated floodplain in this reach traps ca. 10% of the suspended sediment load transported by the river and average rates of deposition of 8 mm year^{-1} have been reported by the same author. However, this deposition is approximately balanced by the remobilization of older sediment deposits by lateral erosion, and Meade (1994) suggests that, taking the floodplain as a whole, the overbank deposits are being recycled within a period of about 2500 years. In many lowland rivers, where channels are less active and rates of channel migration are therefore less, the overbank deposits may represent a more permanent sink. In the case of the Lower Amazon, Meade (1994) estimates that ca. 12.5% of the suspended sediment load passing through the gauging station at Obidos is deposited on the floodplain as essentially permanent storage, before reaching the river mouth.

Concern for the role of fine sediment in the transport of contaminants through river systems (cf. Allan, 1986; Krishnappan and Ongley, 1989; Walling, 1989) has also directed attention to the significance of overbank sediment deposits as sinks for contaminants adsorbed to fine sediment. This significance relates to the potential for both the accumulation of contaminants on areas of the floodplain experiencing overbank deposition and the subsequent remobilization of contaminants stored in overbank sediments by channel migration. Marron (1987, 1989) documents a valuable, if extreme, example of both these facets in her study of the fate of finely milled mine tailings discharged into Whitewood Creek, a tributary of the Belle Fourche River in west-central South Dakota, USA. Gold mining around Lead, South Dakota, during the period extending from the 1870s through to 1978, was responsible for producing large amounts of fine tailings containing high levels of arsenic, and approximately 100 million tonnes of such tailings were discharged into Whitewood Creek during this period. These tailings were transported downstream into the Belle Fourche River and large quantities of arsenic-contaminated sediment were deposited as overbank deposits along this river. In places, these deposits are as much as 2 m thick and Marron has estimated that as much as one third of the discharged tailings are stilled stored within a 164 km reach of the Belle Fourche River extending downstream to its confluence with the Cheyenne River. Approximately half of this storage is associated with overbank deposits. Remobilization of this contaminated sediment occurs through the process of meander migration, and current estimates of its residence time indicate that it will continue to be a source of arsenic to the river for many centuries to come, despite the fact that mining and ore-processing ceased in 1978. Similar problems undoubtedly exist in many European rivers where a long history of mining and industrial activity in the upstream basin has provided a source of contaminated sediment which is now stored in floodplain sediments and which could continue to release polluted sediment into the channel well into the future, even though the original source may have been eliminated (cf. Bradley and Cox, 1987; Leenaers and Rang, 1989; Macklin, chapter 13, this volume). Such deposits have indeed sometimes been referred to as "chemical timebombs". The work of Leenaers and Schouten (1989) in the valley of the River Geul a tributary of the River Meuse, in the Netherlands has, for instance, shown how recent increases in flood magnitude and frequency associated with land-use change within the catchment have initiated a phase of channel migration. The resulting streambank erosion is mobilizing substantial quantities of fine overbank sediment contaminated with lead, zinc and cadmium which was deposited primarily during the peak of ore extraction in the catchment in the nineteenth century.

Against this background there is clearly a need for geomorphologists to direct increased attention to the fine-grained overbank deposits associated with floodplain development and, more particularly, to assemble information on the rates and patterns of sedimentation involved. Existing information indicates that the rates of sedimentation associated with overbank deposits are, as might be expected, relatively low. Wolman and Leopold (1957) suggest a general average of 0.15 cm year^{-1}. Thus, for example, in the case of the investigation undertaken on the floodplain of the River Waal in the Netherlands by Middelkoop and Asselman (1994), which was cited above, average rates of accretion of fine (<0.053 mm) sediment associated with the December 1993–January 1994 flood, measured using sediment traps at several different locations were in the range 0.1–0.33 cm (1.2–3.98 kg m^{-2}) and averaged 0.21 cm. Similarly, the average rate of overbank sedimentation on the floodplain of the lower River Culm estimated by Lambert and Walling (1987) from the measured conveyance losses along the 11 km reach was 0.04 cm year^{-1}. As might be expected, considerably higher values have been reported where rivers have been impacted by greatly increased sediment loads due to land-use change. In his study of Coon Creek, Wisconsin, Trimble (1976) documented total depths of overbank sedimentation in the main river valley of ca. 2 m, which, assuming the deposition occurred during the 85 years between 1853 and 1938, is equivalent to an average rate of ca. 2.35 cm year^{-1}. Knox (1987) reports even greater depths of sedimentation of up to 3–4 m in the valleys of tributaries of the Upper Mississippi in the Lead–Zinc District of Wisconsin and Illinois, which were related to the onset of agricultural and mining activity after about 1820. Similarly, in the case of the Belle Fourche River described above, the large quantities of fine-grained mine tailings introduced into the river caused overbank accretion rates of ca. 2.0 cm year^{-1} (Marron, 1989).

12.2 DOCUMENTING CONTEMPORARY RATES OF FLOODPLAIN SEDIMENTATION

Attempts to assemble further information concerning contemporary rates of sediment deposition on floodplains face many uncertainties and difficulties related to the inherent spatial and temporal variability of such sedimentation and to the operational problems of studying an inundated floodplain. The infrequent and essentially random occurrence of floodplain inundation is itself a major problem, since for most rivers it is extremely difficult to forecast the occurrence of major overbank flood events so as to provide sufficient lead time for deploying monitoring equipment. Equally, because the amounts of deposition associated with individual events will vary according to the magnitude and duration of the flood and its other characteristics, including the suspended sediment concentrations involved, estimates of deposition rates based on a single event or even a few events could prove to be unrepresentative of the longer term.

The methods that have been most widely and successfully used to quantify contemporary rates of overbank sediment accretion can conveniently be grouped into four main approaches, which also involve different time-scales. The first approach represents direct monitoring of sediment deposition, using sedimentation traps. Such measurements are generally event-based and attempt to measure the deposition associated with individual floods by deploying traps on the floodplain in advance of flood events. These traps have commonly been of simple design and have involved the fixing of wooden boards (cf. Mansikkaniemi, 1985; Gretener and Strömquist, 1987) or "plastic grass" mats (cf. Lambert and Walling, 1987; Walling and Bradley, 1989; Asselman and Middelkoop, 1993; Middelkoop and Asselman, 1994) on the floodplain surface and measuring the mass of sediment deposited on the trap during a flood event. The sediment recovered from these traps can also be analysed for its physical and chemical properties. Although such traps have provided valuable information on rates and patterns of contemporary overbank sedimentation (cf. Middelkoop and Asselman, 1994), limitations on the sampling densities that can be realistically employed, questions of representativeness, difficulties of retrieving traps from areas of extended floodplain ponding, uncertainties in the assessment of very low rates of deposition and the difficulty of obtaining long-term data, necessarily restrict the potential of this approach.

The second approach is closely related to the first and again considers individual events. It is, however, primarily applicable to rare high-magnitude floods which result in substantial amounts of deposition and involves post-event surveys of the resultant sediment deposits, aimed at determining their depth and distribution (e.g. McKee et al., 1967; Kesel et al., 1974; Brown, 1983; Marriott, 1992). It is heavily dependent upon the existence of significant depths of sediment and a readily defined interface between the deposits and the underlying surface. Other workers such as Macklin et al. (1992) have extended this general approach by identifying the deposits of known sequences of recent floods within exposed sections of sediment deposits. This strategy is, however, likely to be of limited application under more "normal" conditions, where deposition rates are low and where it is impossible to distinguish the deposits associated with individual events.

The third approach is essentially indirect in nature and involves measuring the decrease in suspended sediment load along a floodplain reach and ascribing the losses to overbank deposition (cf. Lambert and Walling, 1987; Gretener and Strömquist, 1987). However, the results obtained will represent an estimate of the net change in storage along a reach and could underestimate the overbank deposition where significant amounts of suspended sediment are released by bank erosion within the reach or where there are significant inputs of sediment from small tributaries. Equally, rates of overbank deposition could be overestimated where channel storage represents an important sink during flood events. Furthermore, such estimates are heavily dependent upon the accuracy and precision of the sediment load measurements which may themselves be open to question (cf. Walling and Webb, 1989) and can only provide information on the average rate of deposition along the entire reach.

The fourth group of techniques generates estimates of deposition rates averaged over a number of years and essentially relies on establishing the date of a particular level or horizon and calculating the rate of deposition from the depth of material overlying this level. A range of methods have been used to establish levels of known date, and these include previous benchmark surveys (cf. Happ, 1968; Leopold, 1973, Trimble. 1983); the existence of datable surfaces, or material (cf. Costa, 1975, Hupp, 1988); relating trace metal concentrations in the sediment profile to the known history of mining activity in the upstream catchment (e.g. Knox, 1987; Lewin and Macklin. 1987; Popp *et al.*, 1988; Macklin and Dowsett, 1989); and the use of fallout radionuclides such as caesium-137 (Cs-137) (e.g. Ritchie *et al.*, 1975; McHenry *et al.*, 1976; Popp *et al.*, 1988; Walling and Bradley, 1989; Walling *et al.*, 1992; Walling and Quine, 1993; Walling and He, 1994) and unsupported lead-210 (Pb-210) (e.g. Walling and He, 1994; He and Walling, 1996). Although valuable, the first three types of evidence will be limited by the need for prior surveys, the frequent lack of suitable datable material or objects, and the long time-scales and lack of temporal and spatial resolution that are commonly involved. In general they have been used to provide estimates of deposition rates at specific points or exposures, rather than to study the spatial patterns involved. However, the use of the fallout radionuclides Cs-137 and unsupported Pb-210 would appear to offer considerable potential in terms of their general applicability in a wide range of environments, the medium-term time-scales involved (i.e. about 35 years and 100 years for Cs-137 and unsupported Pb-210 respectively) and the possibility of assembling data for a large number of points on a floodplain and thereby investigating the spatial variability of deposition rates and the patterns involved.

12.3 USING FALLOUT RADIONUCLIDES TO INVESTIGATE RECENT RATES AND PATTERNS OF OVERBANK DEPOSITION ON FLOODPLAINS

12.3.1 Caesium-137

Although both Cs-137 and unsupported Pb-210 have been used to estimate recent rates of overbank sedimentation on floodplains (cf. Walling and He, 1994), the former has been more widely used. Caesium-137 is an artificial radionuclide with a half-life of 30.17 years which was introduced into the environment by the atmospheric testing of thermonuclear weapons primarily during the late 1950s and the 1960s. The radiocaesium released into the stratosphere by these weapon tests was dispersed globally and subsequently deposited as fallout on the land surface. Most of the fallout occurred in the decade between 1956 and 1965, with maximum deposition in 1963, the year of the Nuclear Test Ban Treaty. Since the 1970s, rates of Cs-137 fallout have been very low, although in some parts of Europe and adjacent regions an additional short-term input was received in 1986 as a result of the Chernobyl accident, which also released radiocaesium into the atmosphere. In most environments, Cs-137 deposited as fallout is rapidly and strongly fixed by clay particles in the surface soil (cf. Frissel and Pennders, 1983; Livens and Rimmer, 1988) and its subsequent redistribution occurs in association with sediment particles.

In essence, the basis for using Cs-137 to estimate rates of floodplain accretion reflects its accumulation in floodplain sediments as a result of inputs from two primary sources. These sources represent, firstly, direct atmospheric fallout to the floodplain surface and, secondly, the deposition of sediment-associated Cs-137 during the process of sediment accretion. This sediment-associated Cs-137 represents radiocaesium originating as fallout over the upstream basin which has been adsorbed by soil and sediment particles and subsequently mobilized by erosion and transported downstream as an integral part of the suspended sediment load of the river. Both the vertical distribution and total inventory of Cs-137 in floodplain sediments will therefore commonly differ from those in the soils of adjacent undisturbed areas above the level of floodplain inundation, since the latter will have received Cs-137 only from direct fallout (cf. Figure 12.1).

Information such as that presented in Figure 12.1B has been used by a number of workers to estimate the rate of overbank sedimentation over the past 30–40 years using one of two possible approaches (cf. Walling and Bradley, 1989). In the first, the shape of the depth profile is used. If this is compared with that from an adjacent undisturbed reference site (cf. Figure 12.1A), the degree of "stretching" can be estimated and this can be used to estimate the depth of sedimentation. Alternatively, the position of the level with peak activity can be used to estimate the depth of the 1963 surface, or more complex models can be used to ascribe dates to several depths in the profile (cf. Walling and He, 1992). In the second approach, the total inventory of a floodplain core is compared with a reference value representing the local fallout inventory and the value of "excess" inventory is used in combination with an estimate of the mean Cs-137 content of deposited sediment during the period since 1955 to estimate the mass, and thus the depth, of sediment accretion (cf. Walling et al., 1992).

Both the approaches outlined above possess limitations. Use of information relating to the shape of the Cs-137 depth profile necessitates analysis of a large number of samples from each individual sectioned core and this approach is of limited value in studying detailed spatial patterns of floodplain sedimentation where a substantial

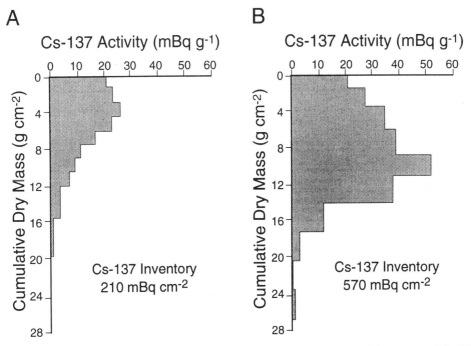

Figure 12.1 Representative examples of the vertical distribution and total inventory of Cs-137 in undisturbed permanent pasture above the level of inundating floodwater (A) and in an adjacent area of the floodplain of the River Culm, Devon, UK (B)

numbers of cores would be required. The "excess" inventory approach possesses a major advantage in that only a single measurement is required on each core, in order to assay the total inventory, and a greater number of cores can therefore be used. However, several uncertainties are involved in converting the value of "excess" inventory to an estimate of sedimentation mass or depth. There is a need to estimate the mean Cs-137 content of deposited sediment and this is usually assumed to be essentially constant within the area studied (cf. Walling *et al.*, 1992). In reality, some variation in this concentration is likely to occur across an area in response to variations in the grain size composition of deposited sediment. Because Cs-137 is known to be preferentially associated with the finer fractions of the deposited sediment (cf. Walling and Woodward, 1992), use of a mean value of Cs-137 content could lead to overestimation of deposition rates for points where the deposited sediment contained an increased proportion of fines and underestimation for points where coarser sediment predominates.

In view of the considerable advantage attached to whole core measurements in terms of the number of sampling points on the floodplain that can be investigated and therefore the increased degree of spatial resolution that can be achieved when studying sedimentation patterns, Walling and He (1993) have developed a procedure for taking account of spatial variations in the grain size composition of suspended sediment within a study area. Using this procedure, the mean sedimentation rate at a specific point on the floodplain can be estimated from the value of "excess" inventory at that point and information on the Cs-137 content of the recently deposited surface sediment (which also reflects the grain size composition of that sediment) provided detailed information on the Cs-137 depth profile is available for at least one representative site. If, as would seem reasonable from the case study reported by Walling and He (1993), it can also be assumed that the grain size composition of the deposited sediment has remained essentially constant through time, the Cs-137 concentration of the recently deposited sediment at each sampling point can be used in a model which has been calibrated using the Cs-137 depth profile, to estimate the mean sedimentation rate for the point at which the core was collected. The "excess" Cs-137 inventory $A_{c,Cs}$(mBq cm^{-2}) of a sediment core is calculated as the difference between the measured total Cs-137 inventory A_{Cs}(mBq cm^{-2}) of the core and the estimated local fallout input $A_{a,Cs}$(mBq cm^{-2}). Because the "excess" inventory represents the Cs-137 associated with the deposition of suspended sediment derived from the upstream catchment, it will be related to the Cs-137 concentration $C_{d,Cs}$(mBq g^{-1}) of deposited sediment and the sedimentation rate R(g cm^{-2} yr^{-1}) according to:

$$A_{c,Cs} = \int_0^t R(t')C_{d,Cs}(t')e^{-\lambda_{Cs}(t-t')} \, dt' \tag{12.1}$$

where t(yr) is the time since 1954 to the present and λ_{Cs}(yr^{-1}) is the decay constant of Cs-137. In order to estimate the mean sediment deposition rate at a sampling point on the floodplain during the past 40 years, Walling and He (1993) have assumed a constant rate of sediment deposition on the floodplain and have modelled the temporal variation of the Cs-137 concentration $C_{c,Cs}(t)$(mBq g^{-1}) in sediment mobilized from the catchment by taking account of the accumulation and behaviour of Cs-137 in different sediment sources and the relative contribution of sediment contributed from these sources. The Cs-137 concentration in deposited sediment can be related to that of the eroded source material using an enrichment ratio $R_{en,Cs}$ to characterize the deposited sediment:

$$C_{d,Cs}(t) = R_{en,Cs}C_{c,Cs}(t) \tag{12.2}$$

The Cs-137 enrichment ratio $R_{en,Cs}$ can be estimated by comparing the grain size composition of recently deposited sediment with that of the source material. Where recently deposited sediment is not available for a sampling point, surface sediment can be analysed for particle size comparison. When the average Cs-137 concentration in suspended sediment is known, the Cs-137 concentration of recently deposited sediment for a sampling position can also be estimated by considering the grain size composition of both the fluvial suspended sediment and the surface sediment. The average sediment deposition rate can be estimated via:

$$R = \frac{A_{c,Cs}}{R_{en,Cs}\int_0^t C_{c,Cs}(t')e^{-\lambda_{Cs}(t-t')}\,dt'} \tag{12.3}$$

12.3.2 Unsupported Pb-210

Although unsupported Pb-210 is also a fallout radionuclide, it differs from Cs-137 in two important respects. First, it is of natural origin, and, secondly, because of its natural origin, the annual fallout may be viewed as being essentially constant through time. It therefore affords a means of estimating deposition rates over somewhat longer time periods (i.e. 50–150 years). Lead-210 is a product of the U-238 decay series with a half-life of 22.26 years. It is derived from the decay of gaseous Rn-222, the daughter of Ra-226. Radium-226 occurs naturally in soils and rocks and diffusion of a small proportion of the Rn-222 from the soil introduces Pb-210 into the atmosphere. Its subsequent fallout provides an input of Pb-210 to surface soils and sediment which is not in equilibrium with its parent Ra-226. This component is termed unsupported Pb-210, since it cannot be accounted for (or supported) by decay of the *in situ* parent. The amount of unsupported or atmospherically derived Pb-210 in a sediment sample can be calculated by measuring both Pb-210 and Ra-226 and subtracting the supported component. As in the case of Cs-137, floodplain surfaces will receive inputs of unsupported Pb-210 as a result of both direct fallout and in association with the deposition of sediment eroded from the upstream basin. The depth distribution of unsupported Pb-210 in a floodplain deposit will therefore again differ significantly from that in adjacent undisturbed pasture above the level of flood inundation (cf. Figure 12.2).

Although most existing work which has made use of measurements of unsupported Pb-210 to establish chronologies for lake sediments has been based on information concerning the depth profile of this radionuclide, He and Walling (1996) have demonstrated how a single measurement of the total unsupported Pb-210 inventory for a bulk sediment core can be used to estimate the average rate of accretion for the point on the floodplain at which the core was collected. Again, therefore, estimates of sedimentation rate can be assembled for a large number of points on a floodplain. In

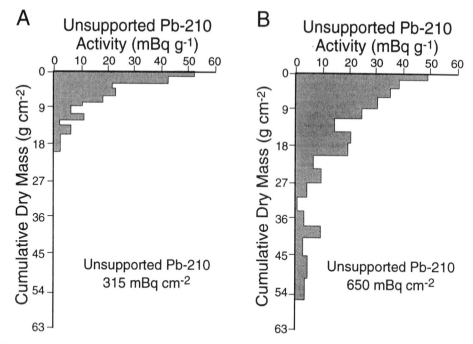

Figure 12.2 Representative examples of the vertical distribution and total inventory of unsupported Pb-210 in undisturbed permanent pasture above the level of inundating floodwater (A) and in an adjacent area of the floodplain of the River Culm, Devon, UK (B)

this case, the total unsupported Pb-210 inventory A_{Pb}(mBq cm^{-2}) of the core is assumed to comprise two components, viz.:

$$A_{Pb} = A_{a,Pb} + A_{c,Pb} \qquad (12.4)$$

where $A_{a,Pb}$ is the direct atmospheric fallout input to the floodplain surface (mBq cm^{-2}), and $A_{c,Pb}$ is the catchment-derived input associated with the deposition of suspended sediment mobilized from the upstream catchment (mBq cm^{-2}).

$A_{a,Pb}$ can be estimated from measurements of the total unsupported Pb-210 inventory of undisturbed pasture sites adjacent to the floodplain, but above the level of inundation. $A_{c,Pb}$ can therefore be calculated by subtracting $A_{a,Pb}$ from the total unsupported Pb-210 inventory of the core. The value calculated for $A_{c,Pb}$ can then be used to estimate the average rate R of sedimentation during the past 100–150 years, using the CICCS (constant initial concentration – constant sedimentation rate) model proposed by He and Walling (1996), i.e.:

$$R = \lambda_{Pb} \frac{A_{c,Pb}}{C_{d,Pb}} \qquad (12.5)$$

where λ_{Pb}(year^{-1}) is the decay constant of Pb-210 and $C_{d,Pb}$(mBq g^{-1}) is the average initial unsupported Pb-210 concentration of deposited sediment associated with the catchment-derived input. Values of $C_{d,Pb}$ for different points on the floodplain can be obtained by analysing the unsupported Pb-210 content of recent overbank deposits collected using sediment traps in the sampling area and establishing a relationship between the unsupported Pb-210 concentration and the grain size composition of the sediment deposits. Alternatively, if the unsupported Pb-210 concentration in fluvial suspended sediment is known, values can be estimated from measurements of the grain size composition of recently deposited sediment by taking account of their relationship to that of the suspended sediment.

Cs-137 and unsupported Pb-210 measurements afford a valuable and essentially unique means of estimating recent rates of overbank sedimentation on floodplains over the past 30–150 years. In both cases it is possible to assemble information for a substantial number of points on a floodplain, which can provide a basis for investigating both the rates and spatial patterns involved. At locations where recent inputs of Chernobyl-derived Cs-137 have complicated the interpretation of radiocaesium inventories or where Cs-137 inventories are low due to the global pattern of Cs-137 fallout (cf. Walling and Quine, 1995), unsupported Pb-210 can provide a valuable alternative to Cs-137. Elsewhere, they can be used in parallel to generate complementary information relating to ca. 35 year and 100–150 year time-scales (cf. Walling and He, 1994). Both procedures generate estimates of average rates of accretion and are most appropriate for overbank deposits on lowland floodplains which are characterized by essentially continuous, albeit sporadic and event-based, sedimentation. Any attempt to use Cs-137 and unsupported Pb-210 measurements for this purpose should involve a preliminary analysis of the depth profiles of the two radionuclides to ensure that the assumption of quasi-continuous sedimentation is not violated. In the case of unsupported Pb-210, the profile should exhibit a steady decline of concentration with depth, whereas for Cs-137 there should be evidence of the typical negatively skewed single peak profile shape for depositional environments (cf. Walling and He, 1992). The potential for using fallout radionuclides to generate spatially distributed information on rates and patterns of overbank floodplain sedimentation can usefully be demonstrated by introducing case studies of two British rivers, namely the River Culm in Devon and the Dorset Stour.

12.4   RATES AND PATTERNS OF OVERBANK SEDIMENTATION ON THE FLOODPLAINS OF THE RIVER STOUR, DORSET, AND THE RIVER CULM, DEVON

The procedures for using bulk core measurements of Cs-137 and unsupported Pb-210 inventories to estimate recent rates of floodplain deposition developed by the authors and described above, have been used to investigate the pattern of sedimentation within two sections of the floodplains of the Dorset Stour and the River Culm. In the case of the Dorset Stour, only Cs-137 measurements were employed, whereas both radionuclides were used in the River Culm study. At both study sites, bulk sediment cores were collected from the floodplain using a motorized percussion corer equipped with a 6.9 cm diameter core tube. A small sample was collected from the base of each core for subsequent radionuclide assay, in order to ensure that the core had penetrated to the full depth of the Cs-137 and unsupported Pb-210 profiles. Samples of surface sediment were also collected from a point immediately adjacent to each core. Where sectioned cores were required, a larger 12 cm diameter core tube was employed and the sediment core was sectioned into 1 or 2 cm depth increments after extrusion. Suspended sediment samples from the two rivers were collected during flood events over a period of 2 years. In the case of the River Culm, overbank sediment deposits were also collected using sediment traps installed at various locations on the floodplain prior to flood events and retrieved immediately after the flood receded. All bulk cores and other samples were air-dried, ground and homogenized prior to measurement of their Cs-137 and unsupported Pb-210 content by gamma spectrometry. In the case of Pb-210, these measurements were undertaken using a high resolution, low background, low energy, n-type coaxial HPGe detector. Samples were sealed for 20 days before gamma assay in order to ensure equilibrium between Ra-226 and Rn-222. Values of unsupported Pb-210 activity were derived from measurements of the total Pb-210 activity by subtracting the Ra-226 supported Pb-210 activity (cf. Joshi, 1987). The Ra-226 activity in the sample was estimated from the measured activity of its short-lived daughter Pb-214 which is also in equilibrium with Ra-226. Count times were typically 10 hours and produced values of unsupported Pb-210 activity with a precision of ca. ±10% at the 90% level of confidence. In the case of samples analysed for their unsupported Pb-210 content, values of Cs-137 concentration were obtained at the same time by measuring the activity at 662 keV. However, because of the limited laboratory capacity for Pb-210 measurements, only Cs-137 activity was determined for the River Stour samples, and in this case p-type coaxial HPGe detectors were employed. Count times were typically 30 000s, providing a precision of ca. ±5% at the 90% level of confidence. Measurements of the grain size composition of samples were undertaken using laser diffraction apparatus (Malvern Mastersizer), after appropriate pretreatment.

12.4.1 The Dorset Stour

The measurements undertaken on the Dorset Stour focused on the 14 km section between Hammoon (GR ST820146) and Blandford Forum (GR ST880077) (Figure 12.3). At Hammoon the Stour drains an area of 523 km². The upstream catchment is primarily occupied by grassland and is underlain by a variety of Cretaceous and Jurassic strata. These include extensive areas of clay, which give rise to low relief clay valves. The mean annual precipitation and runoff for the catchment above Hammoon are 785 mm and ca. 400 mm respectively and the mean annual flood is estimated to be ca. 117.5 $m^3 s^{-1}$. Suspended sediment concentrations in the river during flood events are typically within the range 100–250 mg l^{-1} and the floodplain is inundated on average several times each year.

A nested sampling strategy was employed to provide a general assessment of sedimentation along the section and a more detailed analysis of the pattern of sedimentation within a specific reach. This was achieved by selecting nine representative cross-sections (Figure 12.3), from which cores were collected in order to document lateral variation in sedimentation rates, and a small reach of the floodplain near Chisel Farm (see Figure 12.3) from within which 97 cores were retrieved at the intersections of a 25 m grid. The local Cs-137 reference inventory for

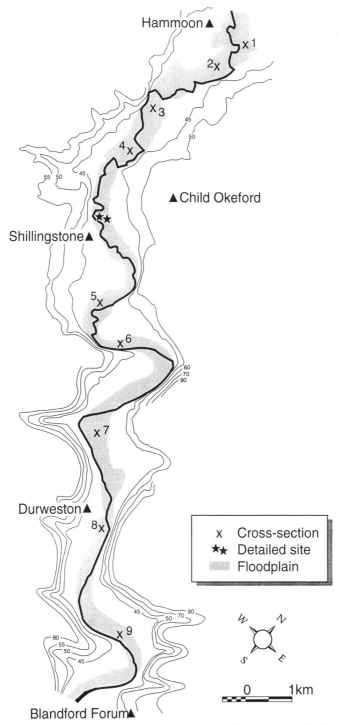

Figure 12.3 The study reach of the River Stour, Dorset, UK

the study area was estimated to be ~230 mBq cm^2 and the values of excess Cs-137 inventory associated with the individual bulk cores were converted to estimates of average deposition rate over the past 35 years using the approach described previously. The mean Cs-137 concentration of suspended sediment was estimated to be ~14 mBq g^{-1}.

For the 99 cores collected from the nine cross-sections established along the reach, the average deposition rate was 0.08 g cm^{-2} yr^{-1}. This value must be seen as relatively low and is likely to reflect, at least in part, the relatively low suspended sediment concentrations and the extremely fine-grained nature of the suspended sediment transported through the reach ($d_{50} \sim 2.9$ μm). Figure 12.4 presents the data assembled for the nine individual cross-sections and provides a means of assessing the variation of deposition rates both laterally across the floodplain and through the study section. All sites evidence a general decline in deposition rate with increasing distance from the channel, which is consistent with the diffusion effect documented by other workers (e.g. Marriott, 1992) and the general reduction in frequency and duration of inundation in the outer zones of the floodplain. On average, sedimentation rates on the floodplain close to the channel are about 2.4 times greater than those encountered towards the outer edge of the floodplain. Figure 12.5 plots the relationship between the estimated sedimentation rates (R) and distance (Y(m)) from the river channel for all 99 cores. In this case, the following generalized function has been fitted to the data to describe the overall trend of the decreasing sedimentation rate with increasing distance from the channel through the reach:

$$R = 0.11 \, e^{-Y/240} \tag{12.6}$$

with $r^2 = 0.22$. Because the rate of suspended sediment deposition will be closely related to the detailed flow pattern on the floodplain and the local microtopography,

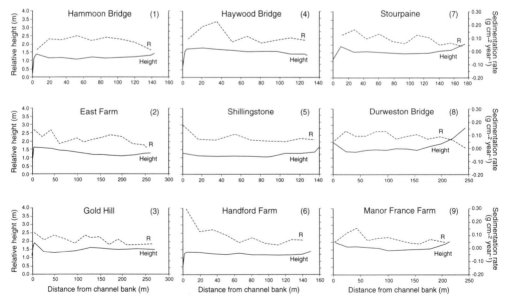

Figure 12.4 Sedimentation rates estimated for the nine cross-sections of the floodplain of the Dorset Stour shown on Figure 12.3 using Cs-137 measurements

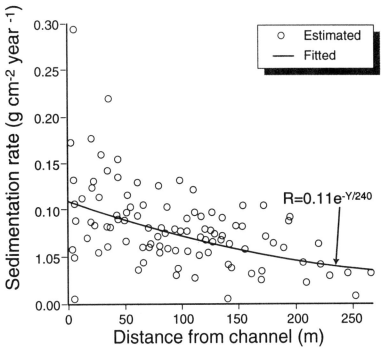

Figure 12.5   The relationship between sedimentation rate and distance from the channel established for the floodplain cross-sections shown on Figure 12.3

there are significant variations in deposition rates between the individual cross-sections (cf. Figure 12.4), but no clear longitudinal trend is apparent.

Figure 12.6 presents the results of the investigation of the detailed pattern of sedimentation in the reach at Chisel Farm based on the grid of 97 cores. The general form of the area of the floodplain investigated is shown in Figure 12.6B. This is relatively simple and consists of a levee or zone of higher land bordering the channel, with the remaining area exhibiting a gentle slope away from the channel. This is, however, complicated by a number of small depressions linked to the channel and by several depressions towards the outer margin of the floodplain. All cores collected within the study reach were characterized by Cs-137 inventories in excess of the local reference (fallout) inventory, indicating that deposition was occurring throughout the reach. The sedimentation rates estimated using the Cs-137 measurements range from $0.01 \, \mathrm{g \, cm^{-2} \, yr^{-1}}$ to $0.66 \, \mathrm{g \, cm^{-2} \, yr^{-1}}$ (cf. Figure 12.6C). The average sedimentation rate estimated for this area ($0.20 \, \mathrm{g \, cm^{-2} \, yr^{-1}}$) is significantly higher than that documented for the individual cross-sections. This probably reflects the increased depth and frequency of inundation in this area, which comprises the inner part of a meander bend where the floodplain narrows and abuts a steep slope on the other side of the river. The pattern of sediment deposition in this reach can be seen to reflect the effects of the local microtopography and the processes associated with sediment transport and deposition. For example, sedimentation rates are relatively high in areas close to the channel bank and evidence a general decrease with increasing distance from the channel. This trend

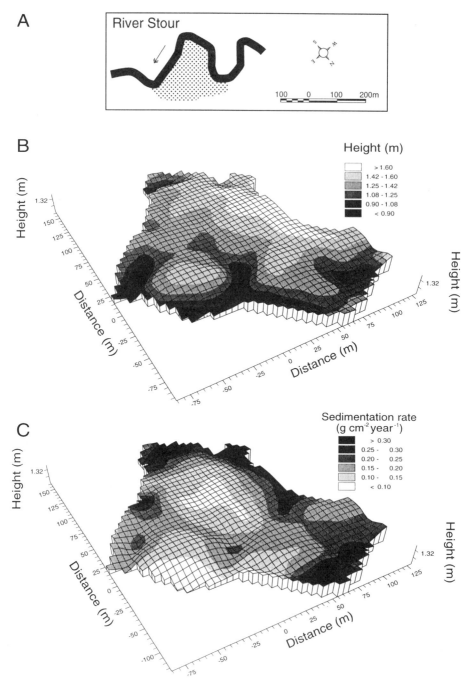

Figure 12.6   The location (A) and local microtopography of the floodplain study site at Chisel Farm (B) and the pattern of deposition rates estimated using Cs-137 measurements (C)

results, at least in part, from the deposition of coarser sediment with higher settling velocities close to the channel banks and from the general reduction in the concentration of coarser particles in flows across the floodplain in response to deposition within the intervening zone. The influence of the local microtopography is also apparent in the high sediment accumulation rates observed in the depression areas close to the channel banks. These reflect the increased depth of floodwater in these areas during inundation, since the amount of sediment deposited per unit area will be proportional to the total mass of sediment in the overlying water column, which is, in turn, proportional to the water depth. The sedimentation pattern in this area is therefore generally consistent with the models proposed by other workers such as James (1985), Pizzuto (1987) and Marriott (1992). An attempt has been made to represent the combined effects of distance from the channel and floodwater depth on the observed patterns of deposition by developing a simple depositional function. Following Howard (1992), floodplain sediment deposition can be viewed as comprising two components. The first represents the deposition of the coarser fraction across the floodplain at a rate proportional to the floodwater depth and its concentration in floodwater, which is assumed to decrease exponentially with increase in distance Y from the nearest channel. The second reflects the deposition of fine particles, for which the rate will be directly proportional to the floodwater depth Z(m) (defined as the difference between the maximum floodplain height and the local floodplain height), since concentrations can be assumed to be constant across the floodplain. For a specific point on the floodplain, it is reasonable to assume that the concentration of the fine sediment particles in the overlying water will be relatively uniform vertically, but the vertical distribution of coarse particles is likely to be characterized by higher concentrations in the water close to the floodplain surface. The following function may be used to represent these mechanisms:

$$R = \alpha Z^{\beta} e^{-Y/\gamma} + \delta Z \qquad (12.7)$$

The first term on the right side of Equation 12.7 represents the deposition of the coarser particles and the second term the deposition of fine-grained sediment. α and δ are the average concentrations of the coarse and fine fractions of suspended sediment respectively. The constant γ represents the general decline in the concentration of the coarse fraction, with increasing distance from the channel. The parameter β acts as a correction factor to account for the non-uniform vertical distribution of the coarse fraction in the floodwater. Fitting Equation 12.7 to the available data provided the following function:

$$R = 0.33 Z^{0.40} e^{-Y/12} + 0.10Z + 0.08 \qquad (12.8)$$

with $r^2 = 0.44$. The constant on the right side of Equation 12.8 takes account of the base level selected for the relative floodwater depth on the floodplain. In Figure 12.7 the fitted function (Equation 12.8) has been superimposed on a plot of observed sedimentation rate versus distance from the channel, with the points classified according to relative floodwater depth. While this simple function accounts for the major trends in the data presented in Figure 12.7, it is clear that the detailed pattern of variation in sedimentation rates within the study reach will also reflect the influence of the local microtopography and the local flow patterns. A more dynamic approach

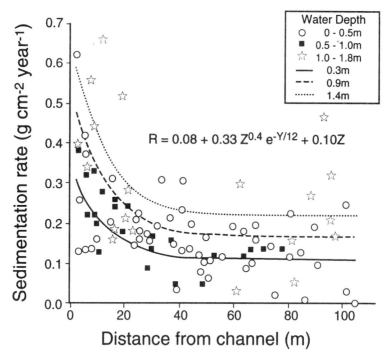

Figure 12.7 The interrelationship between sedimentation rate and distance from channel and water depth demonstrated by the data assembled for the floodplain study site at Chisel Farm

capable of incorporating the interaction of water flow patterns on the floodplain with the local microtopography and the interaction between these flow patterns and sediment transport and deposition is required to provide a more accurate representation of the deposition of suspended sediment on the floodplain.

Figure 12.8 depicts the spatial variation of the relative magnitude of the <63 μm fraction of the surface sediment across the study site at Chisel Farm. Over most of the floodplain surface, there is a well-developed trend for the <63 μm fraction to account for the smallest proportion of the deposited sediment close to the channel bank and for its relative importance to increase towards the outer margins of the floodplain. The planform of the river channel also plays a role in accounting for the pattern of grain size composition of deposited sediment, since an increased proportion of <63 μm sediment is also found in areas close to the channel bank where the river meander changes direction. Reflecting the above trends, correlation analysis indicates that the percentage <63 μm fraction (P) in surface sediment is inversely related to the estimated sediment deposition rate, and the following function was used to represent the relationship:

$$P = -57.3R + 85.2 \qquad (12.9)$$

with $r^2 = 0.26$ (cf. Figure 12.9A). The precise influence of hydrodynamic controls on the spatial pattern of the grain size composition of surface sediment across the floodplain is likely to be obscured by the fact that the suspended sediment is

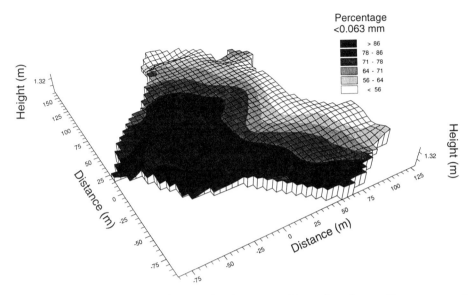

Figure 12.8 Spatial variation of the grain size composition of surface sediment across the floodplain study site at Chisel Farm

transported and deposited as aggregates and not as discrete particles. However, an attempt has also been made to relate the observed distribution of the percentage <63 μm fraction (P) in surface sediment from the study reach to distance from the channel and relative floodwater depth. If it is assumed that deposition of the <63 μm fraction and the >63 μm fraction of the suspended sediment at a specific point can be represented by the two terms on the right side of Equation 12.7 respectively, the following function may be proposed:

$$P = \frac{100}{1 + (\alpha/\delta)Z^{\beta-1}e^{-Y/\gamma}} \qquad (12.10)$$

The relative importance of floodwater depth Z and distance from the channel Y in controlling the value of P depends on the values of β and γ. Equation 12.10 suggests that, in general, water depth can only have a limited influence on P because the concentration of the >63 μm fraction will be primary controlled by distance from the channel. However, areas associated with maximum values of Z are also likely to include areas of the floodplain subject to ponding after recession of the flood peak. The extended period of time available for settling of fine sediment in these areas will permit deposition of the very fine particles and such areas can therefore be expected to be characterized by high values of P. For example, the depression along the outer margin of the study reach holds ponded water after the recession of a flood and this area is marked by high values of P. Based on the observed pattern of the percentage <63 μm fraction in the sediment in the study area and the discussion outlined above, the following function was fitted to the results:

$$P = -29.9\,e^{-Y/12} + 13.3Z + 70.4 \qquad (12.11)$$

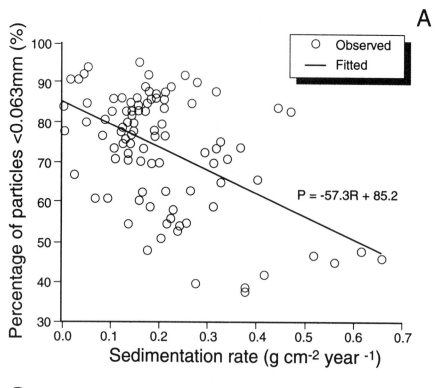

$$P = -57.3R + 85.2$$

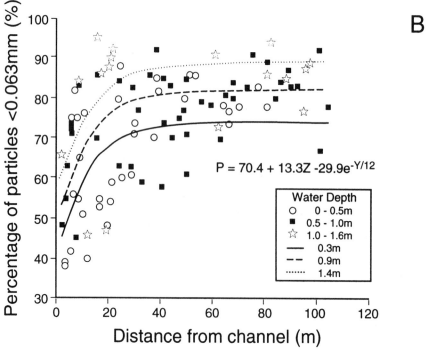

$$P = 70.4 + 13.3Z - 29.9e^{-Y/12}$$

with $r^2 = 0.40$. The first term on the right side of Equation 12.11 reflects the influence of the distance from the channel on the relative magnitude of the >63 μm fraction in the deposited sediment, and the second term represents the effects associated with floodwater depth. The fitted results are also depicted in Figure 12.9B. Again, the observed pattern of grain size distribution in the study area is similar to those documented by other workers (cf. James, 1985; Pizzuto, 1987; Marriott, 1992).

12.4.2 The River Culm

In the case of the River Culm, attention focused on documenting the detailed pattern of sedimentation within a short reach of the floodplain near Silverton Mill (cf. Figure 12.10A). Above this point the River Culm drains an area of ca. 276 km^2 underlain by Permian, Triassic and Cretaceous rocks. The catchment is characterized by mixed land use of pasture and arable, with the proportion of arable land increasing towards its outlet. The mean annual precipitation and runoff for the catchment are ca. 925 mm and 510 mm respectively and the mean annual flood at the Woodmill gauging station, 6 km upstream, has been estimated to be ca. 80 m^{3}s^{-1}. The floodplain bordering the lower reaches of the river is notable for its regular inundation by flood events and such inundation typically occurs on eight to ten occasions per year. The mean annual specific suspended sediment yield of the basin has been estimated to be ca. 32 t km^{-2}yr^{-1} and peak suspended sediment concentrations during flood events are typically in the range 400–800 mg l^{-1}. The suspended sediment load of the river is also notable for its fine-grained composition (cf. Walling and Moorehead, 1989), with a d_{50} averaging ca. 6.5 μm.

Within the study reach, 274 sediment cores were collected at the intersections of a 12 m grid and these were analysed for both their Cs-137 and unsupported Pb-210 inventories. The reference fallout inventories established for the site were ~220 mBq cm^{-2} for Cs-137 and ~300 mBq cm^{-2} for unsupported Pb-210, and these values were used to calculate the excess Cs-137 and unsupported Pb-210 inventories for each core. These in turn were used to estimate the mean annual deposition rates using the procedures outlined previously. Estimates of values for the concentration of Cs-137 and the initial concentration of unsupported Pb-210 in deposited sediment were derived from analysis of overbank sediment deposits. These values of ~12 mBq g^{-1} (for Cs-137) and ~37 mBq g^{-1} (for unsupported Pb-210) were corrected for particle size effects by making use of information on the particle size composition of the surface sediment collected from the vicinity of each coring site.

Figure 12.10B provides further information regarding the nature and topography of the study reach at Silverton Mill. In general terms it comprises a depression within a meander bend, bordered by an elevated bank margin or natural levee, and a linear depression along the outer margin of the floodplain. The mean annual sedimentation rates within the area estimated from the Cs-137 and unsupported Pb-210 measurements

---

Figure 12.9 Relationships between the grain size composition of surface sediment and sedimentation rate (A) and distance from the channel and water depth (B) established for the floodplain study site at Chisel Farm

A

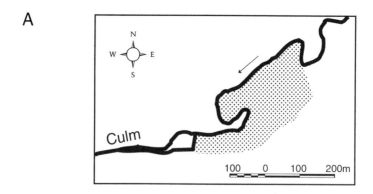

B

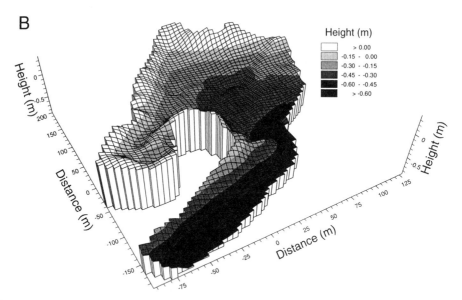

Figure 12.10 The location of the floodplain study site at Silverton Mill (A) and the local microtopography of the site (B)

are depicted in Figures 12.11A and 12.11B. The two sets of estimates are similar in terms of both magnitude and overall pattern, and this in turn suggests that rates of sedimentation have been relatively constant at this site during the past 100 years. There is, however, some evidence that sedimentation rates may have increased slightly since the 1950s, because the estimates derived from the Cs-137 measurements are in places slightly greater than those derived from the unsupported Pb-210 measurements. The rates shown in Figures 12.11A and B are also consistent with independent evidence of short-term rates of overbank sedimentation on the floodplain of the Lower River Culm provided by sediment traps. Simm (1993) reports the results obtained from a series of sediment traps installed on the floodplain at a site with similar physical characteristics and a similar frequency of inundation, at Columbjohn, 4 km downstream. These measurements relate to a period of little more than a year and may therefore be somewhat unrepresentative of the longer-term, but the values, which range from 0.09 to 0.80 $g\,cm^{-2}\,yr^{-1}$, are in close agreement with the longer-term estimates provided by the fallout radionuclide measurements.

In common with the results from the River Stour floodplain reported above, the patterns presented in Figure 12.11A and B show clear evidence of enhanced rates of

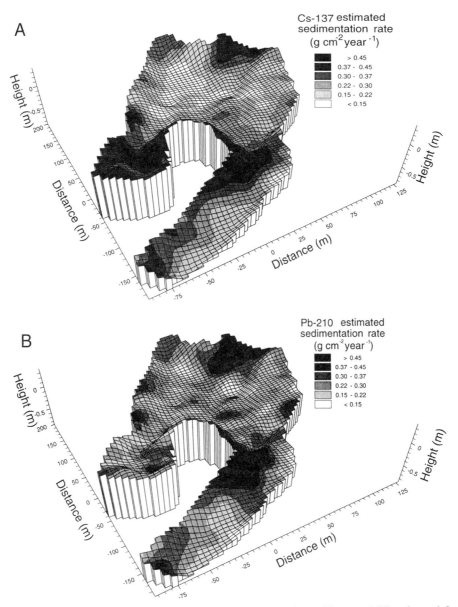

Figure 12.11 Deposition rates across the floodplain study site at Silverton Mill estimated from Cs-137 (A) and unsupported Pb-210 (B) measurements

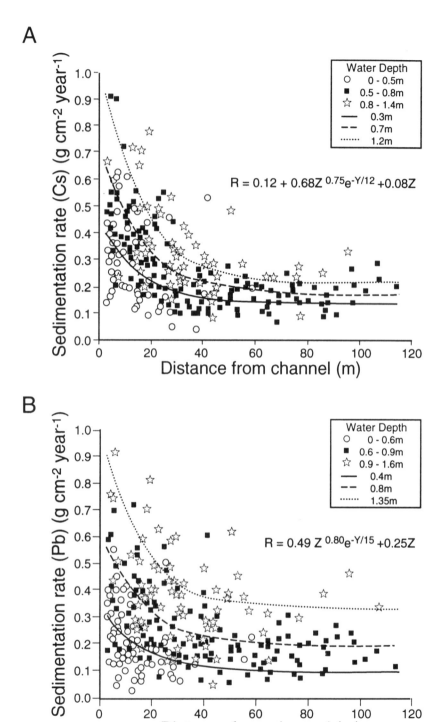

Figure 12.12 The interrelationships between sedimentation rate and distance from channel and water depth established for the floodplain study site at Silverton Mill based on Cs-137 measurements (A) and unsupported Pb-210 measurements (B)

deposition close to the channel, which reflect the transfer of suspended sediment from the channel onto the adjacent floodplain. Both the central depression and the linear depression at the outer margin of the floodplain are characterized by low rates of deposition which further emphasize the importance of transfer from the main channel in accounting for high deposition rates and indicate that water depth *per se* is of less importance in influencing the spatial pattern of deposition. An attempt has again been made to explain the key features of the spatial distribution of deposition rates depicted in Figures 12.11A and 12.11B, in terms of distance from the channel and water depth (i.e. the sediment content of the water column). Fitting the same general function as proposed for the River Stour, the sedimentation rates estimated by the Cs-137 method gave the following results:

$$R = 0.68Z^{0.75}e^{-Y/15} + 0.08Z + 0.12 \tag{12.12}$$

with $r^2 = 0.43$. The sedimentation rates estimated using unsupported Pb-210 measurements gave a similar result viz.:

$$R = 0.49Z^{0.80}e^{-Y/15} + 0.25Z \tag{12.13}$$

with $r^2 = 0.41$. The results presented in Figure 12.12 again confirm the general influence of both distance from the channel and floodwater depth in accounting for the spatial pattern of sedimentation rates. As in the case of the River Stour, however, it is clear that the detailed pattern reflects the additional influence of the local microtopography and its interaction with the flow and sediment transport processes.

Figure 12.13 provides information on the spatial variation of the grain size composition of the surface sediment on the floodplain study area in terms of the

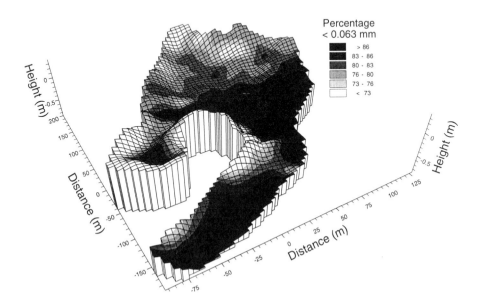

Figure 12.13 Spatial variation of the grain size composition of surface sediment across the floodplain study site at Silverton Mill

percentage <63 μm. In comparison with the results from the River Stour floodplain presented above, there is relatively little variation in this index over most of the area. However, a clearly defined pattern is evident and this is closely related to the microtopography of the area of the floodplain investigated, and more particularly to the spatial distribution of water depths across the floodplain during periods of inundation. Sediment with the greatest proportion of <63 μm particles is found in low-lying areas and more particularly those areas or depressions likely to retain ponded water during the recession of the flood.

For a clearer assessment of the influence of both relative floodwater depth and distance from the nearest channel bank on the grain size composition of surface sediment at this location, Figure 12.14 plots the percentage <63 μm fraction versus distance from the channel with the sample points classified according to water depth. A function similar to that used for the River Stour (i.e. Equation 12.11) was fitted to the data to describe the trends involved, i.e:

$$P = -12.8 \, e^{-Y/15} + 14.4Z + 75.4 \qquad (12.14)$$

with $r^2 = 0.47$. The results are also presented in Figure 12.14. The pattern exhibited by the grain size composition of deposited sediment on the Culm floodplain is more clearly related to the local microtopography than that of the Stour. Conversely, unlike the River Stour (cf. Figure 12.9A), the results from the River Culm show little evidence of a relationship between the grain size composition of deposited sediment (P) and the

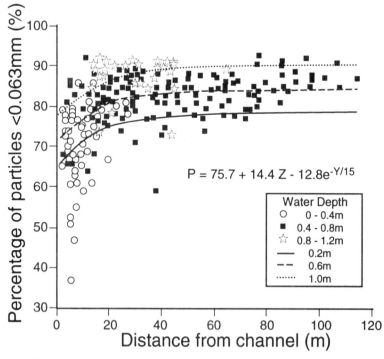

Figure 12.14 The relationship between the grain size composition of surface sediment and distance from the channel and water depth established for the floodplain study site at Silverton Mill

sedimentation rate (R). In this case, the coefficients of determination for the relationships between P and R were 0.17 and 0.08 for the estimates of R based on Cs-137 and Pb-210 measurements respectively.

12.5   THE GRAIN SIZE COMPOSITION OF OVERBANK FLOODPLAIN DEPOSITS

In discussing the results of the field investigations undertaken on the floodplains of the Dorset Stour and the River Culm, information relating to spatial variation in the particle size composition of the surface sediment was presented (cf. Figures 12.8, 12.9, 12.13 and 12.14). Distance from the channel and water depth were shown to exert a significant influence on the grain size composition of overbank deposits. In both study areas, the deposits were dominated by fine-grained (<63 µm) sediment. Existing concepts of sedimentation in shallow waterbodies such as floodplains are founded on the close relationship between fall velocity and grain size, and suggest that overbank accretion should be associated with preferential deposition of coarser particles. Comparison of the grain size distribution of suspended sediment with that of the overbank deposits should therefore evidence preferential loss of the coarser fraction of the suspended sediment load. This preferential loss of coarser fraction should in turn be reflected by a progressive downstream fining of the suspended sediment load.

In Figure 12.15A, representative grain size distributions for a number of surface samples of sediment collected from the study reach of the floodplain of the River Culm at Silverton Mill are compared with equivalent size distributions for the suspended sediment load of the river. In this case, the floodplain deposits are clearly coarser than the suspended sediment load, but there is no clear evidence of preferential loss of the coarser fraction. It is apparent that the floodplain deposits contain a substantial proportion of clay (<2 µm) and silt sized (2–63 µm) particles. Settling of fine sediment from areas of ponded water could account for some of the fine fraction associated with the floodplain deposits, but in view of the limited extent of such areas, it cannot account for the large amounts of fine sediment involved. Recent work by a number of workers (e.g. Droppo and Ongley, 1989, 1994; Walling and Moorehead, 1989; Walling and Woodward, 1993; Phillips and Walling, 1995) has emphasized that a substantial proportion of the suspended sediment load transported by rivers comprises composite particles (aggregates or flocs), which are considerably larger than their constituent primary particles. Figure 12.15B compares a typical *in situ* or *effective* grain size distribution of suspended sediment transported by the River Culm determined shortly after sample collection using laser diffraction equipment (Malvern Mastersizer), with the *ultimate* grain size distribution (chemically dispersed mineral sediment) of the same sediment, measured using the same equipment. Figure 12.15B clearly demonstrates that a considerable proportion of the fine sediment transported by the River Culm is incorporated into large composite particles, and deposition of these composite particles would account for the substantial amounts of clay and fine silt associated with the floodplain deposits.

Figure 12.16A presents the results of a series of measurements of the *in situ* or *effective* grain size distribution of suspended sediment transported by the River Culm at

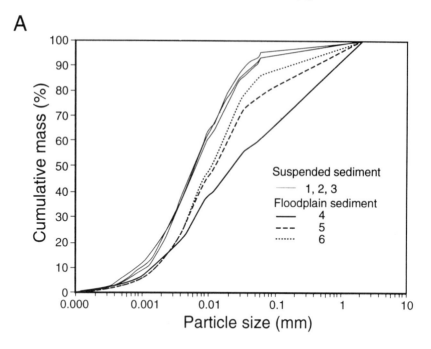

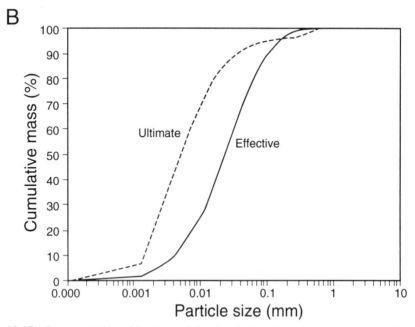

Figure 12.15   Representative ultimate particle size distributions for suspended sediment and surface floodplain sediment for the lower reaches of the River Culm (A) and a comparison of typical ultimate and effective particle size distributions for suspended sediment from the same river (B)

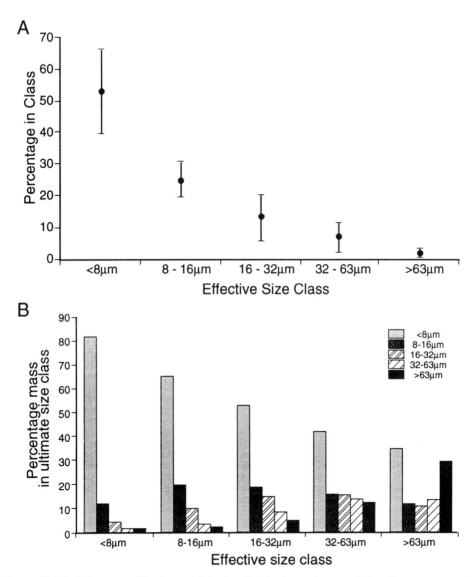

Figure 12.16 The mean effective particle size distributions of suspended sediment transported by the River Culm at Rewe based on 10 elutriation measurements (A) (confidence limits indicate one standard deviation about the means) and the mean ultimate particle size distributions for the sediment contained within each of the five effective size classes (B)

the gauging station at Rewe which is located about 5 km downstream of the study reach. These measurements were made using a field-portable water elutriation system which is capable of fractionating the suspended sediment into a number of effective size classes on the basis of its fall velocity (cf. Walling and Woodward, 1993). Each measurement involved withdrawing water from the river over a period of several hours and fractionating the suspended sediment particles into five effective size classes. The sediment in each effective size fraction was recovered and the *ultimate* grain size composition of these fractions was determined using laser diffraction apparatus (Malvern Mastersizer) after appropriate pretreatment. The results presented in Figure 12.16B again emphasize that each of the effective size fractions contains a substantial proportion of fine sediment irrespective of its effective size class. The data presented in Figure 12.16B demonstrate that if, as might be expected, floodplain sedimentation involves the preferential deposition of the coarser (e.g. 32–63 μm and >63 μm) effective size fractions, the resultant deposits would contain ca. 50% of sediment with an ultimate grain size <16 μm. This value is in close agreement with the absolute size distributions for floodplain surface sediment presented in Figure 12.15A, where the <16 μm fraction accounts for between 46 and 60% of the total sediment mass.

The results presented above indicate that the presence of composite particles within the suspended sediment load exerts an important influence on the grain size composition of deposited sediment and that information on the *in situ* or *effective* grain size distribution of suspended sediment is an important requirement for understanding floodplain sedimentation. It is clear that any attempt to account more fully for the spatial variations in the grain size composition of deposited sediment documented within the two study reaches, in terms of hydrodynamic controls, must also consider the *effective* size distribution of the transported sediment. Equally, floodplain sedimentation models must make use of *effective* rather than *ultimate* grain size data if they are to provide a meaningful representation of sedimentation processes.

12.6   TOWARDS A REALISTIC MODEL OF OVERBANK SEDIMENTATION

12.6.1   The problem

The field investigations of overbank sedimentation reported in this contribution serve to emphasize that any attempt to develop a realistic model of floodplain deposition must incorporate a meaningful representation of the feedbacks occurring between floodplain topography, floodwater hydraulics and suspended sediment dispersion and deposition and take account of the effective size characteristics of the suspended sediment load. Existing models have adopted one of two basic approaches. James (1985) and Pizzuto (1987) examined overbank processes in cross-sections characterized by simplified floodplain topographies. This enabled them to represent overbank flows as steady uniform flow, thus considerably simplifying the hydraulic components of their models. The schemes produced by both workers predict an exponential decrease in deposition amounts with increasing distance from the main channel, with the rate of decrease being positively related to the grain size of the transported sediment. Howard (1992) has considered more complex and hence realistic floodplain configurations and addressed overbank sedimentation in two lateral dimensions. He employed relative floodplain elevation as a surrogate for the key hydraulic variables and determined mean deposition for a sequence of overbank flows using a simple exponential function.

In order to simulate overbank processes during individual flood events, numerical schemes are required which are capable of predicting both hydraulic and sedimentological variables over complex topographic surfaces. Finite difference and finite element schemes have recently been successfully employed to generate predictions of depth-averaged hydraulic variables in two lateral dimensions (Wijbenga, 1985; Gee *et al.*, 1990). However, despite the considerable potential offered by such approaches, a number of problems remain to be overcome, largely associated with the representation of complex floodplain topography (Bates *et al.*, 1992). In addition, hydraulic schemes such as these have yet to be coupled with appropriate sediment transport relations. An alternative strategy for representing overbank flow patterns has involved use of the diffusion wave form of the one-dimensional Saint-Venant momentum equation (Cunge *et al.*, 1980). The relative simplicity of the one-dimensional Saint-Venant momentum equation, compared to the equivalent Navier–Stokes equation, reduces the difficulty in obtaining a stable solution to the latter. The latter approach would therefore appear to offer considerable potential where floodplain geometry is complicated.

12.6.2 A numerical model of overbank processes

A numerical model has been developed which predicts floodwater inundation extent, flow depths and velocities, and patterns of suspended sediment dispersion and deposition, over a two-dimensional finite difference grid capable of representing a complex, realistic floodplain surface (cf. Nicholas, 1994; Nicholas and Walling, 1996a). In order to reduce the difficulty of obtaining stable solutions to the equations of fluid motion employed, and to limit the run-time of the hydraulic component of the model, a simplified hydraulic approach has been adopted. Rather than solving the equations of fluid motion for the given upstream boundary conditions for each time-step of each flood event, these equations are solved initially for a number of upstream water levels, in order to generate a set of flow surfaces. These are then used, in conjuction with a relationship between the water levels at the upstream and downstream boundaries of the finite difference grid, to predict hydraulic patterns during all subsequent simulations. The approach is applicable to short floodplain reaches and its implementation relies upon the validity of a number of assumptions. It is assumed that flood hydrographs can be divided into a number of discrete time intervals, during each of which the discharge crossing the upstream boundary of the solution grid is equal to that crossing the downstream boundary, and that unique relationships exist between stage and discharge at these boundaries. It is also assumed that the pattern of overbank flow within the solution grid, resulting from a given upstream water level, is similar for separate flood events and for conditions of both rising and falling stage. The validity of these assumptions depends largely upon the properties of the reach within which hydraulic predictions are being made.

The relationship between the water levels at the upstream and downstream boundaries of the finite difference grid is established by determining the discharge entering the reach, for a given upstream stage, and calculating the downstream stage required to yield the same discharge leaving the reach. Discharge components normal to the boundary cross-sections are determined at each node in the solution grid from the following equations:

$$h_i = w_b + Sy(y_b - y_i) - z_i \qquad (12.15)$$

$$q_i = \frac{h_i^{5/3} Sx^{1/2}}{n} \Delta y \qquad (12.16)$$

where h_i is the depth of flow at node i, z_i is the bed elevation, n is the Manning roughness coefficient, w_b is the water level at the boundary reference node (i.e. the node at which the boundary water level is specified), y_b and y_i are lateral distances, q_i is the discharge component normal to the boundary of the finite difference grid, Δy is the finite difference space step, and Sx and Sy are the components of the downstream friction slope in the x and y directions of the finite difference grid (the friction slope is assumed to be equal to the water surface slope). The total discharge crossing the reach boundary in the x direction of the finite difference grid is given by the sum of the discharge components at each node across the relevant section. In Equation 12.15 it is assumed that floodplain flow moves in a direction parallel to the longitudinal axis of the valley floor and that the elevation of the water surface is constant along lines running perpendicular to this axis. This enables lateral variations in water levels and hence flow depths to be determined where the y axis of the finite difference grid is not parallel to the cross-stream direction of the valley. As with certain other methods for estimating the relationship between stage and discharge (cf. Ervine and Ellis, 1987), Equations 12.15 and 12.16 take no account of the impact of momentum transfer between adjacent elements of differing depth and velocity, hence they are likely to overestimate the total discharge crossing the boundaries of the model grid (Wormleaton et al., 1982). In addition, the total discharge for a given reference boundary water level is strongly dependent upon the values employed for the roughness coefficients and friction slopes. In this study, friction slopes were approximated using surveyed floodplain and channel bed slopes at the two boundary cross-sections. Although this method will result in discrepancies between predicted discharges and discharges experienced in the field, it is reasonable to assume that the errors involved will be similar at both upstream and downstream reach boundaries. Furthermore, as it is only the boundary water levels that are carried forward to the next stage of the model, the errors in estimated discharges associated with the technique outlined above have no further effect upon the hydraulic calculations.

The next step in the modelling procedure is to generate the flow surfaces needed to make hydraulic predictions for any given upstream stage. These flow surfaces are termed Water Surface Functions because they describe the water surface elevation at each node in the finite difference grid as a function of the specified upstream water level and the corresponding downstream water level determined using the discharge calculation procedure. Water Surface Functions are generated for a number of upstream boundary water levels at 5 cm increments by solving the following continuity of mass and momentum equations over the finite difference grid.

Continuity of momentum:

$$u = \frac{h^{2/3} Sx^{1/2}}{n} \qquad \text{where} \qquad Sx = \frac{\partial h}{\partial x} + \frac{\partial z}{\partial x} \qquad (12.17)$$

$$v = \frac{h^{2/3} Sy^{1/2}}{n} \qquad \text{where} \qquad Sy = \frac{\partial h}{\partial y} + \frac{\partial z}{\partial y} \qquad (12.18)$$

Continuity of mass:

$$\frac{\partial(uh)}{\partial x} + \frac{\partial(vh)}{\partial y} = 0 \qquad (12.19)$$

where h is the depth of flow, z is the bed elevation, n is the Manning roughness coefficient and u, v, Sx and Sy are respectively the velocity components and friction slopes in the x and y directions. Equations 12.17 and 12.18 are a pair of one-dimensional relationships equivalent to the diffusion wave form of the Saint-Venant momentum equation. The derivation of Equations 12.17 to 12.18 neglects both temporal derivatives and convective acceleration terms, so that friction slopes are approximated by water surface slopes.

The differential Equations 12.17 to 12.18 are transformed to a set of difference approximations which are then solved, using Newton Raphson iteration, to yield the depth of flow at each node in the finite difference grid for the given upstream and downstream boundary water levels. When a stable approximate solution is obtained, the value of the Water Surface Function (f_i) at each wet node in the finite difference grid is calculated as:

$$f_i = \frac{w_u - h_i - z_i}{w_u - w_d} \qquad (12.20)$$

where w_u and w_d are the specified upstream water level and the corresponding downstream water level determined using the discharge calculation procedure. Water Surface Function values are stored as a series of arrays which characterize the flow fields for the given upstream boundary water levels. These can then be used to predict flow depths and velocities throughout the finite difference grid for any given upstream water level.

Each Water Surface Function is used to predict hydraulic patterns over a specified interval of upstream water levels, ranging from that for which it was generated down to that for which the next available flow surface was generated. Equation 12.20 can be rearranged to allow the flow depth at each node within the solution grid to be determined for a given upstream boundary water level. The patterns of flow depth generated in this way are then adjusted in certain places using a simple ponding algorithm, details of which are given in Nicholas (1994). This is necessary because certain floodplain regions containing flowing water at the upstream boundary water level for which the Water Surface Function in use was generated, may be dry or contain ponded water at the slightly lower (i.e. 0–5 cm) upstream water level for which the hydraulic calculations are being made. Two types of ponding are identified by the model. First, backwater ponding where a region is supplied with water from a point downstream. Second, recession ponding where floodplain depressions retain ponded water as the floodplain is drained on the falling limb of the flood hydrograph. Having determined the depth of flow at each node within the finite difference grid for the given upstream boundary water level, velocity components are calculated in the x and y directions throughout the solution grid using the finite difference approximations of Equations 12.17 and 12.18.

Hydraulic predictions are carried forward to the sediment transport component of the model which predicts patterns of suspended sediment dispersion and deposition for a range of sediment size fractions. The depth-averaged mass balance equation for suspended sediment transport by convective and diffusive processes in two horizontal dimensions is solved for each of these size fractions, to allow the sediment concentration and amount of deposition for each fraction to be determined at each floodplain node within the solution grid:

$$(uh)\frac{\partial c}{\partial x} - \frac{\partial}{\partial x}\left(\varepsilon h \frac{\partial c}{\partial x}\right) + (vh)\frac{\partial c}{\partial y} - \frac{\partial}{\partial y}\left(\varepsilon h \frac{\partial c}{\partial y}\right) + DR = 0 \qquad (12.21)$$

where c is the depth-averaged suspended sediment concentration, ε is a horizontal mixing coefficient and DR is the net rate of deposition (i.e. deposition less erosion and/ or resuspension). Horizontal mixing coefficients are determined using standard relationships (cf. Pizzuto, 1987):

$$\varepsilon = \lambda h U_* \qquad (12.22)$$

where λ is a constant which is assigned a value of 0.13 after Fischer et al. (1979) and U_* is the shear velocity given by:

$$U_* = (ghS_f)^{1/2} \qquad (12.33)$$

where g is the acceleration due to gravity and S_f is the water friction slope. Deposition rates are predicted from:

$$DR = kcVs \qquad (12.24)$$

where k is an empirical coefficient which is used to calibrate the model with the aid of measured deposition amounts and Vs is the sediment particle fall velocity.

At the boundary between wet and dry nodes, sediment fluxes are set equal to zero to ensure continuity of mass in these regions. Equation 12.21 is written in finite difference form and solved using Newton Raphson iteration. When a stable approximate solution to Equation 12.21 has been obtained, the total amount of deposition at each floodplain node during the given time-step is determined from:

$$D_i^t = k\Delta t \sum_{j=1}^{J} (Vs_j c_{i,j}^t) \qquad (12.25)$$

where D_i^t is the amount of deposition at node i during time-step t, Δt is the duration of the time-step, Vs_j is the fall velocity of the jth size fraction, $c_{i,j}^t$ is the sediment concentration of the jth size fraction at node i during time-step t and J is the number of size fractions employed in the model. For most floodplain nodes this represents the total amount of deposition that they receive. However, for nodes in areas of the floodplain that are susceptible to recession ponding, there is the additional possibility of this deposition being supplemented by material which settles out in trapped stationary water. The extra sediment contributed in this way is given by:

$$Da_i^t = hp_i \sum_{j=1}^{J} c_{i,j}^{t-1} \qquad (12.26)$$

where Da_i^t is the additional deposition at node i resulting from the settling of trapped sediment, hp_i is the depth of ponded water and $c_{i,j}^{t-1}$ is the sediment concentration of the jth size fraction at node i during the time step prior to the isolation of the ponded area from the main flow. Equation 12.26 assumes that 100% of the sediment trapped in the ponded water is deposited. The validity of this assumption is dependent upon the time interval between flood events and the significance of flocculation processes within the ponded water.

12.6.3   Model verification

This model has been calibrated and tested using data collected from a 600 m reach of the River Culm, Devon. The study reach was located downstream of the river

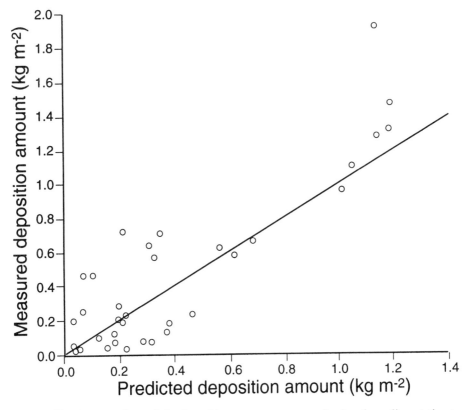

Figure 12.17   A comparison of the deposition amounts measured using the sedimentation traps with those predicted for the same points using the floodplain sedimentation model

monitoring station at Rewe, which provided a continuous record of upstream boundary water level and suspended sediment concentration. These variables were also monitored at a temporary downstream installation adjacent to the downstream boundary of the study reach. Particle fall velocity distributions representative of the effective size distribution of suspended sediment recorded for use in Equation 12.24 were determined using a custom-built water elutriation system (cf. Walling and Woodward, 1993; Nicholas and Walling, 1996b). The study reach was surveyed in detail and a finite difference grid with a nodal spacing of 5 m was constructed. Astroturf sedimentation traps (cf. Lambert and Walling, 1987) were located upon the floodplain to measure amounts of overbank deposition. In total 52 sets of traps (each set consisting of nine traps) were successfully deployed and retrieved over a 14-month period. Measured deposition amounts were employed to calibrate the model parameter k in Equation 12.24. This calibration procedure determined the mean amount of deposition predicted by the model for the 52 measured data points. However, calibration had a negligible effect upon predicted relative spatial patterns of deposition.

Figure 12.17 shows a scatter plot of predicted versus measured deposition amounts. The line in this figure represents the line of perfect agreement between the two. Of the 52 trap sets, the model overpredicts deposition amounts in 26 cases and underpredicts

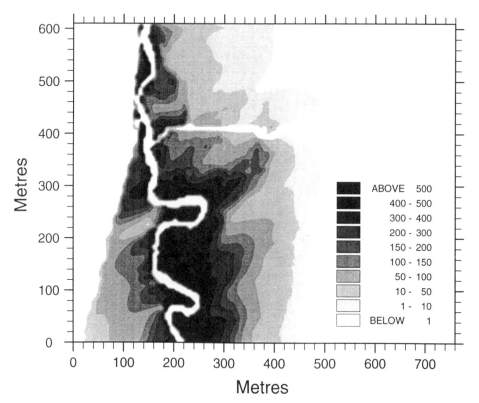

Figure 12.18 The pattern of deposition (g m^{-2}) predicted by the floodplain deposition model for the flood event of 30 November to 5 December 1992

them in 24 cases. The mean absolute error for the 52 data points is 130 g m^{-2} while the mean percentage error is 61%. Although the latter figure may appear high, it should be noted that a large proportion of this error is associated with a small number of data points.

Figure 12.18 shows the predicted pattern of overbank deposition amounts for the flood event of 30 November to 5 December 1992. Predicted amounts of deposition are highest in the low-lying areas close to the main channel where suspended sediment concentrations and durations of inundation are greatest. The extensive areas where deposition amounts exceed 300 g m^{-2} in the downstream portion of the reach result from increased channel sinuosity which stimulates the convective transportation of suspended sediment to areas within the channel belt.

Figure 12.19 shows the predicted *effective*, predicted *ultimate* and measured *ultimate* grain size distributions for three typical flood deposits. These results clearly illustrate that there is a reasonable level of agreement between the measured and predicted

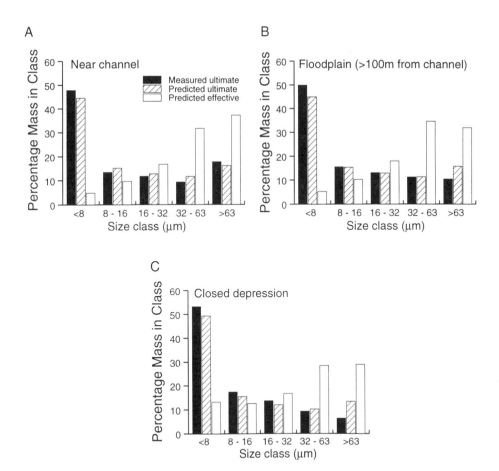

Figure 12.19 A comparison of predicted *effective*, predicted *ultimate* and measured *ultimate* particle size distributions for three representative flood deposits

ultimate grain size distributions, and that there is a sharp contrast between the *ultimate* and *effective* grain size distributions. The former tend to be composed of approximately 50% material in the <8 μm class, with the remaining 50% spread fairly evenly over the other four size classes. In contrast, the *effective* size distributions comprise roughly 65% material >32 μm in diameter, with decreasing proportions in the smaller size fractions. The expected fining of flood deposits with distance from the main channel is also evident in Figure 12.19, although the data presented here indicate the importance of recession ponding in depressions, which results in the deposition of large amounts of fine (<8 μm) material.

12.7   THE PROSPECT

It is suggested that future advances in our understanding of the role and behaviour of river floodplains as suspended sediment sinks will depend upon the integration of recent developments in both mathematical modelling and the field monitoring of overbank sedimentation processes. The simplification of reality provided by laboratory channel facilities undoubtedly offers considerable potential for exploring the process interactions involved, but real-world floodplains are infinitely more complex in terms of their microtopography and hydraulic conditions. Detailed field data are required to assist in the development and verification of two-dimensional distributed models capable of representing this complexity. The field evidence presented in this contribution has, for example, underscored the need to take account of the potential contrast between the ultimate and effective grain size distribution of suspended sediment, a consideration which has been almost totally ignored by existing modelling developments. Equally, the complex microtopography of natural floodplains will exert an important influence on the local hydraulic conditions which in turn control the detailed pattern of sedimentation rates. While event-based measurements will undoubtedly be valuable in promoting the integration of field monitoring and modelling initiation, the medium and longer-term perspectives afforded by fallout radionuclide measurements clearly offer important advantages for any attempt to progress towards realistic models of the evolution of lowland river floodplains. Complex feedbacks will exist between floodplain microtopography and sedimentation rates, and the high spatial resolution distributed data provided by fallout radionuclide measurements can provide a basis for exploring such feedbacks. Collaborative work is currently in progress between research groups at the Universities of Bristol and Exeter aimed at coupling the information provided by the numerical modelling of overbank flow and of overbank sediment deposition on several British lowland river floodplains with the observed field evidence generated from Cs-137 and unsupported Pb-210 measurements.

In terms of the longer-term geomorphological evolution of lowland river floodplains, there is an increasing need to address the apparent dilemma raised by available evidence for contemporary and recent floodplain sedimentation rates. Extrapolated backwards over centuries, such rates are frequently inconsistent with the existing alluvial evidence and it is important to consider the potential magnitude and extent of changes in suspended sediment loads and concentration, in the incidence of floods and floodplain inundation, and in the condition of the floodplain surface. There is, for example, some evidence that the gradual breakdown of the system of leats and distributary channels on the floodplain of the River Culm since the nineteenth century has resulted in reduced channel capacity and therefore increased overbank inundation. Recent rates of overbank sedimentation may thus be unrepresentative of the longer-term. Similarly, forward extrapolation of contemporary sedimentation rates may appear to result in unrealistic increases in channel depth, but it is important to consider whether increases in the elevation of the floodplain surface are coupled with equivalent increases in the elevation of the channel bed or whether a negative feedback will develop. If the latter prevails, an increase in the elevation of the floodplain surface will result in an increased channel capacity and therefore a reduced magnitude and frequency of overbank flows and floodplain inundation and reduced rates of overbank deposition.

REFERENCES

Allan, R. J. (1986) The role of particulate matter in the fate of contaminants in aquatic ecosystems. *Inland Waters Directorate*, Environment Canada, Scientific Series No. 142.

Asselmann, N.E.M. and Middelkoop, H. (1993) Floodplain sedimentation. Quantities, patterns and processes. *Report GEOPRO 1993.03*, Institute of Geographical Research, Rijkuniversiteit Utrecht.

Bates, P.D., Anderson, M.G., Baird, L., Walling, D.E. and Simm, D. (1992) On the potential for using two-dimensional finite element schemes in geomorphological investigations of floodplain environments. *Earth Surface Processes and Landforms*, **17**, 575–588.

Brackenridge, G.R. (1988) River flood regime and floodplain stratigraphy. In: *Flood Geomorphology*, V.R. Baker, R.C. Kochel and P.C. Patton (eds), 139–156. Wiley-Interscience, New York.

Bradley, S.B and Cox, J.J. (1987) Heavy metals in the Hamps and Manifold valleys, North Straffordshire, UK: partitioning of metals in floodplain soils. *The Science of the Total Environment*, **75**, 135–153.

Brown, A.G. (1983) An analysis of overbank deposits of a flood at Blandford-Forum, Dorset, England. *Revue de Geomorphologie Dynamique*, **32**, 95–99.

Costa, J.E. (1975) Effects of agriculture on erosion and sedimentation in the Piedmont Province, Maryland. *Geological Society of America Bulletin*, **86**, 1281–1286.

Cunge, J.A., Holly, F.M. and Verwey, A. (1980) *Practical Aspects of Computational River Hydraulics*. Pitman, London.

Droppo, I.G. and Ongley, E.D. (1989) Flocculation of suspended solids in Southern Ontario rivers. In: *Sediment and the Enrironment* (Proceedings of the Baltimore Symposium) IAHS Publ. no. **184**, 95–105.

Droppo, I.G. and Ongley, E.D. (1994) Flocculation of suspended sediment in rivers of Southeastern Canada. *Water Research*, **28** 1799–1809.

Ervine, D.A. and Ellis, J. (1987) Experimental and computational aspects of overbank floodplain flow. *Transactions of the Royal Society of Edinburgh: Earth Sciences*, **78**, 315–325.

Fischer, H.B., List, E.J., Koh, R.C.Y., Imberger, J. and Brooks, N.H. (1979) *Mixing in Inland and Coastal Waters*. Academic Press, London.

Frissel, M.J. and Pennders, R. (1983) Models for the accumulation and migration of ^{90}Sr, ^{137}Cs, ^{239,240}Pu and ^{241}Am in the upper layers of soils. In: *Ecological Aspects of Radionuclide Release*, P.J. Coughtrey (ed.), 63–72. Blackwell Scientific.

Gee, D.M., Anderson, M.G. and Baird, L. (1990) Large-scale floodplain modelling. *Earth Surface Processes and Landforms*, **15**, 513–523.

Gretener, B. and Strömquist, L. (1987) Overbank sedimentation rates of fine grained sediments: A study of the recent deposition in the Lower River Fyrisan. *Geografiska Annaler*, **69A**, 139–146.

Happ, S.C. (1968) Valley sedimentation in north-central Mississippi. *Proceedings of the Mississippi Water Resources Conference*, 1–8.

He, Q. and Walling, D.E. (1996) Use of fallout Pb-210 measurements to investigate longer-term rates and patterns of overbank sediment deposition on the floodplains of lowland rivers. *Earth Surface Processes and Landforms*, **21**, 141–154.

Howard, A.D. (1992) Modelling channel migration and floodplain sedimentation in meandering streams. In: *Lowland Floodplain Rivers: Geomorphological Perspectives*, P.A. Carling and G.E. Petts (eds), 1–41. John Wiley & Sons, Chichester.

Hupp, C.R. (1988) Plant ecological aspects of flood geomorphology and palaeoflood history. In: *Flood Geomorphology*, V.R. Baker, R.C. Kochel and P.C. Patton (eds), 330–356. Wiley-Interscience, New York.

James, C.S. (1985) Sediment transfer to overbank sections. *J. Hydraulic Res.*, **23**, 435–452.

Joshi, S.R. (1987) Nondestructive determination of lead-210 and radium-226 in sediments by direct photon analysis. *J. Radioanal. Nucl. Chem., Articles*, **116**, 169–182.

Kesel, R.H., Dunne, K.C., McDonald, R.C., Allison, K.R. and Spicer, B.E. (1974) Lateral erosion and overbank deposition on the Mississippi River in Louisiana caused by 1973 flooding. *Geology*, **2**, 461–464.

Kesel, R.H., Yodis, E.G. and McCrow, D.J. (1992) An approximation of the sediment budget of the lower Mississippi River prior to human modification. *Earth Surface Processes and Landforms*, **17**, 711–722.

Knox, J.C. (1987) Historical valley floor sedimentation in the Upper Mississippi valley. *Annals Association American Geographers*, **77**, 224–244.

Krishnappan, B.G. and Ongley, E.D. (1989) River sediments and contaminant transport—changing needs in research. *Proceedings of the Fourth International Symposium on River Sedimentation*, IRTCES, Beijing, 530–538.

Lambert, C.P. and Walling, D.E. (1987) Floodplain sedimentation: A preliminary investigation of contemporary deposition within the lower reaches of the River Culm, Devon, UK. *Geografiska Annaler*, **69A**, 47–59.

Leenaers, H. and Rang, M.C. (1989) Metal dispersal in the fluvial system of the River Geul: the role of discharge, distance to the source, and floodplain geometry. In: *Sediment and the Environment* (Proceedings of the Baltimore Symposium) IAHS Publ. no. **184**, 47–55.

Leenaers, H. and Schouten, C.J. (1989) Soil erosion and floodplain soil pollution: related problems in the geographical context of a river basin. In: *Sediment and the Environment* (Proceedings of the Baltimore Symposium) IAHS Publ. no. **184**, 75–83.

Leopold, L.B. (1973) River channel change with time: an example. *Geological Society of America Bulletin*, **84**, 1845–1860.

Lewin, J. (1978) Floodplain geomorphology. *Progress in Physical Geography*, **2**, 408–437.

Lewin, J. and Macklin, M.G. (1987) Metal mining and floodplain sedimentation in Britain. In: *International Geomorphology, Part I*, V. Gardiner (ed.), 1009–1027. Wiley.

Livens, F.R and Rimmer, D.L. (1988) Physico-chemical controls on artificial radionuclides in soils. *Soil Use and Management*, **4**, 63–69.

Macklin, M.G. and Dowsett, R.B. (1989) The chemical and physical speciation of trace metals in fine-grained overbank flood sediments in the Tyne Basin, north-east England. *Catena*, **16**, 135–151.

Macklin, M.G., Rumsby, B.T. and Newson, M.D. (1992) Historical overbank floods and vertical accretion of fine-grained alluvium in the lower Tyne valley, north east England. In: *Dynamics of Gravel-bed Rivers*, P. Billi, R. Hey, P. Tacconi and C. Thorne (eds), 564–580. Wiley, Chichester.

Mansikkaniemi, H. (1985) Sedimentation and water quality in the flood basin of the River Kyronjoki in Finland. *Fennia*, **163**, 155–194.

Marriott, S. (1992) Textural analysis and modelling of a flood deposit: River Severn, UK. *Earth Surface Processes and Landforms*, **17**, 687–697.

Marron, D.C. (1987) Floodplain storage of metal-contaminated sediments downstream of a gold mine at Lead, South Dakota In: *Chemical Quality of Water and the Hydrological Cycle*, R.C. Averett and D.M. McKnight (eds), 193–209. Lewin Publishers, Chelsea, Michigan.

Marron, D.C. (1989) The transport of mine tailings as suspended sediment in the Belle Fourche River, west-central Dakota, USA. In: *Sediment and the Environment* (Proceedings of the Baltimore Symposium) IAHS Publ. no. **184**, 19–26.

McHenry, J.R., Ritchie, J.C. and Verdon, J. (1976) Sedimentation rates in the Upper Mississippi River. In: *Rivers '76*, Vol. II, Am. Soc. Civ. Eng., 1339–1349.

McKee, E.D., Crosby, E.J. and Berryhill, H.L. (1967) Flood deposits, Bijou Creek, Colorado, June 1965. *Journal of Sedimentary Petrology*, **37**, 829–851.

Meade, R.H. (1994) Suspended sediments of the modern Amazon and Orinoco rivers. *Quaternary International*, **21**, 29–39.

Mertes, L.A.K. (1990) Hydrology, hydraulics, sediment transport and geomorphology of the Central Amazon floodplain. PhD dissertation, University of Washington, Seattle.

Middelkoop, H. and Asselmann, N.E.M. (1994) Spatial and temporal variability of floodplain sedimentation in the Netherlands. Report GEOPRO 1994.05, Faculty of Geographical Sciences, Rijkuniversiteit Utrecht.

Nanson, G.C. and Croke, J.C. (1992) A genetic classification of floodplains. *Geomorphology*, **4**, 459–486.

Nicholas, A.P. (1994) Modelling overbank deposition on floodplains: A case study of the River Culm, Devon. Unpublished PhD thesis, University of Exeter.

Nicholas, A.P. and Walling, D.E. (1996a) Modelling flood hydraulics and overbank deposition on river floodplains. *Earth Surface Processes and Landforms* (in press).

Nicholas, A.P. and Walling, D.E. (1996b) The significance of particle aggregation in the overbank deposition of suspended sediment on river floodplains. *Journal of Hydrology* (in press).

Phillips, J.M. and Walling, D.E. (1995) Measurement in-situ of the effective particle size characteristics of fluvial suspended sediment using a field-portable laser backscatter probe: some preliminary results. *J. Marine and Fresh Water Research*, **46**, 349–357.

Pizzuto, J.E. (1987) Sediment diffusion during overbank flows. *Sedimentology*, **34**, 301–317.

Popp, C.L., Hawley, J.W., Love, D.W. and Dehn, M. (1988) Use of radiometric (Cs-137, Pb-210), geomorphic and stratigraphic techniques to date recent oxbow sediments in the Rio Puerco drainage Grants uranium region, New Mexico. *Environmental Geology and Water Science*, **11**, 253–269.

Ritchie, J.C., Hawks, P.H. and McHenry, J.R. (1975) Deposition rates in valleys determined using fallout cesium-137. *Geol. Soc. Am. Bull.*, **86**, 1128–1130.

Ritter, D.F., Kinsey, W.F. and Kauffman, M.E. (1973) Overbank sedimentation in the Delaware river valley during the last 6000 years. *Science*, **179**, 374–375.

Simm, D.J. (1993) The deposition and storage of suspended sediment in contemporary floodplain systems: a case study of the River Culm Devon. Unpublished PhD thesis, University of Exeter.

Trimble, S.W. (1976) Sedimentation in Coon Creek Valley, Wisconsin. *Proc. Third Federal Interagency Sedimentation Conference*, 5.100–5.122.

Trimble, S.W. (1983) A sediment budget for Coon Creak basin in the Driftless Area, Wisconsin, 1853–1977. *American Journal of Science*, **283**, 454–474.

Walling, D.E. (1989) Physical and chemical properties of sediment, the quality dimension. *International Journal of Sediment Research*, **4**, 27–39.

Walling, D.E. and Bradley, S.B. (1989) Rates and patterns of contemporary floodplain sedimentation: a case study of the River Culm, Devon, UK. *GeoJournal*, **19**, 53–62.

Walling, D.E. and He, Q. (1992) Interpretation of caesium-137 profiles in lacustrine and other sediments: the role of catchment-derived inputs. *Hydrobiologia*, **235/236**, 219–230.

Walling, D.E. and He, Q. (1993) Use of caesium-137 as a tracer in the study of rates and patterns of floodplain sedimentation. In: *Tracers in Hydrology* (Proceedings of the Yokohama Symposium) IAHS Publ. No. **215**, 319–328.

Walling, D.E. and He, Q. (1994) Rates of overbank sedimentation on the flood plains of several British rivers during the past 100 years. In: *Variability in Stream Erosion and Sediment Transport* (Proceedings of the Canberra Symposium) IAHS Publ. No. **224**, 203–210.

Walling, D.E. and Moorehead, P.W. (1989) The particle size characteristics of fluvial suspended sediment: an overview. *Hydrobiologia*, **176/177**, 125–149.

Walling, D.E. and Quine, T.A. (1993) Using Chernobyl-derived fallout radionuclides to investigate the role of downstream conveyance losses in the suspended sediment budget of the River Severn, United Kingdom. *Physical Geography*, **14**, 239–253.

Walling, D.E. and Quine, T.A. (1995) The use of fallout radionuclide measurements in soil erosion investigations. In: *Proceedings International Symposium on Nuclear and Related Techniques in Soil/Plant Studies on Sustainable Agriculture and Environmental Preservation.* IAEA, Vienna, 597–619.

Walling, D.E. and Webb, B.W. (1989) The reliability of rating curve estimates of suspended sediment yield: some further comments. In: *Sediment Budgets.* IAHS Publ. No. **174**, Wallingford, UK, 337–350.

Walling, D.E. and Woodward, J.C. (1992) Use of radiometric fingerprints to derive information on suspended sediment sources. In: *Erosion and Sediment Transport Monitoring Programmes in River Basins,* J. Bogen, D.E. Walling and T. Day (eds), 153–164. IAHS Publ. No. **210**.

Walling, D.E. and Woodward, J.C. (1993) Use of a field-based water elutriation system for monitoring the *in situ* particle size characteristics of fluvial suspended sediment. *Water Research*, **27**, 1413–1421.

Walling, D.E., Quine, T.A. and He, Q. (1992) Investigating contemporary rates of floodplain sedimentation. In: *Lowland Floodplain Rivers: Geomorphological Perspectives,* P.A. Carling and G.E. Petts (eds), 166–184. John Wiley & Sons, Chichester.

Wijbenga, J.H.A. (1985) Steady depth-averaged flow calculations on curvilinear grids. In: *Second International Conference on the Hydraulics of Floods and Flood Control,* 373–387.

Wolman, M.G. and Leopold, L.B. (1957) River floodplains: some observations on their formation. *US Geological Survey Professional Paper* **282**-C, 109 pp.

Wormleaton, P.R., Allen, J. and Hadjipanos, P. (1982) Discharge assessment in compound channel flow. *Journal of the Hydraulics Division. Am. Soc. Civ. Eng.*, **108**, 975–994.

13 Fluxes and Storage of Sediment-Associated Heavy Metals in Floodplain Systems: Assessment and River Basin Management Issues at a Time of Rapid Environmental Change

M.G. MACKLIN
School of Geography, University of Leeds, UK

13.1 INTRODUCTION

Soil and sediment are the ultimate sinks for heavy metals (defined here as metallic elements of density equal to or greater than $6\,\text{g/cm}^3$ after Davies, 1980) in the terrestrial environment, and fluvial processes are the principal mechanism responsible for their transportation and redistribution on the land surface of the earth. Studies of the channel and floodplain sediment system, therefore, assume some considerable importance not only for quantifying fluxes and storage of particulate-bound metals but also for mitigating the worst effects of metal pollutants on plants, animals and humans who rely chiefly on the river environment as a source of water and food. Indeed, it is argued in this chapter that appropriate and effective environmental protection strategies in river basins affected by metal pollution can only be devised with a knowledge, and understanding, of short and long-term fluvial sediment transport and storage processes.

Since the mid-1970s, and throughout the 1980s, geomorphologists in continental Europe (e.g. Rang *et al.*, 1987), the USA (e.g. Graf, 1985; Knox, 1987) and especially in Britain (e.g. Davies and Lewin, 1974; Lewin *et al.*, 1977) have become increasingly involved in metal pollution studies of river systems. This interest arose partly from geomorphologists wishing to apply the findings of their more pure research to environmental management issues, but also with the realization that metal contaminants, uniquely in many respects, can be used in long-term and large-scale sedimentological "experiments" within the fluvial system.

The river metal pollution literature published up until 1985 was reviewed by Lewin and Macklin in a paper published in 1987. This paper, though pertaining largely to Britain, was a benchmark study in many respects as it provided for the first time a generic framework, based on fluvial processes and sedimentation styles, for

Floodplain Processes. Edited by Malcolm G. Anderson, Des E. Walling and Paul D. Bates.
© 1996 John Wiley & Sons Ltd.

investigating and interpreting the dispersal and storage of mining wastes in river basins. It also highlighted three themes which had been prominent in river metal pollution research up to 1985. The first comprised studies of the chemical (Grimshaw *et al.*, 1976; Bradley and Lewin 1982) and physical (Wolfenden and Lewin, 1978) factors that control metal concentration decline downstream from sources of contamination. The second involved the use of sediment-associated metals as tracers (Lewin and Wolfenden, 1978) and as stratigraphic markers (Macklin, 1985). The third represented the distinction between what Lewin and Macklin (1987) termed "passive dispersal" of mining wastes, where mining material is transported as part of the sediment load of the river in a manner that does not disrupt the channel and floodplain, as opposed to "active transformation", where the whole fluvial system is transformed through the input of wastes. The latter situation can arise either through a dramatic increase in sediment supply to the river channel (Gilbert, 1917), or as a result of phytotoxic metals disrupting riparian vegetation leading to reduced bank stability (Lewin *et al.*, 1983). Since 1987, metal-related river research emanating from geomorphological studies in Australia, Europe and North America has burgeoned. Much of this work has developed along the lines of the three broad research themes outlined by Lewin and Macklin (1987), with studies of mining-related metamorphosis of fluvial sedimentary systems (Macklin and Lewin, 1989; Knighton, 1991; Rowan *et al.*, 1995) and the use of metalliferous sediment for fingerprinting (Passmore and Macklin, 1994) and dating purposes (Macklin *et al.*, 1992) being particularly prominent. One major new development, however, has been the emergence of environmental change, resulting from both natural and anthropogenic causes, as one of the central research issues of geomorphology in the 1990s. This chapter focuses on this theme and the implications that rapid environmental change has for the very difficult task of rehabilitating metal-polluted floodplains. It is divided into three parts. In the first, sediment-associated metal dispersal processes in channel and overbank environments are examined with particular reference to the effects of large flood events. Second, the spatial and temporal patterns of metal contaminant storage in floodplains are considered together with their role as secondary sources of metal contamination. In the third and final section, remedial options available for rehabilitating contaminated floodplains are discussed in the context of river instability arising from environmental change, including global warming.

13.2   TRANSPORT AND DISPERSAL OF SEDIMENT-ASSOCIATED METALS IN CHANNEL AND OVERBANK ENVIRONMENTS

Transport rates and dispersal patterns of sediment-associated metals are controlled by four factors (Lewin and Macklin, 1987). These represent: firstly, hydraulic sorting according to differential particle density and size; secondly, chemical sorption–desorption processes related to the formation of Fe and Mn oxides, and organic complexes; thirdly, mixing processes, with additions of sediment from tributaries, or from bank erosion, resulting in an increases or decrease, of trunk river metal levels; and fourthly, floodplain deposition and storage. In many river systems metal levels in fluvial sediments tend to decline away from the point of emission and downstream reductions in metal concentrations have been successfully approximated using a simple decay model of a negative linear, exponential or power form (Wolfenden and Lewin, 1978; Lewin and Macklin, 1987; Marcus, 1987). Departures from modelled dispersion patterns have been used to indicate the relative importance of contaminated floodplain sediments as a source of heavy metals.

Prior to 1987, however, certainly in Britain, the transport and dispersal of sediment-bound heavy metals tended to be studied in relatively small streams, commonly with a single, well-defined point source. More recently, larger more complex river systems with multiple metal input sites have been investigated (e.g. Axtmann and Luoma, 1991; Macklin, 1992), including several studies evaluating metal dispersal during single flood events (e.g. Leenaers, 1989). These latter type of studies are particularly informative as they provide a "snapshot", and some measure of control, of the factors that determine both temporal and spatial patterns of sediment-associated metal dispersal. Investigations in the contrasting environments of the perennial River Tyne (UK) and River Geul (Netherlands–Belgium), northwest Europe, and the ephemeral Puerco River, New Mexico, USA, provide very useful illustrations of the approach.

In the River Tyne (Macklin and Dowsett, 1989) and River Geul (Leenaers, 1989) chemical and physical partitioning of metals was carried out on fine-grained sediments deposited on floodplains by overbank flooding. Results from the River Tyne contradicted the findings of several previous investigations of metal-polluted rivers in three significant respects. Firstly, they showed that only a small proportion (generally less than 10%) of particulate-bound metals were in the form of the original sulphide minerals (galena and sphalerite) and that the majority of heavy metals were present as metal-bearing Fe and Mn hydroxides, found either as coatings on sand-size material or as discrete grains. Secondly, although, on a per unit weight basis, the silt and clay fraction had the highest concentration of metals, it comprised only a tiny (0.05–1.0% by weight) component of the deposited overbank fines and is responsible for very little of the sediment-associated metal fluxes. Thirdly, while (as expected) metal concentrations decreased downstream away from the Northern Pennine mining area the proportion of metals in a potentially exchangeable, bioavailable form increased (Figure 13.1). Overall, in the upper and middle parts of the Tyne basin (closer to the former Pb and Zn mines) physical processes of downstream sediment metal content reduction predominate, while in the lower reaches of the Tyne, particulate-bound metal concentrations are primarily governed by chemical sorption–desorption processes associated with Fe and Mn oxides and organic material. In a more recent investigation of the River Tyne (Hudson-Edwards et al., 1996), however, mineralogical analyses have shown that although hydrodynamic dispersal processes (dilution of contaminated sediment by uncontaminated material, abrasion, hydraulic sorting, resuspension of contaminated sediment) are important, the downstream decrease in sediment-borne metal content also reflects an evolutionary sequence in metalliferous mineralogy. In this sequence, thermodynamically unstable metal-bearing primary minerals (galena, sphalerite) oxidize to form secondary minerals (primary Fe and Mn hydroxides), which contain proportionately lower contents of metal than the original source mineral.

Similar trends were also evident in the chemical partitioning of heavy metals in overbank flood sediments of the River Geul investigated by Leenaers (1989). Although total heavy metal concentrations rapidly decreased with distance from the source of

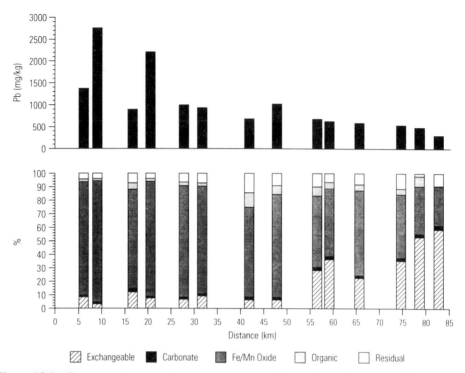

Figure 13.1 Concentrations and chemical speciation of Pb in the medium sand (500–250 μm) fraction of overbank sediment in the River South Tyne (0–52 km) and River Tyne (53–83 km), northern England

contamination, the chemical reactivity of these metals increased in relative terms (Figure 13.2). As a result, even at large distances from the source area, substantial amounts of potentially mobile metals were being supplied to, and stored in, the floodplain. Leenaers (1989) suggested that during transport, metals were chemically released from the residual fractions and subsequently entered the potentially mobile host fractions, although how this related in detail to variations in redox conditions and pH was not considered.

In the River Puerco (Graf, 1990) and River Tyne very detailed surveys of metal concentrations in fine-grained within-channel sediment, as opposed to overbank material, have also been performed. Sediment samples were collected along extended lengths (80 km for the River Tyne and 48 km for the River Puerco) and at relatively frequent intervals (0.5 km in the Tyne, 0.3 km in the Puerco) in both river systems, enabling basin and reach-scale variations in metal concentrations to be determined. In the ephemeral, semi-arid Puerco basin, following the failure of a uranium tailings pond, Graf (1990) found that heavy metal concentrations showed no overall

---

Figure 13.2 (*opposite*) Downstream changes of total and potentially mobile concentrations of Pb, Zn and Cd in flood deposits (<2 mm) of the River Geul, Netherlands–Belgium (after Leenaers, 1989)

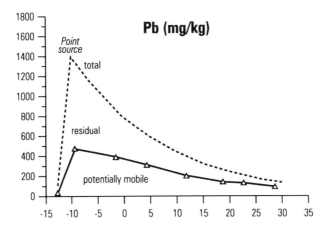

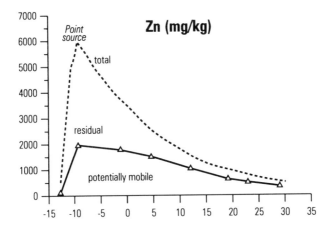

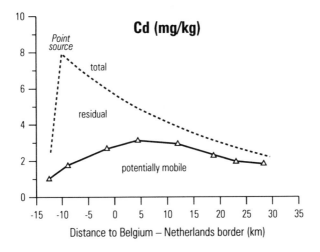

upstream–downstream change and were not related to distance from source by a simple exponential function (Figure 13.3). Instead metal concentrations were found to alternately increase and decrease with metal levels inversely related to unit stream power and the length of time that shear stresses exceeded critical values during the passage of the flood wave. This inverse relationship between stream power and contaminant concentration may have also been enhanced by the mixing of "clean" sediment eroding from channel banks. By contrast, in the Tyne basin, as shown by earlier investigations using less frequent sampling intervals (Macklin and Dowsett, 1989; Passmore and Macklin, 1994), most of the heavy metals analysed decreased down-river and could be modelled at the basin scale using simple decay functions (Figure 13.4). When, however, metal concentrations are examined in individual reaches, they appear to vary in a quasi-periodic manner over distances of between 5–10 km. In some cases, abrupt increases, or decreases, in metal values occur immediately downstream of tributary confluences, though more frequently changes in trunk river sediment metal levels do not coincide with tributary junctions. This wave-like distribution of sediment-associated metals is evident for all of the metals analysed, but is most prominent for Cd, which alternately increases and decreases in a very similar way to metals in the Puerco river reported by Graf (1990). In the case of the Tyne, which unlike the Puerco has a well-developed floodplain, the underlying control appears to be the organization of the valley floor into alternating "transport" reaches and "sedimentation" zones (cf. Macklin and Lewin, 1989). Transport reaches tend to be narrower and steeper, and generally sediment travel distances are larger. Thus, sediment entering the transport reach moves quickly through it with limited storage into the downstream sedimentation zone, where valley floors are wider, channel gradients decrease and sediment storage volumes are larger. It is in these areas where the largest quantities of sediment-associated metals, produced by historic Pb and Zn mining, are currently stored (Macklin and Smith, 1990), and also where present river sediment metal concentrations are generally highest (Macklin and Dowsett, 1989). This association between valley floor geometry and trunk river metal concentrations reflects

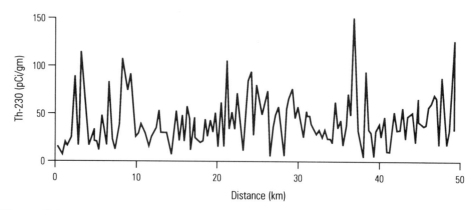

Figure 13.3 The downstream distribution of thorium-230 on the channel floor of the River Puerco, New Mexico, USA at 154 cross-sections through the first 47.4 km below tailings spill (after Graf, 1990)

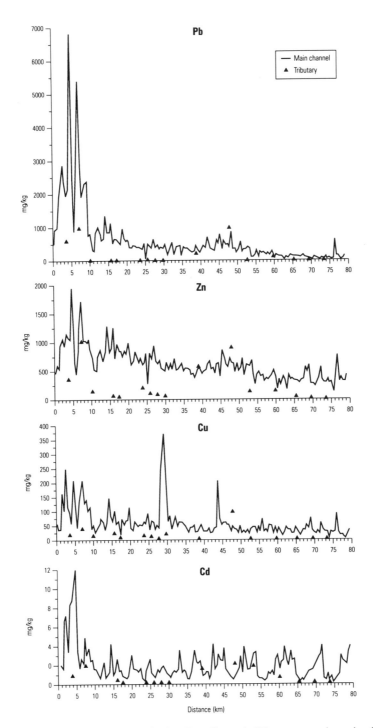

Figure 13.4 Downstream changes of Pb, Zn, Cu and Cd concentrations in fine-grained (<2 mm) channel sediments sampled 0.5 km intervals along the River South Tyne (0–52 km) and River Tyne (53–80 km), northern England. Sediment metal concentrations in major tributaries are also plotted

the remobilization of contaminated mining-age alluvium, by bank erosion and channel entrenchment, and its reintroduction back into the River Tyne. This process is most effective during major flood events.

Data from the metal-polluted River Tyne and River Puerco both indicate marked spatial and temporal variations in particulate-bound contaminant transport rates, partly as the result of intermittent inputs of metal waste into these systems but also due to valley floor configuration and channel characteristics. The River Puerco investigation suggests that for dam-burst events involving single-source spills, in drainage basins unaffected by previous pollution incidents, the hydraulic properties of the flood wave are the primary physical control of downstream metal dispersal rates and patterns. In the case of the River Tyne, however, geographic variations in metal levels are determined primarily by sediment supply through the erosion of alluvium contaminated by historic mining, which is now the principal source of heavy metals in this system. There is also evidence that particulate-bound metals were moved through the River Tyne and River Puerco in the form of a series of sediment pulses, or slugs (cf. Nicholas *et al.*, 1996), which have not resulted in a simple pattern of decreasing metal

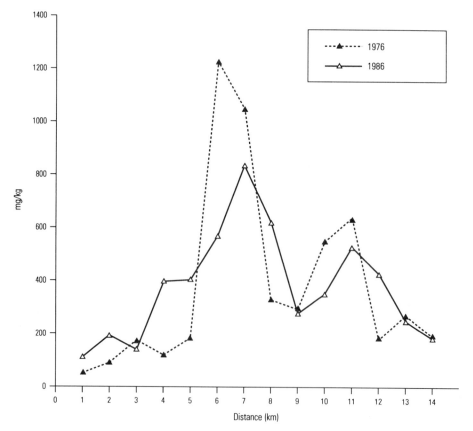

Figure 13.5 Comparison of Zn concentration in fine-gained (<250 μm) channel sediments in the River Teign, Devon, England, sampled in 1976 and 1986 (after Merefield, 1987)

concentration with distance from source as might be predicted by standard diffusion models. This is a phenomenon commonly observed in rivers. For example, Merefield (1987), working in the Teign valley that drains the former Dartmoor orefield in southwest England, resampled channel sites after a 10-year interval and demonstrated the downstream migration of heavy metal anomalies in a wave-like form at a rate of around 100 m per year (Figure 13.5). All of the studies discussed above have assessed variations in river sediment metal levels solely in a downstream, or longitudinal, direction and did not determine the lateral dispersal of contaminants away from the channel across the floodplain and valley floor. By contrast, overbank metal depositional processes have not, to the author's knowledge, been documented during individual flood events and have generally been inferred from the surface and subsurface distribution of metals in floodplain sediments. This topic is considered in the next section which reviews the factors that govern storage of metal-contaminated sediment in alluvial environments.

13.3 SPATIAL AND TEMPORAL PATTERNS OF SEDIMENT-ASSOCIATED METAL DEPOSITION AND STORAGE IN ALLUVIAL ENVIRONMENTS

Sediment-associated metals follow the same transport pathways as any other particulate-bound element and therefore their deposition and storage patterns can be related directly to floodplain geomorphological processes and channel sedimentation styles (Lewin and Macklin, 1987; Macklin, 1992). In British floodplains, for example, strong contrasts in patterns of metal storage have been found between braided-river reaches (Lewin et al., 1983; Macklin, 1986; Macklin and Lewin, 1989) and single-thread meandering rivers (Davies and Lewin, 1974; Wolfenden and Lewin, 1977), and also in sinuous, low gradient systems in which rivers have not changed their courses measurably since the first accurate surveys in the nineteenth century (Macklin, 1985; Lewin and Macklin, 1987; Bradley and Cox, 1990). In laterally mobile braided and meandering river environments, valley floors are reworked and metal-rich increments, to the depth of lateral reworking, are progressively keyed into alluvial bodies (Wolfenden and Lewin, 1977; Lewin et al., 1983; Macklin, 1986; Macklin and Lewin, 1989). This results in considerable variability of metal concentration across the floodplain with metal levels changing abruptly between alluvial units of different age (Figure 13.6). This can be especially marked when inputs of metal waste fluctuated between sedimentation events and where channel incision followed the cessation of mining activity (Lewin et al., 1983; Macklin and Lewin, 1989). By contrast in rivers, or reaches, with relatively low rates of lateral movement, deposition of metals occurs during overbank floods when fine-grained sediment of variable thickness is added across the entire floodplain (Lewin and Macklin, 1987; Bradley and Cox, 1990). Two rather different patterns of metal dispersal, however, can be produced by this process, as can be illustrated by data from the River Derwent, Derbyshire (Bradley and Cox, 1990), and laterally stable reaches of the lower River South Tyne, Northumberland (Macklin, 1988). In the former, metal concentrations show no clear variation in levels across the floodplain, nor vertically in the sediment profile (Figure 13.7). In the latter

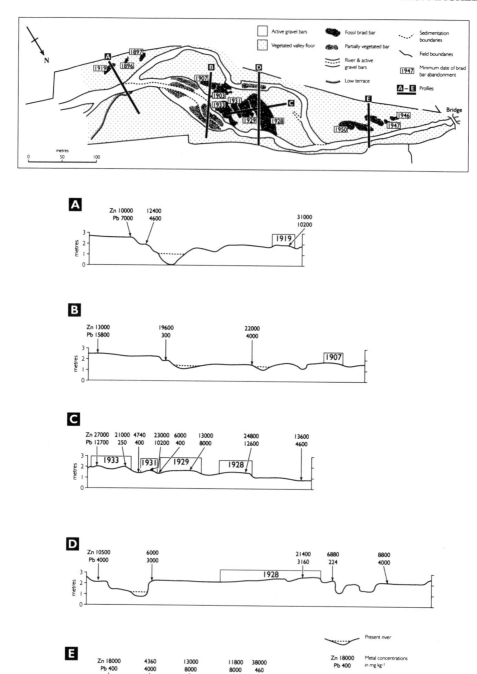

Figure 13.6 Zn and Pb concentrations in alluvial sediments within a historically near-braided reach of the River Nent, Cumbria, England (after Macklin, 1986)

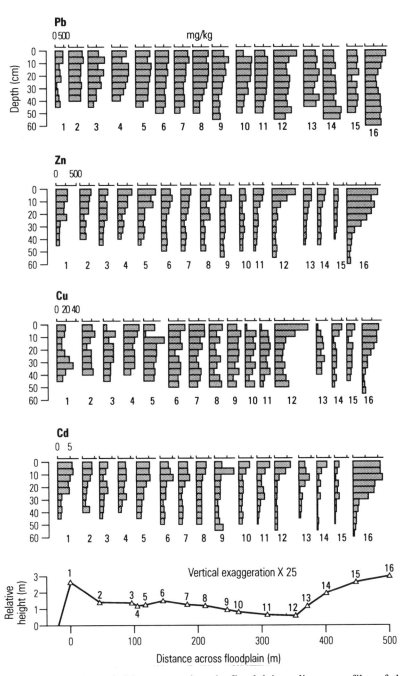

Figure 13.7 Pb, Zn, Cu and Cd concentrations in floodplain sediment profiles of the River Derwent, Derbyshire, England (after Bradley and Cox, 1990)

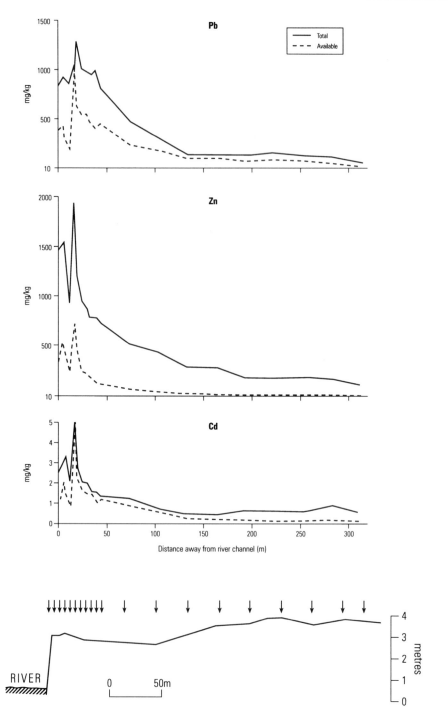

Figure 13.8   Total and available Pb, Zn and Cd concentrations in surface sediments across the floodplain of the River South Tyne, Northumberland, England (after Macklin, 1988)

system, with the exception of sites immediately adjacent to the present channel, metal values decrease with increasing distance from the river bank (Figure 13.8). Lateral variations in surface floodplain sediment metal concentrations in both rivers appear to relate primarily to the particular grain size fractions with which metals are associated. In the River South Tyne, metals are found in the coarser (sand) fraction of the suspended load which tends to be deposited closer to the channel, while in the River Derwent metals are found predominantly in the finer silt and clay fractions, which are deposited by settling from quiescent flows, rather than by diffusion from the main channel as in the case of the South Tyne. This probably explains the more uniform pattern of metal concentrations observed on the floodplain surface of the River Derwent.

Where heavy metals are more evenly distributed over the suspended load grain size fractions, factors such as the frequency of flooding and the rate of sediment deposition are more likely to determine metal concentrations in actively sedimenting overbank environments. This was found in the meandering River Geul, the Netherlands, where sediment-associated metals are not transported in one particular grain size class and metal levels are relatively high nearer the main channel, and in backwater areas, which are characterized by high inundation frequencies and higher rates of sediment deposition (Figure 13.9; Leenaers, 1989). Fluctuating inputs of heavy metals to river

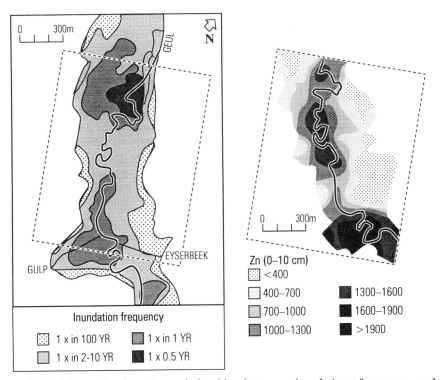

Figure 13.9 Maps showing the relationship between inundation frequency and Zn concentrations (mg/kg) in the River Geul, the Netherlands (after Leenaers, 1989)

systems, together with changes in inundation frequency and deposition rate, are commonly reflected in floodplain sequences by varying amounts of metals in vertical sediment profiles (Rang and Schouten, 1989; Macklin *et al.*, 1992). Floodplains in general, and overbank sediments in particular, can therefore provide an invaluable record of pollution in a river basin (Macklin *et al.*, 1994; Ridgway *et al.*, 1995). But, by acting as a focus for deposition, particulate-bound contaminants can remain stored in these alluvial environments for periods of decades to centuries, particularly in floodplains characterized by low rates of channel migration or those experiencing aggradation. Marron (1992), in an investigation of Pb and As storage along a 121 km reach of the Belle Fourche River, South Dakota, USA, ascertained that the floodplain had stored between 29 and 44% of the metals entering the system from mine tailings, and that about 60% of this material was incorporated into overbank sediments and 40% in point-bar deposits.

Floodplains in many river basins of the world that have been affected by mining or industrial activity are now functioning as major non-point sources of sediment-associated metal contaminants. In the historically polluted River Tyne in northern England (Macklin, 1992), and the River Geul in the Netherlands (Leenaers, 1989) discussed above, alluvial sediments contaminated by former Pb and Zn mining presently constitute the most important single source of metal contamination. For example, in the River Geul (which is a representative of many historically metal-polluted river systems in Western Europe) 66, 47 and 39% of the Pb, Zn and Cd, respectively that enters the channel is supplied by river bank erosion. Although more stringent controls on effluent discharges to rivers in North America and the countries of the European Union since the early 1970s have resulted in reduced metal loads, none of these measures will have any affect on metals currently stored in alluvial sediments on valley floors. Yet, it can be anticipated that in these and other historically polluted river systems the importance of contaminated floodplain sediments as a source of heavy metals is likely to increase significantly in coming years. Of equal, or perhaps greater, concern is the present high levels of metal pollution in rivers of the developing world (e.g. Martinelli *et al.*, 1988) and the former Warsaw Pact countries of Eastern Europe (e.g. Macklin and Klimek, 1992), where in the rush towards industrialization the environment has paid the heavy cost of economic development. Experience in metal-polluted Dutch, British and US rivers would indicate that contaminated alluvium is likely to pose a major environmental hazard well into the twenty-first century and beyond.

Many of the generalizations concerning the behaviour and storage patterns of metal contaminants in river systems outlined above have been developed for streams in humid areas, where organic matter, fine sediment particles and continuous flows are common, and solution–dissolution processes are active. Recent research in semi-arid catchments of the southwest USA polluted by the nuclear industry has shown, however, that many dryland streams lack such properties (Miller and Wells, 1986; Graf, 1990). In the deeply entrenched arroyos, that are typical of this part of the USA, channel-bed and bar sediments are the most critical repository for contaminants rather than the floodplain (Graf *et al.*, 1991). This results from the incised form of these channels which can contain floodwaters and contaminated sediment except during extremely large flood events. The construction of flood protection dykes along many perennial rivers in

Northern Europe, by inadvertently increasing rates of incision and decoupling the channel and floodplain, has had a very similar effect. It is clear that in both dryland and humid regions the geomorphic history of a river system, and the frequency of inundation, are the significant controls of the spatial and temporal distribution of metals in the fluvial sedimentary environment.

13.4   CONTAMINATED FLOODPLAINS AND ENVIRONMENTAL CHANGE: CONTROLLING CYCLING OF HEAVY METALS IN RIVER SYSTEMS

As has been emphasized in this review, it is the protracted residence time of heavy metals in sediment and soil that is one of the most significant aspects of metal contamination in river environments. These are not, however, static or sealed reservoirs and particulate-bound metals can be remobilized if contaminated soil or sediment is disturbed, or if environmental conditions change. Indeed, the fact that many historically contaminated floodplains are now themselves functioning as sources of river pollution is the most clear indication that they are rather "leaky" sedimentary sinks. Any increase in the rate at which metals are remobilized from polluted river sediments will obviously have very serious environmental consequences.

In floodplains, the cycling of heavy metals is controlled by both chemical and physical processes. Changes in the chemical environment (particularly pH and redox conditions) frequently result in the alteration of element forms which can dramatically increase the solubility, mobility and bioavailability of sediment-based metals. A fall in floodplain watertable levels (following, for example, channel-bed incision, over-pumping or hydroclimatic change) can lead to the oxidative remobilization of heavy metals, previously stored in a less labile form in waterlogged anoxic sediments, and their release in solution into ground and surface waters (Forstner and Kersten, 1988). Moreover, this process can be accelerated if floodplain soils become acidified, either through weathering or pedogenic processes, or as the result of atmospheric pollution. Selective chemical extraction techniques, in which operationally defined metal-associated sediment fractions are identified (Tessier *et al.*, 1979), have provided very useful basic information in this context, most notably as regards the manner in which metals are stored in river sediments and how their mobility may change with respect to variations in the ambient physico-chemical environment (Chester, 1988). Bradley and Cox (1987), for example, in a study of the effects of long-term pedogenic processes on the chemical speciation and mobility of metals in floodplain soils, demonstrated that in a little over 100 years significant amounts of Cd were translocated down-profile, in response to a change in its chemical partitioning from an exchangeable to a Fe/Mn oxide form. Despite the well-documented analytical and interpretative problems that exist with selective chemical extraction techniques, they do appear to mimic a general environmental mobility sequence of sediment-bound metals and give a reasonable indication of how they might behave should chemical conditions change. Information on the chemical partitioning of metals in floodplains may therefore be of considerable value in developing assessment and long-term containment strategies in polluted river basins.

The second principal mechanism whereby metals are reintroduced into rivers is by erosion of the latters banks and beds. During floods, contaminated streambank deposits are reworked, resulting in high metal levels in suspended sediment which may not only affect water quality but also be deposited on, and contaminate, floodplains further downstream. An increase in the frequency and magnitude of floods is likely to cause an increase in the proportion of the valley floor that is regularly flooded and, in the case of catchments affected by metal waste, polluted. Changes in flood regime can arise through altered catchment land use (Walling, 1979; Bosch and Hewlett, 1982) but recent research, notably in the USA (Knox, 1984; 1993), Australia (Warner, 1987) and the UK (Higgs, 1987; Rumsby and Macklin, 1994), has established linkages between the frequency of large floods in the last 300 years or so and short-term climatic change. These studies indicate that river corridors are likely to be especially sensitive to abrupt changes in flood hydroclimates anticipated with global warming, and that significant river bed and bank erosion could result in large-scale reworking of metal contaminated alluvium presently stored in floodplains. The serious flooding that occurred across northwest Europe in January/February 1995 may be a harbinger of this.

In view of the diffuse nature of metal contamination in many catchments affected by metal waste, possibly the only economically viable, and practically feasible, remedial solution is to implement a programme of environmental management that seeks to regulate and minimize plant, animal and human exposure to toxic metals. This is an especially difficult task if disposal of metal waste occurred sometime in the past, and also where the location and nature of metal inputs are not precisely known. Therefore, in most river systems affected by historic metal pollution, the first step in environmental assessment is to identify where contaminated sediments are located on the floodplain or valley floor. Unfortunately, this kind of information is not generally available to, or collected by, most regulatory agencies and expensive sediment quality surveys are commonly required after alluvial material is discovered (usually by chance) to be contaminated by heavy metals.

The frequently variable nature of floodplain metal levels, reflecting differences in river sedimentation styles and channel migration rates discussed earlier, also adds considerably to the difficulty of the task. An alternative (and cheaper) option, which can be used where records of metal waste discharge into rivers are available, is to zone alluvial valley floors on the basis of age and delimit river sediments that were deposited during periods of pollution (cf. Davies and Lewin, 1974). These are likely to be those most severely contaminated by heavy metals and can be given precedence for environmental monitoring and surveillance. This approach has been successfully employed in a number of metal-polluted river systems both in Europe (Macklin and Smith, 1990; Macklin and Klimek, 1992) and in the USA (Graf, 1994), with historic maps and aerial photographs being used to date floodplains. This obviates the need for all-inclusive surveys of river corridors and allows alluvium deposited during mining or industrial periods to be recognized and targeted for more detailed field sampling and metal analysis. If a site was found to be polluted, it would then be critical to restrict the remobilization of metals by chemical, physical or biological processes. While there are a range of protective measures that can be undertaken to achieve this, there are three which appear to be of paramount importance. The first involves regulating drainage on floodplains in order to minimize changes in water-table levels that could lead to alterations in the bonding strengths of sediment-associated metals, resulting in their release to surface waters or groundwaters. The second comprises the construction of bank protection structures to prevent erosion of contaminated floodplain material and it re-entering the river system. The third and final measure involves prohibiting the raising of livestock or cultivation of crops on contaminated land. Unfortunately, the last and easiest course of action is rarely taken, most usually because metal contamination in many floodplains has frequently gone unrecognized, especially when metal burdens do not manifestly inhibit plant growth or affect animal health. The development of GIS-based river corridor metal databases, to enable contaminated land to be delineated for environmental protection and planning purposes, would seem to offer a solution to this problem.

13.5 CONCLUSIONS

Over the last two decades or so, interest in sediment-associated heavy metals among fluvial scientists has grown, and diversified, considerably. Fundamental research on metal transport and depositional processes in rivers carried out by geomorphologists in the 1970s and 1980s has produced a wealth of empirical data on the short and long-term behaviour of particulate-bound metals in channels and floodplains. The emphasis of many of these studies lay either in utilizing heavy metals as tracers, to help understand reach and basin-scale sediment fluxes and storage, or in evaluating the impact of mining-related changes of sediment supply on river development. During the 1990s, increasing public and government concern for environmental quality, and more stringent regulatory systems, have necessitated the implementation of monitoring programmes to document environmental contaminants such as heavy metals (Graf, 1994). Unfortunately, very few regulatory agencies concerned with river pollution routinely sample and analyse sediment-bound metals, while those that do usually focus on metal levels in suspended sediment rather than contaminant movement and storage. Yet, as has been highlighted in this review, information about the geographic distribution and long-term storage of heavy metals in floodplain environments is of paramount importance for safeguarding surface water and groundwater quality, and for protecting plant, animal and human health. Geomorphologists are not only well placed to provide such data, but they also can offer useful guiding principles for explaining the dynamics of river systems that transport and store sediment-bound contaminants (Macklin, 1992; Graf, 1994). They have a key role to play in developing remedial measures for contaminated floodplains, especially in the context of global warming predictions of increased flooding and river instability over the next few decades.

ACKNOWLEDGEMENTS

I wish to thank both NERC and EPSRC for their continuing support of my research in the form of postgraduate studentships, fellowships and research grants.

REFERENCES

Axtmann, E.V. and Luoma, S.N. (1991) Large-scale distribution of metal contamination in fine-grained sediments of the Clark Fork River, Montana, USA. *Applied Geochemistry*, **6**, 75–88.

Bosch, J.M. and Hewlett, J.D. (1982) A review of catchment experiments to determine the effect of vegetation changes on water yield and evapotranspiration. *Journal of Hydrology*, **55**, 3–23.

Bradley, S.B. and Cox, J.J. (1987) Heavy metals in the Hamps and Manifold valleys, north Staffordshire, UK: Partitioning of metals in floodplain soils. *The Science of the Total Environment*, **65**, 135–153.

Bradley, S.B. and Cox, J.J. (1990) The significance of the floodplain to the cycling of metals in the river Derwent catchment, UK. *The Science of the Total Environment*, **97/98**, 441–454.

Bradley, S.B. and Lewin, J. (1982) Transport of heavy metals on suspended sediments under high flow conditions in a mineralised region of Wales. *Environments Pollution B*, **4**, 257–267.

Chester, R. (1988) The storage of metals in aquatic sediments. In: Strigel, G. (ed.), *Metals and Metalloids in the Hydrosphere; Impact through Mining and Industry, and Prevention Technology*. Proceedings of an IHP workshop, Bochum, Federal Republic of Germany, 21–25 September 1987, 81–110.

Davies, B.E. (1980) Trace element pollution. In: Davies, B.E. (ed.), *Applied Soil Trace Elements*, John Wiley & Sons, Chichester, 287–351.

Davies, B.E. and Lewin, J. (1974) Chronosequences in alluvial soils with special reference to historical pollution in Cardiganshire, Wales. *Environmental Pollution*, **6**, 49–57.

Forstner, U. and Kersten, M. (1988) Sediment–water interactions: chemical mobilisation. In: Strigel, G. (ed.), *Metals and Metalloids in the Hydrosphere; Impact through Mining and Industry, and Prevention Technology*. Proceedings of an IHP workshop, Bochum, Federal Republic of Germany, 21–25 September 1987, 135–164.

Gilbert, G.K. (1917) Hydraulic mining debris in the Sierra Nevada. *US Geological Survey Professional Paper*, **105**, Washington.

Graf, W.L. (1985) Mercury transport in stream sediments of the Colorado Plateau. *Annals of the Association of American Geographers*, **75**, 552–565.

Graf, W.L. (1990) Fluvial dynamics of thorium-230 in the Church Rock Event, Puerco River, New Mexico. *Annals of the Association of American Geographers*, **80**(3), 327–342.

Graf, W.L. (1994) *Plutonium and the Rio Grande: Environmental Change and Contamination in the Nuclear Age*, Oxford University Press, New York.

Graf, W.L., Clark, S.L, Kamnerer, M.T., Lehman, T., Randall, K. and Schroeder R. (1991) Geomorphology of heavy metals in sediments of Queen Creek, Arizona, USA. *Catena*, **18**, 567–582.

Grimshaw, D.L., Lewin, J. and Fuge, R. (1976) Seasonal and short-term variations in the concentration and supply of dissolved zinc to polluted aquatic environments. *Environmental Pollution*, **11**, 1–7.

Higgs, G. (1987) Environmental change and hydrological response: flooding in the upper Severn catchment. In: Gregory, K.J., Lewin, J. and Thornes, J.B. (eds), *Palaeohydrology in Practice*, John Wiley & Sons, Chichester, 131–159.

Hudson-Edwards, K.A., Macklin, M.G., Curtis, C.D. and Vaughan, D.J. (1996) Processes of formation and distribution of Pb, Zn-, Cd- and Cu- bearing mineral species in the Tyne basin, NE England: implications for metal-contaminated river systems. *Environmental Science and Technology* **30**(1), 72–80.

Knighton, A.D. (1991) Channel bed adjustment along mine-affected rivers of north east, Tasmania. *Geomorphology*, **4**, 205–219.

Knox, J. (1984) Fluvial response to small scale climatic changes. In: Costa, J.C. and Fleisher, P.J. (eds), *Developments and Applications of Geomorphology*, Springer-Verlag, Berlin, 318–342.

Knox, J. (1987) Historical valley floor sedimentation in the upper Mississippi Valley. *Annals of the Association of American Geographers*, **77**, 224–244.

Knox, J. (1993) Large increases in flood magnitude in response to modest changes in climate. *Nature*, **361**, 430–432.

Leenaers, H. (1989) *The Dispersal of Metal Mining Wastes in the Catchment of the River Geul (Belgium – The Netherlands)*. Amsterdam: Geografische Institut, Rijksuniversitat Utrecht.

Lewin, J. and Macklin, M.G. (1987) Metal mining and floodplain sedimentation. In: Gardiner, V. (ed.), *International Geomorphology 1986 Part 1*, John Wiley & Sons, Chichester, 1009–1027.

Lewin, J. and Wolfenden, P.J. (1978) The assessment of sediment sources: a field experiment. *Earth Surface Processes*, **3**, 171–178.

Lewin, J., Davies, B.E. and Wolfenden, PJ. (1977) Interactions between channel change and historic mining sediment. In: Gregory, K.J. (ed.), *River Channel Changes*, John Wiley & Sons, Chichester, 353–367.

Lewin, J., Bradley, S.B. and Macklin, M.G (1983) Historical valley alluviation in mid-Wales. *Geological Journal*, **18**, 331–350.

Macklin, M.G. (1985) Floodplain sedimentation in the upper Axe valley, Mendip, England. *Transactions of the Institute of British Geographers, New Series*, **10**, 235–244.

Macklin, M.G. (1986) Channel and floodplain metamorphosis in the river Nent, Cumberland. In: Macklin, M.G. and Rose, J. (eds), *Quaternary River Landforms and Sediments in the Northern Pennines, England: Field Guide*, British Geomorphological Research Group/Quaternary Research Association, Cambridge, 13–19.

Macklin, M.G. (1988) A fluvial geomorphological based evaluation of contamination of the Tyne basin, north-east England by sediment-borne heavy metals. Unpublished report to the Natural Environment Research Council, 29 pp.

Macklin, M.G. (1992) Metal contaminated soils and sediment: a geographical perspective. In: Newson, M.D. (ed.), *Managing the Human Impact on the Natural Environment: Patterns and Processes*, Belhaven Press, London, 172–195.

Macklin, M.G. and Dowsett, R.B. (1989) The chemical and physical speciation of trace metals in fine grained overbank flood sediments in the Tyne basin, north-east England. *Catena*, **16**, 135–151.

Macklin, M.G. and Klimek, K. (1992) Dispersal, storage and transformation of metal contaminated alluvium in the upper Vistula basin, south-west Poland. *Applied Geography*, **12**, 7–30.

Macklin, M.G. and Lewin, J. (1989) Sediment transfer and transformation of an alluvial valley floor: the river South Tyne, Northumbria, UK. *Earth Surface Processes and Landforms*, **14**, 233–246.

Macklin, M.G. and Smith, R.S. (1990) Historic riparian vegetation development and alluvial metallophyte plant communities in the Tyne basin, north-east England, U.K. In: Thornes, J.B. (ed.), *Vegetation and Erosion*, John Wiley & Sons, Chichester, 239–256.

Macklin, M.G., Rumsby, B.T. and Heap, T. (1992) Flood alluviation and entrenchment: Holocene valley floor development and transformation in the British uplands. *Geological Society of America Bulletin*, **104**, 631–643.

Macklin, M.G., Ridgway, J., Passmore, D.G. and Rumsby, B.T. (1994) The use of overbank sediment geochemical mapping and contamination assessment: results from selected English and Welsh floodplains. *Applied Geochemistry*, **9**, 689–700.

Marcus, W.A. (1987) Copper dispersion in ephemeral stream sediments. *Earth Surface Processes and Landforms*, **12**, 217–228.

Marron, D.C. (1992) Floodplain storage of mine tailings in the Belle Fourche river system: a sediment budget approach. *Earth Surface Processes and Landforms*, **17**, 675–685.

Martinelli, L.A., Feweira, J.R., Forsberg, B.R. and Victoria, R.E. (1988) Mercury contamination in the Amazon: a gold rush consequence. *Ambio*, **17**, 252–254.

Merefield, J.R (1987) Heavy metals in Teign Valley sediments: ten years after. *Proceedings of the Ussher Society*, **6**, 529–535.

Miller, J.R and Wells, S.G. (1986) Types and processes of short-term sediment and uranium tailings storage in Arroyos: an example from the Rio Puerco of the West, New Mexico. In: Hadley, R.F. (ed.), *Drainage Basin Sediment Delivery*. IAHS publication number 159. Albuquerque, New Mexico, 335–353.

Nicholas, A.P., Ashworth, P.J., Kirkby, M.J., Macklin, M.G. and Murray, T. (1996) Sediment slugs: large-scale fluctuations in fluvial sediment transport rates and storage volumes. *Progress in Physical Geography* **19**(4), 500–519.

Passmore, D.G. and Macklin, M.G. (1994) Provenance of fine-grained alluvium and late Holocene land-use change in the Tyne basin, northern England. *Geomorphology*, **9**, 127–142.

Rang, M.C. and Schouten, C.J. (1989) Evidence for historical heavy metal pollution in floodplain soils: the Meuse. In: Petts, G.E., Moller, H. and Roux, A.L. (eds), *Historical Change of Large Alluvial Rivers: Western Europe*, John Wiley & Sons, Chichester, 127–142.

Rang, M.C., Kleijn, C.P and Schouten, C.J. (1987) Mapping of soil pollution by application of classical geomorphological and pedological field techniques. In: Gardiner, V. (ed.), *International Geomorphology 1986 Part 1*, John Wiley & Sons, Chichester, 1029–1044.

Ridgway, J., Flight, D.M.A., Martiny, B., Gomez-Caballero, A., Macias-Romo, C. and Greally, K. (1995) Overbank sediments from central Mexico: an evaluation of their use in regional geochemical mapping and in studies of contamination from modern and historical mining. *Applied Geochemistry*, **10**, 97–109.

Rowan, J.S., Barnes, S.J.A., Hetherington, S.L, Lambers, B. and Parsons, F. (1995) Geomorphology and pollution: the environmental impacts of lead mining, Leadhills, Scotland. *Journal of Geochemical Exploration*, **52**, 57–45.

Rumsby, B.T. and Macklin, M.G. (1994) Channel and floodplain response to recent abrupt climate change: the Tyne basin, northern England. *Earth Surface Processes and Landforms*, **19**, 499–515.

Tessier, A., Campbell, P. and Bission, M. (1979) Sequential extraction procedure for the speciation of particulate trace metals. *Analytical Chemistry*, **51**, 844–851.

Walling, D.E. (1979) The hydrological impact of building activity – a study near Exeter. In: Hollis, G.E. (ed.), *Man's Impact on the Hydrological Cycle in the UK*. Geo Books, Norwich, 135–151.

Warner, R.F. (1987) The impacts of alternating flood- and drought-dominated regimes of channel morphology at Penrith, New South Wales, Australia. *International Association of Hydrological Science Publication*, **168**, 327–338.

Wolfenden, P.J. and Lewin, J. (1977) Distribution of metal pollutants in floodplain sediments. *Catena*, **4**, 309–317.

Wolfenden, P.J. and Lewin, J. (1978) Distribution of metal pollutants in active stream sediments. *Catena*, **5**, 67–78.

14   Linking Hillslopes to Floodplains

T.P. BURT* and **N.E. HAYCOCK†**
*Department of Geography, University of Durham, UK
†Quest Environmental, Harpenden, UK

14.1   INTRODUCTION

> Intuitively, there is an assumption that the condition of the stream and the condition of the
> riparian zone are intimately linked.
>
> (Bren, 1993, p. 286)

This chapter examines the role of floodplains as transitional zones between the catchment and the river channel. By strict definition the riparian zone is that part of the land surface in intimate contact with the river and includes only vegetation along the bed and banks of the watercourse (Tansley, 1911). In recent years, however, the term has come to include a wider strip of land either side of the channel; often the floodplain is seen as being coincident with this more broadly defined area, especially in the USA (e.g. Brown *et al.*, 1978; Anderson, 1987). Riparian zones and floodplains have become of much interest in recent years and many people have an interest in their use. The zone is important for human recreation, as a habitat and corridor for wildlife, and as a major component of the visual landscape. More specialized interest in riparian zones has focused on their function as repositories for sediment, their role as a nutrient sink for the surrounding catchment and their influence on the quality of water leaving the basin. One of the simplest and most effective measures for reducing non-point pollution from agricultural land is believed to be the use of near-stream land as a runoff buffer zone. Vegetated buffer strips in general and riparian forests in particular have been shown to be effective in filtering sediments, nutrients, pesticides, particulate organic matter and bacteria from farm runoff (Phillips, 1989; Gilliam, 1994). Until recently, floodplains were viewed predominantly as a source of good timber or as productive farmland (Lant and Tobin, 1989). Their distinctive biota and ability to protect the stream environment has prompted renewed interest in their broader ecological function.

Riparian zones are landscape boundaries, or ecotones, which physically separate aquatic and terrestrial ecosystems. Ecotones are important regulators of the movement

Floodplain Processes. Edited by Malcolm G. Anderson, Des E. Walling and Paul D. Bates.
© 1996 John Wiley & Sons Ltd.

of energy and material across landscapes and through catchments; they may act both as conduits and barriers. The dynamic nature and management potential of riparian ecotones have recently become subjects of great interest (e.g. Naiman and Decamps, 1990). The ecotone was originally conceived as a spatial transition between two community types (Clements, 1897). Subsequently, it was modified to include edge effects, whereby the number of individuals and species within ecotones often appeared to be greater than in the ecosystems on either side (Leopold, 1933). As the ecotone concept has matured, so has come the realization that ecotones may, in fact, be characterized by their own complex ecological processes and interactions (Odum, 1990). An ecotone is now regarded as a dynamic rather than a static zone, possessing properties of its own which derive from its position in the landscape. The terrestrial–aquatic ecotone provides a connection between a terrestrial ecosystem (often farmland) and the aquatic ecosystem of a river channel or lake. Usually, the fluxes will be from land to water, though overbank flooding or seepage into the channel bank may reverse this. Such connectivity may provide the opportunity for near-stream ecotones to function as natural sinks for sediment and nutrients emanating from farmland. This in turn raises the possibility of establishing and maintaining buffer zones at the bottom of fields adjacent to river channels, to trap both particulate and soluble pollutants and so protect waterbodies from possible eutrophication and minimize pollution of drinking water supplies (Waikato Valley Authority, 1973; Omernik, 1976).

In general, there is agreement that near-stream land should be maintained in a state approaching naturalness (Bren, 1993), with the implicit assumption that the natural processes operating within this ecotone are capable of providing the necessary buffering functions noted above. However, formalized study of such areas is relatively new and it is clear that such systems may not simply act as sinks for pollutants. Processes operating within the riparian zone may transform one pollutant into another (e.g. nitrate to nitrous oxide) or cause an apparently inert substance to be released into the environment (e.g. the increased solubility of phosphorus in anoxic conditions). There is the need, therefore, to establish the conditions under which these various buffer zone processes can operate effectively. Fundamental to this is the need to understand the transmission of hillslope water through the floodplain. The precise pathway taken by water draining from farmland will determine whether that water is resident within the floodplain system for a sufficiently long periods of time to allow internal processes to operate effectively. Accordingly, our first task in this review is to consider the hydrology of floodplains and to examine in particular how hillslope water moves across the floodplain to the river. The water quality functions of floodplains may then be considered within this context; the hydrological pathways govern not only the flux of sediment and nutrients into and across the floodplain but may also be responsible for the concurrent transport of substrates required for their effective removal or transformation. Finally, an integrated approach to catchment management is taken by considering the functional role of floodplains within the entire drainage basin system. Although it is our intention to emphasize the drainage of hillslope waters on to the floodplain and thence to the river, we do not intend to ignore totally the process of bank storage whereby the normal direction of flow is reversed and water enters the floodplain from the river channel.

14.2   RUNOFF PATHWAYS FROM HILLSLOPE TO CHANNEL VIA THE FLOODPLAIN

14.2.1   Storm runoff mechanisims

There are may different pathways by which hillslope runoff can reach a river (Figure 14.1). Some precipitation (or snowmelt) takes a rapid route to the channel and is often described as quickflow or stormflow; quickflow is usually associated with high discharge. Subsurface flow generally moves at much lower velocities, often by longer flowpaths and, although it may contribute to stormflow, its main effect is to maintain streamflow during dry periods through the sustained release of water stored in soil and bedrock. There has been a good deal of research into hillslope runoff mechanisms during the last few decades, both in the field and by the use of computer simulation models. Full details of this work are to be found in Kirkby (1978) and Anderson and Burt (1990); therefore, only a brief review is provided here. A particular emphasis on stormflow is justified here because, in most cases, it transports the largest fraction of solute and sediment export from the catchment to the river.

Infiltration capacity, the maximum rate at which the soil surface can absorb falling rain (or meltwater), decreases asymptotically over time as the upper part of the soil profile becomes progressively saturated from the surface down. Changes in the soil surface (e.g. swelling of clay particles, in-washing of fine particles into pores, compaction by rainbeat) may also reduce infiltration capacity during the course of a storm. Once rainfall intensity exceeds infiltration capacity, surface depressions begin to fill up. When these are full, the excess rainfall then overflows downslope as *infiltration-excess overland flow*. Overland flow may also be generated when the entire soil profile becomes saturated. In this case, rainfall intensity may be well below

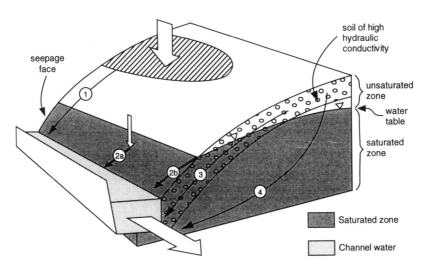

Figure 14.1   The hydrological pathways involved in the delivery of rainfall to a stream channel from a hillslope. 1. Infiltration-excess overland flow. 2. Saturation-excess overland flow: 2a. direct runoff; 2b. return flow. 3. Subsurface stormflow. 4. Groundwater flow

infiltration capacity: the soil becomes saturated either because of prolonged rainfall or because of inflow from further upslope; often both processes operate together. Where rain falls on to a saturated surface, no infiltration can occur and direct runoff takes place. Where too much subsurface flow accumulates at a point on the slope and so exceeds the transmission capacity of the soil, excess flow must exfiltrate the soil and flow across the surface; this is return flow. Direct runoff and return flow together comprise *saturation-excess overland flow*. The source areas for saturation-excess overland flow are those parts of a drainage basin where soil moisture tends to accumulate: at the foot of any slope, especially those that are concave in profile; in areas of thin soils where soil moisture storage is reduced; and in hillslope hollows where convergence of flowlines favours the accumulation of soil water. Drainage from surrounding slopes on to wide, low-angle floodplains may be expected particularly to favour the development of extensive areas of surface saturation. However, there have been few studies of floodplain hydrology from this point of view, most consideration having paid been paid to valleys where steeper slopes abut directly the stream channel. Subsurface flow may also drain rapidly from hillslopes and so contribute to quickflow; this is *subsurface stormflow*. Hillslope hydrologists have, until recently, emphasized flow through the micropores of the soil matrix, but there has been continued interest in pipeflow (Gilman and Newson, 1980; Jones, 1981), and in under-drainage of floodplains and heavy clay soils (Robinson and Beven, 1983). Recently, much attention has been paid to the hydrological effects of rapid drainage through macropores both in relation to stormflow generation and pollutant transport (Germann, 1990).

It is impossible to divorce the generation of subsurface stormflow from the production of saturation-excess overland flow since both depend to large extent on the generation of flow through the soil profile. Subsurface stormflow will be most important where steep slopes with permeable soils are contiguous with the channel. Where surface saturation occurs to any great extent (e.g. where wide valley floors exist), saturation-excess overland flow will dominate the stormflow response with higher peak discharges and less delay between rainfall and runoff (Anderson and Burt, 1990, p. 370). If soils are permeable, the source areas for saturation-excess overland flow form only a small part of the total basin area and storm runoff is usually less than 5% of the total precipitation input. Where soils are less permeable, surface saturation will be more extensive, with proportionately more rainfall converted into runoff. Infiltration-excess overland flow yields the largest amounts of runoff, with least delay between rainfall and runoff; in direct contrast to subsurface stormflow, its sediment load may be high but its dissolved load is almost always low. Saturation-excess overland flow, being a mixture of return flow (old soil water) and return flow (water which has not infiltrated the soil) yields sediment and solute loads intermediate between these two extremes. Groundwater discharge rarely contributes to the storm hydrograph; its long residence time typically yields water of high solute load.

14.2.2 The hydrology of floodplains: a two-dimensional approach

The runoff pathways shown in Figure 14.1 indicate a direct linkage between the hillslope and the river. Indeed, most research on hillslope hydrology, whether by modelling or in the field, has considered this configuration and only a few studies have included the effect of a floodplain. Two questions arise here: the entry of runoff on to the floodplain; and, the hydrology of the floodplain itself.

Freeze and Witherspoon (1967) studied regional flow systems using numerical simulation. The topographic situation considered, a wide upland plateau draining to a valley, provides a direct analogue for a floodplain/channel system (and for this reason the following discussion refers to floodplain and channel rather than to plateau and valley as in the original presentation). In these simulations, the water table is assumed to be the upper level of the saturated groundwater flow system; a two-dimensional section is representative of the three-dimensional flow net if it is taken parallel to the water-table slope. Recharge refers to water that percolates down through the unsaturated zone to the water table and actually enters the dynamic groundwater flow system. In the recharge area, equipotential lines meet the water table obliquely, with the acute angle on the upslope side, and the direction of flow is away from the water table. Discharge refers to water leaving the system via seepage areas and springs. In the discharge area, flow direction is towards the water table; equipotential lines again meet the water table at an oblique angle, with the acute angle on the downslope side. At the hinge line, which separates the recharge and discharge areas, equipotential lines meet the water table at right angles. The length of the groundwater basin (S) is dimensionless with a depth of $0.1S$ at the right hand side; the total relief is $0.0167S$, most of which is the channel. Surface topography and subsurface stratigraphy can obviously exist in an infinite variety. Figure 14.2 shows a selection of flow net simulations; further examples are provided in Freeze and Witherspoon (1967, 1968).

In Figure 14.2a the flow net for a simple system of homogeneous permeability is shown. The water-table configuration, which closely follows the smooth topography, has a gentle slope. The hinge line lies on the edge of the channel and the entire floodplain is a recharge area. Water-table slope and hydraulic gradient are highest in the discharge area close to the channel. Throughout the rest of the floodplain, there is a relatively uniform flow pattern. In Figure 14.2b the existence of a hummocky terrain, as might arise where old channels cross a floodplain, results in numerous sub-basins; water may be discharged locally to the nearest topographic low or may enter the main flow towards the channel.

The effect of introducing a high-permeability layer into the system is to create a conduit for groundwater such that the total amount of flow traversing the area increases. This depends on the hydraulic conductivity of the high-permeability layer compared to the surrounding sediment, and its depth compared to the total available for flow. When an upper layer is more permeable, the flow pattern is almost identical to the homogeneous case; similarly, a less permeable layer below a permeable one has little or no effect on the flow. Often, on a floodplain in England, fine-grained alluvial sediments overlie gravels of early Holocene age. In Figure 14.2c, the effect of introducing a layer at depth with a hydraulic conductivity 100 times that of the overlying beds is shown. The lower layer forms an aquifer in which there is essentially horizontal flow with recharge from above. The vertical component of flow in the upper layer is much more pronounced than it was in the homogeneous case. As the permeability ratio increases, vertical flow (up or down) in the upper layer becomes more pronounced; the hydraulic gradient in the aquifer decreases but flow increases. In Figure 14.2d a lens of permeable material occupies just the upper half of the floodplain.

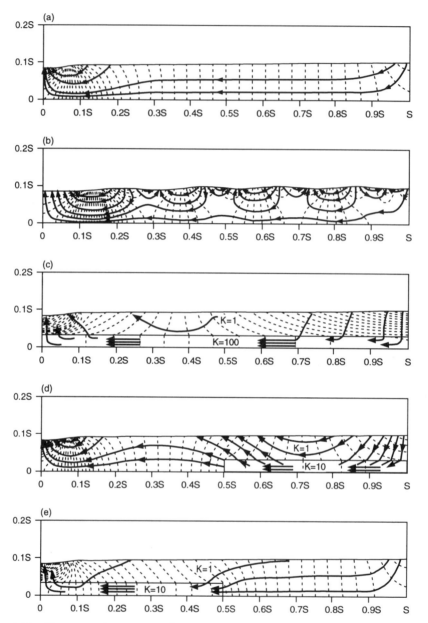

Figure 14.2 Flow nets for a channel/floodplain system. See text for explanation. (Adapted from Freeze and Witherspoon, 1967)

The majority of flow which has entered the upper part of the slope is discharged in the middle of the floodplain; what was originally a single groundwater basin has become two under the influence of a partial layer. In Figure 14.2e the permeable lens is in the downstream half of the floodplain. The central discharge area does not exist and recharge is concentrated over the lens. This situation is analogous to the three-dimensional case discussed below (Figure 14.5), where floodplain drainage becomes concentrated in permeable sediments, in this case a buried channel.

Most field studies of hillslope hydrology have been conducted in landscapes where the influence of floodplains is very small or absent. Studies of groundwater-fed wetlands in humid headwater basins provide a useful analogy, however. Hill and colleagues have published several studies of a riparian wetland near Toronto, Canada (Roulet, 1989; Hill, 1990a,b; Hill and Waddington, 1993; Waddington *et al.*, 1993). Here, a large aquifer occurs in glacial and fluvioglacial deposits; relatively constant groundwater discharge sustains a forested wetland 20–100 m wide along the perennial stream channel. Flow nets indicate a typical outflow pattern with horizontal groundwater flow at the upland perimeter of the wetland with nearly vertical upward movement around the stream channel and within the wetland itself (Figure 14.3). The water table intersects the ground surface at the wetland perimeter and then slopes towards the stream at a shallow depth beneath the surface of the organic deposits. There are three main hydrological pathways for groundwater discharge. Springs emerging at the margin of the wetland produce rivulets which cross the riparian zone to the stream. A second pathway involves upward groundwater flow which forms further springs within the wetland itself. These two pathways together contribute about 60% of the groundwater input to the stream. The third pathway consists of groundwater which reaches the stream directly as bed and bank seepage. Rapid mixing of precipitation with a significantly large pool of pre-event groundwater in the saturated areas of the wetland is responsible for the rapid appearance of pre-event water in surface-generated streamflow. In a similar type of study, Bonell *et al.* (1990) found that the few storm events dominated by event water were also associated with saturation-excess overland flow; in such cases, however, there was mostly direct runoff of precipitation and little contribution from pre-event water. In most events, pre-event water from subsurface runoff was the major contributor to stormflow and saturation-excess overland flow was absent.

In general, permeable substrates favour subsurface stormflow, which in turn provides the opportunity for flow to interact with streamside soil and vegetation. The emergence of groundwater in springs limits this possibility. Hill (1990 a,b) found, in the aerobic environment of surface runoff, that ammonium was rapidly depleted due mainly to microbial immobilization but that travel times were too short and oxygen levels too high to allow denitrification (i.e. the bacterial reduction of nitrate to the gases N_2 or N_2O). Similarly, groundwater which moves beneath a relatively impermeable wetland and flows directly into the stream also has little opportunity for interaction with the riparian system, except briefly at the water–sediment interface of the channel bed. Figure 14.4 summarizes the various hydrological pathways which may link the hillslope to the channel. Most riparian wetlands are maintained by significant groundwater contributions but often, because of their low permeability, most groundwater flow bypasses the wetland, minimizing the opportunity for the riparian zone to function as a

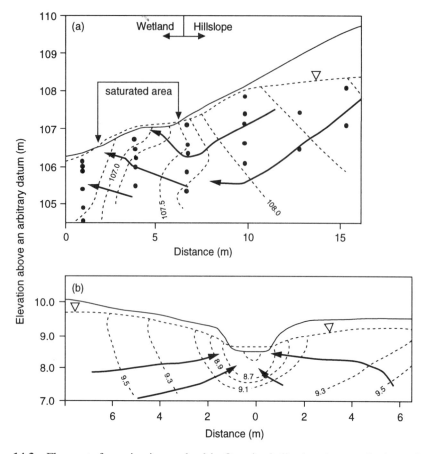

Figure 14.3 Flow nets for a riparian wetland in Ontario, indicating the contributions of return flow, shallow throughflow and deep groundwater flow to stream discharge. (Adapted from Waddington *et al.*, 1993)

buffer zone as far as dissolved load is concerned. The situation is somewhat different for sediment-laden overland flow which reaches the floodplain from upslope. Even if this water does not infiltrate (which would be ideal), the low flow velocities across the floodplain will limit the capacity of the surface runoff to transport the sediment load. Thus the failure of overland flow water to infiltrate the riparian sediments may not be crucial although a buffer zone still needs to be wide enough to trap sediments as they move across the zone. These matters are explained in more detail in Section 14.3.

14.2.3 The hydrology of floodplains: a three-dimensional approach

Complexities of soil and sediment properties and of surface topography complicate the two-dimensional analysis presented above. Discharge to the surface is, *ceteris paribus*, encouraged by the presence of less permeable sediments or depressions; however, lenses of more permeable sediments cause convergence of soil water flow and may

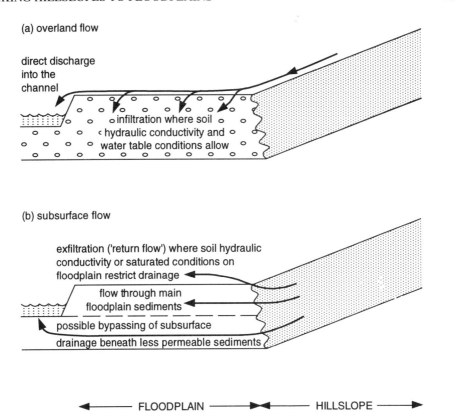

Figure 14.4 Summary of main flow paths by which hillslope discharge moves through a floodplain to reach the stream channel

provide a conduit for subsurface discharge from the floodplain. Most often neglected is the influence of downvalley gradient. Many studies assume that the floodplain is flat and that flow occurs straight across it to the stream. The downvalley slope of a floodplain may, in fact, be several degrees; this can significantly divert hillslope drainage, which eventually flows subparallel to the river channel. Because of this, the residence time of hillslope water within the floodplain may be greatly extended. One important point which follows from this is that field experiments which employ a single line of piezometers orthogonal to the channel may seriously misjudge flow patterns within the floodplain and not be sampling the same body of water.

Anderson and Kneale (1980) studied soil water conditions on a low-angle slope with clay soils; although the slope's planform was only a slight departure from planar, this should have been sufficient to induce convergent flow into the hollow from adjacent spurs. However, results showed that the effect of landslope angle in the downstream direction served to enhance the lateral hydraulic gradient so that soil water drainage was no longer simply determined by hillslope topography. The authors concluded that when the hillslope angle approximates the downvalley floodplain angle, and both angles are low, then the pattern of flow is skewed such that the downslope spur zone has wetter soils than the corresponding upslope spur. Further analysis of this situation by Anderson

(1982), using a simulation model, demonstrated that only in the case of steep slopes and high soil hydraulic conductivities do soil water flowpaths converge into hillslope hollows. For all other conditions simulated (lower slopes and less permeable soils), more complex patterns of cross-slope movements were shown to occur, even though the downvalley tilt used was only 2 °. The implication is that flows on the floodplain may be strongly downvalley in direction; this effect will be compounded where low-angle valley-side slopes exist.

Haycock (1991) studied soil water flow and nitrate leaching on a floodplain in the Cotswold Hills, England (see also Haycock and Burt, 1993). Two sites were instrumented on the floodplain, one grassland and the other a poplar (*Populus italica*) plantation; flow patterns were similar in both cases. During the summer, geological controls meant that flow was from the stream channel into the floodplain sediments and thence into the local limestone aquifer. In winter, hydraulic gradients were reversed once recharge of the aquifer had taken place. Figure 14.5 shows (a) surface topography for the grass floodplain site and (b) mean water-table elevation for the site during January 1990. Direct drainage from the floodplain into the stream takes place only very close to the channel. Most hillslope drainage is diverted down valley along a buried channel filled with permeable sediments, flowing eventually into the stream some 100 m downstream. A two-dimensional view along one flowline would be similar to that shown in Figure 14.2e. In this case the buried channel provides a major conduit for floodplain drainage. This may not always be the case: at another site near Oxford (Matchett, in prep), impermeable sediments in an oxbow divert rather than focus slope drainage.

It is clear that a detailed understanding of drainage on a given floodplain required close scrutiny of the distribution and permeability of the alluvial sediments found as well as an assessment of the connectivity of the floodplain sediments. Porous sediments within the floodplain may be isolated both from a relict channel and the current river, thereby being unable to effectively conduct water through the floodplain. The complication of groundwater movement in floodplains is amply illustrated by Carling and Petts (1992).

14.2.4   Hydrogeological classification

Soil, geology, topography, upslope catchment characteristics, vegetation, soil management practices and hillslope–floodplain coupling influence the degree of surface or subsurface flow at or within riparian systems. In terms of assessing the effectiveness of a riparian buffer strip in controlling the flux of agrochemicals, or catchment sediment/solute loads generally, it is necessary to assess the ability of chemical and biological processes to interact with hillslope runoff. All the combinations of surface to subsurface runoff, plus changes in substrate material have led a number of authors to review the competence of buffer strips on the basis of the strips' broad hydrogeological environment (Haycock and Muscutt, 1994; Lowrance et al., 1995). This systems analysis has its roots in more formal classification procedures of wetland environments. Gilvear et al. (1989) classified the hydrogeological context of wetlands in eastern England both to assess the dominate water bodies maintaining the status of the wetland but also to assess the major hydrological pathways, and thereby pollution

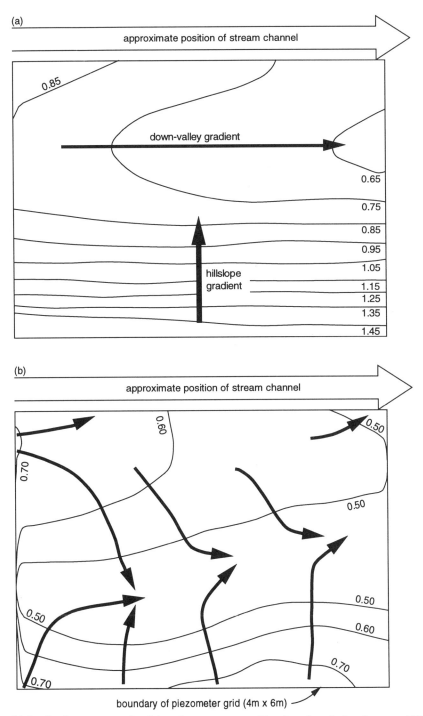

Figure 14.5 Surface topography (a) and mean water table elevation during January 1990 (b) for the grass floodplain of the River Leach studied by Haycock (1991). All units in metres

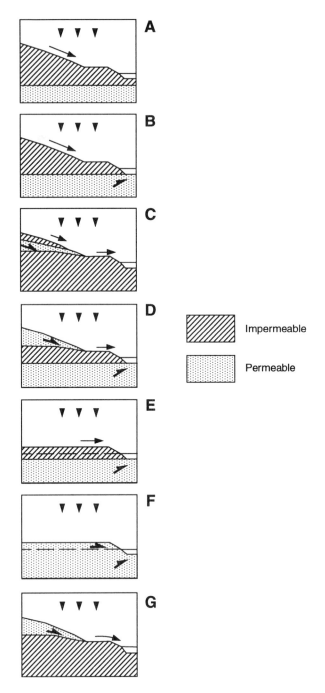

Figure 14.6 A summary of hydrogeological transect types, indicating the influence of permeable and impermeable sediments on runoff pathways. (Adapted from Gilvear *et al.*, 1989)

routes, that may affect the biological quality of the wetland ecosystem. This hydrogeological assessment of wetland vulnerability provides a useful framework when reviewing the competence of riparian buffer strips in attenuating, transforming and storing hillslope nutrients and materials.

Figure 14.6 is a summary of the hydrogeological transect types as defined by Gilvear *et al.* (1989). In the following sections, the fate of nutrients and the limits to the ability of a riparian strip to influence catchment scale fluxes of nitrogen and phosphorus in particular will be reviewed.

14.3   WATER QUALITY FUNCTIONS OF FLOODPLAINS

As outlined in Section 14.2.4, this review of the fate of hillslope nutrients and sediments within riparian buffer strips is made on the basis of the broad hydrogeological setting of the strip. This review will limit its focus to nitrogen and phosphorus dynamics within the strips and examine the hydrological and sediment factors that may enhance or limit the ability of riparian buffer strips to control fluxes of N and P.

14.3.1   Water quality functions of floodplains: an overview

In the past decade considerable research interest has focused on the role of riparian zones in controlling the movement of agriculturally derived nutrients from entering surface water. This research was built upon a growing academic interest in the role of riparian zones in influencing the quality of aquatic habitats, both in relation to fisheries (Vannote *et al.*, 1980; Platts, 1989) as well as the way in which the structure of the riparian fringe influences the diversity of aquatic ecosystems generally (Likens and Bormann, 1974; Newbold *et al.*, 1980). Paralleling this debate has been the increased interest in the broader role of wetland habitats in enhancing the biological diversity of catchments and the role of wetland environments in influencing the quality of surface water (Williams, 1990). Riparian research represents the synergy of these two distinct schools of research, aquatic conservation and wetland research. The difference between riparian research and wetland research groups is a function of the environment both study. Wetland literature suggests that a wetland, placed in an optimum location within a catchment, can fundamentally influence the quality of both the catchment and the water passing through it. The riparian group of researchers share the belief that extensive protection of riparian habitats along the whole course of a riverine/littoral environment throughout the catchment will not only fundamentally influence the quality of both adjacent habitats (e.g. Hehnke and Stone 1978; Decamps *et al.*, 1987; Johnson and Brown, 1990) but will also influence the quality of water before it enters the river network thereby protecting the aquatic ecology of the river at any section of the river.

The development of a branch of science revealing the role of riparian zones in controlling the flux of nutrients is reviewed elsewhere (Haycock *et al.*, 1993; Norris, 1993; Osborne and Kovacic, 1993); in this section the key spatial elements studied by this science will be outlined.

14.3.2   Water quality functions of floodplains: nitrogen

Hillslope–floodplain interactions

In relation to nitrogen, a growing awareness of the importance of protecting riparian zones was first demonstrated in lake protection plans, principally developed in New Zealand (Waikato Valley Authority, 1973). Definitive evidence of floodplain functions began to appear in the late 1970s (e.g. Johnson and McCormick, 1978; McColl, 1978) and early 1980s (e.g. Yon and Tendron, 1980; Schlosser and Karr, 1981). Later articles reported on detailed field observation of groundwater movement through wooded riparian zones, notably in the Coastal Plains region of the Eastern United States of America (Lowrance *et al.*, 1984; Peterjohn and Correll, 1984; Jacobs and Gilliam, 1985). A consistent trend was observed: as groundwater left arable land and entered the edge of a riparian woodland, ranging in width from 20 to 35 m, nitrate concentrations fell dramatically in so much as the water in piezometers 10–19 m into the riparian woodland had little or no nitrate present. This pattern has been observed globally (e.g., Cooke and Cooper, 1988; Pinay and Decamps, 1988; Knauer and Mander, 1989; Smith, 1989; Fustec *et al.*, 1991; Vought *et al.*, 1991; Haycock and Burt, 1993; Haycock and Pinay, 1993; Jordan *et al.*, 1993; O'Neill and Gordon, 1994).

The chemical and biological processes accounting for the reduction of nitrate-rich groundwater as it passes through the zone appear to be to conceptually straightforward (Figure 147). Firstly, there is biological uptake by plants within the riparian zone (Fail *et al.*, 1987; Fustec, 1992), predominantly during the summer months. The role of root biomass accretion versus above-ground accumulation has not been examined in detail,

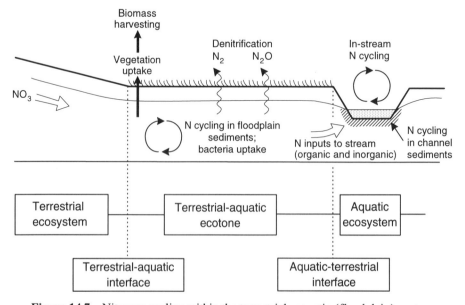

Figure 14.7   Nitrogen cycling within the terrestrial–aquatic (floodplain) ecotone

but in broad terms approximately 40 kg ha^{-1}a^{-1} can be attributed to macrophyte uptake. Secondly, denitrification can take place; this is the bacterial conversion of nitrate to di-nitrogen or nitrous oxide gas within the riparian soils and sediment (Cooper, 1990; Ambus and Lowrance, 1991; Schipper et al., 1993; Hanson et al., 1994). Measuring the rate of denitrification within riparian soil still presents many difficulties primarily in the selection of an appropriate field methodology. For example, Ambus and Lowrance (1991) chose to sample shallow riparian soil cores and used a standard laboratory acetylene inhibition assessment of denitrification potential of the soils. Correll et al. (1992) tested many methods to determine the actual and potential denitrification rates of riparian soils. Denitrification in the top 10–20 cm of a riparian soil was assessed by a complex array of cloches which sample the near-soil atmosphere; these are pumped a short distance to an in-field tunable diode infrared laser that allows continuous monitoring of gas fluxes within the sample areas. The resulting range of actual and potential rate of denitrification within a range of field experiments is of the order of 0.02–20 kg ha^{-1}d^{-1} or 7–7000 kg ha^{-1}a^{-1}. Given such rates, denitrification seems most likely to be the dominant process controlling nitrogen dynamics within riparian zones. One of the more significant limits on denitrification is the availability of soluble carbon which is required by the bacterial community to sustain their metabolic reaction. The source of the carbon is being closely questioned, i.e. whether soluble carbon in runoff from adjacent fields is sufficient or whether riparian soil and vegetation provides the dominant pool of carbon. It has been argued by Haycock and Pinay (1993) that the role of vegetation within the riparian zone is possibly to refresh soil and riparian sediment carbon reserves during the summer months as roots develop into the previously saturated sediments. As the riparian zone wets up during the autumn and winter months, these root systems and their associated fungal communities become submerged, thereby making available a pool of carbon that could be utilized as an anaerobic bacterial community develops. Vegetation, therefore, would appear to play an important role in sustaining the viability of a large soil/sediment bacterial community that has the potential to undertake denitification reactions. This proposal seems to gain support in recent research on poplar vegetation filters undertaken by O'Neill and Gordon (1994).

The debate as to whether denitrification is the dominant process still continues. The principal reason for this debate is the fact that most measurements of denitrification are taken in the top 0–50 cm of a riparian soil, which is invariably unsaturated, although rich in carbon. The saturated sediments where most of the nitrate flow occurs appear to have a low or near-negligible denitrification potential as well as low soluble carbon contents (e.g. Groffman et al., 1992; Matchett, personal communication). These observations lead to considerable difficulty: denitrification does appear to operate within the surface soils of riparian zones, but most nitrate moves through the saturated groundwater zone which may flow at a depth of 0.5–10 m below the riparian soil and yet the concentration of nitrate within the groundwater zone still appears to be reduced (although exceptions are now being reported e.g. Correll et al., 1993). Nitrate-rich water, in some sites at least, would therefore appear to have little opportunity to interact with zones of high denitrification potential. Other possible mechanisms for nitrate reduction are therefore being explored. Observations by Hill (1991) on groundwater fluxes into a riparian wetland (i.e. a permanently saturated riparian zone) suggest that nitrate may be converted into other soluble nitrogen products, e.g. dissolved organic nitrogen, such that the net retention of the riparian wetland is zero (see also Howard Williams, 1984). In many studies of riparian soils, saturation of the riparian sediments is associated with a reduction in nitrate retention. Schipper *et al.* (1991) undertook a series of laboratory experiments whereby riparian soils were saturated to varying degrees and the denitrification potential of the soil assessed. If the soils were 80–100% saturated, nitrate reduction was enhanced. However, field evidence from Haycock (1991) suggests that if riparian soils are 90–100% saturated, or if the water table within the floodplain is within 10 cm of the soil surface, then nitrate reduction within the riparian zone is undermined (Figure 14.8). In sites where groundwater does not form return flow and travels through the riparian zone as surface flow, the formation of soluble organic nitrogen compounds has been reported (e.g. Mander, 1995), although in many cases it may not have been measured in the first place.

An alternative hypothesis that has yet to be explored in detail is the possibility of bacterial communities accumulating nitrogen within their cell structure (nitrogen uptake), but not undertaking denitrification. Experiments by Hughes (personal communication) suggest that bacterial biomass accretion of nitrate can be considerable under optimal conditions (2000 mg N kg^{-1} of dry soil d^{-1}) but is limited by changes in hydrological properties of the soil due to the bio-fouling of the soil micropores which reduces the supply of nitrate to the soil mass. Field observations by Matchett (personal communication) also suggest that biomass accretion may be occurring if carbon : nitrogen ratios within the soil are not optimal for denitrification. The enhanced

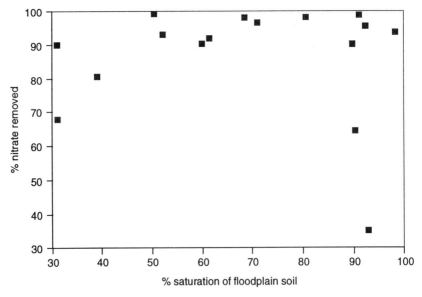

Figure 14.8 The efficiency of nitrate reduction within a floodplain system in relation to the degree of saturation (Haycock, 1991). Note that efficiency is reduced when saturation is low (too little subsurface flow) and when the floodplain is completely saturated (near-surface flow is too rapid to allow optimal conditions for nitrate reduction)

carbon dioxide respiration rate in the presence of nitrate, but with no associated rise in nitrogen gas fluxes within the soil, suggests assimilation of nitrate is occurring within Matchetts' riparian soils. If this hypothesis is correct, then there should be a seasonal trend in riparian soil and deep sediment bacterial biomass with a distinctive vertical distribution of biomass; in particular, excess biomass should occur in sediments that receive high nitrate loadings. Beare *et al.* (1991) undertook work on plant fungal respiration rates and are undertaking work on the distribution of bacterial biomass within riparian woodlands (Beare, personal confirmation). In this unpublished work, microbial biomass has two peaks in biomass with depth through the riparian sediments: the first peak is associated with the riparian soil, the second was associated with a zone where groundwater fluctuates between summer low and winter high. The lower zone (2–3 m deep) appeared to have a higher than normal bacterial biomass and, coincidentally, was associated with a zone in the sediment which would be bathed with nitrate-rich groundwater in the winter months. We have not seen this work repeated elsewhere on riparian sediments, although the distribution of bacteria in geological formations is well reported (Foster *et al.*, 1982; Hiscock *et al.*, 1991). This hypothesis does not undermine the notion that denitrification is the most significant process controlling the mass movement of nitrogen within riparian soil, but it does suggest that the process may not be *directly* involved in nitrate removal from groundwater.

The complex picture of nitrogen retention that appears to be emerging is summarized as follows. Nitrate enters riparian zones predominantly in the winter months, when above-ground biomass is dormant and high water tables within riparian sediments ensure anaerobic conditions. Within a winter runoff event, nitrate entering a riparian zone as subsurface flow may undergo direct assimilation into bacteria biomass, either in the form of nitrate and ammonium (Lewis, 1986) or assimilated into soluble N-rich proteins within the cell cytoplasm. If carbon is not limiting, but not in the right $C:N$ ratio range for denitrification (Sprent, 1987), then bacterial biomass will accumulate and in so doing adsorb more nitrate from groundwater. As the storm flux of nitrogen recedes, the water table will fall, essentially stranding the N-rich bacteria in a range of near-saturated sediments (capillary fringe) immediately above the water table. The N-rich bacteria would have to be digested and mineralized sequentially into ammonium and then nitrate (an oxidization reaction that does require aeration of the soil, which may occur as the water table is drawn down). This nitrate could then be exposed to denitrification within anaerobic microsites of the riparian sediment due to both the lowering of the carbon content of the substrate and higher soil water concentrations of nitrate relative to groundwater concentrations. Denitrification would therefore work at a lower rate, but for a longer period after the storm at reducing the nitrate content of the sediment thereby affecting the mass flux of nitrogen through the riparian zone. This could account for hydrologically-based observation of nitrate being lost at an average rate of 750 $mg\,m^{-2}\,d^{-1}$ versus rates of denitrification only averaging 20 $mg\,m^{-2}\,d^{-1}$. Denitrification would still rely on a source of carbon to sustain the rates of reduction for a longer period; this may accrue through the mineralization of the N-rich biomass. The carbon that enabled the N-rich bacteria to grow must derive from an external source. This may be from the vegetation biomass within the riparian zone, especially woodland, or be derived from dissolved organics leached from adjacent hillslopes. The most likely source, especially in a storm event, is the vertical percolation of dissolved organic carbon from the riparian soil which would initially saturate the riparian soil prior to the passage of the hillslope groundwater flux through the riparian zone. The existence of an accumulation of N-rich biomass at depth within the riparian soil may also encourage above-ground biomass into this soil/sediment zone thereby leading to an accumulation of fine root and root biomass that would be available for bacteria to use as a carbon source in the next autumn/winter. This model of nitrogen reduction within riparian zones requires various conditions. Firstly, stagnation of the water within the riparian zone must not occur and all flow must be subsurface; otherwise the oscillation of wetting and drying would not stimulate the adsorption, mineralization and transformation of nitrate within the riparian soils. Secondly, carbon must be abundant and available at critical times to ensure that denitrification occurs. Thirdly, mineralization of the bacterial organic-N to nitrate must not be hindered by oxygen penetration into the soil mass since this may result in excess accumulation of ammonium within the soil mass. Groundwater ammonium concentrations ($150-2000$ mg NH_4-N 1^{-1}) in riparian soils with a clay alluvial matrix have been observed by Matchett (personal communication) but have been explained by the leaching of urine from livestock held on the land. Since the average concentration of ammonium in human waste water is 34 mg NH_4-1^{-1} (max. approx. 200 mg NH_4-N 1^{-1}) it may be unreasonable to expect concentrations in excess of 260 mg NH_4-N 1^{-1} to be directly associated with livestock urine; this raises the prospect of the ammonium being internally generated within the riparian soil.

Hill (1991) noted the transformation of nitrate into dissolved organic nitrogen when groundwater entered a riparian zone; this response within return flow may be explained with reference to the above model. Nitrate is quickly assimilated into the bacterial biomass, but the aerobic conditions associated with water flowing over the surface of the riparian wetland, plus the small proportion of nitrate that directly interacts with anaerobic sediments capable of denitrification, means that nitrogen would stay in an organic form unless mineralization processes were allowed to occur to convert the organics into ammonium. In surface runoff systems, bacterial accretion of nitrate and equally ammonification is likely to be controlled by temperature, especially in the wetland systems of Canada.

Finally, chemical denitrification, the reduction of nitrate in anaerobic environments facilitated by an accumulation of methane or reduced forms of iron (Moragman and Buresh, 1987), is the subject of continuing research, especially in colder climates where bacterial reduction of nitrate may be hindered (Walker, personal communication) or where the presence of bacteria maybe limited due to the unavailability of carbon (Groffman, personal communication). The difficulty in quantifying chemical denitrification is a problem in common with conventional measurements of biological denitrification.

In summary, observations of nitrate reduction within riparian zones are common. The processes accounting for this change in groundwater concentration were initially stereotyped as being the role of either vegetation or bacteria denitrification. The merging of these two views, which acknowledges the possible interaction of vegetation and bacterial processes, has been further complicated by the fact that direct measurements of denitrification rates within riparian soils cannot account for the hydrological losses observed. A model suggested here introduces the possibility of nitrate being rapidly assimilated in the microbial biomass prior to conversion back into nitrate and subsequent exposure to denitrification. The hydrological regime of riparian sediments ensures that a series of diverse biochemical processes can occur (Haycock *et al.*, 1993). This may account for the resilience of riparian zones in buffering the episodic fluxes of nitrate from arable land that occur in autumn and winter periods (Cooke and Cooper, 1988; Burt and Haycock, 1993; Haycock and Burt, 1993).

Riverine – floodplain interactions

The role of riparian zones in influencing the quality of riverine water once in the river channel itself is a distinct subset of research,. The movement of water from the channel into alluvial sediments, often following relict channel networks or tree root systems, exposes riverine water to reduced sediment conditions that have the capacity to undertake denitrification (Triska *et al.*, 1989). Work on coarse alluvial sediments on the River Garonne, southern France (e.g. Pinay *et al.*, 1993; 1994) demonstrates that the combination of vegetation cover and fine organic mud accumulation on gravels ensure that conditions exist within gravel bar sediments to reduce water that passes through them. The potential of this "hyporheic" flow (Stanford and Ward, 1988) to influence stream water quality can be considerable when the volume of hyporheic flow is also significant. In small a Swedish stream the significance of nitrate reduction due to the domination of hyporheic flow ensured that gross stream nitrogen fluxes were regulated (Vought *et al.*, 1991). Similar observations on sandy alluvial sediments have been made in North America by Duff and Triska (1990). In riverine systems that are ephemeral in nature, subsurface water fluxes below the river bed mean that the role of riverine alluvial sediments in regulating nitrogen dynamics can become critical for the hydro-ecology of the environments as well as the riverine system downstream (Fisher, 1986).

The fate of nitrogen in river flood events is less clearly known, but Brunet *et al.* (1994) have sought to determine lateral deposition of fine sediments and changes in floodplain water quality during large flood events on the Garonne during the years 1991–1994. In large riverine environments (eighth-order in the Garonne example), the physical retention of nitrogen appears to be a significant factor in the overall retention of nitrogen at a catchment scale. The combined storage of particulate organic forms of nitrogen, the adsorption of ammonium and the denitrification of nitrate in shallow standing water means that as much as 2% of the total N budget for a catchment can be lost. Denitrification as a process can influence the concentration of nitrate in standing water, taking place at the water–sediment interface of the pond or shallow lake (Van Kessel, 1977). The diffusion of nitrate into anaerobic sediments is a controlling factor (Reddy *et al.*, 1978). Physical movement of the water within a pond can enhance nitrate reduction by passing more nitrate-rich water over the anaerobic sediments, but if the water velocity exceeds a critical flow the sediment–water interface can become aerated and thus the denitrification potential of the sediment reduced (Van Kessel, 1978; Bencala, 1984).

Sustained flooding of riparian environments may lead to the development of riparian wetland or wetland communities. The fate of nitrogen within wetlands, either man-made or natural, is comprehensively reviewed by other authors (e.g. Brinson *et al.*, 1981; Hook, 1988; Ward, 1989; Ambio, 1994; Mitsch, 1994).

The management of floodplains and the engineering of levées to control flood risk must be considered as having a profound impact on the ability of floodplains to regulate storm water nitrogen loads. Equally, drainage and management of floodplain soils for agricultural purposes has ensured that these floodplain soils, through prolonged aeration, have become a source of nitrogen to riverine environments (Green, 1979; Carling and Petts, 1992; Muscutt *et al.*, 1993).

14.3.3   Water quality functions of floodplains: phosphorus

Hillslope – floodplain interactions

The movement of phosphorus into riparian zones from agricultural land is dominated, in most hydrogeological environments, by surface runoff pathways. Within riparian wetlands the phosphorus content of the standing biomass is on average $10.7 \, \mathrm{g\,m^{-1}\,a^{-1}}$ (Johnston, 1991); as little as $0.00029 \, \mathrm{g\,m^{-2}}$ of phosphorus may be in active circulation within a riparian wetland in any one day. This is approximately two to three orders of magnitude lower than nitrogen circulation within similar riparian wetlands. The phosphorus cycle, in mass flux terms, is therefore considerably smaller than the gross nitrogen fluxes ($0.0027 \, \mathrm{g\,m^{-2}\,d^{-1}}$); therefore the loading of riparian wetlands with external phosphorus may have a considerable bearing on the total P cycle within the riparian zone. The phosphorus cycle, in contrast to the nitrogen cycle, does not have a gaseous phase. Whereas with the nitrogen cycle, any excess loading appears to result in the activation of the denitrification process and therefore the movement of nitrogen from a soil to atmospheric pool, phosphorus will remain within the soil–plant–biomass pool, unless it is dissolved or physically exported from the riparian zone.

The phosphorus cycle in wet soils is summarized by Moore and Reddy (1994). Figure 14.9 outlines the major phases of phosphorus dynamics in wet soils. The controlling influence of the redox state of the soil is demonstrated by the various forms of phosphorus that can occur depending of the presence of various reduced or soluble cations (Fe, Al, Ca). Within riparian zones phosphorus may enter in two primary forms, inorganic (either in solution or bound to eroded soil particles) and organic (dissolved or particulate). The fate of inorganic phosphorus as it enters a riparian zone has not been clearly traced although budgets have been produced (Yates and Sheridan, 1983; Peterjohn and Correll, 1984; Cooper and Gilliam, 1987). It is assumed that the immediate fate of inorganic phosphorus (i.e. Ortho-P, SRP) is that the cation exchange capacity of the soil rapidly adsorbs these fractions. In wet or anaerobic conditions, Fe^{+2} is the principal controlling chemical parameter of the soil. Changes in the oxidation status of the soil ensure that any phosphorus bonded in this way will go through various phases of bonding with Fe^{2+} or Fe^{3+}. Oscillations in wetting and drying render wet soil unstable as a permanent sink of phosphorus. In highly reduced conditions, and where the pH of the soil pore water is alkaline, the bonding of phosphorus with calcium occurs. The interaction of iron and calcium salts can lead to the formation of apatite, a stable Ca–phosphorus crystal. These conditions are difficult to replicate except in extremely static soil water environments. The formation of more soluble Ca–Fe phosphate compounds ensures that, if there is a fluctuation in aeration status of the soil

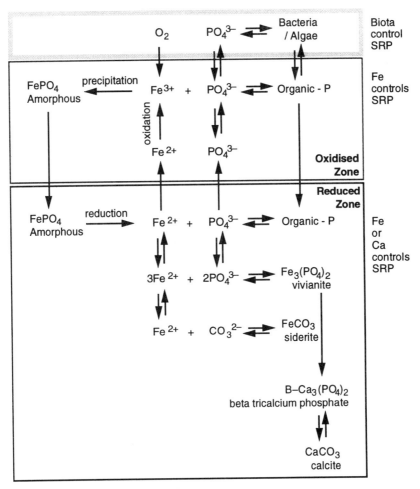

Figure 14.9 Phosphorus cycling within wet soils. (Adapted from Moore and Reddy, 1994)

and swings in the pH of the soil water, then phosphorus can return into solution. As a result of these phenomena, the seasonal dynamics of phosphorus retention can be extreme. Work reported by Smith (1989) and Uusi-kamppa and Ylaranta (1992) indicates high rates of phosphorus retention in the autumn period, but low rates in the spring. In some cases the experimental riparian buffer strips have yielded more phosphorus than they were dosed with, due to changing moisture status of the soil (Uusi-kamppa and Ylaranta, 1992).

The accumulation of inorganic phosphorus (either dissolved or particulate bonded) into the soil organic pool is facilitated by soil microfauna and riparian-zone vegetation. Rates of total phosphorus uptake in riparian woodland have been given by Fail *et al.* (1987) who reported that sites exposed to high phosphorus loading had 2.5 times more phosphorus in the woodland biomass compared to unexposed plots. Equally, Pringle and Triska (1991) noted that phosphorus adsorption within floodplain woodlands, where groundwater was at or near the surface, was strongly influenced by light gaps within the forest canopy. This suggests that soil and/or waterbody organisms were the primary sink for external phosphorus entering the woodland. Figure 14.8 suggests that in aerated systems biomass uptake would be the primary regulatory mechanism in wet soils. The fate of dead biomass itself becomes an important consideration in phosphorus budgets within riparian zones. Longer-term accumulation of organic detritus can store a small, but significant fraction of the phosphorus that may enter a riparian wetland (Mitsch *et al.*, 1979; Howard Williams, 1984). Detritus accumulation may account for as much as 1 $mg m^{-2} a^{-1}$, or, in other words, 10% of the phosphorus in the annual standing biomass (Johnston, 1991).

The fate of the organic-P fraction within a riparian zone is poorly understood. Budgets by Mander and colleagues (e.g. Knauer and Mander 1989; Mander *et al.*, 1991) suggest that organic-P is effectively retained within riparian zones, but there is potential for organic-P to be mineralized within the riparian zone and thus become a source of phosphorus. Organic-P that flows overland into the riparian zone, infiltrating at the edge of the strip, enhances the organic-P concentration of groundwater flowing through the riparian zone. This soluble organic-P, once in the groundwater, seems to change little before it enters the surface water environment (Mander *et al.*, 1991; 1995).

Finally, phosphorus bound to eroded soil particles is a significant source of P when riparian wetlands are adjacent to arable land. The topography and surface roughness of vegetation, and the moisture status of the riparian soil, control the filtration capacity of the riparian zone. Observations by Cooper and Gilliam (1987) and Cooper *et al.* (1987) suggest as much as 80–90% of the sediment exported from arable land is retained with riparian zones. The fate of the phosphorus attached to this sediment is uncertain, although prolonged storage seems viable unless extreme runoff events or gullying of the riparian soil occurs, leading to the deposition of these phosphorus-rich sediments within the riverine environment (McColl, 1978). Modelling of the fate of sedimentary bonded phosphorus within grass buffer strips by Dowon *et al.* (1989) provides a useful tool in terms of sizing of phosphorus filter strips. Some attention has recently been given to the *erosion* of ephemeral gullies within the floors of dry valleys (e.g. Slattery *et al.*, 1994) but to our knowledge the implications of such erosion for phosphorus transport into river channels has not been explored.

From this summary we conclude that phosphorus retention in riparian zones is highly variable, ranging from 6 to 90% retention. There is strong seasonality in phosphorus retention, especially in riparian soils that experience prolonged periods of saturation. In these circumstances riparian zones can become a source and not a sink of phosphorus. Only two papers to our knowledge report total saturation of riparian soils. Vanek (1991) reported that the littoral zones of a lake in Sweden become saturated with phosphorus due to the influence of septic-tank discharges and were no longer were acting as buffer zones for the lake ecosystem. Trudgill *et al.* (1991) studied P absorption in a reedbed receiving river flow and a marsh receiving point-source sewage effluent. Whereas the reed bed was effective in reducing P concentrations, the marsh was a source of P, indicating P saturation of the wetland system. The observations of Mander and colleagues (1995) of significant quantities of dissolved organic phosphorus in groundwater below riparian zones raises many questions as to whether riparian zones control the gross flux of phosphorus or, equally valuably, transform inorganic fractions into organic. The environmental fate of these organic compounds within a riverine ecosystem remains an outstanding question and therefore the eutrophication risk associated with this phenomenon is unclear.

Riverine–floodplain interactions

The deposition of nutrient-rich sediments within floodplain environments is a phenomenon that has been exploited in many agricultural societies, and in some cases flooding of land has been positively encouraged (Mahtab and Karim, 1992). The mass of sediment periodically deposited on floodplains can be considerable, but is obviously dependent on the sediment load of the river at any one point in time as well as being a function of the historical land use of the catchment. Data on sediment deposition rates within floodplain wetlands have been published by Johnston (1991); rates range from $0.6 \ g \ m^{-2} a^{-1}$ to $100 \ kg \ m^{-2} a^{-1}$. Variations in the rate of sediment accumulation locally within a floodplain depend on the degree of connectivity of the floodplain to the main channel as well as more local factors such as floodplain water flow velocity plus the roughness of the vegetation within the floodplain. Floodplains can become a significant sink for sediment-bound phosphorus. The subsequent distribution of phosphorus within the floodplain vegetation community has been documented only at a limited number of sites. Studies by Yarbro (1979, 1983) have examined the fate of phosphorus; the principal conclusion appears to be that plant uptake is limited ($<1 \ g \ m^{-2} a^{-1}$) and the long-term sink remains the soil. Phosphorus concentrations in wetland soil vary with depth (Nixon, 1980), with the greatest proportion of the total phosphorus in the readily available fraction in the organic-rich upper soil horizons. Total phosphorus concentrations decease gradually through wetland systems possibly reflecting the leaching of phosphorus in anaerobic/anoxic wetland sediments in natural sites. The mechanics of phosphorus retention in wetland soils has been reviewed by Moore and Reddy (1994).

The susceptibility of leaching or release of phosphorus from floodplain soils due to the changing redox status of these soils during flood events and fluctuating local groundwater levels means that floodplains can become a potential source of phosphorus within a catchment. Changes in the residence time of floodwater affect the nature of phosphorus cycling. Potential fears in the UK centre on the fact that wetland nature reserves are at risk both from watertable control measures that may lower groundwater levels locally, but also the fact that enhanced aeration of the wetland soils may mineralize organic phosphorus reserves thereby ensuring the soil water within the wetland favours more eutrophic plant communities. In organic-rich wetlands, the rate of mineralization and potential release of readily available phosphorus has stimulated considerable efforts to control water levels at critical periods in the year.

14.4   AN INTEGRATED VIEW

In this chapter our aim was to examine the principal role of hydrology in governing the effectiveness of the the buffering capacity of riparian ecotones. In this section we seek to provide an examination of the potential role of riparian buffer zones within an entire catchment.

In most experimental projects on the nutrient retention capacity of buffer zones there is an implicit assumption that water passes through the buffer zone under investigation. Indeed sites are chosen that meet this criteria. In many catchments this simple assumption does not hold true and the flow of water and nutrients takes a complex pathway, often bypassing riparian zones altogether. In the UK, many floodplains were underdrained in the 1960s and 1970s and farming changed from lightly grazed hay meadows to intensively farmed arable land (Green, 1979). This had two effects: firstly, the floodplains themselves became important sources of agricultural pollution. Rainfall rapidly infiltrates the floodplain soil, enters a tile drain and then passes either to an open ditch or directly into the river. Secondly, and perhaps more importantly, hillslope water is intercepted either by the floodplain drainage or by specially built footslope catch drains; in either case water moves rapidly through the riparian zone without any opportunity to interact with the floodplain soils. Even in headwater tributaries, drainage of the land can fundamentally affect the ability of buffer zones to interact with contaminated water. Often "damp" patches within fields are drained by a single pipe directly to a ditch network. Springs in many headwater catchments occur in inconvenient locations and have been piped to a permanent drainage ditch, sometime as long ago as the eighteenth century (Haycock, 1991). Ephemeral springs in the Cotswolds were seem as an asset when the field where they occurred was a pasture, allowing watering of livestock. As these fields have been converted to arable production, these springs have been piped, again to ditches. The pipes often clog and therefore allow water to return to the soil surface, but the land manager will be quick to re-pipe the spring due to the visible soil erosion problem associated with the water flowing through an often ploughed field in the winter months. Thus from headwater catchments through to large floodplain river environments, man-made ditches often interfere with the natural pathways of water movement within the landscape. Equally, conversion of riparian woodland or grass meadows to arable land use often means that the maintenance of riparian drainage and the piping of springs is perpetuated. Inevitably the impact of drainage is to accelerate the rate of drainage so that the terrestrial and aquatic systems become much more intimately linked. As noted below decoupling farmland from the river environment is vital if pollution risk is to be significantly reduced.

In addition to farm drainage alternative drainage networks have evolved within catchments. Roads and their associated conduits form a major extension to the surface drainage network of many river catchments in the UK. In permeable catchments, with a low density of river channels, road networks can increase the network length by 150%. In clay catchments with a naturally high river network density, roads do not extend the network significantly, but since there is a greater liability for water to pond and flow on the soil surface, the presence of roads can be equally significant. Observations by Hitchman (1993) suggested that up to 50% of the sediment load within the UK Hampshire Avon catchment could have originated from the road drainage network. This proportion was assisted by the fact that secondary roads have a high density of field gateways. The field gateways are often located at the lowest point of the field and therefore any drainage emanating from the field tends to collect at the gateway. The gateway is connected to secondary road drainage, often an annually cut ditch or pipe. This ensures that both the water and its load are effectively transferred to the river network. This phenomenon can be less of a problem in permeable catchments since road engineers tend to favour the use of soak-aways or the discharging water into permanent grass fields. In clay catchments these options are not available meaning that most road drainage would have to be piped to the nearest ditch.

The net result of the changing hydrology of most catchments in the UK is that the location of riparian or buffer zone systems should not be focused solely along the main river channel network; there is also a need to address the buffering of man-made drainage systems, particularly for sediment, phosphorus and pesticide runoff control.

14.4.1 Matching risk with mitigation measures

Within a catchment the risk of nutrient runoff being generated, either as surface or subsurface runoff, is spatially variable. Therefore the risk of either a river or man-made drainage network being contaminated is similarly variable. There is a need to match the risk of a network being contaminated with the location of buffer zones. In many cases the setting aside of large floodplain environments which are either themselves not at risk or are not connected to source areas at risk, fundamentally undermines the ability of these systems to function as buffer systems. On the other hand, the establishment of large floodplain environments as conservation areas relies on them *not* being connected to contaminated water which would undermine the ability of these areas to support a diversity of biotic life. Thus, the role and function of a floodplain environment, be it as a filter or a conservation zone, is dependent on the degree of hydrological interaction with areas where there is a high risk of nutrient runoff. In a policy framework, the successful deployment of buffer zones adjacent to high-risk source areas is required to ensure the competence of this strategy to protect surface water quality.

As a consequence, critical reaches of the drainage network need a higher degree of protection than others. Often, these areas consist of headwater catchments, both along ephemeral flowlines and man-made drainage ditches, where the adjacent land use is arable, usually ploughed annually. The nature of the buffer zone goes beyond the narrow limits of the riparian zone and can include parcels of arable land which were historically "damp" due to either the existence of a spring, or where the topography of the land allows both subsurface and surface flow to concentrate. For both man-made and natural drainage networks, physical alteration in the drainage structure may be required, for example discharging runoff into ponds or establishing grass soakaways (Baade, 1994). The particular issue of field gateways, and the concentration of runoff at these points presents a difficult issue. Physically moving the gateways away from the lowest point of the field and bunding the former gateway leads to the generation of a pond, which is rapidly silted up. The control of runoff gathering at the gateway can often be managed by preventing the farmer from establishing a plough board ditch when ploughing the headland edge of the field. In essence this means that the farmer would not plough any headlands within a field which are physically connected with a gateway or the lowest point of the field.

14.4.2 Towards an integrated approach

The ready supply of wholesome drinking water is now taken for granted in many parts of the world; so, too, is the plentiful supply of food. However, problems can arise when these two basic needs conflict. Most attention has been paid to nitrogen and phosphorus but, pesticide pollution and the off-site impact of eroded soil are also now recognized as important impacts of modern agriculture on the wider environment. Recent legislation, for example the European Union's Nitrate Directive (91/676), has invoked the so-called precautionary principle, seeking to prevent pollution at or near the source, rather than having to find a solution at the point downstream where problems arise. The Nitrate Directive also goes further than previous legislation in recognizing the problem of eutrophication as well as that of drinking water contamination. At the same time, money is now available to compensate farmers for being prevented from farming at the economic optimum in order to effect environmental protection; in Europe this is being achieved via reforms to the EU's Common Agricultural Policy. Controlling pollution loss from farmland can be achieved in two ways: changes to cropping practice and the creation of buffer strips. Strategic planning of entire drainage basins is the logical conclusion. Within such plans, the targeting of floodplains in order to create buffer zones assumes particular importance since this may be the best way to allow modern agriculture and good quality water to co-exist in the same catchment (Burt and Haycock, 1991; Burt, 1993).

It is important, however, to note that an integrated approach to catchment planning extends well beyond the narrow conflict between farming and water. There is growing recognition of the multiple uses of rural land and of the wider range of interests now operating within the countryside. Some of the "users", tourists for example, are consumers rather than producers. It seems clear that the agricultural industry will have to become more benign and farmers more flexible as custodians of the rural landscape. As these various demands on the countryside increase, we can expect to see more diverse rural landscapes, with more visual variety, richer in natural habitats and species, with better access for the public and maximum environmental protection. Once again, the focus on floodplains is crucial, given their high importance as wildlife corridors and their attraction for recreational pursuits. Given this wide range of interests, it seems logical that many government agencies, as well as landowners, farmers and the local community, must become involved in the determination of floodplain use. Already in Australia, the Landcare programme and other related community projects have focused attention on the integrated management of drainage basins; we expect such schemes to become much more widespread. Given their economic value and humanity's liking for streams and riparian areas, it seems likely that conflict about optimal use of these areas will increase in the near future (Bren, 1993).

14.5 CONCLUSIONS

Floodplains are seen as being critical to the control of material and energy fluxes from terrestrial to aquatic ecosystems. In agricultural catchments the role of these buffer zones has been associated with the filtering of contaminated agricultural runoff, thereby protecting the quality of surface water. The ability of riparian buffer zones to control pollution fluxes within a catchment is fundamentally controlled by the hydrology of the buffer. The hydrology of even small floodplains is complex. The sedimentology of the floodplain and the solid geology of both the hillslope and floodplain ensure that water travels through these systems via a variety of pathways. This complexity of flow has ensured that most research to date has been conducted along simple transects where pollutants flow through the riparian zone via subsurface routes.

In relation to nitrogen, the fate of the nitrogen lost from groundwater remains uncertain. Denitrification is attributed as the major biological mechanism reducing nitrate concentration within groundwater. The apparent lack of denitrification activity within saturated groundwater sediments as opposed to high rates in unsaturated riparian soils and sediments within the riparian zone has lead to a variety of other mechanics being explored recently. The fundamental role of hydrology again seems to determine the rate and competence of riparian zones in regulating nitrogen fluxes. In relation to phosphorus, buffer systems appear to offer a limited retention capacity in the long term. Storage of phosphorus within soil organic pools or biomass appears to be the only (semi-)permanent sink, so that concerns have been raised as to whether phosphorus is only temporarily stored in riparian systems. Alternating wetting–drying cycles of the floodplain soil may mean that phosphorus is transformed into dissolved organic fractions and exported from the riparian zone. Limited evidence exists for this hypothesis since complete budgets on phosphorus transport are lacking.

The ability of buffer zones to retain, transform and store nutrient runoff is not only a function of the hydrology of the buffer, but also a function of hillslope and catchment hydrology. Changes in the drainage network within most agricultural catchments have ensured that land which generates nutrient pollution is now physically well connected to drainage ditches and therefore closely coupled to aquatic ecosystems. The management of this risk requires the establishment of buffer zones; in the first instance such approaches should be applied to headwater catchments where the potential benefits are greatest. Those riparian zones which are hydrologically isolated may be designated as major conservation features of the landscape since such areas may be most easily protected.

REFERENCES

Ambio (1994) Special edition: Wetlands as nitrogen traps. *Ambio*, **23**, 320–386.

Ambus, P. and Lowrance, R. (1991) Comparison of denitrification in two riparian soils. *Soil Science Society of America Journal*, **55**, 994–997.

Anderson, E.W. (1987) Riparian area definition – a view point. *Rangelands*, **9**, 70.

Anderson, M.G. (1982) Modelling hillslope soil water status during drainage. *Transactions of the Institute of British Geographers*, NS7, 337–353.

Anderson, M.G. and Burt T P. (1990) *Process Studies in Hillslope Hydrology*. John Wiley & Sons, Chichester.

Anderson, M.G and Kneale, P.E. (1982) The influence of low angled topography on hillslope soil water convergence and stream discharge. *Journal of Hydrology*, **57**, 65–80.

Baade, J. (1994) *Gelandeexperiment zur Verminderung des Schwebstoffaufkommens in land-wirtschaftlichn Einzugsgebieten*. Geographical Institute, Heidelberg University, Germany.

Beare, M.H., Neely, C.L., Coleman, D.C. and Hargrove, W.L. (1991) Characterisation of a substrate induced respiration method for measuring fungal, bacterial and total microbial biomass on plant residues. *Agriculture, Ecosystems and Environment*, **34**, 65–73.

Bencala, K.E. (1984) Interactions of solutes and streambed sediment 2. A dynamic analysis of coupled hydrologic and chemical processes that determine solute transport. *Water Resources Research*, **20**, 1804–1814.

Bonell, M., Pearce, A.J. and Stewart, M.K. (1990) The identification of runoff-production mechanisms using environmental isotopes in a tussock grassland catchment, eastern Otago, New Zealand. *Hydrological Processes*, **4**, 15–34.

Bren, L.J. (1993) Riparian zone, stream, and floodplain issues: a review. *Journal of Hydrology*, **150**, 277–300.

Brinson M.M., Bradshaw, H.D. and Kane, E.S. (1981) Nitrogen cycling and assimilative capacity of nitrogen and phosphorus by riverine wetland forests. *Water Resources Research Institute Report* 167, 90 pp.

Brown, S., Brison, M M. and Lugo, A.E. (1978) Structure and function of riparian wetlands. In: Johnson, R.R. and McCormick, J.F. (eds), *Strategies for Protection and Management of Floodplain Wetlands and other Riparian Ecosystems*, 17–31.

Brunet, R.C., Pinay, G., Gazelle, F. and Roques, L. (1994) Role of the floodplain and riparian zone in suspended matter and nitrogen retention in the Adour River, South-West France. *Regulated Rivers*, **9**, 55–63.

Burt, T.P. (1993) From Westminster to Windrush–public policy in the drainage basin. *Geography*, **78**(4), 388–400.

Burt, TP. and Haycock, N.E. (1993) Controlling losses of nitrate by changing land use. In: Burt, T.P., Heathwaite, A.L and Trudgill, S.T. (eds), *Nitrate, Processes, Patterns and Management*, John Wiley & Sons, Chichester, 341–368.

Carling, P.A. and Petts, G.E. (1992) *Lowland Floodplain Rivers: Geomorphological Perspectives*. John Wiley & Sons, Chichester.

Clements, F.E (1897) Peculiar zonal formations in the Great Plains. *American Naturalist*, **31**, 968.

Cooke, J.G. and Cooper, A.B. (1988) Sources and sinks of nutrients in a New Zealand hill pasture catchment. III Nitrogen. *Hydrological Processes*, **2**, 135–149.

Cooper, A.B. (1990) Nitrate depletion in the riparian zone and stream channel of a small headwater catchment. *Hydrobiologia*, **202**, 13–26.

Cooper, J.R. and Gilliam, J.W. (1987) Phosphorus redistribution from cultivated fields into riparian areas. *Soils Science Society of America Journal*, **51**, 1600–1604.

Cooper, J.R., Gilliam, J.W., Daniels, R.B. and Robarge, W.P. (1987), Riparian areas as filters for agricultural sediment. *Soil Science Society of America Journal*, **51**, 416–420.

Correll, D.L., Mihursky, J.A., Jordon, T.E., Prestegaard, K.L. and Weller, D.E. (1992) Denitrification and the fate of nitrogen inputs to a riparian forest receiving agricultural runoff. A Progress Report (DEB-8917038-02) to The National Science Foundation from Chesapeake Research Consortium, Inc.

Correll, D.L., Jordan, T.E. and Weller, D.E. (1993) Failure of agricultural riparian buffers to protect surface waters from groundwater nitrate contamination. *Groundwater/Surface Water Ecotones*. Proceedings of UNESCO, MAB, IHP Meeting, Lyons, France, July 1993.

Decamps, H., Joachim, J. and Lauga, J. (1987) The importance for birds of the riparian woodlands within the alluvial corridor of the river Garonne, S.W. France. *Regulated Rivers*, **1**, 301–316.

Dowon, L., Dillaha, T.A. and Sherrard, J.H. (1989) Modeling phosphorus transport in grass buffer strips. *Journal of Environmental Engineering*, **115**, 409–427.

Duff, J.H. and Triska, F.J. (1990) Denitrification in sediments from the hyporheic zone adjacent to a small forested stream. *Canadian Journal of Fisheries and Aquatic Science*, **47**, 1140–1147.

Fail, J.L. Jr, Haines, B.L. and Todd, R.L. (1987) Riparian forest communities and their role in nutrient conservation in an agricultural watershed. *American Journal of Alternative Agriculture*, **2**, 114–121.

Fisher, S.G. (198S) Structure and dynamics of desert streams. In: Whitford, W.G. (ed.), *Pattern and Process in Desert Ecosystems*. University of New Mexico Press, Albuquerque, 119–139.

Foster, S.S.D., Cripps, A.C. and Smith-Carington, A. (1982) Nitrate leaching to groundwater in Britain from permeable agricultural soils. *Philosophical Transactions of the Royal Society of London*, Series B, 447–489.

Freeze, R.A. and Witherspoon, P.A. (1967) Theoretical analysis of regional groundwater flow: 1. Analytical and numerical solutions to the mathematical model. *Water Resources Research*, **2**, 641–656.

Freeze, R.A. and Witherspoon, P.A. (1968) Theoretical analysis of regional groundwater flow: 2. Effect of water-table configuration and subsurface permeability variation. *Water Resources Research*, **3**, 623–634.

Fustec, E. (1992) Transfer of chemical contaminants: the role of soils and plant communities in the bottom of valleys. *Comptes Rendus de l'Academie d'Agriculture de France*, **78**(6), 107–116.

Fustec, E., Mariotti, A. and Sujus, J. (1991) Nitrate removal by denitrification in alluvial groundwater: role of former channel. *Journal of Hydrology*, **123**, 337–354.

Germann, P. (1990) Macropores and hydrologic hillslope processes. In: Anderson, M.G. and Burt T.P. (eds), *Process Studies in Hillslope Hydrology*, John Wiley & Sons, Chichester, 327–363.

Gilliam, J.W. (1994) Riparian wetlands and water quality. *Journal of Environmental Quality*, **23**, 896–900.

Gilman, K. and Newson, M.D. (1980) *Soil Pipes and Pipeflow: A Hydrological Study in Upland Wales*. British Geomorphological Research Group Monograph No. 1, Geobooks, Norwich.

Gilvear, D.J., Tellam, J.H., Lloyd, J.W. and Lerner, D.N. (1989) *The Hydrodynamics of East Anglian Fen Systems*. NCC, NRA, Broads Authority Contract Final Report.

Green, F H.W. (1979) In: Hollis, G E. (ed.), *Man's Role in the Hydrological Cycle of the U.K.* Geobooks, Norwich.

Groffman, P.M., Gold, A.J. and Simmons, R.C. (1992) Nitrate dynamics in riparian forests – microbial studies. *Journal of Environmental Quality*, **21**, 666–671.

Hanson, G.C., Groffman, P.M. and Gold, A.J. (1994) Denitrification in riparian wetlands receiving high and low groundwater nitrate inputs. *Journal of Environmental Quality*, **23**, 917–922.

Haycock, N.E. (1991) Riparian Land as Buffer Zones in Agricultural Catchments. Unpublished D.Phil. Thesis. University of Oxford, UK.

Haycock, N.E. and Burt, T.P. (1993) Role of floodplain sediments in reducing the nitrate concentration of subsurface run-off: a case study in the Cotswolds, UK. *Hydrological Processes*, **7**, 287–295.

Haycock, N.E. and Muscutt, A.D. (1994) Landscape management strategies for the control of diffuse pollution. *Landscape and Urban Planning*, **3**, 313–321.

Haycock, N.E. and Pinay, G. (1993) Nitrate retention in grass and poplar vegetated riparian buffer strips during the winter. *Journal of Environmental Quality*, **22**, 273–278.

Haycock, N.E., Pinay, G. and Walker, C. (1993) Nitrogen retention in river corridors: European perspective. *Ambio*, **22**, 370–376.

Hehnke, M. and Stone, C.P. (1978) Value of riparian vegetation to avian population along the Sacramento River system. In: Johnson, R.R. and McCormick, J.F. (eds), *Strategies for Protection and Management of Floodplain Wetlands and Other Riparian Ecosystems*, 228–235.

Hill, A.R. (1991) A groundwater nitrogen budget for a headwater swamp in an area of permanent groundwater discharge. *Biogeochemistry*, **14**, 209–224.

Hill, A.R. (1990a) Groundwater cation concentrations in the riparian zone of a forested headwater stream. *Hydrological Processes*, **4**, 121–130.

Hill, A.R. (1990b) Groundwater flow paths in relation to nitrogen chemistry in the near-stream zone. *Hydrobiologia*, **206**, 39–52.

Hill, A.R. and Waddington, J.M. (1993) Analysis of storm runoff sources using oxygen-18 in a headwater swamp. *Hydrological Processes*, **7**, 305–316.

Hiscock, K.M, Lloyd, J.W. and Lerner, D.N. (1991) Review of natural and artificial denitrification in groundwater. *Water Resources*, **25**, 1099–1111.

Hitchman, J. (1993) A methodology for the identification of extended river networks linking critical source zones to the riverine environment. Unpublished MSc thesis, Silsoe College, UK.

Hook, D.D. *et al.* (1988) *The Ecology & Management of Wetlands*, 2nd edition. Vol. I. *Ecology of Wetlands*; Vol. II. *Management Use and Value of Wetlands*. Croom Helm.

Howard Williams, C. (1984) Wetlands and watershed management: the role of aquatic vegetation. *Journal of the Limnological Society of Southern Africa*, **9**, 54–62.

Jacobs, T.C. and Gilliam, J.W. (1985) Headwater stream loss of nitrogen from two coastal plain watersheds. *Journal of Environmental Quality*, **14**, 467–472.

Johnson, R.R. and McCormick, J.F. (1978) *Strategies for Protection and Management of Floodplain Wetlands and Other Riparian Ecosystems*. Proceedings of the Symposium Dec. 11–13, 1978, Callaway Gardens, Georgia. General Tech. Report W.O.-12. US Forest Service. Washington, DC.

Johnson, W.N. Jr and Brown, P.W. (1990) Avian use of a lakeshore buffer strip and an undisturbed lakeshore in Maine. *Northern Journal of Applied Forestry*, **7**, 114–117.

Johnston, C.A. (1991) Sediment land nutrient retention by freshwater wetlands: effects on surface water quality. *Critical Reviews in Environmental Control*, **21**, 491–565.

Jones, J.A.A. (1981) *The Nature of Soil Piping: A Review of Research*. British Geomorphological Research Group Monograph No. 3, Geobooks, Norwich.

Jordan, T.E., Correll, D.L. and Weller, D.E. (1993) Nutrient interception by a riparian forest receiving inputs from adjacent cropland. *Journal of Environmental Quality*, **22**, 467–473.

Kirkby, M.J. (ed.) (1978) *Hillslope Hydrology*. John Wiley & Sons, Chichester.

Klimas, C.V. (1988) River regulation effects on floodplain hydrology and ecology. In: Hook, D.D. *et al.* (eds), *The Ecology & Management of Wetlands*, 2nd edition, 40–49.

Knauer, N. and Mander, U. (1989) Studies on the filtration effect of differently vegetated buffer strips along inland waters in Schleswig-Holstein 1. Filtration of nitrogen and phosphorus. *Zeitschrift fur Kulturtechnik und Landentwicklung*, **30**, 365–376.

Lant, C.L. and Tobin, G.A. (1989) The economic value of riparian corridors in cornbelt floodplain: a research framework. *Professional Geographer*, **41**, 337–349.

Leopold, A. (1933) *Game Management*. Charles Scribner's Sons, New York.

Lewis, O.A.M. (1986) *Plants and Nitrogen*. Institute of Biology Series 166, Edward Arnold.

Likens, G.E. and Bormann, F.H. (1974) Linkages between terrestrial and aquatic ecosystems. *Bioscience*, **24**, 447–456.

Lowrance, R., Todd, R., Fail, J. Jr, Hendrickson, O. Jr, Leonard, R. and Asmussen, L. (1984) Riparian forests as nutrient filters in agricultural watersheds. *Bioscience*, **34**, 374–377.

Lowrance, R., Altier, L.S., Newbold, J.D., Schnabel, R.R., Groffman, P.M., Denver, J.M., Correll, D.L., Gilliam, J.W., Robinson, J.L., Brinsfield, K.W., Staver, K.W., Lucas, W.C. and Todd, A.H. (1995) Water quality functions of riparian forest buffer systems in the Chesapeake Bay watersheds. U.S. Environmental Protection Agency, Chesapeake Bay Program Report, EPA 903-R-95-004.

Mahtab, F.U. and Karim, Z. (1992) Population and agricultural land use: towards a sustainable food production system in Bangladesh. *Ambio*, **21**, 50–55.

Mander, U., Matt, O. and Nugin, U. (1991) Perspectives on vegetated shoals, ponds, and ditches as extensive outdoor systems of wastewater treatment in Estonia. In: Etnier, C. and Guterstam, B. (eds), *Ecological Engineering for Wastewater Treatment*. Proceedings of the International Conference held at Stensund Folk College, Sweden, March 24–28, 1991, 271–282.

Mander, U., Kuusemets, V. and Ivask, M. (1995) Nutrient dynamics of riparian ecotones: a case study from the Porijogo Catchment, Estonia. *Landscape and Urban Planning*, **31**, 338–348.

McColl, R.H. (1978) Chemical runoff from pasture: the influence of fertilizer and riparian zones. *New Zealand Journal of Marine & Freshwater Research*, **12**, 371–380.

Mitsch W.J. (ed.) (1994) *Global Wetlands: Old World and New*. Elsevier Science BV.

Mitsch, W.J., Dorge, C.L. and Wiemhoff, J.R. (1979) Ecosystems dynamics and phosphorus budget of an alluvial cypress swamp in Southern Illinois. *Ecology*, **60**, 1116–1124.

Moore, P.A. Jr, and Reddy, K R. (1994) Role of Eh and pH on phosphorus geochemistry in sediments of Lake Okeechobee, Florida. *Journal of Environmental Quality*, **23**, 955–964.

Moragman, J.T. and Buresh, R.J. (1987) Chemical reduction of nitrite and nitrous oxide by ferrous iron. *Soil Science Society of America Journal*, **41**, 47–50.

Muscutt, A.D., Harris, G.L., Bailey, S.W. and Davies, D.B. (1993) Buffer zones to improve water quality: a review of their potential use in UK agriculture. *Agriculture, Ecosystems and Environment*, **45**, 59–77.

Naiman, R.J. and Décamps, H. (1990) *The Ecology and Management of Aquatic – Terrestrial Ecotones*. Parthenon Press, UNESCO, Paris.

Newbold, J.D., Erman, D.C. and Roby, K.B. (1980) Effects of logging on macroinvertebrates in streams with and without buffer strips. *Canadian Journal of Fisheries and Aquatic Sciences*, **37**, 1076–1085.

Nixon, S.W. (1980) Between coastal marshes and coastal waters. A review of twenty years of speculation and research on the role of salt marshes in estuarine productivity and water chemistry. In: Hamilton, P. and Macdonald, K.B. (eds), *Estuarine and Wetland Processes*, Plenum Press, New York, 437–525.

Norris, V. (1993) The use of buffer zones to protect water quality: a review. *Water Resources Management*, **7**, 257–272.

O'Neill, G.J. and Gordon, A M. (1994) The nitrogen filtering capacity of Carolina Poplar in an artificial riparian zone. *Journal of Environmental Quality*, **23**, 1218–1223.

Odum, W.E. (1990) Internal processes influencing the maintenance of ecotones: do they exist? In: Naiman, R.J. and Décamps, H. (eds), *The Ecology and Management of Aquatic–Terrestrial Ecotones'* 1, Parthenon Press, UNESCO, Paris.

Omernik, J.M. (1976) The influence of land use on stream nutrient levels, USEPA Report 600/3,76,014.

Osborne, L.L. and Kovacic, D.A. (1993) Riparian vegetated buffer strips in water quality restoration and stream management. *Freshwater Biology*, **29**, 243–258.

Peterjohn, W.T. and Correll, D.L. (1984) Nutrient dynamics in an agricultural watershed: observations on the role of a riparian forest. *Ecology*, **65**, 1466–1475.

Phillips, J.D. (1989) Nonpoint source pollution control effectiveness of riparian forests along a coastal plain river. *Journal of Hydrology*, **110**, 221–237.

Pinay, G. and Décamps, H. (1988) The role of riparian woods in regulating nitrogen fluxes between the alluvial aquifer and surface water: a conceptual model. *Regulated Rivers*, **2**, 507–516.

Pinay, G., Roques, L. and Fabre, A. (1993) Spatial and temporal patterns of denitrification in a river riparian forest. *Journal of Applied Ecology*, **30**, 581–591.

Pinay, G., Haycock, N.E., Ruffinoni, C. and Holmes, R.M. (1994) The role of denitrification in nitrogen retention in river corridors. In: Mitsch, W.J. (ed.), *Global Wetlands: Old World and New*, Elsevier Science BV, Amsterdam, 107–116.

Platts, W.S. (1989) Compatibility of livestock grazing strategies with fisheries. In: Gresswell, R.E., Barton, B.A. and Kershnar, J.L. (eds), *Practical Approaches to Riparian Resources Management. An Educational Workshop*, May 8–11, Billings, MT. US Department of the Interior, Bureau of Land Management, 103–110.

Pringle, C.M. and Triska, F.J. (1991) Effects of geothermal groundwater on nutrient dynamics of a lowland Costa Rican stream. *Ecology*, **72**, 951–965.

Reddy, K.R., Patrick, W.H. Jr. and Philips, R.E. (1978) The role of nitrate diffusion in determining the order and rate of dentrification in a flooded soil. I Experimental Result. *Soil Science Society of America Journal*, **42**, 268–272.

Robinson, M. and Beven, K.J. (1963) The effect of mole drainage on the hydrological response of a swelling clay soil. *Journal of Hydrology*, **64**, 205–23.

Roulet, N. (1990) Focus: aspects of the physical geography of wetlands. *Canadian Geographer*, **34**, 79–88.

Schipper, L.A., Cooper, A.B. and Dyck, W.J. (1991) Mitigating non-point source nitrate pollution by riparian zone denitrification. In: Bogardi, I. and Kuzelka, R.D. (eds), *Nitrate Contamination: Exposure, and Control*. NATO ASI Series G: Ecological Sciences vol. 30, Springer-Verlag, New York.

Schipper, L.A., Cooper, A.B., Harfoot, C.G. and Dyck, W.J. (1993) *Regulators of denitrification in an organic riparian soil. Soil Biology and Biochemistry*, **25**, 925–933.

Schlosser, I.J. and Karr, I.R. (1981) Water quality in agricultural watersheds: impact of riparian vegetation during base flow. *Water Resources Bulletin*, **17**, 233–240.

Slattery, M.C., Burt, T.P. and Boardman, J. (1994) Rill erosion along the thalweg of a hillslope hollow: a case study from the Cotswold Hills, central England. *Earth Surface Processes and Landforms*, **19**, 377–85.

Smith, C.M. (1989) Riparian pasture retirement effects on sediment, phosphorus and nitrogen in channellised surface run-off from pastures. *New Zealand Journal of Marine and Freshwater Research*, **23**, 139–146.

Sprent, J.I. (1987) *The Ecology of the Nitrogen Cycle*. Cambridge University Press.

Stanford, J.A. and Ward, J.V. (1988) The hyporheic habitat of river ecosystems. *Nature*, **335**, 64–66.

Tansley, A.G. (1911) *Types of British Vegetation*. Cambridge University Press.

Triska, F.J., Kennedy, V.C., Avanzino, R.J., Zellweger, G.W. and Bencala, K.E. (1989) Retention and transport of nutrients in a third-order stream in Northwestern California: hyporheic processes. *Ecology*, **70**, 1893–1905.

Trudgill, S.T., Heathwaite, A.L. and Burt, T.P. (1991) The Natural History of Slapton Ley Nature Reserve XIX: A preliminary study on the control on nitrate and phosphate pollution in wetlands, *Field Studies*, **7(4)**, 731–742.

Uusi-kamppa, J. and Ylaranta, T. (1992) Reduction of sediment, phosphorus and nitrogen transport on vegetated buffer strips. *Agricultural Science in Finland*, **1**, 569–575.

Van Kessel, J.F. (1977) Factors affecting the denitrification rates in two water sediment systems. *Water Research*, **11**, 259–267.

Van Kessel, J.F. (1978) The relation between redox potential and denitrification in a water sediment system. *Water Research*, **12**, 285–290.

Vanek, V. (1991) Riparian zone at, a source of phosphorus for a groundwater-dominated lake. *Water Research*, **25**, 409–418.

Vannote, R L, Minshall, G.W., Cummins, K.W., Sedall, J.R. and Cushing, C.E. (1980) The river continuum concept. *Canadian Journal of Fisheries and Aquatic Science*, **37**, 130–37.

Vought, L.B.-M., Lacoursiére, J.O. and Voelz, N.J. (1991) Streams in agricultural landscapes? *Vatten*, **47**, 32–328.

Vought, L.B.-M., Pinay, G., Fugslagg, A. and Ruffinoni, C. (1995) Structure and function of buffer strips from a water quality perspective in agricultural landscapes. *Landscape and Urban Planning*, (in press).

Waddington, J.M., Roulet, N. and Hill, A.R. (1993) Runoff mechanisms in a forested groundwater discharge wetland. *Journal of Hydrology*, **147**, 37–60.

Waikato Valley Authority (1973) *Lake Taupo Catchment Control Scheme, Appendix VI: The History, Principles and Status of the Lakeshore Reserves Proposals – The Taupo County Report*. Waikato Valley Authority, Hamilton.

Ward, J.V. (1989) Riverine–wetland interactions. In: Sharitz, R.R. and Gibbons, J.W. (eds), *Freshwater Wetlands and Wildlife*, CONF 8603101, DOE Symposium Series. No 61. USDOE Office of Scientific and Technical Information, Oak Ridge, Tennessee, 385–400.

Williams, M. (1990) *Wetlands: A Threatened Landscape*. Blackwell, Oxford.

Yarbro, L.A. (1979) *Phosphorus cycling in the creeping Swamp Floodplain ecosystem and export from the Creeping Swamp watershed*. Unpublished PhD thesis, University of North Carolina, Chapel Hill, USA.

Yarbro, L.A. (1983) The influence of hydraulic variations on phosphorus cycling and retention in a swamp ecosystem. In: Fontaine, T.D. and Bartell, S.M. (eds), *Dynamics of Lotic Ecosystems*, Ann Arbor Science, Michigan, 223–246.

Yates, P. and Sheridan, J M. (1983) Estimating the effectiveness of vegetated floodplains/ wetlands as nitrate, nitrite and orthophosphorus filters. *Agriculture, Ecosystems and Environment*, **9**, 303–314.

Yon, D. and Tendron, G. (1980) Etude sur les forets alluviales en Europe, elements du partrimoine naturel international. *Rap. Con Eur.*, **65782**, 1–48.

15 Risks and Resources: Defining and Managing the Floodplain

E.C. PENNING-ROWSELL and S.M. TUNSTALL
Flood Hazard Research Centre, Middlesex University, Enfield, UK

15.1 INTRODUCTION: THE HUMAN DIMENSION OF FLOODPLAIN PROCESSES

The analysis of floodplain processes tends to be dominated by discussions of geomorphological, hydrological and hydraulic processes. This, of course, is the physical background to the evolution of floodplain areas. However, some of the most important processes to be observed on floodplains today – if not the most important processes – are those of development and urbanization. These processes need at least as much attention as the others.

Urbanization and floodplain development reflect the fact that floodplain land in all countries – developed or developing – is a valuable resource. This land has natural attributes that means it provides sites for agricultural production, urban expansion and industrial location, and the pressures for the development of floodplains are substantial. Some of these pressures are related to the recreational opportunities in floodplain locations and the maintenance there of important nature conservation, amenity, cultural and historic values. One only has to observe the circumstances in countries as different as the Netherlands and Bangladesh to show the human importance of floodplain locations, and the pace of development and change there.

This chapter therefore, first identifies and evaluates the process of urban development of floodplain areas, and the implications of this development for flood damage potential and the natural function of such areas for flood conveyance and storage. It also, secondly, examines research aimed at identifying more clearly the areas at risk from flooding and the vulnerability of human populations located there. We also, thirdly, identify a number of cases of policy evolution concerning the interaction of land-use planning and water planning (particularly in respect to flood defence). This analysis is undertaken with an emphasis on the situation in Britain (specifically focusing on England and Wales), but with a strong international comparative theme. This comparison is designed to show how the difficulties of controlling floodplain development in Britain are mirrored in circumstances elsewhere.

Floodplain Processes. Edited by Malcolm G. Anderson, Des E. Walling and Paul D. Bates.
© 1996 John Wiley & Sons Ltd.

15.1.1   The human processes of floodplain change

Human activities affecting floodplains can be differentiated by whether they occur on or off the floodplain (Table 15.1).

New development on floodplains means that flood damage potential is created or enhanced, or floodplain storage is reduced if the property is raised above historic and predicted future flood levels by landfill. In areas where economic activity is growing there are constant pressures to intensify urban development, through increasing the density of land use, again adding to flood damage potential. Modern commercial, residential and industrial premises also tend to have more and more equipment that is vulnerable to flooding, including electronic goods and information handling systems. In many areas of the world there are processes of agricultural intensification that are promoted to produce foodstuffs for a growing population. But all this intensification of activities in flood-prone areas creates increased flood damage potential.

The effects of development off the floodplain can be just as significant in that they can have "downstream effects" on floodplain areas lower down the catchment. Urbanization that is not planned with in-built local storage for flood flows tends to increase both the volume of runoff and the speed of concentration. This means that flood severity can increase in the floodplain storage areas downstream, while at the same time the lead time before the flood waters arrive there is decreased. Again flood damage potential is increased and the opportunities for ameliorative action during flood events are reduced.

These human floodplain processes often result from poor understanding or decision making about the consequences that may ensue if flood-affected areas are developed. Individual floodplain occupants are taken by surprise when the floods come. Governments, faced with a major flood disaster, are also surprised that local and regional authorities have given permission for such developments. The agencies responsible often think that someone else is to blame: the water agencies blame the land-use planners and the land-use planners blame the politicians. After the inevitable flood has subsided, and the damage and disruption have been forgotten, the process of development begins again.

A fundamental problem in this field is therefore how to maintain the memory and vigilance about floods in the many years or decades between serious flood events in

Table 15.1   Processes of urbanization and development on and off floodplains

---

1. On the flood plain
    - New development (on the floodplain floor)
    - New development (raised above flood levels)
    - Intensification of existing development:
        Infilling
        Extensions to existing property
        Increasing investment within the existing fabric
    - Agricultural intensification
2. Off the floodplain
    - Development increasing flood volumes downstream by reducing infiltration
    - Development decreasing flood warning time (increasing runoff velocities)

---

such a way as to prevent unwise floodplain development. This difficulty is a function of the complexity of the periodicity and uncertainty of flooding, which are widely misunderstood.

Centuries ago such matters appear not to have been difficult: towns and villages were built where there was maximum natural protection against floods or buildings were adapted to minimize the impact of expected flooding. But with the more recent rapid development of urbanization away from these relatively safe urban cores, and with a transient modern urban population and modern institutions with their constant reorganizations and ever-changing staff, the maintenance of this memory and the safeguards that it promotes are all too often missing.

This problem is compounded by intersectoral and interdepartmental "friction". Each side of the interface between water and land-use planning may be efficient and competent in its own field, but the interface of these two fields can be highly bureaucratic, poorly integrated and inefficient. Recent research has focused on this interface between land and water management with regard to floodplain areas and has identified the best practices for the control of floodplain development being followed in a number of developed countries (Tunstall *et al.*, 1993; Chatterton *et al.*, 1994). The results of this work are described in this chapter and serve to draw out lessons that may be applicable elsewhere.

15.1.2 What are we managing floodplains for?

In this context an important policy issue, as well as a significant research focus, is determining why and for what we are managing floodplains, and for whom. The answer to these questions will determine the type and extent of policy responses by the various levels of government, and the nature of the research we undertake to investigate this floodplain management.

While at first sight this appears a simple question, in reality it is not. This is because in all countries of the world, rivers and their floodplains create assets as well as liabilities: resources as well as hazards. This is just as true about the floodplains of Britain and Europe now (Penning-Rowsell *et al.*, 1986; Penning-Rowsell and Fordham, 1994) as it was for the ancient river-dependent civilizations of Egypt, or the United States at the formative stages in hazards research when the resource issue was clearly understood to underlie hazards investigations (Burton *et al.*, 1978).

So what are these resources? In Britain today the emphasis is on the environmental assets in river corridors that are seen as particularly valuable. It was (and still is) the agricultural potential of river valleys such as the Mississippi or the floodplain of the River Nile that attracted settlement there. Here the annual sediment deposition has created the fertile floodplain and allowed the agricultural production that sustained the Egyptian economy and empire. Elsewhere it is the floods and the water-dominated environments that they create that leads to the valued wetland habitats and important landscape features, such as on the British floodplains.

However, the first point of departure in terms of the objectives of floodplain management has tended to be the reduction of flood damages. This apparently self-evident objective, being the minimization of hazard impacts, ignores the fact that such an objective should not be pursued regardless of cost (economic or environmental).

Therefore, it is necessary to balance the objective of damage reduction by replacing this with cost and loss minimization: in its simplest form this is the minimization of the aggregate monetary costs and losses in the form of both flood damage and the costs of any flood alleviation scheme. Herein lies the basis of benefit–cost analysis, designed to evaluate the implementation of schemes to a standard which maximizes the difference between benefits and costs, thereby allowing the selection of those investments which show the greatest return to society (Penning-Rowsell and Chatterton, 1977; Parker *et al.*, 1987; Ministry of Agriculture, Fisheries and Food, 1993).

In these respects society and governments are generally attempting to tackle the residue from the mistakes of the past, by designing flood alleviation schemes to protect the considerable amount of unwise development that has already occurred on floodplains. These are existing urban developments which have been allowed to encroach into the flood risk area, owing to a number of factors, and are then retrospectively protected against the flooding likely to be experienced into the future. Floodplain development having occurred, benefit–cost analysis is designed to guide decisions in a way that determines the standard of protection in terms of a balance between the potential damages to be avoided and the costs of so doing.

Such an objective, of course, is insufficient with regard to the future. What we seek to do in this respect is to minimize future flood losses, by controlling the development of floodplains in such a way that land uses there which have high damage potential are discouraged, or prevented. Alternatively, we do not seek to minimize losses, but seek a balance between the adverse and favourable effects of floodplain development: a judgement is made about future damage potential being balanced by the contribution to the local, regional or national economy from the development if it is allowed to proceed in the floodplain area. The latter approach is complex, since it involves a judgement about the national contribution of local economic developments.

All this, however, is highly anthropocentric. A more wide-ranging objective would be the management of rivers and their floodplains within a holistic approach (probably emphasizing the whole catchment) aimed at maintaining and enhancing all assets from which is derived human and environmental benefit (Gardiner, 1994). Policies here would seek to maintain the amount of natural assets and protect the critical natural resources. Here we are defining for this policy field the objectives that would meet the criterion of "sustainability". It necessarily involves the management of both the use of floodplain areas (and those areas contributing flood waters to the floodplain), and at the same time managing the rivers located there such that past urban developments are protected to the extent that this is justifiable. At the same time it emphasizes protecting the particular environmental assets of rivers and river corridors (of which the floodplain constitutes a vital part), given the uniqueness of these environments and the folly of urban development occurring there which could occur elsewhere.

This policy direction, and set of objectives, will lead to a tendency towards retaining and protecting natural and semi-natural floodplains, thus retaining environmental values (because these are particularly important and urban development encroaching on to them is particularly unwise). This in turn will need a framework of land-use planning designed to protect assets rather than encouraging adverse development.

This approach also should have designed within it a system of policy instruments embracing disincentives regarding the use of floodplain land for unsuitable purposes.

This would imply a reduction in the grant-aid subsidy from national governments for flood protection, since this subsidy will encourage unwise use of floodplains over the medium to long term. Other economic instruments designed to protect or compensate those liable to flooding might also be discontinued (e.g. flood insurance and public relief) in an attempt to reduce the likelihood that human occupation of floodplains and river corridors will take place at the expense of the natural values inherently and uniquely located there.

Such a set of objectives – tending towards a more ecocentric approach – are difficult to translate into specific project decision and appraisal criteria, although the developing field of environmental economics can help here in certain highly specific circumstances (Green and Tunstall, 1991). However, these objectives do serve to indicate that the goals of floodplain management are more than just the minimization of flood losses to society, and that they do involve environmental values and should be set within a future-oriented strategy rather than one designed (as now) to solve retrospectively the problems and hazards created by unwise decisions in the past.

15.2   INSTITUTIONAL COMPLEXITY AND THE IMPLICATIONS: THE EXAMPLE OF BRITAIN

Any evaluation of floodplain processes and the consequent policy response must start with an analysis of the institutional context to the interface between land and water planning, since this is the basis for managing the human floodplain processes. We start our analysis with the situation in Britain, and this complements more detailed analysis elsewhere (Penning-Rowsell *et al.*, 1986; Parker, 1995).

The institutional context of floods and flood management in Britain is underlain by a three crucial principles. First, it is dominated by a legal tradition emphasizing common law, which places the first and basic burden of responsibility for the management of floods upon the riparian owner, not government (Wisdom, 1975). This individual is obligated not to exacerbate flooding downstream by any works he or she might implement to protect their property, and can be sued for wrongdoing in this respect. Thus policy is based on precedent, built up through litigation over the years, rather than a codified set of statutes, as in other countries such as those dominated by the Napoleonic legal code. It therefore can evolve and is based on judgement rather than absolutes.

Secondly, floods and floodplain management are tackled using a range of policy instruments, rather than a single approach. Thus we have economic instruments in the form of grant aid from central government to local organizations, and regulatory instruments in the form of land-use control exercised by local authorities over development in flood risk locations. The private sector is involved in the arena of flood insurance, but in general the important policy decisions about the public good of flood alleviation are made by state authorities in order to prevent the situation where individuals opt out of flood alleviation without cost penalty (Parker *et al.*, 1987).

Thirdly, the implementation of policy is the responsibility of a range of institutions operating both at central government level and the local level. The main agencies are described below. This arrangement is necessary because flood problems and their solutions are likely to be highly location specific, but are deemed also to require resources and expertise from higher levels of government, not least because the welfare state approach to social matters – including the protection of people from the impacts of natural hazards – concludes that individual flood victims should not be left to their own devices without effective help from the state. Thus in the field of flood defence, in Britain, there is a range of institutions devised to exercise permissive powers and involving every level of government. No one is clearly overall "in charge".

Highest in the institutional hierarchy of floodplain management in Britain (Penning-Rowsell *et al.*, 1986) is the Ministry of Agriculture, Fisheries and Food (MAFF). The Ministry is the responsible agency for implementing and enforcing legislation in this field (Table 15.2), and in particular the Land Drainage Act 1991, which gives powers to the National Rivers Authority [1] and local authorities to implement flood alleviation schemes.

Grant aid is given on these schemes from central government if certain criteria are met, including economic efficiency (the benefit–cost test), technical proficiency (as judged by MAFF's regional engineers), and environmental soundness (as judged by the approval for schemes required by MAFF of a number of environmental organizations involved in the areas concerned). The Ministry has recently issued a strategy document (Ministry of Agriculture, Fisheries and Food and Welsh Office, 1993), setting out its priorities. These are the implementation of flood warning systems, urban coastal

Table 15.2 Environmental legislation and European Community Directives relevant to inland flood defence works in England and Wales

The key environmental legislation relevant to inland flood defence works in England and Wales include:

Wildlife and Countryside Act, 1981
Water Resources Act, 1991, Sections 2(2), 16 and 17
Land Drainage Act, 1991, Sections 12 and 13
Land Drainage Improvement Works (Assessment of Environmental Effects Regulations SI 1988 No. 1217)
Town and County Planning (Listed Buildings and Conservation Areas) Act, 1990
Ancient Monuments and Archaeological Areas Act, 1979

The European Community Directives which are relevant to the environmental aspects of inland flood defence works include:

Council Directive, 79/409/EEC on the conservation of wild birds
Council Directive 85/337/EEC on the assessment of the environmental effects of certain public and private projects on the environment
Council Directive 92/43/EEC on the conservation of natural habitats and of wild fauna and flora. (This new Directive has had to be implemented into Member States' legislation by 1994. Under its terms, Member States are required to list potential "Special Areas of Conservation" and to designate these sites after clearance with the EC Commission)

---

[1] In April 1996 the National Rivers Authority (NRA) merged with Her Majesty's Inspectorate of Pollution (HMIP) to form the Environment Agency (ENVAGE).

defence, urban flood defence, rural coastal defence and existing rural flood defence and drainage schemes, and new rural flood defence and drainage schemes, in that order of priority.

For the protection against flooding at the coast, the Ministry is encouraging the setting up of coastal defence groups for specified stretches of the coast, and the development of shoreline management plans. In the fluvial environment, the Ministry is also encouraging the development of river catchment plans, and encouraging the integration of flood warning and defence measures within catchment areas. Thus the central government steer in policy direction is towards consistency, environmental protection, and also economic efficiency.

At the same time the Ministry is working with the Department of the Environment to discourage the development of urban areas in flood risk locations (Circular 30/92), by emphasizing that where flood risk arises, this should always be taken into consideration by local planning authorities in preparing their development plans and in determining planning applications (Department of the Environment et al., 1992). Such is the autonomy, however, of local authority land-use planning departments that neither the Ministry nor the Environment Agency can overrule them in planning terms, and cannot require the advice from the Environment Agency to be adhered to with regard to development in flood risk areas. However, the Department of the Environment can intervene and call in a planning application and can overrule the rejection of a planning application, including rejection on grounds of flood risk, on appeal.

The second major player in the flood and floodplain management field in England and Wales is the Environment Agency which also has responsibility for water resources, water quality, fisheries, recreation and navigation and important associated duties to conserve and enhance the water environment. The National Rivers Authority was incorporated in April 1996 into the Environmental Agency, bringing together key regulatory pollution control functions for water, air and land, previously the responsibility of Her Majesty's Inspectorate of Pollution, the National Rivers Authority and the local authorities. Whereas the Agency has the characteristics of a regulator in the pollution control field, in the flood defence field it has an executive responsibility, made most obvious in its promotion and construction of flood alleviation schemes to protect areas which suffer from flood risks. The Environment Agency is a non-departmental public body, answerable to Parliament through the Secretary of State for the Environment (and not through the Ministry of Agriculture, Fisheries and Food). Income for flood defence comes from charges on County Councils for flood defence, as well as grant aid from the Ministry of Agriculture, Fisheries and Food for capital schemes.

Institutionally, as far as flood defence is concerned, the Environment Agency retains many of the pre-existing characteristics of the National Rivers Authority, the Water Authorities and the Catchment Boards that preceded it (Penning-Rowsell et al., 1986). It is organized on a regional and catchment-wide basis and it retains the regional structure of flood defence committees that have executive responsibilities for planning a programme of flood defence works. As a statutory consultee on local authority development plans, the Environment Agency has to be consulted on proposals to allocate land for development at an early stage before plans are deposited for general public consultation.

The Environment Agency's consultee role is of particular significance since development plans are now given greater prominence in the planning system (under the Planning and Compensation Act 1991 and PPG12: Development Plans and Regional Planning Guidance, February 1992). Planning decisions on individual applications are required to accord with development plans unless material considerations indicate otherwise. Consultation between the Environment Agency and the local planning authorities on development plans is now seen as the main mechanism for integrating land-use planning and planning for the water environment. The Environment Agency also provides advice to local authority planning departments about planning applications in areas liable to flooding (and areas liable to exacerbate flooding but not located themselves on the floodplain), although ass indicated above this advice may be rejected by the local authority planning departments.

The Environment Agency also has the responsibility for forecasting floods, and currently is responsible for advising the police about the incidence of future events (although the role of the police in this respect is to change during 1996 in favour of the Environment Agency). The police thereafter have the responsibility of disseminating these warnings to the public, following a national agreement in 1968 to structure flood forecasting and warning systems in this way.

Given the broadening objectives of floodplain management (see above) and the plan-led land-use planning system, the local authority planning departments have an increasing role in this field. There are no statutory planning agencies in England and Wales at regional level, only voluntary groupings of local planning authorities which have responsibility for relatively small areas, usually only part of major flood plains or river catchments. Only the local planning authorities through their allocation of land for development in plans and through decisions on planning applications have the power to allow or prevent urban development in flood risk areas, and this is the vehicle that creates future increase in flood damage potential. Local authority planning departments also have the power, thereby, to retain and even enhance the natural assets in floodplain areas, in terms of amenity and nature conservation values, by not allowing inappropriate development. Only development covered by the General Development Order (principally government and quasi-government initiatives) is outside the control of local authority planning departments, along with decisions about the use of agricultural land and decisions about minor urban development (such as small extensions to individual properties).

In this respect local authority planning departments are part of the environmental protection organization devised in Britain to include the designation by statutory conservation agencies of special areas such as Areas of Outstanding Natural Beauty, Sites of Special Scientific Interest, National Parks and other amenity-based designations. Many of these designated sites involve river valleys and floodplain land. Local authorities have also been given the responsibility for implementing Agenda 21 at the local level (National Rivers Authority, Thames Region, 1994), which will be used to direct the path of floodplain management towards sustainability, as supported by the government at the Rio Convention. The Rio Declaration on Environment and Development has as one of its guiding principles that environmental issues are best tackled with the participation of all concerned citizens at the relevant level, and Agenda 21 also stresses that broad public participation in decision making is fundamental for sustainable development. This is likely to give an increased impetus to the development of participatory approaches to the planning and management of floodplain resources.

From this description of institutional arrangements in England and Wales it can be seen that the administration of the interface between land and water management is crucial to the direction, intensity and impact of human-induced processes on floodplains. Floods and floodplain management are not simply concerned with the

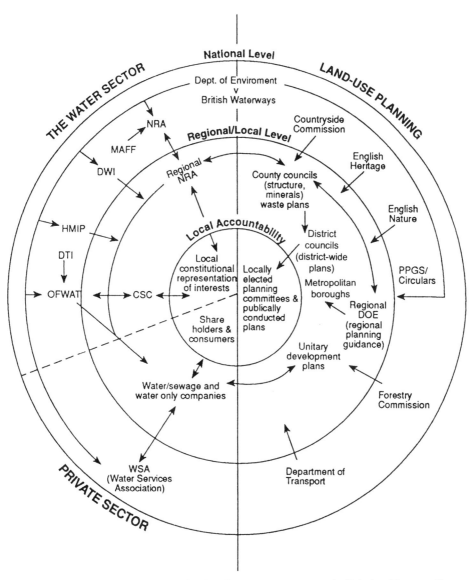

Figure 15.1 The interface between land and water management in Britain. (Source: Slater *et al.*, 1993. Reproduced by permission of Simon Slater). After April 1996 the NRA was replaced by the Environment Agency

management of water, as often it may appear, but the management of the interface between water and land, and the human occupants of that land.

But herein lies the fundamental problem: we are dealing here (as in most countries of the world) with two sets of very different and fragmented organizations that necessarily and inherently only give part of their effort and attention to the way that their policies interact. This can mean that resources such as the floodplain land being managed at this interface are neglected or that unwise decisions result. The general situation is neatly summarized in Figure 15.1 (Slater *et al.*, 1993), which illustrates the complexity of the interface between water planning and land-use planning, and provides also the additional dimension in terms of the juxtaposed position of the private sector and state authorities.

15.3 PROGRESS IN DEFINING TILE FLOODPLAIN AND ITS HAZARDS

15.3.1 Recent advances in defining flood hazards and vulnerability

Whatever the institutional arrangements, and whichever the country or continent, a vital consideration governing the interface between land and water management, in the context of flood defence, is accurate definition of the area at risk from flooding: without this, policy development is stymied. Better information in this area should significantly decrease the uncertainty about the extent of flood problems, and allow the control of floodplain development to proceed on a sounder footing.

Important here are the recent developments in the hydraulic modelling of floods and the extent of the floodplain. The main component of this advance has been the development of mathematical modelling of flood-affected areas, through the development of numerical methods of translating hydrological records, through hydraulic analysis, into predictions of the geographical extents of floods of different return periods. These developments have been a major advance in the last 10 years and have created an important new tool for improved floodplain management; previously reliance had to be put on patchy historical records or expensive physical models.

Table 15.3 gives a list of the available technology in this field in Britain (National Rivers Authority, 1992b). Some of the earliest developments occurred in the United States, with the work of the Corps of Engineers, and the development of the HEC-2 computer models. These have been further developed recently, and have been joined by a number of similar models produced by consulting engineering firms and national hydraulics laboratories for use in the appraisal of flood defence schemes (e.g. Halcrow; Mott MacDonald; Delft Hydraulics; the Danish Hydraulics Institute).

As computer technology has advanced, so these models have become more readily available for all to use, rather than the domain of organizations with large mainframe computer facilities. Also as these mathematical models have become more sophisticated, so the development has occurred of flood damage estimation procedures building on the flood extents predicted by the mathematical models (Penning-Rowsell *et al.*, 1985). Both these aspects are discussed below.

Table 15.3 Availability and summary use features of hydraulic computer programs applicable to flood level estimation (abstracted from National Rivers Authority, 1992b)

Company	Model name	Ease of use
Hydraulics Research, Wallingford	FLUCOMP	Not very user-friendly, but in wide use
Hydraulics Research, Wallingford	SALMON-F (formerly LORIS)	New interface under development
Halcrow	ONDA	Menu system handles input model set-up output, graphics. File editing approach also possible
Mott MacDonald	HYDRO	Menu system for setting up and identifying model
Haestad Methods Inc	HEC-2	Menu input via "The Friend"
Danish Hydraulics Institute	MIKE 11	Menu input
Binnie and Partners	DAMBRK UK	User-friendly input and output
Binnie and Partners	FWAVE	–

Advances in the UK: the application of the ONDA model to the lower Thames catchment

Thames Region of the National Rivers Authority has been appraising its flood problems throughout the Thames catchment, principally focusing on the main river stem.

Following initial analysis at a pre-feasibility scale (Green *et al.*, 1987), full feasibilities studies were undertaken for Maidenhead (Lewin, Fryer and Partners, 1992), and a more major investigation was also undertaken for the reach of the Thames between Datchet and Teddington (to the west of London). This appraisal involved using Halcrow's ONDA model, operating on a personal computer platform, employing hydrological inputs provided by the Institute of Hydrology.

The model used nearly 1000 nodes, and sought to define flood extent both with the existing river geometry and following the implementation of a system of bypass channels, and Figure 15.2 shows the extent of flooding modelled for a major event (the 204-year flood). The river reach was known to have been flooded in 1947, but then there were many fewer properties built in the area at risk. In any case the available records of the detailed extent of flooding in 1947 are few and far between, and therefore the mathematical model greatly facilitated gauging the extent of the hazard the area now faces.

The results of this hydraulic modelling were used to calibrate the flood damage potential in the river reach as a whole, resulting in the estimates given in Table 15.4. This shows that up to 12 500 properties could be affected by a major flood as indicated in Figure 15.2, resulting in the need to evacuate approximately 35 000 people for a period of approximately one week, this being the likely duration of a major Thames flood of that return period.

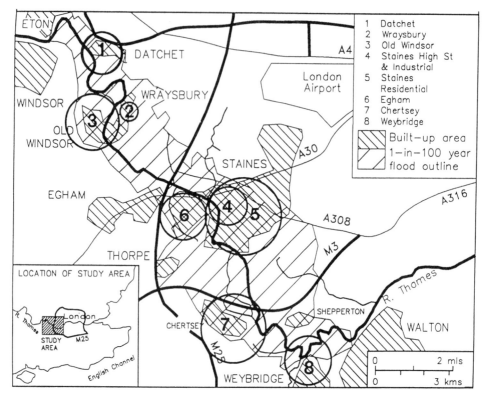

Figure 15.2 Results of modelling the extent of flooding in the Datchet to Teddington area of West London using the Halcrow ONDA model

Table 15.4 The numbers of properties at risk from flooding in the Datchet to Teddington area of West London (source: National Rivers Authority, 1992)

Return period	Number of properties affected
5 years	404
9 years	958
25 years	5258
56 years[*]	8599
101 years	10455
204 years	12458

[*] Approximately the return period of the 1947 flood event.

The major limitation of this mathematical modelling approach to defining areas of flood risk is the intensive nature of data requirements. Extensive field surveys of floodplain cross-sections are needed to complement data on the river channel itself. A large amount of data on the threshold floor heights of large number of properties is required to determine flood damage potential, since until now the mathematical hydraulic models cannot easily be used in conjunction with a digital terrain model to generate flood envelopes and depths. The possibility of linking these hydraulic models through a digital terrain model into a comprehensive hazard assessment model needs to be explored in the future.

Regional-scale flood hazard modelling in Germany

The challenge in this area of research is to devise methods of flood hazard assessment which can operate at the regional rather than just the local scale.

This is because without that kind of regional "overview" there is a danger that analysts will subsequently concentrate on individual projects for appraisal that do not warrant such close attention: we need "to see the wood for the trees" first. With the advent of mathematical hydraulic models, such as the ONDA model described above, it is possible to simulate in detail the extent and depth of flood waters in an area where detailed topographical information is readily available. What is more difficult is to model the likely magnitude of flood impacts at a regional scale and with more modest data inputs, thus building on the work of Green *et al.* (1987). The overview approach will allow us to appraise whether more detailed investigations are worthwhile.

This topic of enhancing regional-scale pre-feasibility hazard appraisal methods has been researched in a module within the EUROflood project (Penning-Rowsell *et al.*, 1992; Penning-Rowsell and Fordham, 1994). The research objective was to devise a regional-scale model which could take as its inputs data that is likely already to be available in the range of different countries of the European Union. Therefore, for example, information on land-use patterns is officially available in mapped form in certain countries (e.g. the United Kingdom and the Netherlands), whereas in other countries information on land-use in floodplain areas is only available by using property numbers and values which are part of a classification of industrial production. In other countries such data is only available in the form of local property tax records. To develop a Europe-wide approach for consistent analysis needs systems that are able to use whatever data is available, rather than incur the expense of field data collection over a large area.

The research has been pursued in the first instance through the development of a case study in Germany, oriented towards developing a prototype model. The case study area is shown in Figure 15.3, forming a large area of the German North Sea coast. The area is partially protected by ring embankments. In this case the smallest administrative unit, termed a "Fluren", was used to input existing data on land-use and industrial production types, and no field surveying of land-use was undertaken. The regional-scale model accumulates potential damage information for the flood-affected areas, using either a sloping hydraulic profile or a flat flood profile as likely to be found in coastal locations. Figures 15.4 and 15.5 show the relationship between potential flood damage and flood scenarios in relation to water levels above sea level, thereby, first, quantifying the

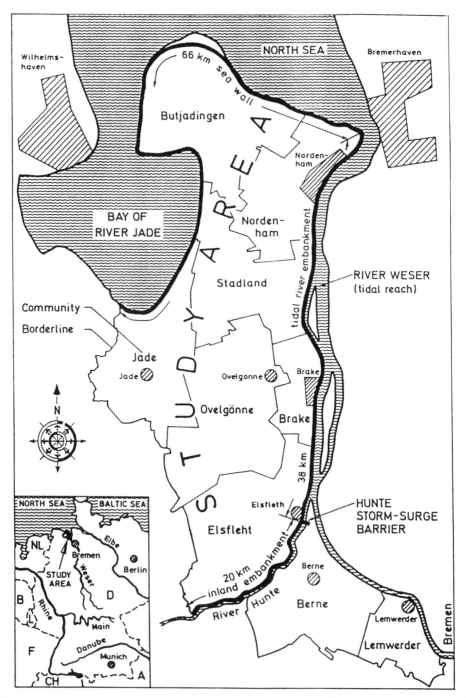

Figure 15.3 Location of the case study for the regional-scale flood hazard impact model within the EUROflood project. (Source: Penning-Rowsell and Fordham, 1994)

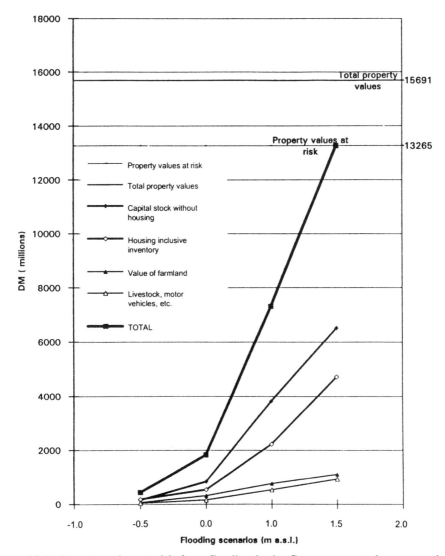

Figure 15.4 Property values at risk from flooding in the German case study concerned with regional-scale flood modelling. (Source: Penning-Rowsell and Fordham, 1994)

nature and extent of the hazard likely to be experienced in this region and, secondly, setting up a data base and model for exploring the enhanced damage potential resulting from different sea level rise scenarios.

The use of such a model is not restricted to project appraisals. One of the main advantages of the development of such a model seen in Germany is to identify areas of particularly high flood vulnerability, so as to control land-use development in these areas. Making the land-use decision takers more aware of the flood damage potential in this way should guide their strategic decisions about particular land uses to locate in the

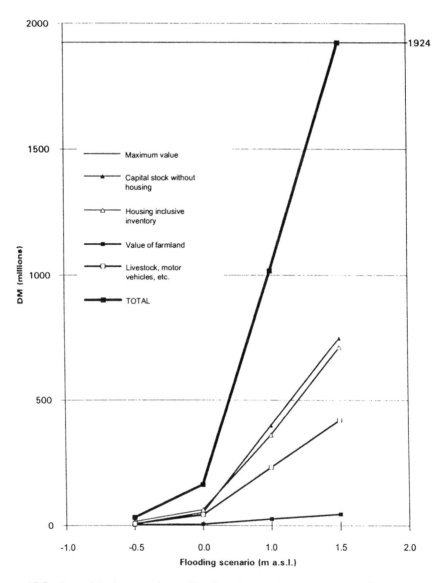

Figure 15.5 Potential damage from flooding in the German case study concerned with regional-scale flood modelling. (Source: Penning-Rowsell and Fordham, 1994)

region, and this is particularly important given the growing disenchantment there with structural flood alleviation policies and the proactive search for an emphasis on preventative action designed to forestall the build-up of flood damage potential into the future. The results from the model could also be useful at a regional scale in setting flood insurance premiums, which hitherto have not been related in any sophisticated way to flood risk but have been set simply and crudely in relation to the aggregate profitability of insurance cover in this field.

Therefore, in summary, we see two parallel advances: more accuracy in defining flood extent and the capability of obtaining an overview of hazardousness. The interface between water and land management is becoming more accurately defined, thus allowing better decision making.

15.3.2 Synthesis of research results: defining flood vulnerability

We have shown above how progress has been made in defining floodplains, in terms of enhanced mathematical modelling to convert flood discharges into flood extents for a range of return periods. Obviously the results of these models need calibration against known events, and can then be used predictively to anticipate the extent of floods of return periods greater than that experienced at any one site, and also to estimate flood extent in areas for which no historical extent records are available.

However, parallel research work has shown the complexity of flood vulnerability, in terms of the effect of floods of different return periods and other characteristics on the population at risk. Figure 15.6 gives an equation relating flood vulnerability to a number of factors (socio-economic; property and infrastructure; flood characteristics; warning variables; and response variables). In turn these factors are broken down into a number of variables, which research over a period of some 20 years (Penning-Rowsell and Chatterton, 1977; Penning-Rowsell and Fordham, 1994) has shown to be important in affecting the vulnerability of householders to the floods that they might experience (Green et al., 1994). What this equation indicates, of course, is that it is not just flood extent that contributes to vulnerability, but that a range of other characteristics concerned with the people at risk and the information they receive about the impending flood is also important.

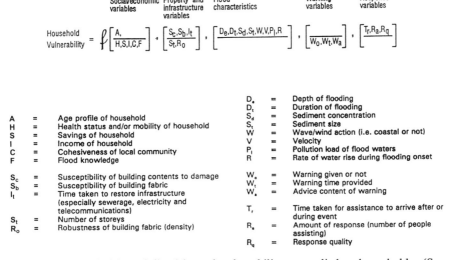

Figure 15.6 The definition of flood hazard vulnerability as applied to households. (Source: Penning-Rowsell and Fordham, 1994)

Social/economic variables

Other things being equal, the research evidence so far has suggested that elderly populations, or elderly individuals within populations, are more vulnerable to floods than the young, especially if the flood event arrives suddenly and with little warning (Green *et al.*, 1985). The elderly are less able to respond with flood-fighting actions, and are more likely to be infirm so that their response is slower and less effective (Parker *et al.*, 1983). Those with most knowledge and experience are correspondingly less vulnerable.

Those householders with a higher health status (i.e. those less unwell) are more able to respond to floods, and those households that have a large family (or family members in the immediate vicinity) are less vulnerable because help can arrive more quickly. A strongly cohesive local community would probably have the same effect, and there is some evidence (Green and Penning-Rowsell, 1989) that those households with higher incomes or more ready access to savings can help themselves more readily in flood events and are therefore less vulnerable to their impacts. Thus, in summary, households composed of young people with financial reserves are less vulnerable to floods than those of the elderly, the poor and the sick.

Property and infrastructure variables

If buildings have more than one storey, their contents are less vulnerable to flood damage. Damage will also be lower if the buildings are robustly built, and the susceptibility of household contents and building fabric damage will depend on their type and age, but in general soft furnishings are more susceptible to damage than wooden furniture, and electrical or electronic goods are generally a total loss when affected by severe and prolonged flooding.

The time needed to restore services to dwellings also affects the damage caused (Parker *et al.*, 1987) because without electricity the drying of flooded goods is difficult and the lack of telecommunications means that assistance cannot be organized quickly, including in the clean up and repair phase of recovery from flood events.

The impact of the flood's characteristics

On average the deeper the floods and the higher the flood water velocity the greater will be the damage, which will also be exacerbated by high speed of onset and long flood duration (Penning-Rowsell and Chatterton, 1977). In some circumstances the duration of flooding is more important than its depth, especially if this duration has major effects on communication systems (Parker *et al.*, 1987).

The character of the flood waters will also affect the damage caused. Pollution in flood water will can have serious long-term impacts, including adverse effects on the health of the people affected. Almost all floods in urban areas will be contaminated with sewage (Green and Tunstall, 1991), and damage will be greater where sediment concentrations are high, as will be the vulnerability of households affected in these circumstances. The greater the rate of rise of flood waters the less warning can be given – other things being equal – and therefore the more vulnerable will be the populations at risk (Torterotot and Roche, 1990).

Warning variables

Warnings can reduce both the tangible and intangible damages caused by floods (Parker, 1991) and thus warnings can reduce flood vulnerability.

Much research has been done on the many variables affecting the quality of the warning, both in terms of lead time and the process of warning dissemination (Handmer and Penning-Rowsell, 1990). In general, the greater the advice content of the warning message the greater its effectiveness, and therefore the lower the vulnerability of the population being warned. However, this is a complex matter and the quality of both the forecast of a flood and the warnings as to what response to take is affected by a large number of variables that will help to determine aggregate vulnerability (Parker *et al.*, 1994).

Response variables

Warning and response are interconnected. The time taken for assistance to arrive at the scene of a flood will affect the vulnerability of the most vulnerable in the flood-affected community. A swift response can save lives and reduce the health impacts of the flood on its victims, as well as helping to save damage (especially to heritage values).

In addition, the amount of response (the number of helpers, etc.) will affect the degree of responsiveness, if not its quality. Too many observers of the flood will not help the flood victims, although greater media attention to a flood event may mean that the media's audience can respond in different ways, for example by donating relief.

The quality of the long-term response will affect the way in which the community returns to a normal – post-flood – state of affairs. Trained social workers can assist in its process (Parker and Handmer, 1992), as can the attention of interviewers from research organizations visiting flood victims sometime after the flood has occurred: research shows that this type of attention and concern of itself helps the process of post-flood recovery (Parker *et al.*, 1983).

More research is needed to calibrate more fully the equation in Figure 15.6, which has developed inductively over a number of years and builds on the experience of a number of disparate case studies; particular attention needs to be devoted to the impact of warnings on flood vulnerability. What is suggested overall, however, is that the immediate and self-evident impact of a flood (in terms of damage to property and broken communications) is directly related to the flood's physical characteristics, and this is well known, whereas the longer-term impacts in terms of individual losses and community recovery are more related to the socio-economic characteristics of the individual and the community at risk. If this hypothesis is correct, then we need to give more attention to data on socio-economic aspects in our analysis of flood events, rather than adopting the less comprehensive approach of defining vulnerability in the conventional way in terms of flood extent, depth, duration and flood-water velocity.

15.4   LAND-USE PLANNING AND FLOOD RISK IN ENGLAND AND WALES: DEVELOPING A STRATEGIC APPROACH

Knowledge of flood extent, vulnerability and flood risk is not a sufficiently sound basis on which to advance policy in the field of floodplain development control. We also need to know more about the institutions involved and the ways that different countries have approached this thorny problem.

A research project sponsored by the National Rivers Authority and undertaken in 1993/94 was designed to develop a strategic approach in England and Wales to furthering the link between land and water planning, particularly in the area of flood defence. The definition of the strategic approach involves four attributes: it is proactive, it takes a long-term view, it requires coordinated decision making embracing all principal interests, and it is flexible and adaptive.

The research was designed to evaluate current practice and policy, assess the potential for change in this practice and policy, identify possible and appropriate modifications to the law and government guidelines, and to provide guidance to the National Rivers Authority on heightening awareness of the problems of interfacing land-use planning and flood defence (Tunstall *et al.*, 1993). The conclusions are equally relevant for the body which replaced the Authority, the Environment Agency.

The research methodology comprised a literature review, a series of interviews with key "stakeholders", representing statutory and non-statutory agencies involved in the town and country planning process as it relates to flood defence, and the identification of recommendations. More details are to be found in Tunstall *et al.* (1993); summarized below are the positions of the three key sets of organizations concerned.

15.4.1   National Rivers Authority policies and practice

The policies and practice of the National Rivers Authority and its forerunners have evolved since 1947 in response to various government circulars which address the question of development in flood risk areas or runoff from new development. The latest circular is Circular 30/92, issued on the 16 December 1992.

However, the development of a strategic forward-looking approach has been slow to develop in the National Rivers Authority. In terms of the organization of planning liaison, a National Planning Liaison Group was established in 1991, and this Group tackled all flood defence regulation and enforcement matters which included the planning and control of urban development in flood risk areas. At the same time, there were marked differences between the National Rivers Authority's regions in the way that strategic planning liaison was organized, and day to day consultations on planning applications remained a dominant activity, rather than strategic forward thinking. There were also wide variations in practice on consultation at the regional level with regional planning conferences.

In terms of the views of National Rivers Authority staff, the potential for improvement in integrating land and water planning was widely appreciated. Catchment management planning was seen as a focus for that improvement process, although some staff within the National Rivers Authority felt that the plans so far produced were too internally oriented, too technical and give insufficient attention to flood defence planning and development control issues and solutions. Furthermore, the pace of plan production was too slow to meet the needs of local land-use planning authorities. There was uncertainty within the National Rivers Authority about the relationship of catchment management planning to land-use planning.

With regards to Circular 30/92, concern was expressed by flood defence staff within the National Rivers Authority about the government's assumption that the National Rivers Authority had flood risk information available and in a form that could be communicated easily to outside bodies. There was also concern about whether the National Rivers Authority had sufficient resources to be able to identify ways of overcoming the development constraint created by flood risk, and to what degree it should do this.

Nevertheless, Circular 30/92 was viewed as a significant advance over Circular 17/82 issued a decade earlier. However, it still leaves the National Rivers Authority and it successor, the Environment Agency without a mechanism for tackling phased development involving multiple developers over a long period of time: major urban flood risk areas can be created through piecemeal processes. Another concern was the lack of recognition in government guidance of the National Rivers Authority's nature conservation and related duties and expertise in relation to the water environment. In this respect the work of the National Rivers Authority in the area of river restoration exemplifies the care that the Authority gave to enhancing the floodplain environment in Britain (Tapsell, 1995).

15.4.2   Central government departments

Central government departments were quite clear about the primacy of local planning authorities in development control matters. The National Rivers Authority was expected to adopt a proactive approach to planning liaison, and was expected to advise local planning authorities using all of the opportunities available within the development plan process.

In this respect, a key requirement is that the National Rivers Authority, now the Environment Agency, should improve its databases in order to be better able to advise local land-use planning authorities during the development plan process, rather than after plans have been agreed. National Rivers Authority "model policies" or guidance on the integration of land and water planning were viewed by the Department of the Environment as an appropriate way of promoting flood defence objectives. However, more importance is attached to specific information being made available on flood risks and the areas to which flood risk policies are applicable, including catchment management plans and the results of the surveys which the National Rivers Authority, now the Environment Agency, is required to carry out under Section 105 of the Water Resources Act 1991 in relation to its flood defence function (Table 15.5).

15.4.3   Local land-use planning authorities

Most critical decisions about floodplain urbanization processes are made locally. The local land-use planning authorities surveyed in this research considered that the National Rivers Authority's views received sufficient weight in planning decisions, and most felt that the National Rivers Authority made sufficient inputs to the statutory planning process.

Table 15.5 Specification and time-scale for the implementation of the Section 105 surveys of flood extent in England and Wales (from National Rivers Authority, 1994b)

Time-scale (0 = start)	Phase		Details
0 →9 months	1	(a)	Issue of existing S24(5) survey data to each LPA; obvious updates to be included if possible; intended to impart general but representative nature of flooding to LPAs; LPAs must be made aware of limitations.
		(b)	Agreement of a programme of work with LPAs, including, where possible, the appropriate level of survey data for each location (Parameters A and B, or other) beyond that required for NRA use.
0 →2 years (1996)	2	(a)	Completion of 50% of accurate flood envelope delimitation work; map scales as specified below.
		(b)	Surveys to include agreed Primary data, which may be selected from Parameters A below.
2 →5 years (1999)	3	(a)	Completion of full work programme to phase 1 standards above.
		(b)	All surveys to include agreed Secondary data, which may be selected from Parameters B below.
1999 onwards	4	(a)	Inclusion of remainder of watercourse lengths (essentially rural watercourses with little/no development potential); degree of data and accuracy to be subject of future agreement if surveys required.

Types of data which may be applicable

Physical or Primary Data (Parameters A)	Calculated or Secondary Data (Parameters B)
• Min. map scales (i) Rural 1 : 10 000 (ii) Urban 1 : 2500 or 1250 • Flood envelopes: at least 100 yr, with consistently agreed flow data. • Relevant boundaries (+SSSIs etc.). • NRA assets + details + areas protected. • Non-NRA defences + details + areas protected • Floodplain contours + plus survey (→0.25 intervals in sensitive areas). • Channel cross-sections + survey data. • Strategic flood storage areas (boundaries). • Land use + SoS data. • Natural defences. • Tidal inundation limit (200 yr SWL min.) • Vulnerable areas: esp. from wave action/surges. • Maintenance requirements (accesses) • Threshold levels of endangered properties.	Essentially supporting data, e.g.: • Storage/discharge details. • Surface water profiles. • Channel spillage regime. • "End dates" of assets. • Blockage possibilities (via modelling runs). • Erosion/accretion details. • River corridor survey data. • Managed retreat/set back. • Coastal erosion. • Development constraints. • Measures to overcome flood risk constraints. • Urban "green corridors" Application beyond NRA use to be agreed with LPA for each location.

However, all was not entirely well at this level. In many cases there appeared to be no standard procedures or mechanisms for contact between National Rivers Authority planning liaison staff and local authority forward planning staff. Overall, local authorities also considered that such liaison could be improved through more personal contact and with better National Rivers Authority guidance documents.

The land-use planners looked to the National Rivers Authority to provide reliable data and planners were critical of the quality of information available from the Authority on floodplains, flood risks and runoff problems. They also believe that catchment management plans would be more helpful if they addressed such matters in sufficient detail for the authorities to use them in development plans and planning application decisions.

15.4.4   The development of a strategic approach

The development of a strategic approach to the interface between land-use planning and flood defence in England and Wales will require greater central coordination than has hitherto been the case. Policies and procedures need to be standardized, or at least coordinated, and an overall methodology needs to be adopted. Currently, a number of problems need to be tackled which act as impediments to this process, including the absence of data, disagreement as to communication processes, and the inadequacy of existing legal instruments.

In this respect, the research indicates that the National Rivers Authority and now the Environment Agency must improve its data on floodplains in order to exert more influence on land-use planning decisions. Section 105 surveys offer a basis for doing this, and Circular 30/92 sets out data requirements which include the identification of areas for managed retreat (i.e. abandoning areas with particularly severe flood problems). In this regard, a high priority should be placed on rapidly improving the floodplain mapping and associated nature conservation and other database systems. Progress has been made in agreeing a timetable and priorities for floodplain mapping, as set out in a Memorandum of Understanding between the National Rivers Authority and local planning authority associations (National Rivers Authority, 1994b, and Table 15.5). But databases also need to be created to identify, record and monitor the production of statutory development plans and other relevant information. These plans need to be identified at an early stage, so that the Environment Agency can influence their content, and this requires wide liaison and consultation with external bodies.

Having an adequate database leads to the need for effective communication between the Environment Agency, as the flood defence authority, and the many local authority planning departments responsible for development decisions. Many vehicles for better liaison need to be explored, including the employment of planners by the Environment Agency, regular meetings between planners and flood defence staff, liaison at national level to agree systems of liaison at regional and local levels, and the development of informal communication systems as well as formal processes. The Environment Agency needs also to develop national policy documents targeted at planners and developers, which should be regularly updated. In terms of guidance, the National Rivers Authority

(1994a) has provided this for planning departments (Table 15.6), but the adoption of this guidance remains voluntary, as before. With regard to changes in the law, only minor modifications are needed, but such topics as whether a statutory basis to catchment management plans would be useful need to be explored.

Table 15.6 Guidance statements issued in 1994 by the National Rivers Authority (NRA) to local planning authorities (LPAs) in relation to flood defence (source: National Rivers Authority, 1994a)

1. **Strategic/county-level concerns**
    "The LPA, after consultation with the National Rivers Authority, should normally resist allocation of land where such development would be at direct unacceptable risk from flooding (including tidal inundation) or likely to increase the risk of flooding elsewhere to an unacceptable level".

2. **Local/District level concerns**

    2.1 **Protection of the floodplain**
    "Within the identified floodplain or in the areas at unacceptable risk from flooding the LPA should resist new development, the intensification of existing development or land raising. Where it is decided that development in such areas should be permitted for social or economic reasons, then appropriate flood protection and mitigation measures, including measures to restore floodplain or provide adequate storage, will be required to compensate for the impact of development. At sites suspected of being at unacceptable risk from flooding but for which adequate flood risk information is unavailable, developers will be required to carry out detailed technical investigations to evaluate the extent of the risk. In all cases, developers will be required to identify, implement and cover the costs of any necessary measures. In some cases the elements of the necessary measures may be such that they are best undertaken by the National Rivers Authority itself, but in these cases the cost would be covered by the potential developers".

    2.2 **Surface water runoff**
    "The LPA should normally resist development which would result in adverse impact on the water environment due to additional surface water run-off. Development which could increase the risk of flooding must include appropriate alleviation or mitigation measures, defined by the LPA in consultation with the National Rivers Authority and funded by the developer. Developers will be expected to cover the costs of assessing surface water drainage impacts and of any appropriate mitigation works, including their long-term monitoring and management".

    2.3 **Tidal and fluvial flood defences (1)**
    "Development should not normally be permitted which would adversely affect the integrity and continuity of tidal and fluvial defences. Access to existing and future tidal and fluvial defences for maintenance and emergency purposes will be protected, and where appropriate, improved. Where development relating to tidal and fluvial defences is permitted, the LPA will, in consultation with appropriate bodies including the National Rivers Authority, require appropriate measures to be incorporated in order to ensure that the stability and continuity of the defences is maintained. Developers will be expected to cover the costs of any appropriate enhancement and mitigation works, including their long-term monitoring and management".

(continued)

Table 15.6 *Continued*

2.4 **Tidal and fluvial flood defences (2)**
"In order to minimise the effects of tidal flooding, the LPA should resist development on land to the seaward side of sea defences, including the siting of temporary holiday chalets and caravans. On land between a first line sea defence and the main defence, the siting of holiday chalets, caravans, and camping sites will not normally be permitted. If after consultation with the National Rivers Authority and other interested bodies the LPA decides that the risk of flooding is sufficiently low to permit certain types of use, time limited occupancy conditions will need to be imposed preventing occupation during the period from November–March inclusive when the risk of tidal inundation is greatest. The development permitted in any area of land subject to a flooding risk must be in line with the level of protection provided by the sea defences which exist. A change in the type of development permitted could result in a need for increasing the level of protection afforded and if so the cost of such provision should be borne by the developer".

2.5 **River corridors and coastal margins (1)**
"By emphasizing the importance of river corridors and coastal margins, the National Rivers Authority aims to promote these aspects of the river environment. Such a corridor is a continuous area of land which is physically and visually linked to the watercourse itself. A coastal margin is similarly an area of land physically and visually linked to the coast and any coastal defences. Studies have shown that there is a high correlation between river corridors in England and existing environmental designations, notably SSSIs and Areas of Outstanding Natural Beauty. In urban areas, the importance of river corridors is even more pronounced since they represent one of few remaining features which link areas of open space. Such links are significant for amenity and recreation, but also for wildlife, allowing otherwise isolated areas to be interconnected and more viable in terms of animal and plant populations and habitat types. These factors suggest that river corridors warrant reference in land use plans as important elements which link areas of open space".

2.6 **River corridors and coastal margins (2)**
"The LPA, in consultation with the National Rivers Authority, should seek to promote river corridors and coastal margins as important areas of open land by:

- conserving existing areas of value and wherever possible seeking to restore the natural elements within the corridors and margins; promoting appropriate public access;
- identifying appropriate locations for water-related recreation;
- protecting and improving access for operational and maintenance purposes, including the provision of maintenance strips where practical;
- resisting development which would have an adverse impact on nature conservation, fisheries, landscape, public access or water-related recreation".

2.7 **River corridors and coastal margins (3)**
"The LPA, in consultation with the National Rivers Authority, should seek to ensure that all works in, under, over and adjacent to watercourses, water bodies and the coast are appropriately designed and implemented and that the likely impacts of development proposals have been adequately assessed by means of a formal Environmental Assessment, where appropriate. In all cases proposals will need to be accompanied by an environmental report so that environmental impacts can be appraised".

The development of a strategic approach to the integration of land-use planning and flood defence appears therefore to constitute an important way forward in the process of reducing flood vulnerability in England and Wales. What is needed is a positive, coordinated and proactive approach, supported by much improved data (especially on the extent of floodplains), and also supported by catchment management plans. In this process the Environment Agency needs to consult widely with both statutory and non-statutory agencies. A number of other conclusions emerged from the research, particularly about the production of flood risk information and processes of consultation, indicating the blockages to a more integrated approach revolve around inter-agency linkages, rather than deficiencies in the law. Such inter-agency problems result from a number of features of environmental agencies currently in Britain, namely the rapid turnover of staff, the move towards the creation of autonomous agencies with narrow specifications (including semi-privatized agencies), and the generally low level of resources available to public bodies concerned with environmental regulations.

This analysis indicates, in summary, that there are innumerable problems at the interface between land-use and flood defence planning, as is shown by the detail of the recommendations from the research described above. This is despite the fact that Britain has had a standardized land-use planning system for nearly 50 years and there is generally good information about the extent of floods (even if this information is not collated into a systematic database). Many of the current problems result from intense development pressure in areas of high population density, and the willingness of local authorities to "bend the rules" in order to attract developers into areas where development might be unwise but the pressures to increase employment opportunities are intense. Some of these circumstances, of course, contrast markedly with the situations in a number of other countries, discussed below, where there may be no tradition of systematic land-use planning, weak enforcement of the plans that exist, little information on flood extent, and little willingness to plan proactively to prevent flood hazards increasing in the future.

15.5 ENHANCING FLOODPLAIN MANAGEMENT: INTERNATIONAL COMPARISONS OF BEST PRACTICE

15.5.1 Research methodology

In addition to the work on Britain, described above, research has also been undertaken to investigate the relationship between land and water planning in a number of European countries and in the United States of America (Chatterton *et al.*, 1994). The objective of this research was to inform the evolution of British practice in this sphere, by looking at contrasting situations elsewhere. In this respect, situations of high development pressure were evaluated, particularly in France and in the Netherlands, and these were contrasted with the situation in the United States of America, where there are a variety of levels of pressure and several different institutional characteristics, notably the federal system of government in the country.

The methodology adopted in this research was, first, a comprehensive review of policy documents, followed by, secondly, in-depth interviews with a variety of "actors" in the land and water planning agencies of the countries sampled. These interviews concerned, for the flood defence field, the institutional framework within the country concerned, the framework of laws and responsibilities at different levels of government, the policies and policy instruments deployed, and the effect of these different arrangements on the implementation of land and water planning in particular case studies.

In addition to these case studies, an analysis has been undertaken of land and water planning problems in Argentina (Penning-Rowsell, 1996), to illustrate the problems of integrating land and water planning in a country undergoing rapid change which has physical, socio-economic and governmental dimensions.

15.5.2   France: centralization and devolution in transition

Historically, France is a highly centralized state, but one where politicians maintain a strong local power base. Currently there are moves in France to devolve responsibilities from the centralized state to local communes and these moves potentially could have significant implications for land and water policy. At the same time new water and catchment planning laws and regulations are being proposed and these could prove very powerful in the planning process, although it is too early to monitor their full effect.

In terms of institutional arrangements in France (Figure 15.7), there was new legislation in 1992 and thereby the creation of the Agences de l'Eau as mechanisms for integrated water quantity and quality management, land-use planning and environmental protection on a catchment-wide basis; this is seen as an innovative step to achieving "all in" planning. The French approach, thereby, is developing from the organization of loose consensuses and coalitions towards statutory Master Planning. In this model, communes and inter-communal committees have planning, finance and regulatory functions under their remit. Basin agencies have a statutory responsibility for the development of catchment Master Plans, giving directions for all users of water, with local planning committees coordinating procedures at a local planning level.

With regard to flood risk, several procedures have been used in the past to define areas of hazard vulnerability. The Plans for Flood Prone Areas (PSS) uses an old procedure dating from a law of 30 October 1935 and considers two main types of zones but does not give any precise information on quantitative criteria to delimit these zones, in terms of water level, depth and velocity. More recently, the definition of Risk Perimeters (PR) is directly related to land-use planning in flood prone areas. It is initiated by the Departmental Prefect and is used by the departmental bodies. The procedure is only applied to new land-use planning and is unable to impose restraints on existing development. The Risk Perimeter method is applied independently of communal boundaries, and this is a determining factor for its likely success.

The most recent tool (law of 13 July 1982) is the development of Risk Exposure Plans (PER). Although this is a sophisticated and much needed approach, the system is cumbersome to set up (taking 2 years on average between the specification of a PER and its approval). It is managed at departmental level, and it is of interest for two reasons: firstly, it is applied to both present and future town planning or development projects, and, secondly, it establishes a relationship between compliance with a specification laid down in the PER and the reimbursement for damage by insurers in the case of natural catastrophes: there is an in-built economic instrument.

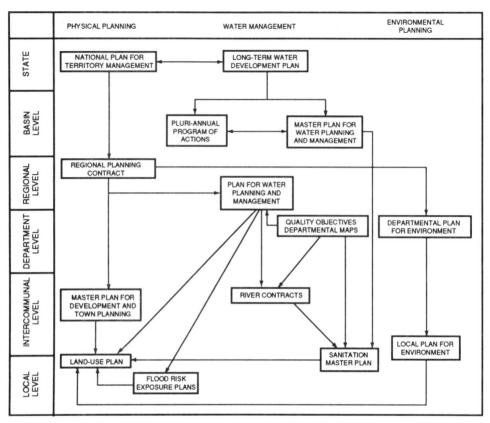

Figure 15.7 Institutional arrangements for land and water management in relation to flood defence in France. (Source: Chatterton *et al.*, 1994)

Despite this progress in refining the definition of flood risk areas, and discouraging development there, it appears to be true that since de-centralization of political power within France, the constraints imposed by the state for controlling the use of flood-prone areas have been very badly received by locally elected representatives. Development often proceeds regardless, with a lack of control and an absence of monitoring by the state. Insurance companies also pay little regard to the risk exposure plans.

Therefore, the French situation is still in a transitional phase. Arrangements appear well-organized in theory in that numerous mechanisms exist to address the problem of flooding and its impacts. Nevertheless, a tension exists between local and central efforts. Procedures that have been set up by the local organizations have enhanced the compatibility between legislative and regulatory measures relating to land use and those related to flood risks. In this way the relationships between flood control and water management and between flood control procedures and flood relief measures have been recognized. There is also an emphasis on the development of public information and training, and sharing the community financing of these schemes.

Nevertheless, the danger of local control of the integration of land and water planning is that an idiosyncratic solution is developed in each administrative area, reflecting local interests and circumstances. The lack of standardization can mean that problems are passed from one area to another, and major risks remain the subject of a patchwork quilt of responses in different local authority areas. Reorganizations in the 1990s in France are attempting to counter the dangers of this situation occurring, through the development of standardized approaches to flood hazard delimitation, and a more standardized approach to local control over the interface between land and water management. Only time will tell whether this balance between central needs and local expectations will produce an enhanced performance. There appears to be good multi-level coordination between the national state apparatus, the Ministry of the Environment and local land-use planning, but this does not yet cover the whole country and securing complete coverage will be an important test for the success of the strategy and policy as a whole.

15.5.3 The United States of America: federal power and local weakness

In the United States a federal system of government prevails, with major power vested in the individual states which in some fields have had power devolved to them and which is now entrenched in constitutional arrangements at the lower level. The highly fragmented local government arrangement below state level has evolved with multiple local government structures (Chatterton *et al.*, 1994). Consequently, there is much emphasis on incentives to encourage action at the lower level of government, strong emphasis on involving stakeholders in decisions, and on conflict resolution and consensus building. There is a strong tradition of compensating flood victims through grants and loans, somewhat modified by the Federal Insurance Programme.

In terms of flood defence policies, emphasis is placed on providing early flood warnings and local preparedness, as well as municipal enforcement of floodplain zoning regulations to control land-use in areas of risk. This process has been assisted by establishing and enforcing state floodplain encroachment permit programmes, with federal financial assistance, and with public participation. Flood hazard areas have been transferred into planning authority development zoning maps, thus combining flood hazard management with other land-use management objectives.

Strong community involvement is an element in attempts at integrated land and water planning in the United States. For example, there has been the creation of private citizen groups, such as the Flood Land Action Committees (FLAC) in the Chicago Metropolitan District. These are successfully embracing community involvement and the FLAC require residential developments of more than 5 acres or commercial developments of more than 2.5 acres to provide 100% retention for runoff from a 100-year storm before construction has begun. This approach to flood source control attempts to tackle the fundamental origin of flood problems – increased runoff from urbanization – with a policy instrument that is nevertheless locally based.

Floodplain legislation, controlling land-use, is based on the National Flood Insurance Programme, which aims to shift the cost of occupying floodplains to the occupants themselves. Subsidised federal flood insurance can only be offered in communities which adopt floodplain ordinances to reduce future flood losses (e.g. flood-proofing to 0.3 metres above the 100-year flood level).

In this respect developing Flood Insurance Rate Maps on a community basis has become a priority. Such a move will allow the production of flood-prone street address directories to allow streamlined insurance rating procedures and, hopefully, increased awareness of flood problems. In the same way the Disaster Relief Act, 1974, requires recipients of disaster relief to evaluate the natural hazards in the area in question and take appropriate action to mitigate such hazards in the future. Other economic approaches involve the acquiring of undeveloped floodplain land with federal assistance by local communities under systems such as the Land and Water Conservation Fund. In this way federal subsidy is being provided to encourage wise practice in terms of floodplain development (or prohibiting development), thus reducing the chance that local communities will choose to develop areas which put themselves at risk.

In summary, in the United States a type of market approach has been adopted, as might be expected in that country. The use of cost sharing (principally federal incentives) is designed to encourage local communities to conform towards wise use of floodplain areas, with a particular emphasis on retaining wetlands and nature conservation aspects of river corridors. This has the advantage of bringing the weight of the federal government to bear upon the problem, combined with local decision making about particular circumstances and flood threats. The worst excesses of development in floodplain areas that have been experienced in the past are thus being avoided, although the cost to the federal budget is substantial.

15.5.4   Land and water management in the Netherlands: a national priority

In a way that is untypical of almost all other nations, the Netherlands is wholly dependent upon the successful management of water. The very existence of land within the country depends on a complex system of dykes designed to prevent flooding, and agricultural production is intimately related to an intricate network of drainage channels and pumping water from low-lying land. The philosophy dictating flood defence policy in the Netherlands is one of zero risk, hence the shock caused by the floods in the Meuse and other areas in 1995 and the orderly response to the need for the evacuation of 250 000 people that ensued. The policy imperative is that the dykes must be managed effectively so that failure does not occur. Flood defence and disaster management take precedence over all else, but at the same time the Dutch system demonstrates the value of community involvement, with environmental concerns now becoming paramount.

In terms of institutional arrangements for flood defence, the Netherlands is a strongly centralized state where water institutions, managed through the state-controlled Rijkswaterstaat, have a powerful political influence. Indeed the country's historic administrative and legislative tradition is strongly influenced by the needs of water management, and many of the water management institutions active today can trace their origins back several centuries, and even beyond the arrangements of modern democratic government.

As Figure 15.8 shows, water management in the Netherlands is formulated and enforced through the governmental institutional framework at national, provincial, and municipal authority levels (Chatterton et al., 1994). Land and water management are intimately integrated, and the Rijkswaterstaat provides the direction to Dutch water management by formulating policy and developing the means for enforcement through legislation at the national level. The provincial councils, in turn, are responsible for the execution of nationally approved laws, and they direct and interpret water law with

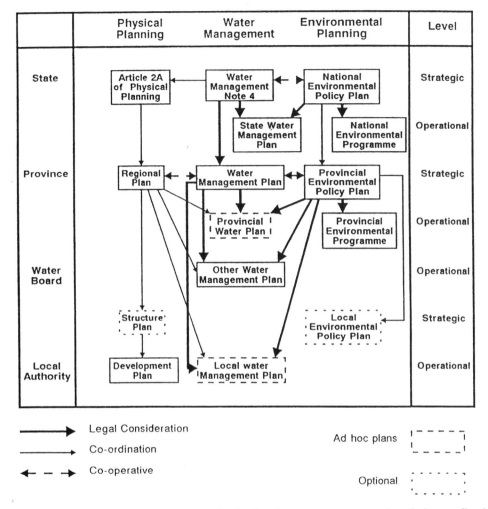

Figure 15.8 Institutional arrangements for land and water management in relation to flood defence in the Netherlands. (Source: Chatterton *et al.*, 1994)

regard to the water boards' activities. Provincial government therefore plays a supervisory role in water management, while regulatory and financial aspects are under central government authority. Water boards are decentralized functional forms of government and are geographically defined by natural and artificial drainage systems. Some, but not all, are multi-functional (and therefore, for example, are responsible for water supply or pollution control), and the national government's push for functional consolidation is behind moves for the reduction in the number of these boards in recent years.

In terms of land and water policy, Dutch water management is very progressive in seeking to redress the ecological balance of water, land and natural values which are seen to have been knocked out of balance in the recent past, for example by the increasing pollution of water and land degradation over the last century of industrialization and population growth. Effective land-use planning has been achieved by allowing water management to remain in the public domain, and "vertical" integration of the management of land and water allows comprehensive planning and the operation of best management practices at all government and functional levels. Strong national plans have been developed, cascaded down to provincial and municipal levels, for the integration of land and water management. Integrated land-use planning has created the opportunity for open debate about policy and the free dissemination of information at every stage of the decision-making process. This has strengthened cooperation between water managers and other bodies, and has been promoted by the separation of the technical design work from the decision-making process. This has been helped by the development of a complex and sophisticated system of impact assessment, through Environmental Impact Assessment Agencies in newly created Special Protection Areas, in accordance with European Commission directives.

Land-use planning for the floodplain is not separated from the integration of land and water management elsewhere in the country. Protecting the country against tidal storm surges has always been the primary area of water management, and municipalities work with provincial government authorities to ensure that dyke maintenance is given the highest priority and that land-use is controlled within dyke ring areas. The integration of development control with Environmental Impact Assessment has resulted in a move away from primary concern for the physical mechanisms of flooding, to a multi-objective evaluation of biological and ecological goals within the process of integrating land and water management.

In terms of policy instruments, legislation and jurisdiction relating specifically to the floodplain includes the development of by-laws ("Keur") for the operational management of water laws in compliance with provincial systems, the issuing of a technical register ("Legger") at the provincial level for upholding maintenance standards for dyke systems, and the provision in the water board by-laws for regular on-site maintenance checks for all watercourses. This strong regulatory regime, together with fixed flood protection standards as the basis of the construction and maintenance of dykes, has provided an important way forward and strictly defines floodplains within which land-use is controlled. The use of some economic instruments within floodplains has introduced some mechanisms to protect against unwise floodplain development, including reducing price support mechanisms for arable production, encouraging set-aside and the withdrawal of field drainage programmes. Thus there is a trend towards agricultural de-intensification, but this has to be seen within a national context of highly intensive agricultural production.

In summary, the Netherlands approach to integrating land and water management is through a complex "bottom-up"/"top-down" consensus approach, involving strong water planning at local and national levels. Policy development is guided through four year National Plans, formulated at state level and coordinated at the local planning level. The control mechanism passes through provincial governments and is implemented by water boards under provincial control. However, the Dutch also lead the way in local community involvement in flood management issues, particularly through their unique use of floodplain boards, with both a water and a land-use planning role. In this way a balance is struck between strong central oversight, local participation, controlled land-use development and flood risk elimination.

15.5.5   The innovative "merit approach" in New South Wales, Australia

Australia is not a country that is associated with major floods, but in reality flood events in New South Wales can be serious. The east coast of Australia is dominated by fast-flowing rivers affected by cyclonic storms, which can bring rapid rise in flood waters and severe devastation. For example, in 1955, a major flood in the Hunter Valley resulted in 14 deaths, 5000 houses being flooded, and totally destroyed some 160 homes.

A number of individual responses to floods have been taken (Penning-Rowsell and Smith, 1987), which date back to the early days of the European settlement of the country when governments were weak. Much more recently a new approach has been developed in New South Wales, emphasizing a careful evaluation of each development in a floodplain on its own merits, guided by a standard system of flood hazard appraisal and different standards of flood defence for different land uses. This approach is termed the merit approach (Government of New South Wales, 1984).

In this respect it should be noted that Australia is a federal country, but all responsibilities in the water area are devolved to states. Within states, local authority councils have responsibilities for land-use control, while state government authorities have responsibilities for water resources management, including flood alleviation. State water resources programmes, including flood alleviation projects and plans, are subsidized by the federal government, which provides funds for selected works through the national Water Resources (Financial Assistance) Act.

Within this context, in 1984 the New South Wales government reviewed its policies and procedures in providing resources to assist flood-affected areas. Previously, there had been designated a 1 : 100-year flood standard for the provision of urban flood alleviation schemes, but this was withdrawn and it was proposed instead to adopt a flexible merit approach primarily to be implemented by local government, but backed up by state government technical and financial support. This policy was set within the context of structural works being used to reduce the flood hazard in areas of existing development, the removal of unnecessary development and the voluntary government acquisition of property in some high hazard locations.

Policy for new urban development in floodplains involved the development of a manual and guidelines for assessing the flood hazard in a particular location, together with advice on decisions on development and building applications and the preparation of local environmental plans in conjunction with floodplain management programmes. With this approach decisions on development depend on the severity of the flood in a particular location (Figure 15.9), the type of development envisaged (i.e. its vulnerability), and the implementation of associated plans to minimize the hazard, such as by flood-proofing, raising property and providing evacuation arrangements.

In this way flooding considerations are systematically weighed along with other planning factors in the decision-making process. A limited amount of development would be permitted in areas of high hazard, but in low hazard areas development would be allowed, subject to conditions. Developers are provided with detailed guidelines about what is likely to be permitted in each type of hazard zone and, crucially, decisions are made locally so that individual circumstances can be judged on their own merits.

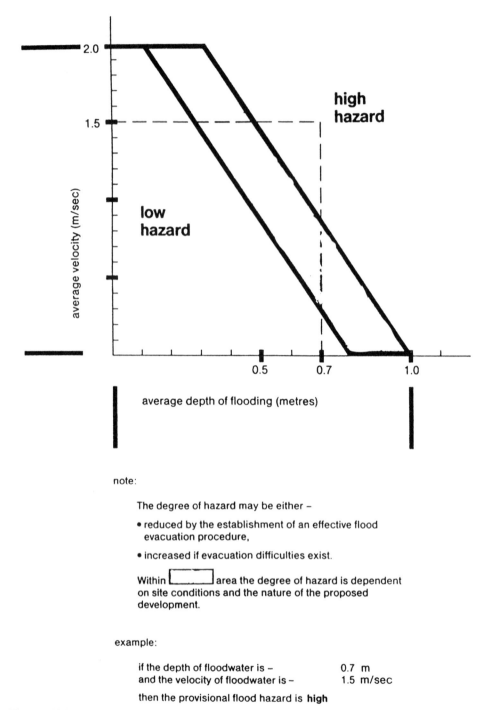

note:

The degree of hazard may be either –

• reduced by the establishment of an effective flood
  evacuation procedure,

• increased if evacuation difficulties exist.

Within ⬚ area the degree of hazard is dependent
on site conditions and the nature of the proposed
development.

example:

if the depth of floodwater is –   0.7 m
and the velocity of floodwater is –   1.5 m/sec

then the provisional flood hazard is **high**

Figure 15.9 The categorization of flood hazard severity involved in the New South Wales "merit" approach to floodplain development decisions. (Source: Government of New South Wales, 1984)

In summary, the Australian situation in New South Wales is one which is moving away from set standards towards appraisal of each case in the day-to-day integration of land and water planning. Technical assistance is provided by the higher levels of government, together with financial assistance to induce action. There is nothing to prevent local communities making unwise decisions with regard to development in high hazard areas, but there is a presumption that the technical advice and guidelines provided will make this less likely than was hitherto the case. The strictness of the planning system in the Netherlands is avoided, and this is appropriate in the context of a country such as Australia where land is plentiful and alternative locations for most development can readily be found.

15.5.6   The changing situation in Argentina

The examples above of practices in different countries in managing the interface between land and water management have concerned situations where the physical environment was more or less in a stable state, albeit in a state of dynamic equilibrium. The situation in Argentina appears to be somewhat different, with the processes of climatic change leading to worsening flood severity (Penning-Rowsell, 1996).

This can be demonstrated with respect to the Paraná/Paraguay/Uruguay river system. This system has generated major floods in recent years, notably in 1983 and 1992. Major floods have also occurred throughout the period of human record, notably in 1812, 1858, 1878 and 1905 (Anderson *et al.*, 1993). The river regime, however, appears either to have changed or to be changing, given the incidence of the these two floods with return periods greater than 100 years (based on the previous hydrological record) in the last 15 years: 1983 and 1992.

This situation of rising flood severity, illustrated in Figure 15.10, appears to be related to increased precipitation which in turn is related to "El Nino" periods when warm equatorial waters in the Pacific Ocean encroach much further to the east than is normal (Burgos *et al.*, 1991). This occurred to a major extent in 1982/83 and to a lesser extent in 1991/92. Such geographical shifts in climatic factors could be being exacerbated by global warming, which appears to be occurring in the Southern Hemisphere (Burgos *et al.*, 1991).

In addition, the main river channels of the Paraná and its tributaries appear to be being affected by geomorphological changes. There is some evidence (Anderson *et al.*, 1993, Annex A, 10) that overgrazing of agricultural land in the areas on the west bank is increasing sediment yields into the main branch of the river, thus reducing its carrying capacity at times of flood. This non-stationarity, in hydrological terms, clearly has significant implications for the control of land-use within the floodplain areas. In Argentina there is pressure to develop these areas, since most of the main towns in the river basin are sited close to the major rivers. These rivers have been active navigation routes, mainly for the export of agricultural products, and the valley floor provides some of the country's best agricultural land. The towns adjacent to the river locations have provided communication foci, focusing on river crossing points. It should be understood that these are large rivers; at the town of Paraná (350 km from the estuary) the normal width of the braided River Paraná amounts to some 4 km, and for a flood event with a return period of 1 in 10 years, the width of the flood-affected area is more than 15 km.

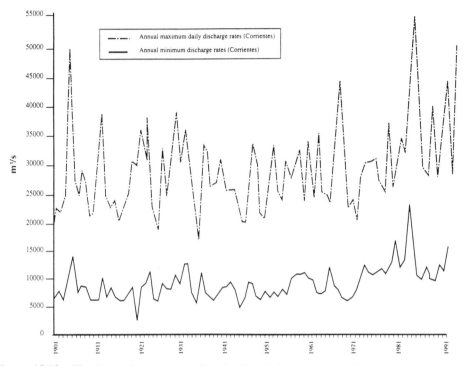

Figure 15.10 The increasing severity of major floods in the Paraná river system since 1940. (From Penning-Rowsell, 1996)

The population located in the floodplain of this large basin is also increasing. This is a product of migration to the urban areas within Argentina, coupled with rural depopulation resulting from a relative decline of the agricultural sector. In addition, the reduction in the importance of the old heavy industrial sector within the country also leads population to migrate to the urban service centres, many of which are adjacent to the floodplain or located on it.

In Argentina, there is a very weak or non-existent tradition of controlling urban development. Building regulations apply to metropolitan and large city areas, but the development of private land by private developers is allowed, and floodplain locations are no exception to this. Local authorities have some jurisdiction over the location of major infrastructure developments, including public works, but the location of residential dwellings tends to be outside the authority of such organizations.

As a result there has been massive expansion on to the floodplains of Argentina since 1905, particularly in the period between 1905 and 1966 which was characterized by a lesser incidence of major floods (Figure 15.10). The situation is compounded by the fact that the main organization responsible for flood alleviation in Argentina, the Subunidad Central de Coordinacion para la Emergencia (SUCCE), is an organization with a prime function in the area of emergency relief. Such relief has been the backbone of flood-alleviation policy in Argentina in recent decades, and the difficulty of moving away from this and towards a more sustainable policy based on land-use control is made the more problematic because the emergency relief distributed by the SUCCE and other organizations has been an important contribution to local income in many areas.

Thus the incentive to move development out of the floodplain (and to discourage movement into the floodplain) is low. Provincial budgets – and Argentina is a strongly federal country – are supplemented after times of flood, to compensate provinces for the flood losses that they have suffered, both institutionally and by individuals. The prospect of climate change inducing greater flood severity has caused concern to the government, and also to the World Bank which has funded much post-flood renovation and evacuation work in the past. To give some indication of the order of magnitude of the problem, available data show that some 120 000 people were evacuated from the River Paraná floodplain in the 1992 flood, which lasted over 6 weeks (Anderson *et al.*, 1993).

Thus, in the recent past, the interface between land and water planning has not been managed in Argentina. The World Bank, fortunately, is promoting a change in policy. The Bank sees the need to integrate land and water management, in order to minimize the losses caused by unforeseen events creating damage in areas developed unwisely. They are promoting projects (Sir William Halcrow and Partners, 1994) to gauge the full extent of the floodplain, using a combination of satellite imagery and hydraulic modelling. Within the areas thereby identified, zones are being established where development will be controlled, based on minimizing flood loss potential. This process is under way, but it is likely to be quite protracted since the lack of a tradition of development control within Argentina makes the imposition of land-use regulation systems within flood risk areas all the more difficult to implement.

The lesson from Argentina, therefore, is that in the absence of a tradition of development control, managing the interface between land and water planning for flood defence is difficult. It is exacerbated at the current time in Argentina by the prospect and the reality of climate change. Only with the incidence of two major floods recently, and the pressure from the World Bank, has the impetus been created to shift policy away from a dependence upon evacuation and post-flood compensation, and towards more integrated and sustainable policies. It is apparent, in this respect, that without such external impetus from the Bank, it is unlikely that greater integration of land and water management would have occurred in this country where the institutional status quo is difficult to shift owing to the corruption and fraud endemic in the government system.

15.6   ASSESSMENT: MANAGING RISKS AND RESOURCES

This chapter has reviewed a large field: the relationship between land-use planning on floodplains and the processes of urbanization and development there. It has drawn on examples in England and Wales and several other countries to show the complexity of these processes.

The research reported in this chapter indicates the difficulty of controlling development in areas of flood risk. Sometimes these difficulties are a function of data deficiencies and the lack of information about the extent and severity of floods in particular locations. Sometimes, however, the difficulties relate to poor institutional structures, and lack of liaison between different arms of government (in this case the arms concerned with land-use planning and those concerned with water management). In many cases arrangements at the interface between land and water management vary significantly within countries, particularly in those with a federal structure and tradition.

Our conclusions from this analysis are relatively simple, but have far-reaching implications (Table 15.7). The management of floodplain resources involves a complex mixture of technical, political and moral issues, and cannot be simplified into any formulaic approach. Always there will be a struggle between competing interests, and always the problem will be to maintain the assets that the floodplain brings while minimizing the impacts of flooding on the developments that have occurred in the floodplain in the past. An integrated approach is the aim, and sustainable floodplain policies are the goal.

However, in few countries, yet, is the aim of integrating land and water management firmly targeted on the development of a more sustainable future, with the full incorporation of environmental values into decisions about the future use of floodplain

Table 15.7 Key principles and rules to guide the management of human-induced floodplain processes

1. Take a broad policy perspective and a long time horizon.
    Floodplain management needs to be considered with sustainability and sustainable development as the key guiding principle. There is therefore a need to define what are the critical natural assets and the constant environmental resources in relation to floodplains.

2. Take a regional view, and a multidisciplinary perspective.
    Floodplain management is best considered in terms of integrated multifunctional catchment-wide management and planning which seeks to take into account and balance all the functions and uses of the water environment and associated land as well as flood defence: water quality and resources, nature and other conservation, amenity, fisheries and recreation are all important.

3. Work with sound and appropriate data.
    Good databases on flood risk areas, and on other floodplain resources, are fundamental to floodplain and river catchment management.

4. Policy development at only one level of government is not enough.
    It is important to develop mechanisms for integrating land-use planning and planning for the water environment at all the relevant levels: national, regional and local level action and coordination are essential requirements for floodplain management. It is also important to recognize that managing floodplains is a political process, rather than a technical one.

5. Involve the public in the decisions that affect them.
    Mechanisms are needed to inform the general public about flood risk and floodplain resources. These are vital requirements as are mechanisms for involving local people in decision making regarding river catchments in order to engender "local ownership" of plans and proposals. The management and planning of flood plain resources should be a "bottom-up" as well as a "top-down" process, as recommended in Agenda 21.

6. Use appropriate techniques as decision aids, but also use wisdom and human judgement about policies and priorities.
    Where consistent with sustainability and sustainable development principles (i.e. where critical assets and constant resources are not threatened by a decision and where economic analysis is not incompatible with a future-oriented approach), economic appraisal techniques such as benefit–cost analysis provide a valuable aid to rational decision making.

areas. In most cases, in contrast, there is a relatively simplistic emphasis on reducing the potential flood damages that might occur in the future – if there is consideration for the future at all – rather than a more holistic approach.

In general, these research results present us with gloomy conclusions. Only in the cases of the Netherlands, New South Wales (Australia) and aspects of the situation in England and Wales do we see coherent systems in place for tackling the processes of urbanization and development on floodplains. In other areas these processes are either treated separately (and flood defence works retro-fitted when problems emerge; land-use planning is undertaken without regard to flood defence matters), or the systems of land-use and flood defence planning are so poor and fragmented that local interest groups and political processes dominate the development of policy, rather than a strategic or coherent view.

The results from this research, and the implications from these conclusions, are that the processes of development and urbanization will continue to exacerbate flood hazard problems into the future. They will also affect the geomorphological evolution of river valleys and floodplains, by constraining rivers with the urbanization of floodplain surfaces, and will thereby profoundly affect their hydrological and hydraulic characteristics. In this respect it will not be sufficient, in the future, for researchers only to consider the physical processes of floodplain and river channel modification, when evaluating floodplain processes generally, without due regard for the significant impacts of human-induced change.

REFERENCES

Anderson, R.J., dos Santos, N. da R. and Diaz, H.F. (1993) *An Analysis of Flooding in the Paraná/Paraguay River Basin*. LATEN Dissemination Note 5. Washington DC, The World Bank (Latin America and Caribbean Technical Department, Environment Division).

Burgos, J.J., Ponce, H.F. and Molion, L.B. (1991) Climate change predictions for south America. *Climate Change*, **18**, 223–239.

Burton, I., Kates, R. and White, G. (1978) *The Environment as Hazard*. New York: Oxford University Press.

Chatterton, J.B., Correia, F.N., Green, C.H., Hubert, G., Penning-Rowsell, E.C., Saraiva, M. and Torterotot, J. (1994) *Strategic Land Use Planning in Europe and USA: Review of Best Practice for the NRA*. Bristol, National Rivers Authority.

Department of the Environment, Ministry of Agriculture, Fisheries and Food, and Welsh Office (1992) *Development and Flood Risk: Circular 30/92*. London, HMSO.

Gardiner, J. (1994) Sustainable development and its implications for flood defence and the water environment. MSc course portfolio, Department of Civil Engineering, Bristol University, UK.

Government of New South Wales (1984) *Flood Development Manual*. Sydney, Government of New South Wales.

Green, C.H. and Penning-Rowsell, E.C. (1989) Flooding and the quantification of "intangibles". *J. Inst. Water Engrs and Env. Man*, **3**(1), 27–30.

Green, C.H. and Tunstall, S.M. (1991) The evaluation of water quality improvements by the Contingent Valuation Method. *Applied Economics*, **23**, 1135–1146.

Green, C.H., Emery, P.J., Penning-Rowsell, E.C. and Parker D.J. (1985) *The Health Effects of Flooding: A Survey at Uphill, Avon*. London: Middlesex University Flood Hazard Research Centre.

Green, C.H., N'Jai, A. and Neal, J. (1987) *Thames Overview Pre-feasibility Study: Report to the Thames Water Authority*. London, Middlesex University Flood Hazard Research Centre.

Green, C.H., van der Veen, A., Wierstra, E. and Penning-Rowsell, E.C. (1994). Vulnerability refined: analysing full flood impacts. In: Penning-Rowsell, E.C. and Fordham, M. (eds), *Floods Across Europe: Hazard Assessment, Modelling and Management*. London, Middlesex University Press.

Handmer, J.W. and Penning-Rowsell, E.C. (eds) (1990) *Hazards and the Communication of Risk*. Aldershot, UK, Gower Technical Press.

Lewin, Fryer and Partners (1992) *Maidenhead Flood Alleviation Scheme*. London: Lewin, Fryer and Partners.

Ministry of Agriculture, Fisheries and Food (1993) *Flood and Coastal Defence: Project Appraisal Guidance Notes*. London, Ministry of Agriculture, Fisheries and Food.

Ministry of Agriculture, Fisheries and Food, and Welsh Office, (1993) *Strategy for Flood and Coastal Defence in England and Wales*. London, Ministry of Agriculture, Fisheries and Food and Welsh Office.

National Rivers Authority (1992a) *Datchet, Wraysbury, Staines and Chertsey Flood Study: Project Report*. Reading, National Rivers Authority.

National Rivers Authority (1992b) *Techniques for Identification of Floodplains*. R&D Digest 14. Bristol, National Rivers Authority.

National Rivers Authority (1994a) *Guidance Notes for Local Planning Authorities on the Methods of Protecting the Water Environment through Development Plans*. Bristol, National Rivers Authority.

National Rivers Authority (1994b) *Memorandum of Understanding: Development and Flood Risk*. Bristol, National Rivers Authority.

National Rivers Authority, Thames Region (1994 presumed) *Thames 21 – A Planning Perspective and a Sustainable Strategy for the Thames Region*. Reading, National Rivers Authority, Thames Region.

Parker, D.J. (1991) *The Damage Reducing Effects of Flood Warning*. London, Middlesex University Flood Hazard Research Centre.

Parker, D.J. (1995) Floodplain development policy in England and Wales. *Applied Geography*, **15**(4), 341–363.

Parker, D.J. and Handmer, J.W. (1992) *Hazard Management and Emergency Planning: Perspectives on Britain*. London, James and James.

Parker, D.J., Penning-Rowsell, E.C. and Green, C.H. (1983) *Swalecliffe Coast Protection Scheme: Evaluation of Potential Benefits*. London, Middlesex University Flood Hazard Research Centre.

Parker, D.J., Green, C.H. and Thompson, P.M. (1987) *Urban Flood Protection Benefits: A Project Appraisal Guide*. Aldershot, UK, Gower Technical Press.

Parker, D.J., Fordham, M. and Torterotot, J.-P. (1994) Real-time hazard management: flood forecasting, warning and response. In: Penning-Rowsell, E.C. and Fordham, M. (eds), *Floods Across Europe: Hazard Assessment, Modelling and Management*. London, Middlesex University Press.

Penning-Rowsell, E.C. (1996) Flood hazard response in Argentina: changing context and changing policies. *Geographical Review* (in press).

Penning-Rowsell, E.C. and Chatterton, J.B. (1977) *The Benefits of Flood Alleviation: A Manual of Assessment Techniques*. Aldershot, UK, Gower Technical Press.

Penning-Rowsell, E.C. and Fordham, M. (eds) (1994) *Floods Across Europe: Hazard Assessment, Modelling and Management*. London, Middlesex University Press.

Penning-Rowsell, E.C. and Smith, D.I. (1987) Self help flood hazard mitigation: the economics of house raising in Lismore, N.S.W., Australia. *Tijdschrift voor economische en social geografie*, **78**(1), 176–189.

Penning-Rowsell, E.C., Chatterton,. J.B., Day, H.J., Ford, D.T., Greenaway, M., Smith, D.I., Wood, R. and Witts, R. (1985) Comparative aspects of computerised floodplain data management. *J. Water Res. Plan. and Man, Am. Soc. Civ. Engrs*, **113**(6), 725–744.

Penning-Rowsell, E.C., Parker, D.J. and Harding, D.M. (1986) *Floods and Drainage: British Policies for Hazard Reduction, Agricultural Improvement and Wetland Conservation*. London, Allen and Unwin.

Penning-Rowsell, E.C., Peerbolte, B., Correia, F.N., Fordham, M., Green, C.H., Flugner, W., Rocha, J., Saraiva, M., Schmidtke, R., Torterotot, J. and Van der Veen, A. (1992) Flood vulnerability analysis and climate change: towards a European methodology. In: Saul, A.J. (ed.), *Floods and Flood Management*. Dordrecht, Germany, Kluwer Academic Publishers.

Sir William Halcrow and Partners (1994) *Estudio de regulacion del valle aluvial de los Rios Paraná, Paraguay y Uruguay para el control de las inundaciones*. Buenos Aires, Sir William Halcrow and Partners.

Slater, S., Marvin, S. and Newson, M. (1993) *Land Use Planning and the Water Sector: A Review of Development Plans and Catchment Management Plans*. Newcastle, UK, Department of Town and Country Planning, Newcastle University.

Tapsell, S. (1995) River restoration: what are we restoring to? A case study of the Ravensbourne River, London. *Landscape Research*, **20(3)**, 98–111.

Torterotot, J. and Roche, P.A. (1990) Evaluation socio-economiques pour la gestion du risque d'inondation. Paper given at the European Conference of Water Managers, Paris, 4–6 December 1990. In: *Gestion de l'eau*, 481090. Paris, Presses de l'Ecole Nationale des Ponts et Chaussees.

Tunstall, S.M., Parker, D.J. and Krol, D. (1993) *Planning and Flood Risk: A Strategic Approach for the National Rivers Authority*. Bristol, National Rivers Authority.

Wisdom, A.S. (1975) *The Law of Rivers and Watercourses*, 4th edition. Croydon, UK, D.R. Publications.

16 Sustaining the Ecological Integrity of Large Floodplain Rivers

G.E. PETTS

School of Geography, University of Birmingham, UK

16.1 INTRODUCTION

> The threat to aquatic biodiversity is more serious than threats to terrestrial diversity or even the integrity of tropical rainforests
>
> (Naiman *et al.*, 1995, p. 59)

Following a long history of ecological degradation as a result of human impacts, environmental restoration is becoming the symbol of advanced societies concerned with wealth creation and the quality of life through the sustainable use of natural resources and the conservation of biodiversity. River systems play an important role in these processes, not only providing vital routeways for migratory species, but also sustaining a high diversity of habitats for a wide range of flora and fauna. Natural river corridors have particularly high species diversity and productivity (Gregory *et al.*, 1991), and function as centres for the dispersal of organisms throughout drainage basins. Large river ecosystems in many regions of the world have been severely degraded by human impacts. Most attention has focused on impacts at the catchment scale that have changed flow, water quality and sediment transport regimes, with major repercussions for fluvial landforms and biological communities (e.g. Petts, 1984; Petts *et al.*, 1989; Calow and Petts, 1994). However, along large floodplain rivers, local impacts of flood control, land drainage and navigation works have also had dramatic effects upon the river ecosystem.

As identified by Naiman *et al.* (1995), the biological impoverishment of aquatic systems has been dramatic. In the United States, for example, the proportion of aquatic biota classed as rare to extinct ranges from 34% for fish to 75% for unionid mussels – the proportion of terrestrial vertebrates (birds, mammals and reptiles) similarly classed ranges from 11 to 15% (see Naiman *et al.*, 1995). Many factors, often acting synergistically, explain the decline and extinction of aquatic fauna. The most frequently cited factors are: overharvesting (15%), hybridization (38%), chemical alteration of habitat (38%), and detrimental effects of introduced species (68%), but the dominant factor is the loss and fragmentation of physical habitat (73%) (Miller *et al.*, 1989). The isolation of large rivers from their floodplains has been a major factor in physical

Floodplain Processes. Edited by Malcolm G. Anderson, Des E. Walling and Paul D. Bates.
© 1996 John Wiley & Sons Ltd.

habitat degradation. Thus, channelization, embanking, and lowering of the channel bed for navigation, flood control and land drainage have contributed significantly to biological impoverishment not only locally but also at the landscape scale (Naiman and Decamps, 1990; Large and Petts, 1994; Petts, 1996).

Recent disastrous floods in Europe, the United States and Bangladesh have led to the questioning of traditional flood-control and floodplain management policies and have focused attention on the values of preserving and restoring large river–floodplain ecosystems (Sparks, 1995). This chapter examines the role of floodplains in sustaining the ecological integrity of large rivers as a basis for programmes of river restoration and management. The "integrity" of a system includes elements (e.g. genes, species, populations and landscapes) and the processes that generate and sustain the elements (e.g. selection, succession, disturbance, sediment transport and nutrient cycling). "Integrity" may be defined as: "the capability of supporting and maintaining a balanced, integrated, adaptive community of organisms having a species composition, diversity, and functional organization comparable to that of natural habitat of the region" (Angermeier and Karr, 1994). However, a more pragmatic definition is "the maintenance of the community structure and function characteristic of a particular locale or deemed satisfactory to society" (Cairns, 1995).

16.2   THE NATURE OF LARGE FLOODPLAIN RIVERS

Traditionally, running waters have been the focus of study by two different and distinct groups of disciplines. First, hydrologists and fluvial geomorphologists, with geographical, geological or engineering backgrounds, have investigated the ways in which flows, sediment loads and channel forms vary along a river from headwaters to mouth and with time over periods ranging from hours (during individual floods), to one year (seasonal patterns), to 100 years (the main period of human impacts), and to 10 000 years (Petts, 1995). Secondly, ecologists, biologists and fisheries scientists have examined downstream patterns in the structure and function of biological communities – algae, macrophytes, macroinvertebrates and fish – over scales ranging from microhabitats (e.g. the individual stone) to whole rivers, including the nature of food webs and nutrient fluxes (see Calow and Petts, 1992). These classic approaches to the study of rivers emphasize two dimensions: longitudinal and temporal.

The classic high-gradient upland stream with small channel, cold temperatures and highly oxygenated water, dominated by fastwater habitats, and high flow variability, contrasts with low-gradient sectors of large floodplain rivers with a diversity of channel forms and floodplain waterbodies, and characterized by a predictable flood regime. This view of the river as a longitudinal continuum emphasizes downstream transfers of energy and matter (Vannote et al., 1980) and that ecosystem-level processes in downstream areas are linked to those in upstream catchments (Minshall et al., 1985). However, the ecological integrity of large floodplain rivers is not only a function of longitudinal processes but also of lateral connectivity (Welcomme, 1979; Junk et al., 1989).

Over the past decade, increasing attention has been given to the linkage between the river and its margins (Gregory et al., 1991; Petts and Decamps, 1996). River margins may be defined as land–water ecotones (Naiman and Decamps, 1990): the areas of land adjacent to a river channel within which habitat patches, their composition and arrangement, are determined by fluvial processes (flooding, erosion, deposition); where biotic communities relate to the dynamic interaction of fluvial and terrestrial processes, and comprise a continuum of dependencies from terrestrial species requiring regular inundation to aquatic species requiring occasional desiccation, as well as purely aquatic and purely terrestrial taxa. Recognition of the importance of ecotones for the structure and function of river ecosystems has led to a new understanding of the processes that sustain large floodplain rivers.

Thus, large floodplain rivers differ from other river systems in two ways. First, large rivers tend to be characterized by more regular and predictable variations of the abiotic variables than headwater streams. Large basins display a range of weather conditions, lithologies and land uses, so that flow and water quality reflect the mixing of water from the different headwater catchments, the routing of flows through the channel network, and channel and floodplain storage. While headwater streams are highly sensitive to catchment changes, downstream such effects may be buffered by mixing of runoff from unaffected, or differently affected, catchments, and by lateral fluxes between the river and its floodplain and vertical fluxes between the river and the alluvial aquifer. Secondly, large floodplain rivers are characterized by strong lateral gradients and linkages, in contrast to rivers lacking floodplains which are dominated by longitudinal processes, especially over short time-scales (<100 years). The complex ecosystems of large floodplain rivers comprise a range of aquatic habitats (flowing channels, backwaters, lakes, and springs), wetlands, and terrestrial patches, in different successional stages. The structure of this patchwork of habitats reflects the fluvial dynamics. High and predictable connectivity between the different patches during the flood season characterizes large floodplain rivers. They display a wide range of habitats and food sources and this is associated with high biotic diversity and a wide range of trophic adaptations (e.g. as shown by fish, Welcomme, 1979, 1995).

A model of the downstream ecological patterns along large rivers may be proposed that views the "river" as comprising both the active channels and their ecotones, which can extend the "*lit majeur*" (the French concept of a river's major bed) for several kilometres beyond the banks of the main channel. In this model (Figure 16.1), the primary variables are (a) the diel and annual temperature ranges, (b) channel stability, and (c) the flood regime (especially the predictability of the flood season), all of which are seen to increase downstream. Also shown are (d) the proportion of the total organic matter that is in the coarse particulate form (CPOM) and (e) the influence of the land–water ecotones for sustaining the integrity of large river ecosystems. The interaction of these factors explains the relatively high biodiversity (f) found along large floodplain rivers.

The model river is seen to comprise three zones (following Schumm, 1977) based on the variation of channel properties through a drainage basin: (i) the headwater (production) zone with steep slopes and channel confined between valley slopes; (ii) a relatively stable (transfer) zone, with a sinuous, single-thread channel; and (iii) a depositional, meandering (storage) zone, with very low slope. A highly unstable, braided zone may be added (iia), classically occurring at the interface between zones i and ii. Headwater channels are laterally constrained by hillslopes and are largely

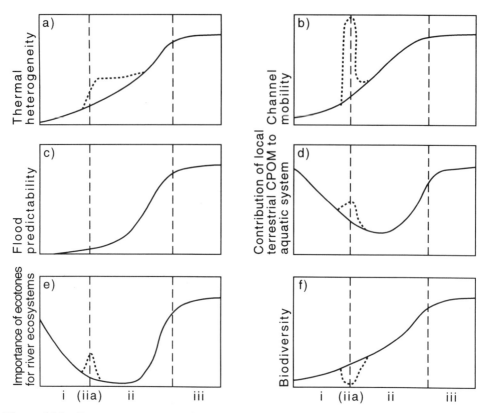

Figure 16.1 Downstream patterns (low to high) along an idealized river contrasting the large floodplain river (iii) with the middle river (ii) including braided sections (iia), and headwater streams (i) (developed from Ward and Stanford, 1995)

dependent upon inputs of organic matter from the riparian zone, especially in heavily shaded streams. Coarse woody debris plays an important role in sustaining a diversity of habitats within these streams (e.g. Gurnell *et al.*, 1995). The middle reaches of rivers are dominated by downstream transfers of organic matter and nutrients, and the role of the ecotone is relatively low. However, progressively, as the structure of the channel margin becomes increasingly complex, the main channel becomes little more than a routeway for the movement of organisms and the channel margin becomes the important zone for both productivity and diversity. In large floodplain rivers, the ecological integrity of the river is dependent upon the predictable connection to the diverse range of floodplain habitats. Most of the animal biomass in large rivers derives directly or indirectly from production within the floodplain and not from downstream transport of organic matter (Junk *et al.*, 1989). This is explained by the large amounts of CPOM derived from the floodplain: wood, bark, leaves, flowers, fruit and terrestrial arthropods; the herbaceous vegetation that shows high productivity during the dry season; and the abundant hydrophyte growths during the wet season. It is within zone iii that the strongest links occur between the river channel and floodplain systems.

Although floodplains may occur in any of the other zones, it is the high connectivity, wide range of habitats created and rejuvenated by channel instability, and high predictability of the flood regime, that gives this zone the highest productivity and biodiversity.

16.3 FLOODPLAIN PATCHES

Fluvial processes create the template of physical features and balance the successional process of sediment filling and terrestrialization with scour and channel migration. Periodic rejuvenation of the system by erosion and deposition sustain the health of the ecosystem (Bayley, 1995). Within each sector of a floodplain river, a quasi-equilibrium condition may be defined involving hydrological, geomorphological and ecological interactions. Each sector can be described by a particular diversity and arrangement of aquatic and terrestrial habitat patches. Floodplain patches must be defined, however, not only by type, for example, sand bars, levees, cutoff channels and backswamps, but also by age (reflected by successional criteria for each type), and by the frequency and duration, or probability, of connectivity. Along the Amazon, for example, Salo et al. (1986) demonstrated that channel migration is the principal process explaining the shifting successional forest mosaic.

The relative abundance of different patch types, the complexity of the mosaic, the arrangement of patches, and the dynamic character of this arrangement, relates to the channel pattern (meandering, braided, anastomosed) and the degree of channel mobility over time-scales of $10 < 1000$ years. Thus, the diversity of vegetation patches on floodplains is related to the rejuvenation of successions associated with channel erosion and deposition so that the patch mosaic reflects age structure and patch type (Petts, 1990a, b). Thus, along the River Rhone, France, the floodplain patch mosaic reflects the varying morphological stability of the different channel sectors (Roux et al., 1989). In the rapidly shifting braided sectors, frequent disturbance inhibits long plant successions and pioneer populations dominate. Less active sectors, in contrast, are characterized by hardwood forests.

The arrangement of patches within a sector changes over time in response to successional processes and disturbance (by erosion and deposition). Thus, the optimal areas for particular fauna shift to different parts of a sector in response to the build-up of sediments (e.g. Bayley, 1991). Although the spatial arrangement of patches may change, the proportions of different patches within each sector tend to remain relatively stable, about an average condition. The changing patch arrangement is a form of disturbance and plays a critical role in organizing communities and ecosystems (Townsend, 1989; Reice et al., 1990).

Thus, a genetic floodplain may be deemed as the area of land adjacent to a river channel within which habitat patches, their composition and arrangement, are determined by the contemporary regime of fluvial processes (flooding, erosion, deposition); where biotic communities relate to the dynamic interaction of fluvial and terrestrial processes, and comprise a continuum of dependencies from terrestrial species requiring regular inundation to aquatic species requiring occasional desiccation, as well as, often opportunistic aquatic and terrestrial taxa. This distinguishes "floodplain" from valley floor situations that are inundated frequently or rarely, by overbank floods but where the structural characteristics of the inundated valley floor are not conditioned by the contemporary regime of fluvial processes.

16.4 FLOOD REGIME

While acknowledging the importance of channel migration and the fluvial processes of erosion and deposition for sustaining a diversity of habitat patches along river corridors, over short time-scales (<100 years), it is the seasonal flood regime that drives the large floodplain–river ecosystem. It is seasonal flooding that promotes the exchange of materials and organisms among the mosaic of habitats and it is the hydrological regime that plays the key role in determining the level of biological productivity and diversity (Bayley, 1995). Within large floodplain rivers, the principal process that has affected the evolution of biota is the dynamic interaction between water and land (Junk *et al.*, 1989). The regular, predictable flood pulse should not be regarded as a disturbance but as a resource. This may be illustrated by three examples, illustrated in Figure 16.2:

1. Flooding adds to the detrital and mineral nutrient pools of both the terrestrial and aquatic systems: during rising stages silts and nutrients replenish the floodplain pool while the drowning of mineral and organic matter, added to by migratory animals and birds, releases nutrients into the water; primary production and decomposition rates are high but production tends to outstrip decomposition. Once flood peak is reached – i.e. once the aquatic littoral has stopped advancing across

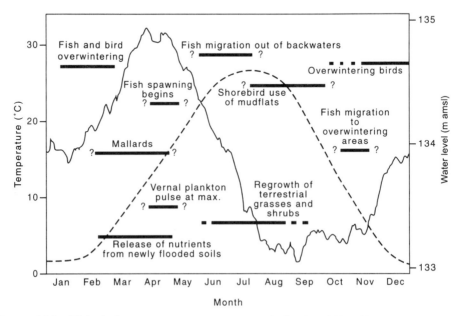

Figure 16.2 Biological responses to average annual flood (1960–1993) and temperature (1989–1993) cycles at Havanna, on the Illinois River, USA (based on Sparks, 1995)

the floodplain – decomposition processes increase relative to production, leading to low oxygen levels. During receding flows, nutrients may be returned to the river, vegetable and animal material will be stranded, to be consumed by terrestrial animals and birds, or decompose, and exposed fluvial deposits are stabilized by the rapid germination of terrestrial plants. Many persistent "weeds" characterize the early successional stages of such highly disturbed patches, reflecting their r-strategy traits.

2. The land–water ecotone comprises a gradient of plant species adapted to seasonal degrees of inundation, nutrients, and light. Thus, many species of river-margin plant, such as the red gum (*Eucalyptus camaldulensis*) forests along the River Murray, Australia (Bren, 1988), and the cypress (*Taxodium distichum*) wetlands of southeastern USA (Mitsch *et al.*, 1979), are dependent upon fluvial flooding. Although every plant has an optimum position on the inundation gradient (see Table 16.1), species may be displaced by many other factors: patch stability; fertility and groundwater level; and biogenic processes such as organic matter accumulation, nitrogen fixation and interspecific competition.

3. Most higher biota, including mammals, fish and invertebrates, are adapted to the flood cycle and particularly high diversity characterizes the more mobile organisms, such as fish (Welcomme, 1979). The seasonal change from terrestrial to aquatic phases imposes a severe stress on organisms but many have developed adaptations that not only enable them to survive during the adverse period of drought or flood but also to benefit from it (Junk *et al.*, 1989). Thus, many species of river fish rely on the annual inundation of the river margin for reproduction and feeding. Spawning before or during flood rise allows the use of floodplain nursery habitats and the high production ensures the fish grow quickly during the flood period to reach sufficient size to reduce predation losses or overwinter mortalities. For tropical and temperate rivers, fish diversity and productivity are directly linked to the scale of the annual (predictable) inundation of the floodplain ecotone (Welcomme, 1995), and Bayley (1991) discusses this as the "flood advantage" for fish production.

The flood-pulse concept implies that biological production is enhanced through a variety of processes during the flooding cycle (Bayley, 1995). However, the role of the flood pulse depends upon both hydrological and seasonal temperature changes. The flood pulse may be described by its duration, amplitude, frequency, timing and predictability. Short-duration floods with rapid rise and recession, or irregular and unpredictable events, offer little resource value and disturb floodplain environments (Petts, 1996). There has been little quantitative research on the relaǔve benefits of different flood hydrographs (see Petts and Decamps, 1996) but Bhowmik and Demissee (1982) have shown that most river floodplains act as storage reservoirs for flows at or below the 40-year return-period event. For higher flows, the floodplain and river act more or less as a homogeneous conveyance channel, rather than as a storage reservoir. Nevertheless, disturbance can play a positive role because extreme floods can "reset" floodplain patches, destroy floodplain forest, flush pools, and rejuvenate successions.

Large floodplain rivers characterized by regular, predictable and slow inundation show greatest benefits of the flood pulse where the flood pulse is coupled with the seasonal temperature rise. Thus, in the tropics, consistently high temperatures favour high production and rapid processing of organic material throughout the year. The flood-pulse advantage (Bayley, 1991) is lost if the hydrological and temperature regimes are not coupled, although the floodplain ecosystem may benefit from the distribution of water and nutrients prior to the growing season and the flushing of detritus to the main channel during the spring may provide a valuable food source for aquatic biota. In temperate maritime climates, summer rainfalls benefit the production of plant communities, as well as directly and indirectly influencing the life cycles of many animals. In cold temperate and polar rivers, where the timing of the flood is predictable because of massive snowmelt – sometimes extended by ice melt – the high sediment loads, cold water and ice floes present an inhospitable environment. Thus, the "flood advantage" for the large river ecosystem may be greatest in tropical, semi-arid and temperate continental climatic regions.

The 1993 flood on the Mississippi and Illinois rivers, USA, provided an opportunity to investigate the effects of an extreme flood with unusual timing and duration. The annual flood usually occurs in April–May (Figure 16.2) but in 1993 above-normal river stages persisted throughout the growing season (Figure 16.3), peak flow occurred more than 100 days later than normal. Data on fish abundance (Mather, 1994) for the lower Illinois, demonstrated that the flood supported a high diversity and abundance of fishes. The high abundance of young-of-the-year fishes supports the hypothesis that ideal conditions for spring spawners occur during years in which the flood and temperature

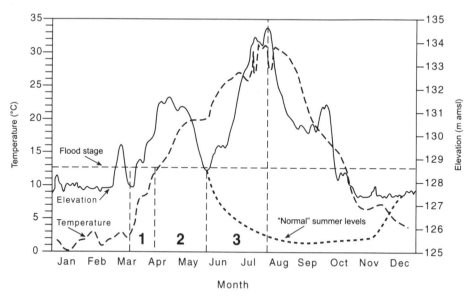

Figure 16.3 Flood hydrograph, temperature pattern, and groups of fish spawners, in the Illinois River at Grafton, during 1993 (after Mather, 1994). Fish groups: (1) early spring spawners (grass pickeral *Esox vermiculatus*; and gizzard shad *Dorosoma cepedianum*), (2) spring spawners (bigmouth buffalo *Ictiobus cyprinellus*; largemouth bass *Micropterus salmoides*; black crappie *Pomoxis nigromaculatus*; and white bass *Morone chrysops*), (3) summer spawners (golden shiner *Notemigonus crysoleucas*; bluegill *Lepomis marrochirus*; and western mosquitofish *Gambusia affinis*)

Table 16.1 Mortality of common tree species caused by the 1993 flood along the Mississippi River through missouri, USA (after Yao Yin *et al.*, 1994)

Species	Common name	Flood tolerance*	Tree mortality (%)†	Sapling mortality (%)†
Acer negundo	Box elder	1–2	43	62
Acer saccharinum	Silver maple	2	28	88
Celtis occidentalis	Hackberry	4	92	99
Cornus spp.	Dogwood	2	∅ (0%)	70
Foresttera acuminata	Swamp privet	1	∅ (0%)	52
Fraxinus pennsylvanica	Green ash	1–2	18	47
Morus alba	White mulberry	4	∅ (75%)	90
Populus deltiodes	Cottonwood	2	22	∅ (100%)
Quercus palustris	Pin oak	2–3	26	∅ (60%)
Salix nigra	Black willow	1	8	56
Ulmus americana	American elm	2–3	20	61
Ulmus rubra	Slippery elm	2	19	65

* 1 = very tolerant – 4 = intolerant.
† ∅ = less than 20 tress sampled.

rise are coupled (Junk *et al.*, 1989), floodplain inundation functioning as important spawning and/or nursery areas for many fish species. Moreover, April–May and summer spawners also benefited from the late flood: the large amounts of nursery and refuge habitat, and high primary production, increased the survival of these species. Although the year class produced in 1993 was unusually large, whether or not the young-of-the-year are recruited into the adult breeding population remains to be seen; if not, the lack of overwintering habitat or overharvest may be limiting factors.

Extreme flooding in a single growing season can also cause severe disturbance to the large river ecosystem. The effect of the 1993 flood on the alluvial forest along the lower section of the upper Mississippi was to cause mortality rates of 18–37% for trees and 70–80% for saplings (Yao Yin *et al.*, 1994). Smaller juvenile trees were nearly completely wiped out. Mortality rates increased downstream and were positively correlated with flood duration and amplitude, although mortality rates varied markedly among species (Table 16.1). However, the flood may also be seen to have had a benefit for the sustainability of the habitat diversity provided by the alluvial forest, by "resetting" the floodplain ecosystem. The flood created new opportunities for pioneer species to regenerate. In areas where all layers of above-ground vegetation had been killed and new sediments deposited during the flood, first-year seedlings of black willow, cottonwood and sycamore (*Platanus occidentalis*) were frequently encountered during the summer of 1994 (Yao Yin *et al.*, 1994).

16.5 THE HISTORICAL LEGACY

The 1960/1970s was the era of the mega-project in river works (Cosgrove and Petts, 1990) but palaeoenvironmental reconstructions for some regions, such as western

Europe (Petts *et al.*, 1989; Starkel *et al.*, 1991) have demonstrated human impacts extending over a period of about 4000 years. These began with floodplain siltation induced by deforestation. Progressive aggradation and increasing flood levels resulting from expanding agriculture and catchment disturbance during the eighteenth and nineteenth centuries led to the intensification of river regulation works to improve navigation, to utilize water power, and to drain floodplain lands. The impacts of such regulation works on the integrity of large floodplain rivers have been dramatic, and examples emphasize the importance of (i) lateral connectivity, (ii) the flood pulse, and (iii) fluvial processes. Many floodplain rivers are now hydraulically optimal but their once ecologically diverse and productive corridors have been eliminated as a direct result of dam construction (Petts, 1984) and channelization (Brookes, 1988), and indirectly by land-use developments throughout the drainage basins (Lauga *et al.*, 1986). On many large rivers throughout Europe and North America, and increasingly on other continents, the area subject to flooding has been drastically reduced. The contrasting characteristics of natural and regulated floodplain rivers are summarized in Table 16.2. Ward and Stanford (1995) discuss the effects of dams on abiotic patterns and processes, incorporating interactions between rivers and floodplains. Of particular importance is the source of organic matter, because organic matter that is transported long distances tends to be refractory and have fewer attached microbes, and, thus, it contributes less to secondary production than does material contributed locally from the floodplain or in-river sources (Thorp and Delong, 1994).

The morphological processes of floodplain rivers determine the dynamics of the habitat mosaic for biota. Thus, the biota of large floodplain rivers may be seen as comprising those organisms that are able to track the shifting spatial mosaic of suitable environmental conditions. Along regulated rivers, floods are controlled and the morphological dynamics are stabilized. In these situations metapopulation dynamics may be especially critical for two reasons: (i) lateral connectivity between patches is severely restricted and (ii) given that local extinctions will take place (succession, wash out, etc.) stochastic local extinctions must be balanced by recolonization. Especially for primary patch types (e.g. gravel bar, mudflat and cutoff channel), along regulated rivers, even the largest habitat patches may be so small and/or widely spaced that local populations have a distinct risk of extinction. Early successional stages and pioneer communities have been most severely affected (Bravard *et al.*, 1986; Amoros *et al.*, 1987a, b).

The ecological impact of "river training", isolating them from their natural floodplains and fixing the channel course, is illustrated particularly well by research on fish communities. In Europe, Copp (1991) attributed the limited recruitment by limnophilic plant spawners (e.g. *Blicca bjoerka* and *Tinca tinca*) to the drastic reduction in the number of lentic resource patches – semi-abandoned and abandoned channels – consequent upon the construction of embankments along the Bedford Ouse; on the Danube, Schiemer and Spindler (1989) highlighted the importance of a structurally diverse littoral zone to meet the trophic requirements of juvenile fishes, and Löffler (1990), in comparing two floodplains along this river, demonstrated that the loss of lateral connectivity had caused a dramatic reduction in the number of fish species from 30 to 4; and Jurajda (1995) demonstrated the isolation of backwaters from the main channel of the Morava River, Czech Republic, where the prevention of

Table 16.2 Contrasts between natural and regulated (canalized and embanked) large floodplain rivers having late spring/early summer flood in temperate regions. P = production; R = respiration; CPOM = coarse particulate organic matter; FPOM = fine particulate organic matter

	Natural	Regulated
Productive processes	High P/R ratio due to high *in situ* production on the inundated floodplain	Low P/R ratio due to processing of material from upstream and low *in situ* production
	High CPOM/FPOM ratio – floodplain contributes wood, bark, leaves, fruit, flowers, terrestrial arthropods. High productivity of herbaceous vegetation during dry phase and hydrophytes during wet phase	Low CPOM/FPOM ratio – dependent upon downstream flux of organic matter from headwaters
Dominant role of floods	Resource: life cycles of many aquatic and terrestrial organisms are adapted to the flood regime	Disturbance: *frequent* flushing "resets" physical and biological environments and limits successions
Dominant role of main channel	Routeway for gaining access to adult feeding areas, nurseries, spawning grounds and refuges	Supports whole life cycle
Bulk of riverine animal biomass derives from:	Production within floodplains	Downstream transport of organic matter from headwaters
Biodiversity	Very high	Limited by homogeneous substratum (bed and banks), turbidity, oxygen deficits
Sensitivity of biota to headwater catchment "events" (major floods or pollution incidents)	Low	High

flooding led to the decline in abundance, and in some cases local extinction of species that require access to floodplains for spawning or nursery habitat.

16.6 OPPORTUNITIES FOR RESTORATION

Restoration of large rivers to a pristine or virgin state is a complex and difficult task because fluvial ecosystems are so intimately tied to physical, chemical and biological processes that occur throughout the catchment. Indeed, restoration is probably incompatible with present human population levels (Welcomme, 1992) and past environmental change (Petts *et al.*, 1989). Thus, for many rivers, an end point other than pristine may be the best option. Indeed, such end points may be considered "normal and good" (Regier, 1993). To many people, the alternative states are not only acceptable and normal, but are also desirable for aesthetic, recreational or commercial reasons. Cairns (1995) has argued strongly that an anthropocentric view of "ecological integrity", incorporating the maintenance and sustainable use of desirable ecosystem services, is required if major progress in environmental conservation and restoration is to be achieved. Humans are an integral part of the large floodplain river ecosystems that we are seeking to restore and rehabilitation programmes must seek to achieve a functioning ecosystem that is sustainable (i) under current hydrological, water quality and sediment transport regimes (determined by catchment conditions), (ii) in the context of balancing the social, economic and environmental needs, and (iii) with a minimum level of maintenance (i.e. artificial influence). In the long term, major advances in rehabilitation can be realized through integrated catchment management and local (sector-scale) management. However, in the short term, considerable benefits – environmental and socio-economic – can be achieved by sector-scale management alone.

At the sector scale, restoration of the annual flood pulse supported by habitat management on the floodplain could offer a cost-effective way to increase production of a diverse array of species *"with nature paying a much larger proportion of the bill than taxpayers in the long run"* (Bayley, 1991). Certainly, there is evidence to suggest that reinvasion of macroinvertebrates and fish species would be relatively rapid (Milner, 1994). Because of the historic legacy of development, restoration of the capacity for geomorphic renewal (Bravard *et al.*, 1986; Amoros *et al.*, 1987a) may be a realistic goal in only a few cases. Rehabilitation requires the provision of a new physical structure to enhance the formation of the biotic community (Gore and Shields, 1995), but this may have to be achieved artificially by designing a floodplain morphology to create a predetermined "target" structure. More general opportunities may exist to restore the flood storage function of floodplains by purchasing land, removing embankments and modifying reservoir, lock and dam operations along discrete sectors (Bayley, 1991).

Other human-caused changes to river ecosystems that are irreversible in absolute or practical terms, will also influence restoration programmes. Changes in biological community composition by extinction and the replacement of native woodland by urban surfaces are two examples. Invasion by exotic species and hatchery-based stocking programmes require urgent control to reduce the decline of genetic diversity (Dodge and Mack, 1994). The list of introduced species is considerable and invasive weeds (Brock *et al.*, 1993) may be a particular problem for the restoration of floodplain systems. However, introduction of exotic species is often cited as a factor contributing to the decline or extinction of fish species (Miller *et al.*, 1989). For example, brown trout (*Salmo trutta*), a polymorphic species having wide-ranging habits, was once known as the "European Fish", the limits of its native distribution being Iceland and the Barents Sea in the north, the Atlas Mountains of North Africa and the Mediterranean Basin to the south, and the Ural Mountains and the Caspian Sea to the east, with an isolated population around the Aral Sea. Since the discovery of the means of transporting ova in 1852, it has been introduced into 24 countries (Elliot, 1989), often to the detriment of native species.

This historic legacy of human interference demands that restoration policies are supported by sustained maintenance and management programmes. The legacy is manifest in terms of the physico-chemical environment and in the composition of the biota. Thus, for example, on the Illinois River, USA, increased turbidity from agricultural runoff and boat traffic caused a drastic decline in macrophytes in backwaters from 1958 to 1961 (Sparks *et al.*, 1990). Without plants to break the wind in the backwaters, wave energy increased, sediments were resuspended and the shoreline eroded; sedimentation reduced the depth of the backwaters and increased wind-driven suspension of sediments, further increasing turbidity, and reducing the ecological value of these backwater areas. Reduction of sediment loads, through improved catchment management and/or in-river works and maintenance programmes, is required to mitigate these impacts.

In one important case, on the upper Colorado (including the Green and Gunnison rivers), USA, aimed at preventing the extinction of the razorback sucker (*Xyrauchen texanus*), restoration involves reinstating the historical flooding regime and the lateral connectivity by removing some levees, lowering floodplain levels and manipulating floodplain water levels artificially (P. Nelson, personal communication). The razorback sucker depends on the spring flood. Spawning on the rising limb of the hydrograph, the embryos hatch within one or two weeks and newly hatched larvae drift downstream during the high spring flows into floodplain nursery areas where the water is warmer and calmer, food more plentiful and inundated vegetation provides refuge from predation. Naturally, these habitats flooded two out of three years and remained flooded for several weeks. Today, dams, diversions and levees regulate the river such that the floodplain is flooded rarely and then only for very short durations. The aim of the restoration programme is to establish a complex of bottomland habitats to which drifting razorback larvae have access, and within which a percentage of the larvae can survive to recruit into established populations. However, competition and predation by species introduced into the basin may limit the success of the programme. The razorback's list of predators includes birds (e.g. herons, cormorants and eagles), mammals (otters and racoons), and a long list of piscvorous fishes including many exotic species. Survival of the razorback sucker may depend upon the successful control of these populations.

Political concern for river systems has arisen mainly in response to pollution problems. Although ecosystem sustainability is attracting increasing attention, we must fully recognize that traditional human needs still remain: for protection against floods, for clean water supplies, and for hydroelectric energy and irrigation. However, following recent catastrophic floods in the USA, there has developed a new vision of the twenty-first century floodplain to meet the needs of both human and natural systems (IFMRC, 1994; Table 16.3). Over the past three decades, the natural resources of floodplains have been recognized: biodiversity, scenic beauty, groundwater recharge and flood storage. This vision seeks to optimize these natural values as key components of a policy to reduce the vulnerability of urban areas, industry and agriculture to flooding.

As scientists, we must seek to define the limits of system integrity with special regard to diversity and connectivity; but from an anthropocentric perspective, we must seek to optimize ecosystem resiliance *and* the acceptable risk for human social,

Table 16.3   Selected key features of the twenty-first century floodplain (based on IFMRC, 1994)

Flow regulation for ecological benefit
At the upstream end of levees, water-control structures will permit controlled passage of river waters to keep wetlands moist throughout the year and sustain perennial lentic bodies to maintain and restore aquatic habitat for fisheries, wildfowl and other wildlife.

Levees
Some will be removed or relocated, others will be modified to provide for controlled overtopping.

Management of locks and dams
The navigation system will be operated to enhance riverine ecosystems by water-level adjustments and control.

Land-use change
Some floodplain land will be converted to more appropriate land use or returned to a natural state under federal or state easements.

Catchment management
Programmes will be introduced to improve the treatment of lands, control runoff, and restore wetlands.

economic, cultural and political systems. Current scientific knowledge is sufficient to allow progress in sustaining and restoring the ecological integrity of large floodplain rivers, provided that management incorporates a precautionary approach including the necessary monitoring programmes, and the degree of flexibility necessary to allow response to both successful and unsuccessful actions. The main lesson of the past 100 years during which we have sought to control rivers, is that to sustain the integrity of large river ecosystems, river regulation schemes must find a place for change: changing flows (certainly) and moving channels (ideally).

ACKNOWLEDGEMENTS

I am grateful to many people for opportunities, over the past two years, to discuss a range of issues related to the management of large floodplain rivers, especially: Ken Lubinski and colleagues at the National Biological Service, Environment Technical Center, Onalaska, Wisconsin; Rip Sparks of the Illinois Natural History Survey; Jim Sedell, Stan Gregory, Peter Bayley and colleagues at Oregon State University, Corvallis; Pat Nelson, National Biological Service, Denver, Colorado; and Robin Welcomme, FAO, Rome.

REFERENCES

Amoros, C., Rostan, J.-C., Pautou, G. and Bravard J.-P. (1987a) The reversible process concept applied to the environmental management of large river systems. *Environmental Management*, **11**, 607–617.

Amoros, C., Roux, A.L., Reygrobellet, J.L., Bravard, J.-P., and Pautou, G. (1987b) A method for applied ecological studies of fluvial hydrosystems. *Regulated Rivers: Research and Management*, 1, 17–36.

Angermeier, P.L. and Karr, J.R. (1994) Biological integrity versus biological diversity as policy directives. Protecting biotic resources. *BioScience*, 44, 690–697.

Bayley, P.B. (1991) The flood-pulse advantage and the restoration of river–floodplain systems. *Regulated Rivers: Research and Management*, 6, 75–86.

Bayley, P.B. (1995) Understanding large river–floodplain ecosystems. *Bioscience*, 45, 153–158.

Bhowmik, N.G. and Demissie, M. (1982) Carrying capacity of floodplains. *Proc. ASCE, J. Hydraulics Division*, 108(HY3), 443–452.

Bravard, J.P., Amoros, C. and Pautou, G. (1986) Impacts of civil engineering works on the succession of communities in a fluvial system: a methodological and predictive approach applied to section of the Upper Rhone River. *Oikos* 47, 92–111.

Bren, L.J. (1988) Effects of river regulation on flooding of a riparian red gum forest on the River Murray, Australia. *Regulated Rivers*, 2, 65–78.

Brock, J.H., Childs, L.E., de Waal, L.C. and Wade, P.M. (eds) (1993) *Ecology and Management of Invasive Weeds*. John Wiley, Chichester.

Brookes, A. (1988) *Channelized Rivers*. John Wiley, Chichester.

Cairns, J. Jr (1995) Ecological integrity of aquatic systems. *Regulated Rivers*, 11, 313–324.

Calow, P. and Petts, G.E. (eds) (1992) *The Rivers Handbook*, volume 1. Blackwell Scientific, Oxford.

Calow, P. and Petts, G.E. (eds) (1994) *The Rivers Handbook*, volume 2, Blackwell Scientific, Oxford.

Copp, G.H. (1989) The habitat diversity and fish reproductive function of floodplain ecosystems. *Environmental Biology of Fishes*, 26, 1–27.

Copp, G.H. (1991) Typology of aquatic habitats in the Great Ouse, a small regulated river. *Regulated Rivers*, 6, 125–134.

Cosgrove, D. and Petts, G.E. (eds) (1990) *Water Engineering and Landscape*. Belhaven, London.

Dodge, D.P. and Mack, C.C. (1994) Direct control of fauna: role of hatcheries, fish stocking and fishing regulations. In: Calow, P. and Petts, G.E. (eds) *Rivers Handbook*, 2, 386–400.

Elliot, J.M. (1989) Brown trout: *Salmo trutta. Journal of Fish Biology*, 34.

Gore, J.A. and Shields, F.D. Jr (1995) Can large rivers be restored? *Bioscience*, 45, 142–152.

Gregory, S.V., Swanson, F.J., McKee, W.A. and Cummins, K.W. (1991) An ecosystem perspective of riparian zones. *BioScience*, 41, 540–551.

Gurnell, A.M., Gregory, K.J. and Petts, G.E. (1995) The role of coarse woody debris in forest aquatic habitats: implications for management. *Aquatic Conservation*, 5, 143–166.

IFMRC (1994) *Sharing the Challenge: Floodplain Management into the 21st Century*. Report of the Interagency Floodplain Management Review Committee to the Administration Floodplain Management Task Force, Washington, DC, June.

Junk, W.J., Bayley, P.B. and Sparks, R.E. (1989) The flood pulse concept in river–floodplain systems. *Special Publication of the Canadian Journal of Fisheries and Aquatic Sciences*, 106, 110–27.

Jurajda, P. (1995) The effect of channelization and regulation on fish recruitment in a floodplain river. *Regulated River*, 11, 207–216.

Large, A.R.G. and Petts, G.E. (1994) Rehabilitation of river margins. In: Calow, P. and Petts, G.E. (eds), *The Rivers Handbook*, volume 2. Blackwell Scientific, Oxford, 401–418.

Lauga, J., Decamps, H. and Holland, M.M. (eds) (1986) *Landuse Impacts on Aquatic Ecosystems*. Proceedings of the MAB-UNESCO and PIREN-CNRS Workshop. CRDP, Toulouse.

Löffler, H. (1990) Danube backwaters and their response to anthropogenic alteration: In: Whigham, D.F., Good, R.E. and Kvet, J. (eds), *Wetland Ecology and Management: Case Studies*, Kluwer, Dordrecht, 127–139.

Mather, R.J. (1994) Observations of fish community structure and reproductive success in

Amoros, C., Roux, A.L., Reygrobellet, J.L., Bravard, J.-P., and Pautou, G. (1987b) A method for applied ecological studies of fluvial hydrosystems. *Regulated Rivers: Research and Management*, **1**, 17–36.

Angermeier, P.L. and Karr, J.R. (1994) Biological integrity versus biological diversity as policy directives. Protecting biotic resources. *BioScience*, **44**, 690–697.

Bayley, P.B. (1991) The flood-pulse advantage and the restoration of river–floodplain systems. *Regulated Rivers: Research and Management*, **6**, 75–86.

Bayley, P.B. (1995) Understanding large river–floodplain ecosystems. *Bioscience*, **45**, 153–158.

Bhowmik, N.G. and Demissie, M. (1982) Carrying capacity of floodplains. *Proc. ASCE, J. Hydraulics Division*, **108(HY3)**, 443–452.

Bravard, J.P., Amoros, C. and Pautou, G. (1986) Impacts of civil engineering works on the succession of communities in a fluvial system: a methodological and predictive approach applied to section of the Upper Rhone River. *Oikos* **47**, 92–111.

Bren, L.J. (1988) Effects of river regulation on flooding of a riparian red gum forest on the River Murray, Australia. *Regulated Rivers*, **2**, 65–78.

Brock, J.H., Childs, L.E., de Waal, L.C. and Wade, P.M. (eds) (1993) *Ecology and Management of Invasive Weeds*. John Wiley, Chichester.

Brookes, A. (1988) *Channelized Rivers*. John Wiley, Chichester.

Cairns, J. Jr (1995) Ecological integrity of aquatic systems. *Regulated Rivers*, **11**, 313–324.

Calow, P. and Petts, G.E. (eds) (1992) *The Rivers Handbook*, volume 1. Blackwell Scientific, Oxford.

Calow, P. and Petts, G.E. (eds) (1994) *The Rivers Handbook*, volume 2, Blackwell Scientific, Oxford.

Copp, G.H. (1989) The habitat diversity and fish reproductive function of floodplain ecosystems. *Environmental Biology of Fishes*, **26**, 1–27.

Copp, G.H. (1991) Typology of aquatic habitats in the Great Ouse, a small regulated river. *Regulated Rivers*, **6**, 125–134.

Cosgrove, D. and Petts, G.E. (eds) (1990) *Water Engineering and Landscape*. Belhaven, London.

Dodge, D.P. and Mack, C.C. (1994) Direct control of fauna: role of hatcheries, fish stocking and fishing regulations. In: Calow, P. and Petts, G.E. (eds) *Rivers Handbook*, **2**, 386–400.

Elliot, J.M. (1989) Brown trout: *Salmo trutta*. *Journal of Fish Biology*, **34**.

Gore, J.A. and Shields, F.D. Jr (1995) Can large rivers be restored? *Bioscience*, **45**, 142–152.

Gregory, S.V., Swanson, F.J., McKee, W.A. and Cummins, K.W. (1991) An ecosystem perspective of riparian zones. *BioScience*, **41**, 540–551.

Gurnell, A.M., Gregory, K.J. and Petts, G.E. (1995) The role of coarse woody debris in forest aquatic habitats: implications for management. *Aquatic Conservation*, **5**, 143–166.

IFMRC (1994) *Sharing the Challenge: Floodplain Management into the 21st Century*. Report of the Interagency Floodplain Management Review Committee to the Administration Floodplain Management Task Force, Washington, DC, June.

Junk, W.J., Bayley, P.B. and Sparks, R.E. (1989) The flood pulse concept in river–floodplain systems. *Special Publication of the Canadian Journal of Fisheries and Aquatic Sciences*, **106**, 110–27.

Jurajda, P. (1995) The effect of channelization and regulation on fish recruitment in a floodplain river. *Regulated River*, **11**, 207–216.

Large, A.R.G. and Petts, G.E. (1994) Rehabilitation of river margins. In: Calow, P. and Petts, G.E. (eds), *The Rivers Handbook*, volume 2. Blackwell Scientific, Oxford, 401–418.

Lauga, J., Decamps, H. and Holland, M.M. (eds) (1986) *Landuse Impacts on Aquatic Ecosystems*. Proceedings of the MAB-UNESCO and PIREN-CNRS Workshop. CRDP, Toulouse.

Löffler, H. (1990) Danube backwaters and their response to anthropogenic alteration: In: Whigham, D.F., Good, R.E. and Kvet, J. (eds), *Wetland Ecology and Management: Case Studies*, Kluwer, Dordrecht, 127–139.

Mather, R.J. (1994) Observations of fish community structure and reproductive success in flooded terrestrial areas during an extreme flood on the Illinois River. In: National Biological Service, Illinois Natural History Survey, Iowa Department of Natural Resources and Wisconsin Department of Natural Resources, *Long Term Resource Monitoring Program 1993 Flood Observations*. National Biological Service, Environment Technical Center, Onalaska, December. LTRMP 94-SO11, 95–115.

Meffe, G.K. (1992) Techno-arrogance and halfway technologies: Salmon hatcheries on the Pacific Coast of North America. *Conservation Biology*, **6**, 350–354.

Miller, R.R., Williams, J.D. and Williams, J.E. (1989) Extinctions of North American fishes during the past century. *Fisheries*, **14**, 22–38.

Milner, A.M. (1994) System recovery. In: Calow, P. and Petts, G.E. (eds), *The Rivers Handbook*, volume 2, Blackwell Scientific, Oxford, 76–98.

Minshall, G.W., Cummins, K.W., Peterson, R.C., Cushing, C.E., Burns, D.A., Sedell, J.R. and Vannote, R.L. (1985) Developments in stream ecosystem theory. *Canadian Journal of Fisheries and Aquatic Sciences*, **42**, 1045–1055.

Mitsch, W.J., Dorge, C.L. and Wiemhoff, J.R. (1979) Ecosystem dynamics and a phosphorous budget of an alluvial cupress swamp in southern Illinois. *Ecology*, **60**, 1116–1124.

Naiman, R.J. and Decamps, H. (eds.) (1990) *The Ecology and Management of Aquatic–Terrestrial Ecotones*. UNESCO, Paris and Parthenon, Carnforth, UK.

Naiman, R.J., Magnuson, J.J., McKnight, D.M. and Stanford, J.A. (eds) (1995) *The Freshwater Imperative: A Research Agenda*. Island Press, Washington, DC.

Petts, G.E. (1984) *Impounded Rivers*. John Wiley, Chichester.

Petts, G.E. (1990a) Forested river corridors: a lost resource. In: Cosgrove, D.E. and Petts, G.E. (eds), *Water, Engineering and Landscape*. Belhaven, London, 12–34.

Petts, G.E. (1990b) The role of ecotones in aquatic landscape management. In: Naiman, R.J. and Decamps, H. (eds), *The Ecology and Management of Aquatic–Terrestrial Ecotones*. UNESCO, Paris, and Parthenon, Carnforth, UK, 227–260.

Petts, G.E. (1995) Changing river channels: the geographical tradition. In: Gurnell, A.M. and Petts, G.E. (eds), *Changing River Channels*, John Wiley, Chichester (in press).

Petts, G.E. (1996) The scientific basis of managing biodiversity along river margins. In: LaChavanne, J.-B. (ed.), *Biodiversity and Land–Water Ecotones*, UNESCO, Paris (in press).

Petts, G.E. and Decamps, H. (1996) European river margins as indicators of global change. In: Barth, H. (ed.), *Proceedings of the Special Session on Environment and Climate*, SETAC 1995, European Commission, Brussels (in press).

Petts, G.E., Moller, H. and Roux, A.L. (eds) (1989) *Historical Analysis of Large Alluvial Rivers*. John Wiley, Chichester.

Regier, H.A. (1993) The notion of natural and cultural integrity. In: Woodley, S., Kay, J. and Francis, G. (eds), *Ecological Integrity and the Management of Ecosystems*, St Lucie Press, 3–18.

Reice, S.R., Wissmar, R.C. and Naiman, R.J. (1990) Disturbance regimes, resilience, and recovery of animal communities and habitats in lotic ecosystems. *Environmental Management*, **14**, 647–660.

Roux, A.L., Bravard, J.P., Amoros, C. and Pautou, G. (1989) Ecological changes of the French Upper Rhone River since 1750. In: Petts, G.E., Moller, H. and Roux, A.L. (eds), *Historical Change of Large Alluvial Rivers: Western Europe*. John Wiley, Chichester, 167–182.

Salo, J. (1990) External processes influencing origin and maintenance of inland water–land ecotones. In: Naiman, R.J. and Décamps, H. (eds), *The Ecology and Management of Aquatic–Terrestrial Ecotones*, UNESCO, Paris, Parthenon, Carnforth, 37–64.

Salo, J., Kalliola, R., Häkkinen, I., Mäkinen, Y., Niemelä, P., Puhakka, M. and Coley, P.D. (1986) River dynamics and the diversity of Amazonian lowland forest. *Nature*, **322**, 254–258.

Schiemer, F. and Spindler, T. (1989) Endangered fish species of the Danue River in Austria. *Regulated Rivers*, **4**, 397–408.

Shumm, S.A. (1977) *The Fluvial System*, Wiley, New York.

Sparks, R.E. (1995) Need for ecosystem management of large rivers and their floodplains. *BioScience*, **45**, 168–182.

Sparks, R.E., Bayley, P.B., Kohler, S.L. and Osborne, L.L. (1990) Disturbance and recovery of large floodplain rivers. *Environmental Management*, **14**, 699–709.

Starkel, L. Gregory, K.J. and Thornes, J.B. (1991) *Temperate Palaeohydrology*. John Wiley, Chichester.

Thorp, I.H. and Delong, M.D. (1994) The riverine productivity model: an heuristic view of carbon sources and organic processing in large river ecosystems. *Oikos*, **70**, 305–308.

Townsend, C.R. (1989) The patch dynamics concept of stream community ecology. *Journal of the North American Benthological Society*, **8**, 36–50.

Townsend, C.R. and Calow, P. (1981) *Physiological Ecology. An Evolutionary Approach to Resource Use*. Blackwell Scientific Publications, Oxford.

Vannote, R.L., Minshall, G.W., Cummins, K.W., Sedell, J.R. and Cushing, C.E. (1980) The river continuum concept. *Canadian Journal of Fisheries and Aquatic Sciences*, **37**, 130–137.

Ward, J.V. and Stanford, J.A. (1995) The serial discontinuity concept: extending the model to floodplain rivers. *Regulated Rivers*, **11**, 105–120.

Welcomme, R. (1992) River conservation future prospects. In: Boon P.J., Callow, P. and Pears G.E. *River Conservation and Management*, Wiley, Chichester, 453–62.

Welcomme, R. (1979) *Fisheries Ecology of Floodplain Rivers*. Longman, London.

Welcomme, R. (1995) Relationships between fisheries and the integrity of river systems. *Regulated Rivers*, **11**, 121–136.

Yao Yin, Nelson, J.C., Swenson, G.V., Langrehr, H.A. and Blackurn, T.A. (1994) Tree mortality in the upper Mississippi River and floodplain following an extreme flood in 1993. In: National Biological Service, Illinois Natural History Survey, Iowa Department of Natural Resources and Wisconsin Department of Natural Resources, *Long Term Resource Monitoring Program 1993 Flood Observations*. National Biological Service, Environment Technical Center, Onalaska, December. LTRMP 94-SO11, 39–60.

17 Floodplain Restoration and Rehabilitation

A. BROOKES
Environment Agency, Kings Meadow Road, Reading, UK

17.1 INTRODUCTION

To date there have been impressive restoration efforts related to the water quality improvement of some catchments (Cairns *et al.*, 1977; Jordan *et al.*, 1987) and the re-creation of natural morphological characteristics in channels previously managed for drainage or flood control (Brookes, 1988, 1995a,b,c). However the majority of these projects are only partial solutions when viewed in context of the wider floodplain and catchment. Kern (1992) believes that rehabilitation of channels must include the floodplain. Particularly in lowland rivers, inundation of the floodplain is essential for the development of specific habitats. In ecological terms the floodplain is now regarded as an essential component of the system, without which production is drastically reduced and community composition and energy pathways are radically changed (Odum, 1979; Junk *et al.*, 1989; Ward, 1989). However, the lack of baseline data for large river–floodplain systems, particularly in the temperate climates of developed countries, means that an enormous problem is posed for river managers attempting large-scale restoration (Bayley, 1991). Of all the political, economic and technical constraints on floodplain restoration, perhaps the most significant is the lack of scientific knowledge. In comparison with river channel restoration, there have been relatively few attempts at restoration of an entire floodplain (National Research Council, 1992). For example, floodplain restoration is not a topic specifically dealt with in a key volume on river restoration published in 1985 (Gore, 1985).

This chapter demonstrates that a return to a truly pristine or pre-disturbance condition is not usually a viable option for river managers. In practice only partial solutions are realistically attainable and a range of alternative options for floodplain enhancement are reviewed. A primary concern of many river managers has been the preservation of the terrestrial–aquatic interface by setting aside riparian land corridors, recognized as critical to restoration of river channels themselves (Osborne *et al.*, 1993). Research on riparian zones may be easier, because they are smaller than floodplains. Projects which attempt restoration of the floodplain are most likely to arise from opportunities provided by land-use change, while holistic integrated approaches are likely to be site

Floodplain Processes. Edited by Malcolm G. Anderson, Des E. Walling and Paul D. Bates.
© 1996 John Wiley & Sons Ltd.

specific and geographically more restricted. However, it is argued that for all projects impacts should be anticipated as far as possible for a whole range of environmental attributes prior to implementation and that post-project monitoring and appraisal should be an integral component of a sample of projects. It is only through conducting appropriate scientific monitoring that knowledge can be improved and that river managers can be certain that even partial restoration projects are sustainable in the context of longer-term planning and catchment influences.

17.2 DEFINITION OF TERMS

When considering restoration or rehabilitation of floodplains it is necessary to be clear about the definition of terms. A simple definition of a floodplain is the relatively level valley floor formed of sediment deposits which extends to the valley walls. This is the area adjacent to a river channel which becomes periodically inundated at high flows (Bren, 1993). To hydraulic engineers and river managers it may have a more precise definition as the area covered by, say, the $1:100$ year flood. A more detailed geomorphological definition of floodplain is provided by Nanson and Croke (1992) where the genetic association between the river type and the floodplains they construct is stressed. These floodplains are high-energy non-cohesive floodplains in confined valleys, which erode during extreme flood events; medium-energy non-cohesive floodplains formed by regular flow events in relatively unconfined valleys with lateral migration of channels; and low-energy cohesive floodplains formed by regular flow events along laterally stable single-thread or anastomosing low-gradient channels.

In practice many restoration projects have concentrated on only part of the floodplain, typically a narrower riparian zone or river corridor in closest proximity to the river or stream channel itself. There are also a large number of terms used to describe intervention by river managers to improve the riverine environment and a widely accepted definition for these has not yet been devised. Measures are increasingly implemented to mitigate, or compensate directly for, damage caused by a recent or current development. Practice with such measures has often provided insight into how disturbed systems may be restored. However, there are perhaps four distinct management terms which can be used to describe the restoration of disturbed river systems namely, restoration, rehabilitation, enhancement and creation. Table 17.1 provides some definitions of the term "river restoration" which can be applied to the floodplain. Perhaps one of the most widely accepted is that of Cairns (1991), which describes restoration "as the complete structural and functional return to a pre-disturbance state". This could be regarded as the ultimate objective to which river managers should strive (National Research Council, 1992). Both natural recovery or enhanced recovery could be regarded as processes leading to restoration.

By contrast "rehabilitation" has been defined as "the partial structural and functional return to a pre-disturbance state" (Cairns, 1982). It involves the selection of a limited number of attributes and is a pragmatic option in the management of many disturbed systems. "Enhancement" is regarded as "any improvement of a structural or functional attribute" (National Research Council, 1992) and is not taken to refer to the pre-disturbance condition. For example, instream flow devices may artificially improve the

Table 17.1   Definition for restoration

Author	Definition
Gore (1985)	In essence, river restoration is the process of recovery enhancement. Recovery enhancement enables the river or stream ecosystem to stabilize (some sort of trophic balance) at a much faster rate than through the natural physical and biological processes of habitat development and colonization. Recovery enhancement should establish a return to an ecosystem which closely resembles unstressed surrounding areas.
Herricks (1985) and Osborne	Implicit in the concept of water quality restoration is some knowledge of the undisturbed or natural state of the stream system. Restoration of water quality can be defined as returning the concentration of substances to values typical of undisturbed conditions.
Cairns (1991)	The complete structural and functional return to a pre-disturbance state.
Osborne *et al.* (1993)	Restoration programmes should aim to create a system with a stable channel, or a channel in dynamic equilibrium that supports a self-sustaining and functionally diverse community assemblage.

river environment. Finally, "creation" is seen as "the birth of a new (alternative) ecosystem that previously did not exist at the site" (National Research Council, 1992).

This chapter is concerned with some of the technical and scientific issues surrounding floodplain restoration, rather than economic and political constraints which are covered elsewhere (see Penning-Rowsell and Tunstall, Chapter 15, this volume).

17.3   THE IDEAL SOLUTION: RESTORATION OF ENTIRE FLOODPLAINS

Petts (Chapter 16, this volume) has demonstrated the need for restoration of floodplains and the requirement for historical analysis to reconstruct hydrosystems of the pre-industrial era. Kern (1992) advocates a German concept known as "Leitbild". This is the ideal solution involving restoration of all natural stream and floodplain properties, not considering the economic or political aspects that potentially limit the adoption of this option. Kern (1994) accepts that only in a very few cases would it be possible to carry out the Leitbild option. Osborne *et al.* (1993) conclude that systems in pristine condition should serve as a reference point and not as a goal for most restoration projects. A truely pristine state is not attainable in catchments where humans have modified the land use and cover. However, a holistic view should be taken: reconstructing topography without using the appropriate soils or plant materials is unlikely to recreate the plethora of functional values of the natural or predisturbed aquatic ecosystem (National Research Council, 1992). Sedell and Froggatt (1984) stated that the relationship of floodplain and main-stem in large rivers in North America and Europe no longer exists and is rapidly disappearing in Africa, South America and Asia. Indeed most river valleys in the developed world are little more than wide corridors of intensive land and water uses (Yon and Tendron, 1981). Development on floodplains

Table 17.2   Some potential issues to consider in floodplain rehabilitation

- compression of river corridors by development
- flood control measures such as embankments
- drainage of floodplains
- destruction of soil structure by farming practices
- filling of floodplain lakes and other topographical features
- removal of woodlands and other riparian vegetation leading to accelerated erosion
- mineral extraction
- waste disposal and groundwater contamination
- barriers to groundwater movement
- upstream impoundment for water supply or navigation
- point and non-point source pollution affecting water quality and sediments

has meant the subsequent need for intervention to control floods and to enhance agricultural production. However, although restoration may be demanded in developed countries, it is, ironically, because of the long history of development that there is often a lack of baseline data from which to reconstruct the morphology and ecology of such systems. Table 17.2 shows some of the typical impacts which have occurred on floodplains, particularly in developed countries, and which may constrain restoration.

17.4   RECOVERY AS AN OPTION?

Bayley (1991) suggested that restoration of the river–floodplain and the hydrological regime of most large temperate systems might take upwards of 100 years of sustained effort affecting the entire catchment. As an interim approach, Bayley (1991) recommended that the "natural flood pulse" approach can be adopted to improve the hydrograph in a section of a river to improve overall production. Brookes (1995c) recommends that the most pragmatic approach for widespread restoration of modified channels would be to allow natural recovery to occur where practicable, and where the process is negligible, or slow, to install low-cost devices to enhance recovery. For example, a channel can be locally narrowed using deflectors or groynes, thereby inducing the deposition of sediment which further narrows the channel by natural processes of recovery over a considerable length. In this way many thousands of kilometres of watercourse could be improved over relatively short time-scales, perhaps ranging from 1 to 150 years (Brookes, 1992). In terms of the morphology of floodplains, Kern (1992) believes that evolution of natural morphology may take several hundreds of years and therefore the floodplain must be reshaped by man. In other cases, alluvial soil textures may have been irreversibly destroyed and therefore rehabilitation of the floodplain ecosystem may be limited. Based on experience of revegetation of part of the Illinois River, USA, Sparks *et al.* (1990) question the effectiveness and localized nature of projects including physical habitat restoration. Although field research has provided some promising preliminary results, many questions remain unanswered. They suggest that a quicker way to recover floodplain backwaters is to purchase these areas and re-flood them.

Bayley (1991) suggested that enhanced recovery is appropriate to river floodplains and that restoration based on the "natural flood pulse" should achieve long-term high productivity of a wide variety of native biota for recreational and commercial use in an aesthetically pleasing environment at minimum cost. This involves manipulating the hydrograph in a river, which in practice requires purchasing land and removing embankments. There is evidence along the upper Mississippi River that some parts of the floodplain are exhibiting recovery through natural processes of sedimentation and plant succession, although it is likely to take in excess of 100 years before the process is complete and a new equilibrium established (Bhowmik et al., 1986). Large et al. (1994) suggest that zones of transition (ecotones) remaining on the floodplain of the River Trent, UK, may potentially provide a pool for expansion of species, given improved ecological conditions following restoration. In practice the functional aspects of the newly created floodplains will differ from the old as a direct consequence of regulation of the river for navigation. Sparks et al. (1990) also felt that human intervention and continued maintenance are likely to be required to maintain habitat diversity and desirable functional system characteristics. For example deflectors or artificial islands may be required to divert some flow from the main channel to create and maintain side channels by scour, instead of a uniform floodplain. Long-term monitoring is required to assess recovery in floodplains which are influenced by a variety of factors including yearly variations in flow patterns.

17.4.1   Case study: Kissimmee River, central Florida

The Kissimmee River Demonstration project represents an attempt at restoring a river system previously impacted by channelization for flood control (Toth et al., 1993). Channelization between 1962 and 1971 involved diversion of flow from nearly 170 km of meandering river channel, and a floodplain of between 1.5 km and 3.0 km in width, to a 9 m deep, 64–105 m wide, 90 km long canal. This destroyed most of the fish and wildlife habitats once provided by the river and floodplain wetlands. Initial restoration involved re-introducing flow along remnant river channels and led to increased habitat diversity and biological improvement (Toth, 1991). However, Toth et al. (1993) recognize that, although the diversion of flow improved river channel habitat diversity, regulation of the flood flow upstream by a series of lakes limits or precludes the restoration of biological components. The pre-channelization discharge data show that the river had continuous flow and that baseflow exceeded 11 $m^3 s^{-1}$ during at least 90% of this time interval. As a consequence of upper basin flow regulation, no flow occurred almost 50% of the time between 1970 and 1987. This led to artificial discharges which are pulse-like, with extended periods of low or no flow and a seasonal pattern totally out of phase with that experienced prior to channelization. A major impact of this regulation is the effect of discontinuous flows on floodplain inundation and a seasonal reversal of high and low flow periods. Modelling studies showed that even with extensive backfilling of the artificial channel there would be inadequate inundation to restore the floodplain wetlands. Restoration of wetlands requires the re-establishment of stage hydrographs that existed prior to channelization. Records showed that 94% of the floodplain was inundated over 52% of the pre-channelization period of record (1942–1967) and that the river has a characteristic subtropical wet–dry cycle.

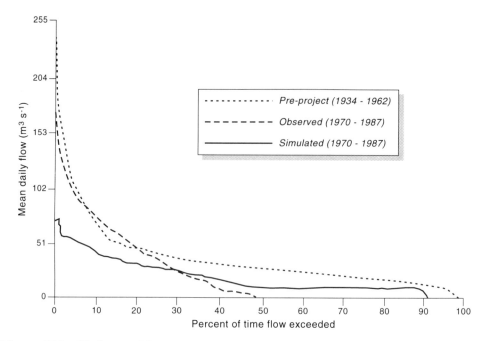

Figure 17.1 Kissimmee River: comparison of actual flow duration curves for the post-channelization period (1970–1987) with the pre-project records (1934–1962) and simulated flows for the new schedule and operation (from Toth *et al.*, 1993)

Using simulation modelling a schedule was therefore developed for regulation of the lakes which re-establishes the seasonal flow pattern and maintains baseflows for a greater proportion of the year. Figure 17.1 compares actual flow regimes for the post-channelization period (1970–1987) with the pre-project (1934–1962) records and the simulated flows for the new schedule and operation.

The plan involves backfilling of 46 km of canal, re-establishing continuous flow through 90 km of the original river channel and inundation characteristics to 11 000 ha of floodplain. Unfortunately, opportunities for restoring more natural flow regimes to river systems are likely to be limited. The need for flood control, impoundment for navigation and reservoirs for water supply are all likely to preclude the complete restoration of a flood-pulse, principally through their effects on water levels and transport of sediment. Similarly below urban areas, the impact of paved areas and associated channelization may radically increase the magnitude and frequency of overbank flows on floodplains (see Penning-Rowsell and Tunstall, Chapter 15, this volume).

17.5 CONSTRAINTS

Perhaps the largest technical constraint to large-scale floodplain restoration is the lack of scientific knowledge. Some of these problems are potentially more reversible than others. Clearly scale considerations are important: it is likely to be more difficult to restore a large-scale floodplain than a small confined floodplain in the headwaters.

There are other various constraints to consider when appraising the need for floodplain restoration or rehabilitation. These include the limitations of scientific understanding concerning river–floodplain systems, catchment influences, landownership and potentially irreversible floodplain development.

17.5.1 Scientific knowledge

The lack of baseline data for large river–floodplain systems in the temperate climates of developed countries presents an enormous problem to those attempting to restore such systems (Bayley, 1991). It is almost impossible to find detailed morphological and ecological information for large river floodplains in developed countries that approximate pristine conditions. The lack of knowledge of floodplain hydrology and sediment transport characteristics is a particular constraint (see Chapters 5, 8, 11, 12, 13, 14, this volume). Geomorphological surfaces form the template for the development of riparian plant communities (Gregory *et al.*, 1991; Gurnell, 1995). The key to ecologically sound management of floodplains is to provide for sufficient overflow habitat and period of inundation to sustain the biota in the river (Bryan, 1991). However, groundwater issues may be important: interactions between surface water and the alluvial aquifer are not well understood (Dister *et al.*, 1990; Gilbert *et al.*, 1990; Plenet *et al.*, 1992). Some individuals have assumed that features in upstream reaches may be used to reconstruct downstream reaches (Cummins *et al.*, 1984). These are very difficult questions to answer. If incorrect evaluations and judgements are made in extrapolating features from relatively undisturbed upstream reaches then the consequences could be both expensive and ineffective.

For example, it is known that geomorphological processes control the spatial pattern and successional development of floodplain vegetation (Tabacchi *et al.*, 1990; Hupp, 1992). The riparian zone appears to play a major role in sediment retention (Schlosser and Karr, 1981; Cooper *et al.*, 1987; Brunet *et al.*, 1994). In a study of the Adour River in southwest France, Brunet *et al.* (1994) found that although the riparian zone was 15 times smaller than the overall floodplain, the total suspended matter and particulate nitrogen deposition were 50 and 70 times greater, respectively, in the riparian zone. The proportion of sediment in different density fractions differs by location within a given community of floodplain succession (Sollins *et al.*, 1985; Pinay *et al.*, 1992). Due to the complexity of processes, restoration of a floodplain as part of the river ecosystem can be very complex. For example, topographical differences brought about by land use change may be difficult to rectify. This will inevitably affect the distribution of flood flows across a floodplain. A note for optimism is that the river corridor environment in the 1980s and 1990s has become a major theme of interdisciplinary research (Malanson, 1993; Gurnell, 1995; Peterken and Hughes, 1996).

17.5.2 Catchment influences

The role of flooding in sustaining the ecological integrity of large rivers is demonstrated by Petts (Chapter 16, this volume). The flood pulse is particularly important for river ecology (Welcomme, 1979; Junk *et al.*, 1989). Indeed many floodplains have degenerated due to a stable water level arising from upstream flow regulation for flood control, navigation and water resources (Sparks *et al.*, 1990; Bayley, 1991). For example, along the lower Illinois River, USA, degeneration of ecologically important areas has been attributed to a combination of increased sediment impact due to farming changes but principally due to the absence of a natural flood-pulse (Bayley, 1991).

Removal of extensive upstream floodplains also alters the flood pattern and sediment budgets of downstream floodplain systems and affects river functioning (Sparks *et al.*, 1990). Excessive sediment yield arising from land-use change in an upstream catchment means that in practice many restoration projects must make allowance for high turbidity and sediment concentrations in the main-stem river and its backwaters. Any significant reduction of sediment loading is likely to take decades and may in the majority of cases be unattainable. This usually means that not all aspects of the river–floodplain system can be restored simultaneously. The National Research Council (1992) described a series of restoration case studies from the USA which illustrate this problem. Roelle *et al.* (1988) state that where former levee districts have been purchased for wildlife areas, the intention is to retain the high levees, maintaining a barrier to the river and therefore excluding sediment-laden water which would degrade the new areas. It is widely accepted in the USA that breaching the levees in conservation areas must not take place until such a time that the sediment loadings in the main-stem river have been reduced (Roelle *et al.*, 1988). In the interim, the main benefit of these new areas will be for migratory animal and bird populations rather than fish and other aquatic species that depend on the link between channels and backwaters. Along the Illinois River there have been some attempts to reduce sediment loading by reducing soil erosion in the drainage basin and along the tributaries (National Research Council, 1992). However, it is recognized that it may take decades before measures to control soil and bank erosion substantially reduce sediment loading of the river because of the large-scale nature of the problem, limited resources and lag effects.

Few studies have specifically focused on the rehabilitation and restoration of urban catchments (CIRIA, 1992; Hall *et al.*, 1993; Ellis and House, 1994). Stormwater treatment and management should also play a more vital role in future urban land-use planning to prevent further degradation of urban rivers and streams. In agricultural catchments nutrient limitation measures such as phosphorus stripping in sewage treatment works may be necessary (Turner, 1987).

17.5.3 Landownership

The majority of floodplains which have been developed over any significant length are likely to be under the ownership of various individuals and organizations. In practice this may present the need for involved negotiation with more than one person when promoting proposals for floodplain restoration. In Denmark, this involves a variety of processes, including compensating individual owners and the exchange of riparian parcels of land with state-owned land elsewhere in the locale (Iversen *et al.*, 1993). Figure 17.2 shows the parcels of land involved in one project in southern Denmark. However, even in Denmark, where watercourse restoration legislation exists, the conflict between agricultural and environmental interests means that re-establishing

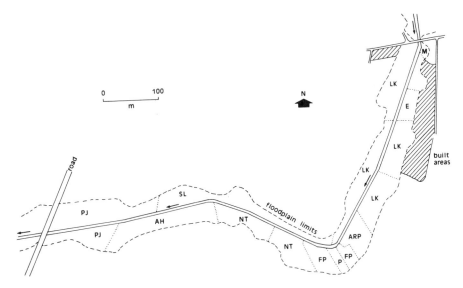

Figure 17.2 Ownership of land for one restoration project (River Brede) in southern Jutland Denmark (each parcel of land is given the initials of the landowner). The straightened channel shown (1991) was meandered across the floodplain in 1992 (based on the data compiled by Southern Jutland County)

extensive riparian areas has been rare. River and floodplain restoration efforts depend very much on the negotiating ability of individual river managers and the cooperation and goodwill of landowners.

Elsewhere in the world, publicly supported habitat rehabilitation projects can only be implemented on community land. In the Upper Mississippi, lands are gradually being acquired, when there are willing sellers, by organizations such as the Nature Conservancy, local park districts and the Illinois Department of Conservation (National Research Council, 1992).

Schemes promoting change in land use are particularly important for restoration. These offer major opportunities for restoration of floodplain habitats for which financial assistance may be available under a number of schemes. For the UK, they include "set-aside" established by the Ministry of Agriculture, Fisheries and Food to reduce agricultural production by removing land from agricultural use and a "Countryside Stewardship Scheme" promoted by the Countryside Commission, which includes incentives for the reversion of arable land to grassland in waterside landscapes in certain areas. These schemes usually include agreements covering periods up to 10 years (see River Restoration Project, 1994).

17.5.4 Floodplain development

The potential for floodplain restoration is most restricted in urban areas, where in practice there may only be a very constrained residual corridor. Objectives for integrated urban river corridor management include the need to provide an effective, efficient and safe drainage system and protection against increased flow velocities, sediments and pollutants associated with urbanization (Ellis and House, 1994). The design goal may therefore be to create an environmental and recreational amenity for local communities that will at the same time provide a significant improvement in stormwater quality as well as fulfilling the prime flood control function. Marginal river wetlands are of particular value in urban areas because they provide important corridors for wildlife, often providing a connection between isolated undeveloped areas and therefore vulnerable habitats (Ellis and House, 1994; Brookes, 1995b). River corridors in urban areas also have particular aesthetic values.

Hydraulic modelling may therefore be necessary to determine the extent of inundation needed to restore lost floodplain wetlands and associated ecological values and to avoid flooding built areas. Even in predominantly rural areas the floodplain may have been developed in such a way that larger scale restoration is impossible. Figure 17.3 illustrates the River Wye at Llangurig in mid-Wales, UK. It is a typical example of a floodplain with a former braided channel system, evident on the 1845 tithe map. The river has been straightened and channelized over a period exceeding 100 years. The 1980s single-thread channel is also shown. To re-create the floodplain would involve major disruption to at least three roads and several built areas.

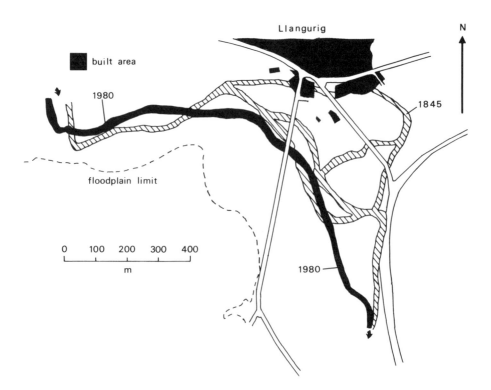

Figure 17.3 Constraints on floodplain restoration on the River Wye at Llangurig, mid-Wales

17.6   PRACTICAL DESIGN CONSIDERATIONS

In addition to the more general constraints described above, there are also more specific issues which might need to be considered during the detailed design stage of a project. Perhaps one of the most contentious technical issues raised by river channel and floodplain restoration proposals is estimating and assigning roughness values to areas proposed for revegetation. These problems have been discussed for south-central England (Brookes, 1995a) and California (Riley, 1989). The roughness value of floodplains is generally larger than that of the channel proper. Higher values are obtained for trees and shrubs than for agricultural crops or grass. Issues surrounding modelling of floodplain hydraulics are discussed by Knight and Shiono, (Chapter 5, this volume).

Floodplain restoration for locations where there are particular land-use issues, may need to consider the appropriate level or standard of service for flood defence purposes. Concerns, particularly in the Netherlands, have arisen over problems of flood threat to property and high-grade agricultural land arising from restoration. This concern appears to have delayed restoration in that country (Aukes, 1992; Gerritsen, 1992).

Structural techniques for restoring floodplain habitats include increasing the hydrological link between the floodplain and river by raising bed-levels by backfilling and/or substrate placement and drop structures or removing obstructions such as embankments; and eliminating and controlling contaminants and re-establishing native vegetation. In contrast to channel restoration, reinstatement of riparian zones and floodplains relies heavily on non-structural methods, leading to natural or enhanced recovery. In small riparian zones this may require relieving grazing pressure to allow recolonization and recovery of natural vegetation (Platts and Rinne, 1985). However, in some instances where the prospects for recolonization by plants are low then selective replanting may be necessary (Anderson and Ohmart, 1985; Baird, 1989). In larger floodplains re-shaping of the surface may be necessary to create a diversity of water levels and habitats (Amoros et al., 1987).

It may be desirable to locally flood riparian areas by raising a channelized river bed to a more natural invert level. This will restore a more natural bankfull level, above which the river flows onto the floodplain. However, the feasibility of locally raising the bed of a river will depend on constraints in adjacent reaches, particularly upstream, which in developed catchments might include bridge inverts, weirs with an industrial or heritage value, and outfalls. As part of appraising the River Cole in south-central England, one of the River Restoration Demonstration Projects in the UK, planned to be constructed during 1995, such constraints have inevitably meant that the bed-level cannot be raised to a more natural level (River Restoration Project, 1993a; Brookes, 1995c). The frequency and extent of overbank flooding is therefore considerably less than might naturally be expected.

17.7   ENVIRONMENTAL IMPACT ASSESSMENT

Rhoads and Miller (1990) argue that because the construction of wetland environments is still very much in its infancy, such projects must be monitored closely to determine

Table 17.3 Some potential impacts of floodplain rehabilitation and restoration: operational phase

Sources of impact	Potential effects
	1. Surface water hydrology/hydraulics
Re-instatement or lowering of ground levels	Changed flow velocities and flow distribution
Replanting	Increased hydraulic roughness
	Increased storage of water
	Changed distribution of flood flows
	Decreased flooding downstream
Re-meandering of channel	Increased hydraulic roughness
	Changed flow velocities
	Increased frequency of flooding
	2. Channel morphology
Re-instatement or lowering of ground levels	Changed stability
	Deposition/siltation
Replanting	Increased stability
	Reduced downstream erosion
Re-meandering of channel	Reduced channel size
	Increased stability
	Erosion of bed and/or banks
	Change of slope
	Change of planform/pattern
	Decreased bedload
	3. Groundwater hydraulics
Re-instatement or lowering of ground levels	Rise in water table
Re-meandering	Rise in water table
	4. Surface water quality
Re-instatement or lowering of ground levels	Changed quality
	Improved quality
Replanting	Decreased sediment load
Re-meandering	Oxygenation/eutrophication
	5. Aquatic and terrestrial ecology
Re-instatement or lowering of ground levels	Improved habitat
	Increased species diversity
Replanting	Improved habitat
Re-meandering	Improved habitat
	Increased fish biomass
	Increase of invertebrates
	Increased plant biomass
	Change in the fish community
	Increased species diversity
	Effects on fish spawning

their total impact on the environment. Coats *et al.* (1989) argued for post-construction monitoring of biological and ecological systems. Construction often involves changing the topographic, biological and hydrological character of a site. In practice many enhancement schemes espoused by river managers through mitigation and enhancement (see Schnick *et al.*, 1981; Ward *et al.*, 1994) are localized, piecemeal and are not adequately monitored and do not necessarily represent a step towards whole-system restoration (Bayley, 1991). Modifications can have unanticipated negative impacts on the geomorphological, hydrological and ecological systems. Some potential impacts are shown in Table 17.3. Longer-term impacts may generally be beneficial, although in the construction phase there is a risk that considerable volumes of sediment could arise from erosion of the floodplain surface prior to the vegetation becoming established. Other impacts which should be addressed include those on recreation, land use, heritage, archaeology and landscape. Rhoads and Miller (1990) outline a framework for evaluating construction and end-state impacts based on the water/sediment budget concept. This is based on the principles of conservation of mass: that is that the total amount of sediment and water moving through a river reach must be conserved. Long-term measurement of channel sediment storage and other water/sediment budget components provide the basis for distinguishing between project-related impacts and those arising from other causes. Changes of channel sediment storage that occur as a result of changes in internal inputs of water or sediment indicate an impact related to the project, while those associated with changes in upstream or tributary inputs denote a change of conditions elsewhere in the catchment (Rhoads and Miller, 1990). The Des Plaines River Wetlands Demonstration Project near Chicago, Illinois, USA, was assessed by this geomorphological approach: channel sediment storage changed little during the initial construction phase, suggesting minimal impact on channel stability.

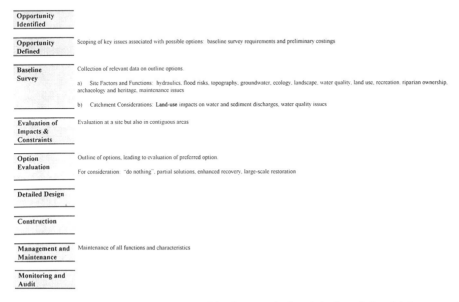

Figure 17.4 Recommended steps involved in the appraisal and design of floodplain restoration or rehabilitation projects

Figure 17.4 illustrates some of the issues which should be considered in the appraisal and design of floodplain restoration or rehabilitation projects. These follow the logical steps which might be considered as essential for any project, ranging from the scoping of key issues following outline definition of a project, collection of relevant baseline data, identification of potential impacts, option evaluation and detailed design. Following construction, monitoring and post-project appraisal are important requirements. Monitoring the outcome of past, existing and planned restoration projects is required for information on the feasibility of alternative techniques and approaches. However, experience has show that finance is rarely available or appropriated for monitoring most projects (Osborne *et al.*, 1993). The options evaluated should include consideration of the "ideal solution" of large-scale floodplain and channel restoration, the "do nothing option" and a range of alternative "partial" solutions. Management of restored floodplains should be directed towards monitoring all functions and characteristics. In practice this may be a difficult objective to achieve (National Research Council, 1992).

17.8 PRAGMATIC APPROACHES TO THE RESTORATION OF FLOODPLAINS

In addition to the ideal solution of major floodplain restoration and the option of recovery, several additional types of partial floodplain restoration and enhancement can be visualized. In the context of experience in rehabilitation of streams in southwest Germany, Kern (1992) states that "the expectation of self-purification in a rehabilitated stream must never be a substitute for further efforts in pollution control". Petersen *et al.* (1987) highlighted the importance of the riparian zone to river managers. A "holistic, ecosystem approach" relying on riparian control rather than a catchment approach is the first goal in saving huge costs associated with river purification. Although watershed management is the goal, riparian control is the starting point. Petersen *et al.* (1992) outline the importance of a series of measures or "building blocks" in stream restoration, including buffer strips, revegetation and small wetlands for the treatment of point-source agricultural discharges. Clearly the type, or combination of types, adopted for a particular location may be determined by economic, political, legal and physical constraints. The option involving greatest land take may generally be envisaged to be the most difficult solution to attain. The final design Kern (1992) admits, will have to be a compromise depending on economic or political constraints.

The first type involves the creation of a single or multi-stage channel cut into the contemporary or historical floodplain and may be the most pragmatic solution in many rural and urban locations where land-use change on the floodplain may only be locally or partially reversible. A significant problem encountered in the restoration of Danish streams, for example, is the extremely incised nature of many streams which have been artificially straightened and have adjusted in erodible sediments (Brookes, 1987a,b, 1990). For the Elbaek in central Jutland, Denmark, a small stream with an existing bank-top width of 2 m was widened into a two-stage channel with a total width of between 8 and 10 m. The second stage was formed by pulling back the steep banks of the incised channel and now provides capacity for floodwaters when the natural "in banks" capacity of the low-flow channel is exceeded. The remainder of the floodplain prior to channelization, which existed about 150–200 years previously, was estimated to be about 25–35 m, representing a total floodplain width of about 35–45 m. The remainder of the floodplain was therefore left in agricultural production.

As well as in Denmark, the author has also negotiated many similar solutions for urban and rural channels in south-central England. For example, the Bear Brook downstream from Aylesbury in Buckinghamshire presented an opportunity to partially restore about 7 km of floodplain as a direct consequence of a development application for residential use. Figure 17.5A shows that while the wider floodplain is to be used for housing development, a residual river corridor of 60 m is retained with a naturally sinuous low-flow channel. The river corridor also accommodates the 100-year flood discharge with a multi-stage channel planted with appropriate vegetation (Brookes, 1995c). In this example the channel was previously degraded by straightening and deepening as a consequence of an agricultural drainage scheme, allowing the adjacent land to be used for arable production. Two-stage or multi-stage channels, have now been used for many kilometres of watercourse in the UK, particularly over the last 10–15 years or so. They are not a panacea and present problems of maintenance of standards of service for flooding, as a consequence of morphological adjustment and recovery of vegetation (Brookes, 1988). Knight and Shiono (Chapter 5, this volume) present an overview of recent advances in the understanding of flow processes in such compound channels. However, if allowance is made at the design stage for future adjustments, and, if necessary, the channel is over-designed, then in practice this concept of limited floodplain creation is likely to be one of the most widely applicable to developed and developing countries.

Within many urban situations, restoration must be conducted within a very constrained corridor. Such constraint means that a full range of wildlife habitats, restoration and landscaping may not be possible (Ellis and House, 1994). In such channels, shallow marginal shelves can be weaved between wet banks to stimulate both emergent and wetland plants in addition to the provision of linear ponds within the flood corridor. An objective of urban corridor restoration is often to create an environmental and recreational amenity for local communities in addition to providing a significant improvement in stormwater runoff quality and fulfilling the prime function of flood control (Ellis and House, 1994). A plan for the Wildcat and San Pablo Creeks in North Richmond, California, USA (Figure 17.5B), illustrates how ecological, recreational and landscape objectives can be integrated with the function of flood control (Riley, 1989). The project included riparian vegetation such as trees next to the channels. There is a meandering, low-flow channel of 3–4 m width, designed to carry the average flow, and floodplains where higher flows can spread, reduce velocity and deposit sediment. Elsewhere work has shown that the inclusion of sports fields, open spaces and parklands adjacent to the river can lead to a higher standard of visual amenity. Figure 17.5C depicts the concept of secondary floodplain creation in an agricultural area based on work in Northern Germany (Kern, 1992).

Figure 17.6 shows secondary floodplain creation in south-central England negotiated by the author as part of a road development. The photograph was taken

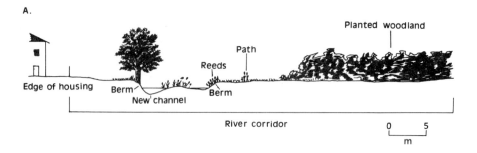

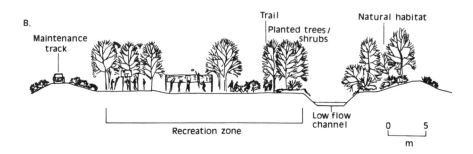

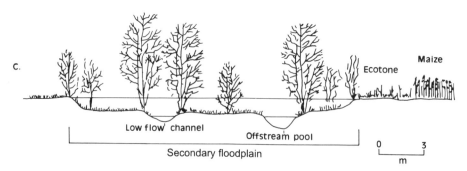

Figure 17.5 (A) Multi-stage channel creation in central England. (B) Example of multi-functional use of floodplains in North Richmond, California (after Riley, 1989). (C) Riparian strips in rural catchments (from North Germany) (after Kern, 1992)

in 1992 during construction. The wider floodplain (to the right of the photograph) is now occupied by the road. While this may only be a partial solution, research has proved that the riparian strip adjacent to the river channel may be the most important in terms of geomorphological processes. Petersen *et al.* (1987) recommended that rehabilitation of the buffer capacity, that is the capacity to

Figure 17.6 Secondary floodplain creation – River Blackwater, Hampshire, UK (1991)

retain suspended material in river systems, should commence with the reclamation of floodable riparian vegetative strips.

A further concept is to allow a river to run wild. Palmer (1976) recommended that for rivers which are actively changing their courses, a sufficient corridor should be provided to allow continued change (Figure 17.7). This is known as the "streamway concept". While a river might move across the entire floodplain in a period of several hundred years, migration of individual bends may occur on a time-scale of only a few tens of years. For the purposes of floodplain management, the meander belt may be confined to a fixed position on the floodplain by localized bank protection at the boundary of the streamway. A similar technique has been applied in Bavaria, Germany, by Binder *et al.* (1983). Geomorphologists can attempt to determine the width and location of a river corridor that can be maintained for planning objectives (say 50 to 100 years).

A previously managed or maintained river can be allowed to run wild. Accommodation of sediment deposition can be achieved by extending the floodway. Figure 17.8 shows the example of Cherry Creek, an urban creek in Denver, Colorado. The channel was extensively modified following catastrophic floods last century and earlier this century. In 1950 a flood control reservoir was completed upstream (Costa, 1978). In this way a previously overwide and deepened flood channel may be allowed to recover through erosion and deposition and the overbank flooding caused by the loss

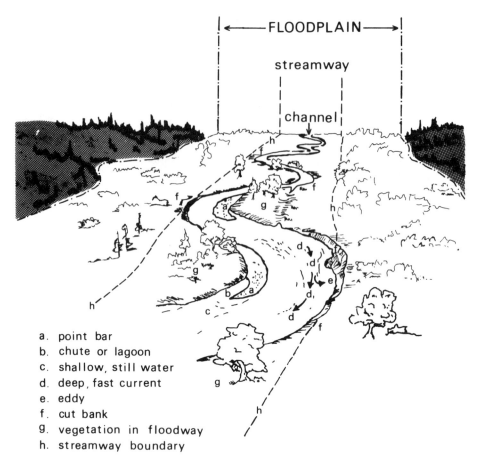

a. point bar
b. chute or lagoon
c. shallow, still water
d. deep, fast current
e. eddy
f. cut bank
g. vegetation in floodway
h. streamway boundary

Figure 17.7 "Streamway concept" (after Palmer, 1976)

of conveyance is accommodated within flood embankments. This technique is perhaps most applicable to gravel-bed rivers and other rivers with a high sediment load to facilitate recovery.

A final option to consider is the "do minimum". Clearly, with all the obstacles posed to larger-scale floodplain restoration, restoration or enhancement of habitats closer to the channel may be more appropriate. For example, along the Lower Mississippi River artificial dykes used to train the river flow create a series of pools, providing habitat for numerous species of fish and invertebrates. Baker *et al.* (1990) suggest that because of the high primary productivity of these pools, compared with other main-stem habitats, they may function similarly to floodplain habitats. This production may be particularly important when that from floodplain habitats is unavailable due to the emplacement of levees or land-use change on the floodplain (Shields, 1995).

Figure 17.8 An urban creek, Denver, Colorado (1991), showing adjustment in the absence of maintenance (photograph by A. Brookes)

17.9 CONCLUSIONS AND RECOMMENDATIONS

In 1990 Berger made a statement which is particularly valid when considering the restoration or rehabilitation of floodplains:

> It is axiomatic that no restoration can ever be perfect; it is impossible to replicate the biogeochemical and climatological sequence of events over geological time that led to the creation and placement of even one particle of soil, much less to exactly reproduce an entire ecosystem. Therefore all restorations are exercises in approximation and in the reconstruction of naturalistic rather than natural assemblages of plants and animals within their physical environments.

Clearly scientific knowledge of floodplains is imperfect, particularly in temperate developed countries. Restoration is very much in its infancy. A variety of different scientific disciplines including river–floodplain geomorphologists, hydrologists and plant ecologists should study temperate systems that are unmodified, modified or under restoration (Bayley, 1991). However, a substantial investment of resources will be required to achieve this if floodplain restoration is to be accelerated and applied to other areas. There is some evidence that relatively undisturbed tropical systems may be tentatively used as a template for re-creating temperate floodplains (Bayley, 1991). Enhanced recovery may be a valid option. However, in terms of the morphology of floodplains Kern (1992) believes that evolution of natural morphology make take several hundreds of years and therefore the floodplain must be re-shaped by man.

As a consequence of the relatively few examples of large-scale floodplain restoration or enhanced recovery, there are relatively few general guidelines produced for restoration of floodplains and riparian zones. Work to date has therefore mainly been "trial and error" and at the most can be described as an art form rather than a science. Although re-establishment of floodplains has been achieved and they at least aesthetically resemble natural systems, little is known of whether appropriate functions have been re-established (National Research Council, 1992). Although some scientists question whether or not more natural floodplains are reproducible, this cannot currently be resolved, and should not preclude consideration of large-scale restoration when evaluating options for a particular site. A holistic approach equivalent to the German Leibild concept (Kern, 1992, 1994) may be a valid starting point. Clearly it may only be as more projects are built that opportunities are provided for monitoring and comparing results. Evaluation of restoration efforts has been limited because, for political, economic or legislative reasons, many agencies have not been able to fund post-project appraisal. However, within the UK, at least, the need to increase scientific knowledge is demonstrated by the River Restoration Project (RRP) which aims to establish demonstration projects which show how state-of-the-art restoration techniques can be used to re-create natural ecosystems in damaged river corridors, to improve understanding of the effects of restoration work on nature conservation value, water quality, visual amenity, recreation and the views of the public, and to encourage others to restore streams and rivers (River Restoration Project, 1993a,b, 1994).

While there are major gaps in scientific understanding, there may be reasons other than ecological ones for carrying out restoration at particular sites: aesthetics and recreation may be equally valid. Hollis and Acreman (1994) suggest that floodplain wetlands may also be important for groundwater recharge and sediment and nutrient retention. However, while suggesting that large-scale attempts at restoration should be made, assessment of the potential impacts is an important consideration. This should not be restricted to an assessment of site issues. Strategic considerations should be addressed, including the effects of catchment land-use changes and repercussive impacts of the project itself on upstream and downstream reaches. Restoration of isolated and selected river–floodplain segments can only be an interim solution. Wider issues such as controlling catchment land and water use and reduction of chronic pollution should be appraised. While isolated interim restoration provides a microcosm of what can be achieved in the longer-term, and will serve as a catalyst for future public support, it is likely that such schemes will remain geographically restricted for the foreseeable future.

The most pragmatic approach from a river manager's viewpoint may be a partial solution and there are many examples from river agencies around the world which have been built for ecological, landscape and/or recreational purposes. Scientists rightly argue that attempts at restoration through mitigation or enhancement are piecemeal. Such projects can result in a change to another unnatural state and are not necessarily a step towards restoration of a whole system (Schnick et al., 1981; Bayley, 1991).

Although understanding of many of these attempts is still largely in the experimental phase, there is little published evidence to suggest that major adverse impacts arise from such solutions.

ACKNOWLEDGEMENTS

The views expressed are those of the author and do not necessarily reflect the views of his employer, the Environment Agency. The permission of the Regional General Manager Thames Region NRA to publish this paper is gratefully acknowledged.

REFERENCES

Amoros, C., Roux, A.L., Reygrobellet, J.-C., Bravard, J.P. and Pautou, G. (1987) A method for applied ecological studies of fluvial hydrosystems. *Regulated Rivers: Research and Management*, **1**, 17–36

Anderson, B.W. and Ohmart, R.D. (1985) Riparian revegetation as a mitigation process in stream and river restoration. In: Gore, J.A. (ed.), *Restoration of Streams and Rivers: Theories and Experience*, Butterworths, 41–79.

Aukes, P. (1992) Criteria for ecological rehabilitation. *Contributions of the Workshop of the Ecological Rehabilitation of Floodplains*, RIZA, CHR KHR, VITUKI, Arnhem, Netherlands, 89–96.

Baird, K. (1989) High quality restoration of riparian ecosystems. *Restoration and Management*, **7**(2), 60–64.

Baker, J.A., Kilgore, K.J. and Kasul, R.L. (1991) Aquatic habitats and fish communities in the lower Mississippi River. *Aquatic Sciences*, **3**(4), 313–356.

Bayley, P.B. (1991) The flood pulse advantage and the restoration of river–floodplain systems. *Regulated Rivers: Research and Management*, **6**, 75–86.

Berger (1990) quoted on page 18 of National Research Council (1992) *Restoration of Aquatic Ecosystems: Science, Technology and Public Policy*, National Academic Press, Washington, DC.

Bhowmik, N.G., Adams, J.R. and Sparks, R.E. (1986) Fate of a navigation pool on the Mississippi River. *Journal of Hydraulic Engineering*, **112**(10), 967–970.

Binder, W., Jurging, P. and Karl, J. (1983) Natural river engineering – characteristics and limitations, *Garten und Landschaft*, **2**, 91–94.

Bren, L.J. (1993) Riparian zone, stream and floodplain issues: a review. *Journal of Hydrology*, **150**, 277–299.

Brookes, A. (1987a) Restoring the sinuosity of artificially straightened stream channels, *Environmental Geology and Water Science*, **10**, 33–41.

Brookes, A. (1987b) The distribution and management of channelised streams in Denmark, *Regulated Rivers: Research and Management*, **1**, 3–16.

Brookes, A. (1988) *Channelised Rivers: Perspectives for Environmental Management*. John Wiley, Chichester, 336 pp.

Brookes, A. (1990) Restoration and enhancement of engineered river channels: some European experiences. *Regulated Rivers: Research and Management*, **5**, 45–56.

Brookes, A. (1992) Recovery and restoration of some engineered British river channels. In: Boon, P.J., Calow, P. and Petts, G.E. (eds), *River Conservation and Management*, John Wiley, Chichester, 337–352.

Brookes, A. (1995a) The importance of high flows for riverine environments. In: Harper, D. and Ferguson, A. (eds), *The Ecological Basis for River Management*, John Wiley, Chichester, 33–49.

Brookes, A. (1995b) Design practices for channels receiving urban runoff: examples from the River Thames catchment UK. In: Herricks, E.E. (ed.), *Stormwater Runoff and Receiving Systems: Impact, Monitoring and Assessment*, CRC Lewis Publishers, Florida, USA, 307–317.

Brookes, A. (1995c) River channel restoration: theory and practice. In: Petts, G.E. and Gurnell, A.M. (eds), *Changing River Channels*, John Wiley, Chichester, 369–388.

Brunet, R.C., Pinay, G., Gazelle, F. and Roques, L. (1994) Role of the floodplain and riparian zone in suspended matter and nitrogen retention in the Adour River, south-west France. *Regulated Rivers: Research and Management*, **9**, 55–63.

Bryan, C.F. (1991) Preface: Flood River Symposium. *Regulated Rivers: Research and Management*, **6**, 73–74.

Cairns, J. (1982) Restoration of damaged ecosystems. In: Mason, W.T. and Iker, S. (eds), *Research and Wildlife Habitat*, Office of Research and Development and US Environmental Protection Agency, Washington, DC.

Cairns, J. (1989) Restoring damaged ecosystems: is predisturbance condition a viable option? *The Environmental Professional*, **11**, 152–159.

Cairns, J. (1991) The status of the theoretical and applied science of restoration ecology. *The Environmental Professional*, **13**, 186–194.

Cairns, J., Dickson, K.L. and Herricks, E.E. (eds) (1977) *Recovery and Restoration of Damaged Ecosystems*. University Press of Virginia, Charlottesville, USA.

CIRIA (1992) *Scope for Control of Urban Runoff*. Vols 1–4. Construction Industry Research and Information Association, London.

Coats, R., Swanson, M. and Williams, P. (1989) Hydrologic analysis for coastal wetland restoration. *Environmental Management*, **13**, 715–727.

Cooper, J.R., Gilliam, J.W., Daniels, R.B. and Robarge, W.P. (1987) Riparian areas as filters for agricultural sediment. *Soil Sci. Soc. Am. J.*, **51**, 416–420.

Costa, J.E. (1978) *Hydrologic and Hydraulic Investigations of Cherry Creek. Denver. Colorado*. Technical Paper 78-1, Department of Geography, University of Denver, 130 pp.

Cummins, K.W., Minshall, G.W., Sedell, J.R., Cushing, C.E. and Peterson, R.C. (1984) Stream ecosystem theory. *Proc. Internat. Assoc. Theor. Applied Limnol.*, **22**, 1818–1827.

Dister, E., Gomer, D., Obrdlik, P., Petermann, P. and Schneider, E. (1990) Water management and ecological perspectives of the Upper Rhine's floodplains. *Regulated Rivers: Research and Management*, **5**, 1–15.

Ellis, J.B. and House, M.A. (1994) Integrated design approaches for urban river corridor management. In: Kirby, C. and White, W.R. (eds), *Integrated River Basin Development*, John Wiley, Chichester.

Gerritsen, G.J. (1992) Ecological rehabilitation of the River Ijssel. *Contributions of the Workshop of the Ecological Rehabilitation of Floodplains* RIZA, CHR KHR, Arnhem, the Netherlands, 97–102.

Gilbert, J., Dole-Olivier, M.J., Marmonier, P. and Vervier, P. (1990) Surface water/groundwater ecotones. In: Naiman, R.J. and Decamps, H. (eds), *Ecology and Management of Aquatic–Terrestrial Ecotones, Man and the Biosphere Series*, UNESCO, Paris, and Parthenon Publ., Carnforth, 119–225.

Gore, J.A. (ed.) (1985) *The Restoration of Rivers and Streams: Theories and Experience*. Butterworth Publishers, Boston, 280 pp.

Gosz, J.R. (1978) Nitrogen inputs to stream water from forest along an elevational gradient in New Mexico. *Water Research*, **12**, 725–734.

Gregory, K.J., Lewin, J. and Thornes, J.B. (1987) *Palaeohydrology in Practice: A River Basin Analysis*. John Wiley, Chichester.

Gregory, S.V., Swanson, F.J., McKee, W.A. and Cummins, K.W. (1991) An ecosystem perspective of riparian zones. *Bioscience*, **41**, 540–551.

Gurnell, A.M. (1995) Hillslopes, floodplains and the riparian zone. In: Petts, G.E. and Gurnell, A.M. (eds), *Changing River Channels*, John Wiley, Chichester, 237–260.

Hall, M.J., Hockin, D.L. and Ellis, J.R. (1993) *Design of Flood Storage Reservoirs*. Butterworth-Heinemann, London.

Herricks, E.E. and Osborne, L.L. (1985) Water quality protection and restoration in streams and rivers. In: Gore, J. (ed.), *Restoration of Rivers and Streams*. Ann Arbor Press, Ann Arbor, Michigan, 1–20.

Hollis, G.E. and Acreman, M.C. (1994) The functions of wetlands within integrated river basin development. In: Kirby, C. and White, W.R. (eds), *Integrated River Basin Development*, John Wiley, Chichester, 351–365.

Hupp, C.R. (1992) Riparian vegetation recovery patterns following stream channelization: a geomorphic perspective. *Ecology*, **73**, 1209–1226.

Iversen, T.M., Kronvang, B., Madsen, B.L., Markman, P. and Nielsen, M.B. (1993) Re-establishment of Danish streams: restoration and maintenance measures. *Aquatic Conservation: Marine and Freshwater Ecosystems*, **3**, 1–20.

Jordan, W.R., Gilpin, M.E. and Aber, J.D. (eds) (1987) *Restoration Ecology*, Cambridge University Press.

Junk, W.J., Bayley, P.B. and Sparks, R.E. (1989) The flood pulse concept in river–floodplain systems. In: Dodge, D.P. (ed.), Proceedings of the International Large River Symposium, *Canadian Special Publication of Fisheries and Aquatic Sciences*, **106**, 110–127.

Kern, K. (1992) Rehabilitation of streams in south-west Germany. In: Boon, P.J., Calow, P. and Petts, G.E. (eds), *River Conservation and Management*, John Wiley, Chichester, 321–335.

Kern, K. (1994) *Grundlagen naturnaher Gewässergestaltung: Geomorphologische Entwicklung von Fließgewassern*. Springer-Verlag, Berlin, 256 pp.

Large, A.R.G., Prach, K., Bickerton, M.A. and Wade, P.M. (1994) Alteration of patch boundaries on the floodplain of the regulated River Trent, UK. *Regulated Rivers: Research and Management*, **9**, 71–78.

Malanson, G.P. (1993) *Riparian Landscapes*. Cambridge University Press.

Nanson, G.C. and Croke, J.C. (1992) A genetic classification of floodplains. *Geomorphology*, **4**, 459–486.

National Research Council (1992) *Restoration of Aquatic Ecosystems: Science, Technology, and Public Policy*. National Academy Press, Washington, DC.

Odum, E.P. (1979) Ecological importance of the riparian zone. In: Johnson, P.P. and McCormick, J.F. (eds), *Strategies for Protection and Management of Floodplain Wetlands and other Riparian Ecosystems*, US Forest Service General Technical Report, WP-12, Washington, DC, 2–4.

Osborne, L.L., Bayley, P.B., Higler, L.W.G., Statzner, B., Triska, F. and Iversen, T. (1993) Restoration of lowland streams: an introduction. *Freshwater Biology*, **29**, 187–194.

Palmer, L. (1976) River management criteria for Oregon and Washington. In: Coates, D.R. (ed.), *Geomorphology and Engineering*, Allen & Unwin, London, 329–346.

Peterken, G.F. and Hughes, F.M.R. (1996) Restoration of floodplain forests. In: *Forests and Water*, Proceedings of the 1994 Discussion meeting of the Institute of Chartered Foresters, ICF, Edinburgh.

Petersen, R.C., Madsen, B.L., Wilzback, M.A., Magadzan, C.H.D., Paarlberg, A., Kullberg, A. and Cummins, K.W. (1987) Stream management: emerging global simulations. *Ambio*, **16**(4), 166–179.

Petersen, R.C., Petersen, L.B.-M and Lacoursiere, J. (1992) A building-block model for stream restoration. In: Boon, P.J., Calow, P. and Petts, G.E. (eds), *River Conservation and Management*, John Wiley, Chichester, 293–309.

Pinay, G., Fabre, A., Vervier, P. and Gazelle, F. (1992) Control of C, N, P distribution in soils of riparian forests. *Landscape Ecology*, **6**, 121–132.

Platts, W.S. and Rinne, J.N. (1985) Riparian and stream enhancement. Management and Research in the Rocky Mountains, *North American Journal of Fisheries Management*, **5**(A), 115–125.

Plenet, S., Gibert, J. and Vervier, P. (1992) A floodplain spring: an ecotone between surface water and groundwater. *Regulated Rivers: Research and Management*, **7**, 93–102.

Rhoads, B.L. and Miller, M.V. (1990) Impact of riverine wetlands construction and operation on stream channel stability: conceptual framework for geomorphic assessment. *Environmental Management*, **6**, 799–807.

Riley, A.L. (1989) Overcoming federal water policies: The wildcat–San Pablo Creeks case, *Environment*, **31**.

River Restoration Project (1993a) *Phase I. Feasibility Study: Summary*. The River Restoration Project, PO Box 126, Huntingdon, UK, 15 pp.

River Restoration Project (1993b) *Business Plan*. The River Restoration Project, PO Box 126, Huntingdon, UK, 23 pp.

River Restoration Project (1994) *Partnership for River Restoration, Part I: Institutional Aspects of River Restoration in the UK*, Summary, March 1994. The River Restoration Project, PO Box 126, Huntingdon, UK, 11 pp.

Roelle *et al.* (1988) quoted on page 424 of National Research Council (1992) *Restoration of Aquatic Ecosystems: Science, Technology and Public Policy*. National Academy Press, Washington, DC.

Roux, A.L., Bravard, J.P., Amoros, C. and Pautou, G. (1989) Ecological changes of the French Upper Rhone River since 1750. In: Petts, G. (ed.), *Historical Change of Large Alluvial Rivers: Western Europe*, John Wiley, Chichester, 323–350.

Schlosser, I.J. and Karr, J.R. (1981) Riparian vegetation and channel morphology impact on spatial patterns of water quality in agricultural watersheds. *Environmental Management*, **5**, 233–243.

Schnick, R.A., Morton, J.M., Mochalski, J.C. and Beall, J.T. (eds) (1981) *Mitigation/ Enhancement Handbook for the Upper Mississippi River System (UMRS) and other Large River Systems*. Technical Report E. Comprehensive Master Plan for the Management of the Upper Mississippi River System, Upper Mississippi River Basin Commission, Minneapolis, MN, 702 pp.

Sedell, J.R. and Froggatt, J.L. (1984) Importance of streamside forests to large rivers: the isolation of the Willamette River, Oregon, USA, from its floodplain by snagging and streamside forest removal. *Verh. Internat. Verein Limnol*, **22**, 1828–1834.

Shields, F.D. Jr (1995) Fate of Lower Mississippi River habitats associated with training dikes, *Aquatic Conservation: Marine and Freshwater Systems*, **5**, 97–108.

Sparks, R.E., Bayley, P.B., Kohler, S.L. and Osborne, L.L. (1990) Disturbance and recovery of large floodplain rivers, *Environmental Management*, **14**, 699–709.

Tabacchi, E., Planty-Tabacchi, A.M.,and Decamps, O. (1990) Continuity and discontinuity of the riparian vegetation along a fluvial corridor. *Landscape Ecology*, **5**, 9–20.

Toth, L.A. (1991) Environmental responses to the Kissimmee River Demonstration Project. *South Florida Water Management District Technical Publication*, No. 91–02, 96 pp.

Toth, L.A., Obeysekera, J.T.B., Perkins, W.A. and Loftin, M.K. (1993) Flow regulation and restoration of Florida's Kissimmee River. *Regulated Rivers: Research and Management*, **8**, 155–166.

Turner, R.K. (1987) Environmental management in the Broadland region. *Regulated Rivers: Research and Management*, **4**, 287–378.

Ward, D., Holmes, N. and Jose, P. (1994) *The New Rivers and Wildlife Handbook*. Royal Society for the Protection of Birds, NRA and Royal Society for Nature Conservation, RSPB, Sandy, Bedfordshire, UK, 426 pp.

Ward, J.V. (1989) Riverine–wetland interactions. In: Sharitz, R. and Gibbons, J. (eds), *Freshwater Wetlands and Wildlife*, CONF-8603101, DOE Symposium Series No. 61, US Department of Energy, USDOE Office of Scientific and Technical Information, Oak Ridge, TN, 385400.

Welcomme, R.L. (1979) *The Fisheries Ecology of Floodplain Rivers*. Longman, London, 317 pp.

Yon, D. and Tendron, G. (1981) Alluvial forests of Europe. *European Committee for the Conservation of Nature and Natural Resources and Environment Series No. 22*.

18 The Effects of River Management on the Hydrology and Hydroecology of Arid and Semi-Arid Floodplains

I.D. JOLLY
CSIRO Division of Water Resources, Canberra, Australia

18.1 INTRODUCTION

Arid and semi-arid regions constitute approximately one third of the world's land area (Rogers, 1981). Development in these regions has been limited in many cases by lack of available water resources as the rivers and streams of these areas usually have flows which are highly variable (in both space and time) and therefore of insufficient reliability to support large populations and their accompanying industries. To overcome the highly variable flow in arid and semi-arid areas, rivers and streams have been regulated by weirs and storages to provide year-round water supplies, transportation, waste assimilation, recreation, generation of electricity and other uses.

In many cases river regulation has had profound hydrological and ecological side-effects on the rivers themselves and their adjacent floodplains. In general, floodplains support higher amounts of biomass, have greater primary production and more dynamic nutrient cycling than surrounding upland areas. This is particularly the case in arid and semi-arid areas where potential evapotranspiration is greater than rainfall for most months and so flooding provides an additional source of water not available to non-floodplain areas (Hollis, 1990). The unique dynamics, structure and composition of floodplain forests are a direct result of episodic flooding (Brinson, 1990).

There is little doubt that river regulation leads to the disturbance of floodplain ecology. Sparks *et al.* (1990) define ecological disturbance in a river–floodplain system as "an unpredictable, discrete or gradual event (natural or man-made) that disrupts structure or function at the ecosystem, community, or population level". Walker and Thoms (1993) contend that regulation of rivers results in ecological disturbance because it provides a regime which is more predictable than that under natural conditions. It is a matter of debate as to whether the effects of flow regulation are greater for arid/semi-arid rivers than those in more temperate regions. Walker (1992) argues that it is in arid/semi-arid areas that the greatest disparities between natural and regulated regimes are found, whereas Poff and Ward (1990) contend that

Floodplain Processes. Edited by Malcolm G. Anderson, Des E. Walling and Paul D. Bates.
© 1996 John Wiley & Sons Ltd.

effects are comparatively minor as ecosystems in these areas have exceptional capacities to absorb the effects of change. These arguments notwithstanding, it is clear that development has significant negative environmental impacts on floodplains and wetlands in arid and semi-arid regions (Hollis, 1990).

In this chapter we summarize the effects of flow regulation on the hydrology and hydroecology of rivers and floodplains in arid and semi-arid regions. We concentrate on the effects on vegetative hydroecology but do not address processes in the near-stream aquatic (littoral) zone. By way of example, we present a case study from a semi-arid floodplain of the River Murray in southern Australia which illustrates how river management has impacted on floodplain hydrology with adverse consequences for the native riparian vegetation communities.

Throughout the chapter the following terms are widely used and are defined thus:

hydroperiod	the duration, frequency, depth and season of flooding (Lugo *et al.*, 1990)
vegetative hydroecology	the ecology of floodplain vegetation communities in relation to available water
groundwater recharge	the addition of water derived from rainfall or flooding to groundwater storage (aquifer)
groundwater discharge	the loss of water from groundwater storage (aquifer) to the atmosphere by evaporation and/or transpiration

18.2 SUMMARY OF THE HYDROLOGICAL AND HYDROECOLOGICAL EFFECTS OF RIVER REGULATION

The hydrological and ecological consequences of river regulation in temperate, humid and tropical areas have been widely studied. Indeed, theory exists which attempts to describe the dependence of river–floodplain ecosystems on flooding (Junk *et al.*, 1989). The effects of river regulation in arid and semi-arid areas are much less studied, particularly in relation to riparian vegetation. In this section we summarize the available literature on the impacts of flow regulation on both the hydrology of the rivers themselves and their floodplains. The consequences of hydrologic change on native riparian vegetation communities are then described.

18.2.1 Impacts on river and floodplain hydrology

Rivers

The nature of changes in flow regime varies widely throughout the world. In almost all cases it is governed primarily by the demands on the water and the hydraulic structures (and their operation) used to meet the demands. It is not possible to generalize the range of responses which occur so we present a number of examples to illustrate the forms of behaviour commonly observed.

The San Joaquin River in southern California (USA) has been progressively regulated since large-scale irrigation projects commenced in the catchment during the late 1940s (Orlob and Ghorbanzadeh, 1981). Since this time flows in the river have decreased by as much as 35% with much of this reduction occurring during the period April–September, the principal irrigation season. Prior to regulation the majority of the water yield from the catchment was derived from snowmelt in spring and early summer. This runoff is now mostly transferred out of the river system or diverted for consumptive use. Flow during the late summer, autumn and winter periods has been much less affected by regulation. Accompanying the changes in flow are significant increases in the salinity of the San Joaquin River due to reduced diluting flows of natural runoff and increases in salt accessions to the river due to irrigation drainage. River salinities are highest from April to July, when regulation has had the greatest impact on flows and the crops irrigated with river water are most sensitive to high salinity.

Diversion of water from the Arkansas River in Colorado (USA) for irrigation has also resulted in substantial reductions in downstream flows. Miles (1977) estimated that as much as 98% of all river water passing Canon City in the upper reaches of the river is consumed before it reaches the Colorado–Kansas state border approximately 300 km downstream. Approximately 60% of this is directly attributed to crop production, while livestock, industrial and domestic use accounts for the rest. Data presented in Konikow and Person (1985) show that for a 12-month period in 1971/72 average river salinity for the same reach increased from 170 mg L^{-1} at the upstream end to 3810 mg L^{-1} at the border. As in the previous example, irrigation drainage entering the river is thought to be mostly responsible for the increases in salinity moving downstream.

The Colorado River in the southwestern USA is one of the most altered and intensively controlled rivers in the United States (Carlson and Muth, 1989). It supplies more water for consumptive use than any other river in the United States in spite of having the lowest discharge per unit area of any basin in the country (Pillsbury, 1981). To meet the burgeoning demands over the last 90 years, the Colorado River and its tributaries have been increasingly modified by the construction of large dams, alteration of channels, and diversions of water. Reservoir construction and diversions of water in open canals have increased water losses due to evaporation by ~15% (Graf, 1985). River regulation has reduced spring flows and increased summer flows. Less than 1% of the river's natural flow now reaches its mouth. Greatly increased salinity levels are the most serious water quality issue arising from the intensive regulation and diversion of the river. Graf (1985) reported that at Lee Ferry (a location about mid-distance along the river) natural salt levels (~250 mg L^{-1}) had doubled by 1957. On a more local scale, water quality problems such as high heavy metal concentrations, radioactive materials and acid mine drainage have arisen as a result of increased industrialization and urbanization in the catchment.

The River Murray in southeastern Australia has been highly regulated by a series of weirs and storages since the early 1920s. Present-day diversions of water for agricultural, urban and industrial uses mean that outflow to the sea is now only 36% of what it was under natural conditions (Close, 1990a). The frequency of medium-size floods in the lower reaches is now a third of that under natural conditions (Ohlmeyer, 1991). The seasonality of flow has also been greatly altered by regulation. Under natural conditions, flows were at a maximum in spring and then rapidly returned to low levels in late summer and autumn. Under current management, flow in the upper reaches of the river system now peaks in late summer and is lowest during winter. This is due to diversion of winter flows into storages and releases to meet peak demands in summer and autumn (the main growing season for the large areas of irrigated agriculture along the river). In the lower reaches the changes in seasonality of flow are less pronounced (Walker and Thoms, 1993) due to the inflows of unregulated tributaries and because the water released from upstream is diverted along the way for irrigation use (Close, 1990a).

Flood behaviour has also been greatly altered by regulation. Most notable is that the recession limb of the flood hydrograph is now much steeper because the weir pools represent a significant loss of floodplain storage capacity. Recession of a flood which would have taken up to two months under natural conditions now occurs in a matter of days (Walker and Thoms, 1993). Somewhat related are the changes in channel morphology (depth, width, bed slope) which have occurred along the lower reaches of the river due to modifications in sediment supply and redistribution (Thoms and Walker, 1993). As in the previous example, flow regulation has led to increases in River Murray salinity (Morton and Cunningham, 1985) due to irrigation drainage and increased accessions of saline groundwater (Close, 1990b).

While not located in an arid or semi-arid region, the effects of regulation on the Mississippi River provide an interesting contrast to the above examples. Belt (1975) described how a large flood on the middle Mississsippi in 1973 was augmented by levees and other river confinement structures that constricted flood flows, thereby preventing the river from utilizing the water storage capacity of the floodplain. More recently, Grubaugh and Anderson (1989) reported the long-term effects of one of the navigation dams on the flow of the upper Mississippi River. They showed that water levels in the lock (weir) pool upstream of the dam had increased at ~2.5 cm yr^{-1} over the 74 years since installation. They attributed this behaviour to the progressive loss in channel volume due to sedimentation caused by installation of the dam and lock (weir). Both the frequency and number of annual flood days was found to have increased significantly as a result of the flow regulation. They concluded that similar behaviour was likely for the other 26 lock pools of the upper Mississippi River.

While it is difficult to generalize the effects of regulation on the hydrology of the rivers themselves, a number of conclusions can be reached. Firstly, flow regulation almost always leads to changes in hydroperiod. In all of the above examples depth, duration and frequency of flooding was severely altered by regulation, with reductions in all three components the most common effect. In the majority of cases seasonality of flooding was also greatly altered. Secondly, in arid or semi-arid areas, flow regulation often leads to increased salinities of the river. In arid and semi-arid areas both groundwaters and soils are often naturally saline. Increased accession of saline groundwater to the rivers is usually caused by (a) alterations to the natural groundwater flow regime (by the hydraulic structures used for regulation and their consequent effects on stream–aquifer interactions), (b) saline drainage beneath irrigation areas developed close to the rivers, and (c) reductions in dilution flows.

Floodplains

Floodplains typically have very low relief and so minor alterations to the heights and lengths of floods have dramatic effects on the extent, depth and duration of floodplain inundation. Regulation of rivers can also affect the flow and chemistry of shallow groundwater systems often found beneath floodplains.

Salinization of floodplain soils is the most commonly observed consequence of changes in hydroperiod. The soils of arid and semi-arid floodplains often contain naturally high levels of salt due to the aridity of the climate. Moreover, many floodplains are underlain by saline groundwater at shallow depths and so groundwater discharge driven by evapotranspiration leads to salt movement up into the soil profile (Thorburn *et al.*, 1992). Salt which accumulates in the near surface can be leached back down the profile during floods and so salt levels in the plant root zone are kept low enough to allow riparian vegetation communities to thrive. While the salt content of a given soil may vary throughout the year in response to flooding, rainfall, water extraction by vegetation, and groundwater fluctuations, over the long-term a dynamic salt "balance" generally exists in which there is no net accumulation or leaching of salt. When changes in flooding regime occur this "balance" can be disturbed, leading to salt accumulation.

Salt accumulation can occur indirectly in response to irrigation of floodplain soils (e.g. Sokolovskiy, 1967) or directly due to changes in the flooding regime, as described above (e.g. Jolly *et al.*, 1993). It is interesting to note that in the former case, the collapse of ancient civilizations such as those of the Tigris and Euphrates rivers (in present-day Iraq) occurred primarily as a result of the salinization of the floodplain soils that were irrigated (Pillsbury, 1981). In the latter case, an example is discussed in the case study presented below which shows that complete salinization of the soil profile can occur within one or two decades following regulation.

18.2.2 Impacts on the vegetative hydroecology of floodplains

As discussed above, changes in hydroperiod are the most direct consequence of flow regulation. Given that the structure, composition and dynamics of riparian vegetation communities are directly related to the long-term natural flooding regime (Breen *et al.*, 1988), it is not surprising that the consequences of regulation for these communities can in many cases be extremely severe. To paraphrase Lugo *et al.* (1990), "hydroperiod (and its modification) can function as both a subsidy and a stress for wetlands". In this section we summarize the range of impacts commonly observed.

Drowning

River regulation is most often carried out by the construction of weirs and locks along the length of the stream. These structures are used to maintain constant minimum river levels in the pools created between successive weirs. Vegetation dies due to permanent inundation (causing waterlogging) which, prior to regulation, would have only occurred intermittently. The effects are greatest immediately upstream of the weir as the water level changes are greatest in this zone. Moreover, drowning only occurs in the lowest lying parts of the floodplain which are in direct hydraulic connection with the weir pool. Higher areas of the floodplain which are hydraulically isolated from the weir pool remain unaffected. In floodplains underlain by shallow saline groundwater the effects can be exacerbated by salinization as groundwater levels in the immediate vicinity of the weir rise toward the surface (due to the higher hydraulic head exerted by the weir pool) causing saline groundwater seepages characterized by areas of efflorescing salt (Walker *et al.*, 1992).

Water stress

As highlighted by Kondolf *et al.* (1987) reduced river flows due to regulation may result in (a) lowering of the stream water surface below levels that littoral vegetation require for direct contact or spray, (b) lowering of underlying water tables below the rooting depths of some species (in reaches where the stream is losing water to the groundwater), and (c) reductions in annual high flows that otherwise would recharge bank sediments as well as soils of the wider floodplain. In areas where the stream is gaining water from the groundwater, the effects of reduced streamflow may be offset by this additional source of water, provided it is of sufficient quality and the vegetation has (or can develop) root systems able to exploit this source.

Channel modifications are often performed on rivers to improve downstream conveyance of water (Brinson, 1990). This can result in water stress of riparian vegetation as flows which otherwise would have flooded an area of floodplain are now constrained within the main stream channel. Furthermore, drainage lines which supply water to regions of the floodplain remote from the main channel can be be blocked off or destroyed. As shown by Buchholz (1981), minor drainage lines can significantly affect the distribution of floodplain vegetation communities. Channelization can also result in the lowering of underlying water tables further reducing water supplies. Other side-effects include increased erosive potential of floods (Rhoads, 1990) and reduced sediment delivery to the floodplain (Brinson, 1990).

The effects of reductions in water availability on riparian vegetation include changes in the area, density, composition and species diversity. The study of Johnson *et al.* (1976) provides a good example of changes in community structure caused by reduced water availability.

Salt stress

As discussed above, salinization of floodplain soils can occur in areas underlain by shallow saline groundwater or subject to irrigation. High soil salt levels causes dieback of riparian vegetation by both ion toxicity and osmotic effects. The former is likely to occur in species which have poor salt exclusion mechanisms and therefore have difficulty in regulating the amount and type of salt taken up through the roots. Species subject to the latter suffer stress and dieback due to physiological drought. Even though adequate water may be available in the soil profile, it is of too high salinity (and hence too low osmotic potential) to be available to the plant. The lower limit of osmotic potential at which water can be drawn varies greatly with floodplain/swamp species (e.g. Mensforth and Walker, 1994). An example of dieback due to osmotic effects is given in the case study below.

Nutrient supply

Movement of sediment and nutrients within a catchment occurs primarily by river transport. While the main channels of floodplain rivers are important in downstream transport of materials, the majority of biological activity and physical storage or retention of nutrients and organic matter occur in lateral areas (Sparks *et al.*, 1990). As floodplains receive all forms of nutrients from the main channel it is expected that the basic nutrient status of the floodplain should correspond to that of the river (Junk *et al.*, 1989). Moreover, Junk *et al.* (1989) contend that flooding is the major driving force responsible for the exchange of nutrients, organic matter and organisms between the river and the floodplain. Mitsch *et al.* (1979) present an example of the influence of flooding on the phosphorus budget and tree growth of a floodplain in a temperate region (Illinois, USA).

It is therefore probable that changes in flooding regime which result in reduced silt inputs to floodplains will have adverse consequences for the vegetation (Breen *et al.*, 1988). However, Brinson (1990) claims that nutrient limitation has never been reported in arid riverine forests, and implies that it may be a minor effect when compared with that of water stress. We contend that too few data exist to conclude one way or the other.

Invasion of exotic species

Invasion of *Tamarix* (salt cedar) sp. into native riparian communities on arid floodplains in the southwestern USA is a widespread problem (Horton, 1972). Reductions in river flow by regulation and the lowering of groundwater levels by pumping has resulted in dieback of the native riparian vegetation due to reduced water availability. These conditions are conducive to invasions of the aggressive *Tamarix* which are able to survive in the drier conditions. In addition to the ecological problems of an exotic invasion, the phreatophytic water use of *Tamarix* has caused great concern as it is widely seen as a "nonbeneficial water-wasting plant" (Horton, 1972). There have been numerous studies to estimate the groundwater use of salt cedar (e.g. van Hylckama, 1970; Gay and Fritschen, 1979). Graf (1985) reports that millions of dollars have been spent clearing *Tamarix* communities in the southwestern United States over the last 40 years. *Tamarix* invasions of arid floodplains have also been reported in central Australia (Griffin *et al.*, 1989).

Of more limited extent is the invasion of native *Eucalyptus camaldulensis* (red gum) into adjoining natural grassland (predominantly *Pseudoraphis spinescens*, moira grass) areas in the Barmah–Millewa Forest in southern New South Wales, Australia (Bren, 1992). This extensive floodplain forest resides on the middle reaches of the highly regulated River Murray. Changes in flooding frequency are also responsible for this invasion, although the major factor in this case is seasonal modification in which there is increased summer flooding combined with reduced winter–spring flows.

18.3   CASE STUDY: THE CHOWILLA ANABRANCH OF THE LOWER RIVER MURRAY, SOUTH AUSTRALIA

18.3.1   Introduction

General

The Murray–Darling Basin (Figure 18.1) is Australia's largest river system, occupying an area of some 10^6 km^2, or one seventh of the continent. While the combined length (~5500 km) of the major tributaries, the River Murray and the Darling River, makes it one of the largest river systems of the world, the mean annual runoff from the basin (14 mm) is extremely low compared to that of other great rivers such as the Amazon (620 mm), the Yangtsze Kiang (523 mm) or the Mississippi (176 mm) (River Murray Commission, 1970). This is the direct result of the aridity of much of the catchment. For example, the headwaters in the Australian Alps (southern–eastern New South Wales/ northeastern Victoria) contribute, on average, 37% of the total River Murray flow from an area less than 1.5% of the total catchment (Eastburn, 1990). In contrast, the Darling River, whose tributaries traverse the arid western portion of the basin, contributes only 12% of the flow from an area representing 60% of the catchment.

Despite its low runoff, the Murray–Darling system is of considerable importance, providing approximately 73% of all water used in Australia (Fleming, 1982). In terms of agriculture, almost three quarters of the land irrigated in Australia is located within the basin and results in the production of ~90% of Australia's irrigated cereals, ~80% of

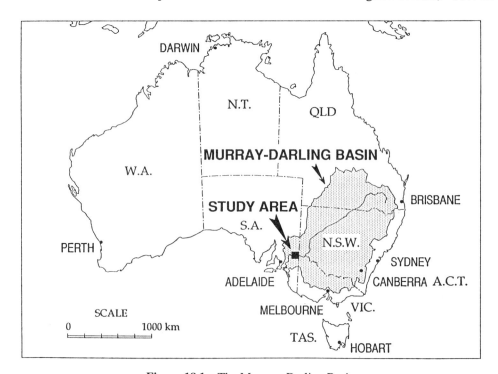

Figure 18.1   The Murray–Darling Basin

its pasture, ~65% of its fruit and ~25% of its vegetables (Murray–Darling Basin Ministerial Council, 1987). The river system supplies water to 16 large urban centres within the basin (including the Federal Capital, Canberra). In addition, the basin supplies water to Adelaide and other South Australian regional centres. On average, ~50% of South Australia's water supply is derived from the River Murray, and in dry years as much as 90% is supplied from this source (Close, 1985). The wetlands of the basin are internationally significant in terms of habitat and breeding for wildlife. Moreover, the riverine landscapes of the basin are of great aesthetic and recreational value to the 1.6 million people who reside in the basin (and to the Australian community at large).

The flow of the River Murray (and to a lesser extent, that of the Darling River) and its tributaries is highly regulated by four major water storages and 16 locks and weirs. Construction of the regulation structures commenced in the early 1920s, primarily to provide year-round navigation and to overcome the effects of water shortages during droughts (Jacobs, 1990). Today, following intensive development over the last 70 years, regulation is essential for the supply of water to the agricultural and industrial industries and urban communities both within and outside the basin.

River salinization

Given the importance of the River Murray, there is much concern over the increasing river salinity which has been observed in recent times. Morton and Cunningham (1985) reported that salinity increases of >2% per year have occurred in the lower reaches since 1970. The extent of the problem is illustrated by Figure 18.2 which

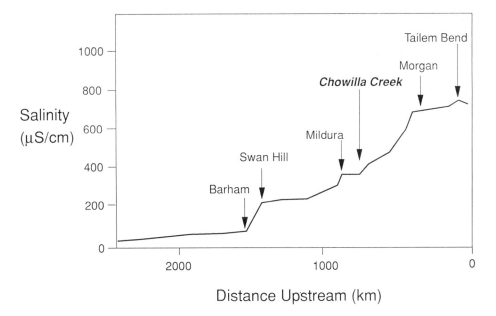

Figure 18.2   Longitudinal salinity (as electrical conductivity) profile of the River Murray

shows the increase in median river salinity with distance downstream from the headwaters. Of note is that the largest increases occur in the lower reaches. On average, 1.1 million tonnes of salt enters South Australia annually, and another 0.5 million tonnes is added in the South Australian reach (Mackay *et al.*, 1988). The large salinity increases in this region are due to the flow of saline groundwater to the river (Figure 18.3). As the groundwater adjacent to the River Murray is saline (up to

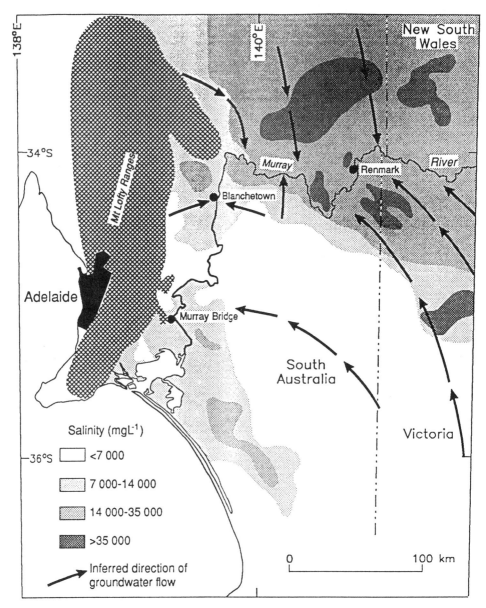

Figure 18.3 Regional groundwater flow and salinity in the the unconfined aquifer of the western Murray Basin

35 000 mg L^{-1}) in this area, this leads to the accession of large amounts of salt to the river. In many instances this natural process has been exacerbated by the establishment of large irrigation areas adjacent to the river. Irrigation in excess of crop requirements, often combined with a lack of adequate subsurface drainage, has resulted in the formation of local groundwater mounds in these areas which displace large volumes of saline groundwater into the river (Close, 1990b). Furthermore, there is great concern over the long-term future impacts of the clearing of native deep-rooted *Eucalyptus* vegetation in the western portion of the basin and its replacement with shallow-rooted crops and pastures. As shown by Allison *et al.* (1990), this has caused increases in groundwater recharge of up to two orders of magnitude resulting in rises in the water table beneath the region, hence increased flow of saline groundwater to the River Murray. Allison *et al.* (1990) predict that this will cause salinity in the River Murray to steadily increase over at least the next 200 years.

The flow of saline groundwater to the lower reaches of the River Murray is complicated by the presence of floodplains of up to 10 km in width. In these areas, the water table is generally less than 4 m below the soil surface, leading to significant loss of groundwater by evapotranspiration. The presence of shallow saline groundwater in a semi-arid climate creates a hostile environment for the riparian vegetation which is mitigated only by the occasional floods which inundate these areas. Flooding provides a source of water for the vegetation, leaches salt from the soils, and freshens the groundwater beneath the floodplain. It also displaces additional salt into the river,

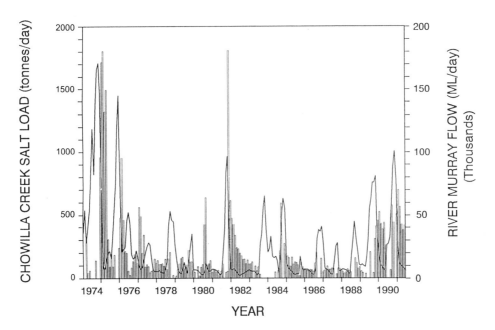

Figure 18.4 River Murray flow into South Australia (solid line) and salt load from the Chowilla floodplain (bar graph) for 1974–1991

causing an increase in the salt loads which can last for several months after the flood (Figure 18.4).

Effects of regulation

In its natural state, flow in the River Murray varied greatly from year to year. It also exhibited a distinct seasonal pattern with peaks in winter and spring and lows in late summer and autumn (Figure 18.5). Under the present regulated conditions flow is much less variable, with reduced flood flows and increased flows in droughts. Outflow to the sea is now only 36% of what it was under natural conditions due to large diversions of water for agricultural, urban and industrial use. In the upper reaches flows are now greatest in summer and winter (Close, 1990a), but in the lower reaches the seasonality has not changed (Walker and Thoms, 1993). Ohlmeyer (1991) has estimated that the frequency of major floods (i.e. return period >1 in 5) in the lower reaches has decreased by a factor of approximately three (Figure 18.6).

Another significant side-effect of river regulation is the rise in water-table levels beneath the floodplains due to the installation of locks and weirs along the river, evaporation basins on the floodplain and irrigation on the floodplain and areas adjacent to the river (National Environmental Consultancy, 1988). As will be shown later, this has led to increases in the discharge of saline groundwater with adverse consequences for the health of the riparian vegetation communities.

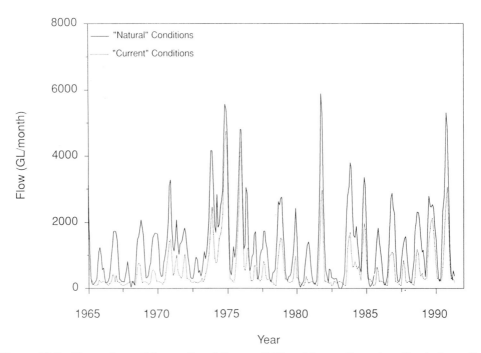

Figure 18.5 Comparison of "natural' and "current" River Murray flows into South Australia for the period 1965–1991 predicted using a monthly simulation model (after Close, 1990a)

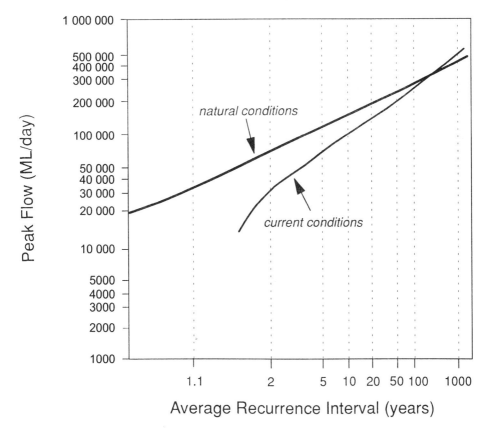

Figure 18.6 River Murray flood recurrence intervals under natural and regulated conditions (after National Environmental Consultancy, 1988)

18.3.2   The Chowilla anabranch

General

The Chowilla anabranch, a 200 km^2 floodplain situated near the junction of the South Australian, Victorian and New South Wales borders (Figure 18.7), is the second largest natural source of salt to the River Murray in the lower reaches. This system of floodplain creeks contributes an average salt load of 140 tonnes day^{-1} to the River Murray, which represents approximately 10% of the salt accession to the South Australian reach of the river. Moreover, inundation of the floodplain causes large volumes of saline groundwater to flow into the creeks following a major flood resulting in salt loads as high as 1800 tonnes day^{-1} (Figure 18.4). This process can continue for up to two years after the flood event (National Environmental Consultancy, 1988; Figure 18.8).

The Chowilla region is the largest remaining area of natural riverine forest along the lower River Murray. It has a high biodiversity of both flora and fauna and is an

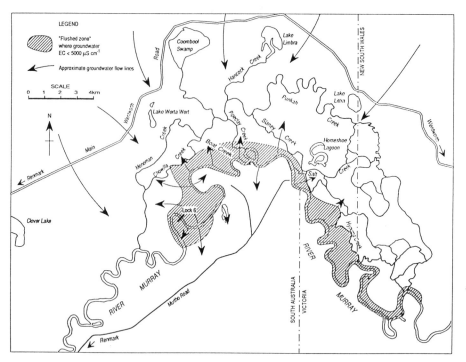

Figure 18.7 The Chowilla anabranch system. The flow of groundwater into and within the floodplain is indicated, as is the area of low salinity groundwater referred to as the "flushed zone"

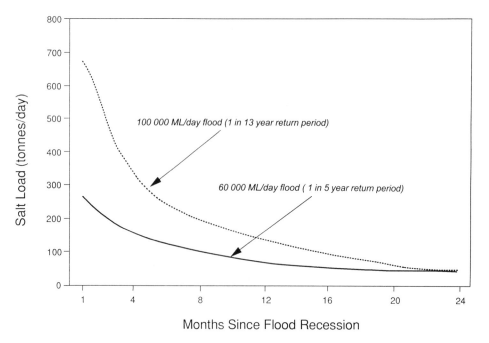

Figure 18.8 Salt load recession curves for the Chowilla anabranch for a medium and large size flood (after R. Ebsary and R. Newman, personal communication)

important native fish nursery and breeding area for water birds (O'Malley and Sheldon, 1990). The ecological significance of the Chowilla region is highlighted by its listing under the UNESCO Ramsar Convention (Section 14.5) as a Wetland of International Importance (National Environmental Consultancy, 1988). In the light of these facts, it is of considerable concern that dieback of this and other native riparian forests, has started to occur over the last decade. Black box, *Eucalyptus largiflorens*, is the species mostly affected on the floodplains of the lower River Murray.

While salt accession to the floodplain streams and riparian vegetation dieback are seemingly disparate problems, as will be shown below, both are related to the flow regulation of the River Murray.

Hydrology and climate

The Chowilla anabranch consists of a network of streams which flow from the River Murray upstream of Lock 6 (one of the major flow control weirs) and eventually join together to form Chowilla Creek which flows back into the River Murray downstream of Lock 6. The Chowilla anabranch streams were ephemeral and flowed only during the periods of flood prior to the construction of Lock 6. The lock creates a permanent upstream weir pool over 70 km long which drives water continually through the Chowilla floodplain streams (40–80% of the River Murray flow is now diverted through the creek system). The streams intercept the saline regional groundwater which flows into the region (Figure 18.7).

The Chowilla region has a semi-arid climate characterized by mild winters and long hot summers. Mean annual rainfall is about 260 mm and potential evaporation is about 2000 mm. Rainfall is highly variable both within and between the seasons with extended dry spells being common. The distribution of mean monthly rainfall has a slight winter dominance (Jolly *et al.*, 1994a).

Hydrogeology

As discussed above, the Chowilla region acts as a natural sink for the regional groundwater systems. A typical hydrogeological cross-section through the Chowilla anabranch is shown in Figure 18.9. The main aquifers are the Monoman Formation, the Pliocene Sands aquifer and the Murray Group aquifer.

The Monoman Formation underlies the floodplain and consists of alluvial sands approximately 30 m thick. The water table formerly resided in this aquifer, but due to the construction of Lock 6 it has risen 2–3 m (Chowilla Working Group, 1992) and is now often found within the overlying Coonambidgal Clay causing the aquifer to be semi-confined. The salinity of this aquifer ranges from 20 000 to 60 000 mg L^{-1}, with the highest values at the base of the aquifer (Jolly *et al.*, 1992a). However, the weir pool upstream of Lock 6 has enhanced groundwater recharge in the immediate vicinity and created a zone of low salinity (<3000 mg L^{-1}) groundwater known locally as the "flushed zone" (Figure 18.7). The Monoman formation is in direct hydraulic contact with the regional Pliocene Sands aquifer and also has good hydraulic connection to the creeks and the River Murray. The hydraulic conductivity of this aquifer is approximately 10 m day^{-1}.

592

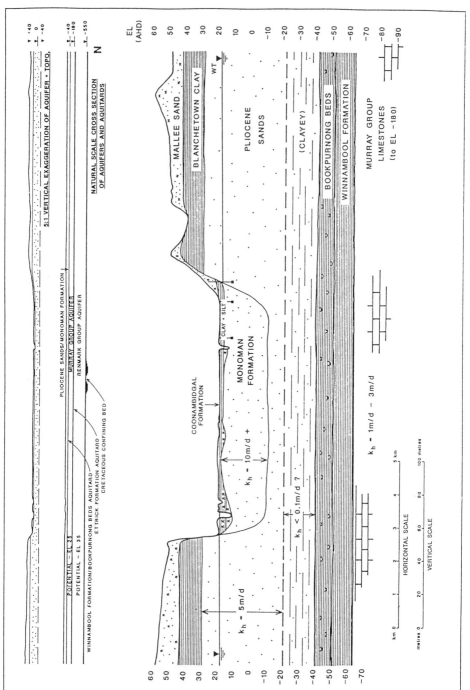

Figure 18.9 Typical hydrogeological cross-section through the Chowilla anabranch (after Waterhouse, 1989)

The Pliocene Sands aquifer consist of estuarine fine, medium and coarse sand of two formations, the Parilla Sand and the underlying Loxton Sand. They contain some clayey layers, and have a clay content of about 10%, which is present as a coating on the grains. The clay content increases towards the base of the aquifer. The aquifer is unconfined and has salinities in the range 40 000–90 000 mg L^{-1}. The hydraulic conductivity is lower than that of the Monoman Formation – approximately 5 m day^{-1}.

The underlying Murray Group aquifer consists of carbonated sandstone and limestone. The aquifer is confined by the Bookpurnong Beds (micaceous and glauconitic sands and marls) and the underlying Winnambool Formation (marls, glauconitic marly limestone, and marly clays) which act as aquitards. Salinities in the Murray Group aquifer are lower then those of the Pliocene Sands and Monoman Formation (15 000–30 000 mg L^{-1}). The hydraulic conductivity of the Murray Group aquifer is approximately 1–3 m day^{-1}.

Regional groundwater flow occurs laterally from the Pliocene Sands aquifer towards the floodplain, entering the Monoman Formation from the north. While the Murray Group aquifer has a potentiometric head approximately 5 m above that of the Pliocene Sands aquifer, the available data suggest that it contributes less than 10% of the water flowing into the Monoman Formation (Vader et al., 1994).

Soils

The soils of the Chowilla region generally consist of a micaceous cracking clay of alluvial origin (Coonambidgal Clay) which has minimal profile development except for occasional gypsum, halite or calcium carbonate horizons. The Coonambidgal Clay can be up to 5 m in thickness, with the deepest deposits occurring close to existing or prior creeks. In some locations (most notably the higher elevation land units, and in limited areas, the beds of the floodplain creeks) the Coonambidgal Clay is absent. As shown by Jolly et al. (1994a), the dispersive nature of this soil results in very low hydraulic conductivities, especially when inundated by low salinity floodwaters. The Coonambidgal Clay overlies the Monoman Formation.

Vegetation

The distribution of vegetation within the floodplain is directly related to flood frequency (and hence surface elevation) and groundwater salinity. It is important to note that small differences in surface elevation correspond to large differences in flooding frequency. Figure 18.10 shows a typical distribution pattern in the vicinity of a stream on the floodplain of the lower River Murray. Black box communities occupy ~40% of the Chowilla anabranch, red gum (*E. camaldulensis*) ~12%, and lignum (*Meuhlenbeckia florulenta*) ~10% (Noyce and Nicolson, 1992).

18.3.3 Salt accession to the River Murray

General

As described above (Figure 18.4), salt loads to the River Murray from the Chowilla anabranch system are highly variable and directly related to river flow. The mean salt

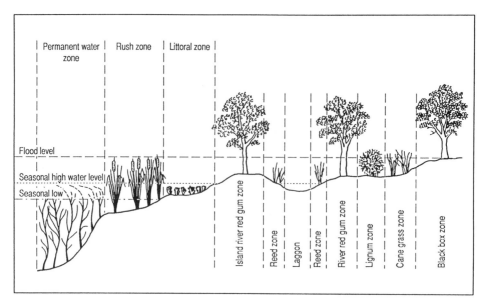

Figure 18.10 Typical vegetation distribution in the the vicinity of a floodplain stream (after National Environmental Consultancy, 1988)

accession from the area of 140 tonnes day $^{-1}$ is comprised of a non-flood component of approximately 40 tonnes day $^{-1}$ and post-flood contributions of up to 1800 tonnes day $^{-1}$. Elevated salt loads originating from the Chowilla anabranch can continue for up to two years after a large flood (Figure 18.8). While it is uncertain as to whether the total volume of salt transported to the River Murray has increased due to regulation, it is probable that the timing and impact of the peak loads have been greatly modified. Prior to regulation, the floodplain streams only flowed during floods and would have ceased to flow soon after the recession of the river (because the river level at the end of its recession would have dropped below the height of the stream offtakes) – a period likely to be only a month or two. While it is probable that the streams would have transported large amounts of salt during this time, the relatively high flows would have provided considerable dilution. Today, streamflow is perennial resulting in continual transport of salt to the river. Moreover, the very high salt loads which enter the streams and river for months after a flood do so at a time when river flow has returned to non-flood levels. It is therefore likely that there is there is much less dilution than would have occurred in pre-regulation times.

Salt sources

Figure 18.11 shows a summary of the contributions of each major stream to the total flow and salt load to the River Murray during flood recession periods. These data are averages calculated from stream flow and salinity measurements made at 17 post-flood samplings during the period 1973–1991. While the inner streams (Slaney, Hypurna, Pipeclay and Boat Creeks) account for the majority of the flow (64%), they contribute only 20% of the salt. The large salt contributions from the outer streams (Punkah, Salt,

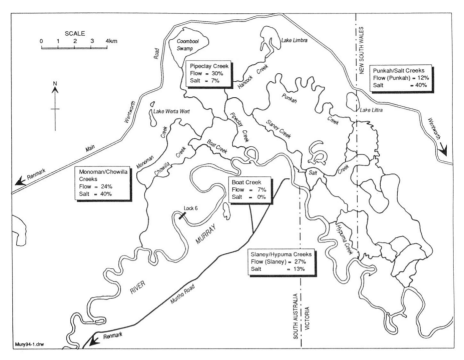

Figure 18.11 Mean flow and salt loads (in percentage of totals from the entire anabranch system) from the major streams based on post-flood flow and salinity measurements carried out at selected sites and times during 1973–1991 (after Sharley and Huggan, 1995)

Monoman and Chowilla Creeks) suggest that they intercept regional groundwater flowing from the north.

Mechanisms

Jolly *et al.* (1994a) studied recharge processes during a flood in 1990 which inundated much of the floodplain. They measured soil and groundwater salinities before and after the flood, and recorded groundwater levels throughout the flood. Very little leaching of salt stored in the soil profile was observed (except in the top 1 m), even at sites inundated for as much as 130 days. Groundwater levels (Figure 18.12) did not rise to the surface at sites which were flooded (i.e. no hydraulic connection was established), except in some areas directly adjacent to streams. Freshening of the aquifer by floodwaters only occurred at some locations in the immediate vicinity of streams. These data, combined with analytical modelling of the water-table behaviour during the flood (Narayan *et al.*, 1993), led them to conclude that vertical diffuse recharge of floodwater did not occur to any great extent due to the highly dispersive nature of the sodic surface soils. They concluded that the unsaturated zone was not the source of salt which entered the floodplain streams following flood recession.

While diffuse vertical recharge of floodwaters was found to be unimportant, the changes in salinity of some of the piezometers adjacent to streams indicated that flow of

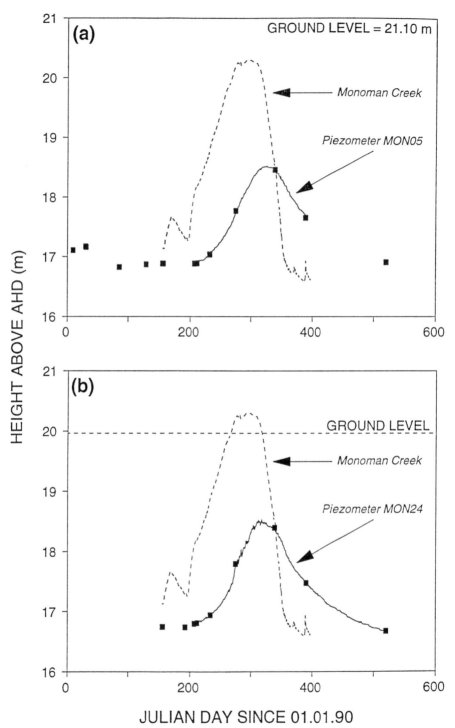

Figure 18.12 Hydrographs of piezometers (solid lines and squares) (a) MON05 (site not flooded) and (b) MON24 (site flooded for 55 days). Ground level at each site is also depicted, as is the hydrograph of Monoman Creek (dotted line)

stream water into the aquifer occurred during the flood. Given that the groundwaters and floodwaters have vastly different salinities (and hence densities), it is probable that mixing of the two waters occurred within bank storage. Hence, water flowing from the aquifer back into the streams during and after the flood recession was likely to have an enhanced salt concentration. Stream salinity data collected during a flood in 1989/90 clearly show that saline groundwater enters the base of streams immediately following flood recession (Figure 18.13). However, these data, and those from some simple analytical modelling analogous to non-steady-state diffusion, indicate that salt loads generated by this mechanism return to non-flood levels within six months of the flood peak. Hence, this mechanism cannot explain the elevated salt accessions which continue for up to two years after a flood (Figure 18.8).

A diagram depicting a possible explanation for the long salt recessions is presented in Figure 18.14. The relatively impermeable surface clay (Coonambidgal Clay) is absent in some areas within the floodplain leaving the underlying Monoman Sand exposed. The saturated hydraulic conductivity of the Monoman Sand is reasonably high (~10 m day^{-1}) and so significant groundwater recharge may occur when these areas are inundated during a flood. If the recharge is high enough then localized groundwater mounds may form beneath these areas. Dissipation of a mound over the following months causes displacement of *in situ* groundwater stored in the region between the mound and the nearest stream. If the displaced groundwater flows into the stream and is saline, then increased stream salinity will result. Using a diffusion-type approach, Jolly *et al.* (1994a) estimated that a mound located 2.5 km from a stream would take about 1.5 years to begin displacing groundwater into the stream.

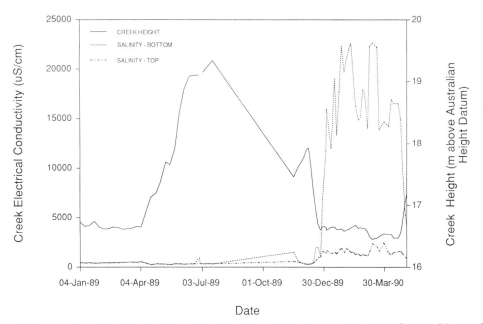

Figure 18.13 Water level and salinity (as electrical conductivity) at the surface and base of Monoman Creek during the 1989/90 flood (data provided by the South Australian Engineering and Water Supply Department)

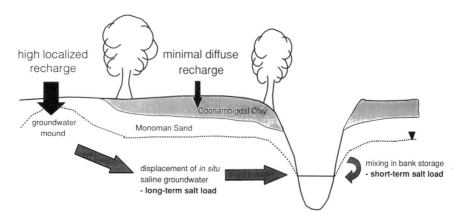

Figure 18.14 Mechanisms which lead to the accession of saline groundwater to floodplain streams following major floods

To further understand the processes which lead to the high salt accessions following a flood, Charlesworth *et al.* (1994) carried out cross-sectional groundwater modelling of a transect across the floodplain passing through Punkah/Slaney Creeks. They used a solute transport model (SUTRA; Voss, 1984) to simulate density-dependent groundwater flow and solute transport within the floodplain aquifer during and after a major flood event. Assuming no floodwater recharge, they estimated that peak salt load increases of five-fold to Punkah Creek and three-fold to Slaney Creek immediately following the recession of a 1 in 13 year flood. These were in good agreement with data collected following floods of a similar magnitude in 1973 and 1986. However, the predicted length of the salt recessions (six months) were considerably shorter than those measured. Repeat simulations which included either diffuse or localized recharge regimes predicted that salt loads returned to non-flood levels after about 10 months, still shorter than those periods observed.

18.3.4 Dieback of riparian vegetation

General

Dieback of native riparian vegetation is a problem along much of the River Murray. In a survey of the status of the riparian vegetation communities of the River Murray floodplains, Margules and Partners *et al.* (1990) estimated that approximately 18 000 ha of vegetation is severely degraded. In particular, they noted that *Eucalyptus largiflorens* F. Muell., the most abundant species in the lower reaches of the River Murray, is suffering severe decline. More recently, Taylor *et al.* (1996) estimated that ~40% of the *E. largiflorens* communities on the Chowilla floodplain are in poor health.

Causes

Margules and Partners *et al.* (1990) identified a number of causes of degradation, the most notable being: saline groundwater, drowning or waterlogging, water stress, logging or clearing, and grazing. They concluded that the issue of greatest concern for the long-term health of riparian vegetation along the River Murray is soil salinization. Subsequent work by Eldridge (1991) and Jolly *et al.* (1992b) suggested that the dieback of *E. largiflorens* is due primarily to water stress caused by a combination of osmotic effects and a reduction in water supply. They found that the osmotic effects were caused by high soil salinities and the reduction in water supply was a result of the lower frequency of flooding.

Thorburn *et al.* (1993) showed that, in order to survive the long dry periods between floods and rains, *E. largiflorens* obtained up to 100% of its water requirements from the moderately saline (14–20 g L^{-1} Total Dissolved Solids, TDS) groundwater. While this enabled it to survive the dry conditions, transpiration rates were low (~0.3 mm day^{-1}) compared to potential evaporation (1–10 mm day^{-1}). Streeter *et al.* (unpublished data, 1994) studied water uptake of *E. largiflorens* at a site of considerably higher groundwater salinity (35 g L^{-1} TDS) and found that transpiration rates were an order of magnitude lower, and that water was mostly derived from the upper 50 cm of the soil profile and was of rainfall origin. They concluded that at sites of this seawater-level salinity, *E. largiflorens* was able to survive by excluding salt from the shoots and lowering its transpiration through reduced stomatal aperture.

In order to generalize the results described above, Thorburn *et al.* (1995) presented a model describing the salinity-limited uptake of groundwater by plants. This analytical model is based on unsaturated zone water and salt balances during dry periods when groundwater discharge and uptake of groundwater by vegetation is the only significant loss of water from the unsaturated zone. Model predictions were compared with measurements of groundwater uptake made over a 15-month period at three *E. largiflorens* sites and at two sites supporting *E. camaldulensis*. Depth and salinity of groundwater, and soil type varied greatly between sites. Good matches between the predicted and measured transpiration rates of *E. largiflorens* were obtained. Sensitivity analysis showed that groundwater depth and salinity were the main controls on uptake of groundwater, while soil properties appeared to have a lesser effect.

It is is clear from these studies that both soil and groundwater salinity are the controlling influences on the water uptake, and therefore the health, of *E. largiflorens*.

Salt accumulation

Jolly *et al.* (1993) investigated the mechanisms leading to salt accumulation within the soils of the Chowilla floodplain and concluded that the salinization was a direct result of the regulation of the River Murray. The higher water tables beneath the floodplain (a consequence of maintaining higher average river levels) results in greater discharge of the saline groundwater, causing increased upward transport of salt into the soil. The decreased incidence of large floods results in less downward leaching of salt from the soils by floodwater recharge. They concluded that prior to river regulation, these processes were "balanced" (over the long term) such that net accumulation of salt did not occur and so vegetation generally had a sufficient depth of low salinity (and hence, low osmotic suction) soil from which to draw water (Figure 18.15). The disruption of this "balance" by river regulation has resulted (in many cases) in the movement of a "salt front" from the water table up through the soil profile. As the "salt front" moves

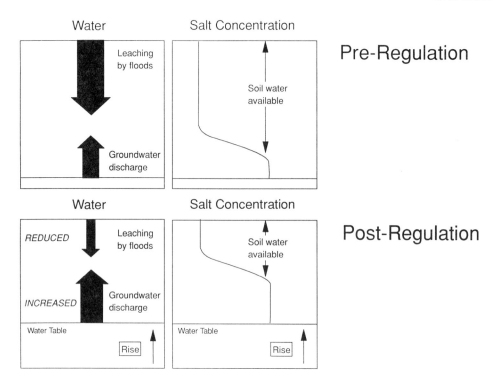

Figure 18.15   Salt accumulation mechanisms in floodplain soils

towards the surface (over a number of years), the depth of low salinity soil water available to the vegetation decreases (Figure 18.15).

Figure 18.16 depicts a soil chloride profile which shows the "salt front" at the surface, and salt accumulating at concentrations much greater than that of the groundwater (the source). Not surprisingly, the *E. largiflorens* community at this site is completely dead. Using an analytical model of groundwater discharge, Jolly *et al.* (1993) predicted that it would have taken approximately 17 years for the "salt front" to reach the soil surface at this site, and a further 20 years for the additional salt to accumulate at the surface. This site has been flooded only once (in 1956) since river regulation commenced. Even if complete leaching of the profile took place at this time, it can be seen that there has been adequate time for the observed accumulation to take place. It is interesting to note that if the non-flood water-table depth was 5 m prior to river regulation then it would have taken approximately 50 years for the "salt front" to rise to the surface. Conversely, if the non-flood water table had risen to 3 m below ground then it would only take six years for the "salt front" to reach the soil surface.

In subsequent work, Jolly *et al.* (1994b) presented a model which described the "balance" between salt transport up into the soil profile by groundwater discharge and downward leaching by flooding. Steady-state soil-limited groundwater discharge (q) is directly proportional to the power of the depth of the water table below an evaporating

SITE BD

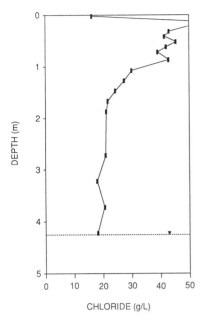

Water table depth ~ 4 m

Groundwater discharge ~ 10 mm/yr

Time for salt front to reach surface (t) ~ 17 yrs

Time for additional salt to accumulate ~ 20 yrs

Last time flooded - 1956

Figure 18.16 Soil salt profile (as chloride ion) from a site where *E. largiflorens* has died as a result of soil salinization

surface (either the soil surface or the base of a plant root zone), z, and soil texture, thus (Warrick, 1988):

$$q = Az^{-n} \qquad (18.1)$$

where A and n are soil physical constants related to the soil texture.

Water available for leaching of salt from the soil (L) is related to the proportion of time the site is flooded (W) and the saturated hydraulic conductivity of the surface soil (K_s) thus:

$$L = K_s W \qquad (18.2)$$

So for a "balance" between salt accumulation and leaching at a given site, Equations 18.1 and 18.2 are equated and rearranged to define a "critical depth", z, thus:

$$z = (A/K_s W)^{1/n} \qquad (18.3)$$

The "critical depth" is interpreted as the minimum water-table depth (below the evaporating front) required to maintain the "balance" and thus prevent long-term net accumulation of salt.

Jolly *et al.* (1994b) compared the "balance" line for a soil texture of loamy clay ($n = 3$) with *E. largiflorens* health (on a scale 0 (dead) to 5 (healthy)) at 22 sites where water-table depth and flooding information were available. They found that the predicted "balance" line reasonably separated the sites which were unhealthy (i.e. the

water-table depth is insufficient to compensate for the reduced flooding) from those
which were in good health (Figure 18.17).

18.3.5 Proposed management scheme

From the above discussion it should be clear that, while seemingly unrelated, floodplain
stream salinization and riparian vegetation dieback are closely linked. In recognition
that both result from the effects of river regulation, a management plan has been
proposed which aims to address both issues (Chowilla Working Group, 1992). The plan
involves the installation of a groundwater pumping scheme to reduce water table levels,
and the enhancement of flood flows to provide more water for the floodplain
vegetation.

A network of 15 pumping bores (tubewells) on the outer northern portion of the
floodplain is proposed (Figure 18.18). Saline groundwater pumped from these bores
would be transported by pipeline to a remote basin where it would be evaporated and
the salt harvested. It is predicted that within 10 years the water table in the immediate
vicinity of pumping bores would be lowered by up to 1.5 m, and consequently flow of
saline groundwater into the outer streams from the regional unconfined aquifer to the
north would be reduced to negligible amounts (Charlesworth et al., 1994). It would also
reduce the rates of groundwater discharge and therefore assist in restoring the "balance"
described by Equation 18.3. Some freshening of the surface of the aquifer is anticipated
which would provide a further supply of low salinity water for the vegetation. In the
light of the finding that diffuse floodwater recharge is low (Jolly et al., 1994a), surface

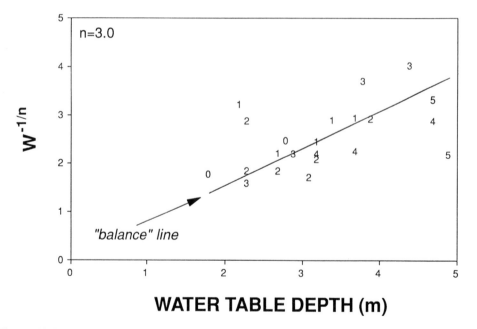

Figure 18.17 E. largiflorens health (on a scale of 0 to 5) versus groundwater depth ($\approx z$) and
$W^{-1/n}$ Also shown is the "balance' line for a loamy clay ($n = 3$)

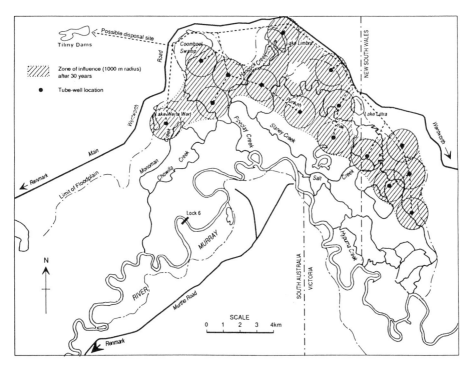

Figure 18.18 Groundwater pumping scheme

freshening of the aquifer will require some form of recharge enhancement (e.g. recharge bores/pits).

The second component of the management plan is the enhancement of flood flows to provide more water to the vegetation. Controlled releases of water from a nearby upstream storage (Lake Victoria) would be carried out at or near the peak of floods. The water needed for the releases would be obtained by partially emptying (if required) Lake Victoria prior to the flood and filling it up on the rise of the flood. Two strategies have been proposed: (1) releasing 7 GL day $^{-1}$ at flows in the range 35–160 GL day $^{-1}$ (return periods of 1 in 2 and 1 in 20 respectively); and (2) releasing 16 GL day $^{-1}$ at flows greater than 60 GL day $^{-1}$ (1 in 5 year return period). The second would require extensive modifications to the flow regulator at the outlet of Lake Victoria This, and possible legal actions from downstream communities inconvenienced by high flows, dictate that strategy (2) is probably not feasible.

As part of the development of the management plan a Geographical Information System (GIS) of the Chowilla anabranch was produced (Noyce and Nicolson, 1992). The GIS contained overlays such as topography, riparian vegetation distribution, water-table depth, groundwater salinity, flooding extent, and surface soil type. Hodgson (1993) used the GIS to develop a spatial model which categorized the occurrence of *E. largiflorens* into six health classes based on the critical parameters of groundwater salinity, flooding frequency and water-table depth (Table 18.1). Using comparisons with site specific data and aerial photography, Taylor *et al.* (1996) found that the six

Table 18.1 Parameters, thresholds, areas and predicted *E. largiflorens* health of GIS classes defined by Hodgson (1993)

GIS class	Groundwater salinity (dS m^{-1})	Flooding frequency (years)	Groundwater depth (m)	Area (ha)	Predicted vegetation health
1	>40	>1 in 13	<4	1046	Poor
2	>40	>1 in 13	>4	764	Good
3	>40	1 in 10–13	–	1724	Poor
4	>40	<1 in 10	–	2038	Good
5	<40	<1 in 10	–	377	Good
6	<40	>1 in 10	–	564	Good

classes could be divided statistically into either good health (four of the classes) or poor health (remaining two classes). Using the GIS, Overton *et al.* (1994) showed that in the flow range 70–100 GL day^{-1} (1 in 7–13 return period, the range over which most of the *E. largiflorens* communities are found), a release of 7 GL day^{-1} would result in an 10–20% increase in area of *E. largiflorens* flooded. For example, a 82 GL day^{-1} flood occurs every 1 in 10 years. By increasing a 75 GL day^{-1} flood by 7 GL day^{-1} this would produce a 82 GL day^{-1} flood at approximately a 1 in 8 year return period. This would convert some of GIS class 3 to GIS class 4, changing the prediction of health from poor to good.

While both the groundwater pumping scheme and flood enhancement will clearly provide a more favourable environment for the riparian vegetation, it is unclear as to the magnitude and nature of the benefits that may accrue. The simple GIS model described above provides some indications but is limited by its discretization of the floodplain into large "homogeneous" areas and its inability to model the mechanisms leading to the observed health patterns. For example, large stores of salt have accumulated in the floodplain soils in the 70 years since river regulation commenced. Lowering the water table will obviously reduce the rate of salt transport up into the profiles. However, it is uncertain as to how long it will take under the improved flooding regime for large amounts of the stored salt to be leached downwards. Moreover, it is uncertain as to how a given degraded vegetation community will respond to a greater availability of flood water, an increase in depth of low salinity soil, and a possible additional supply of low salinity groundwater. Similarly, it is not clear whether the releases from Lake Victoria should be optimized to maximize the flood height (to cover greater areas of *E. largiflorens*) or prolong the flood length (to provide more water and leaching to those areas which would have been flooded anyway). These questions cannot be addressed with the presently available data and GIS model.

Future studies should concentrate on measuring the field response of *E. largiflorens* in a range health conditions to a range of flooding regimes. This will require measurements of the physiological responses of the trees (i.e. transpiration, stomatal conductance, ion balance, etc.) to the prevailing environmental (soil, groundwater, climate, canopy structure, etc.) conditions before, during and after flooding. These measurements would provide valuable validation data for vegetation response models such as TOPOG-IRM (Hatton *et al.*, 1992) which can be used to model the likely outcomes of a wide range of management scenarios. Linking of this understanding with existing or derived overlays in the GIS will provide a means of assessing the management requirements for other floodplains where less data is available.

18.4   CONCLUDING REMARKS

The effects of flow regulation on arid and semi-arid rivers and floodplains have been little studied, particularly in relation to riparian vegetation communities. From the case study and other detailed work (e.g. Kondolf *et al.*, 1987) it is clear that the effects are complex and can take considerable time to become apparent.

Regulation almost always leads to lower river flows and significant changes in hydroperiod. Consequently, increasing river salinities due to alterations to the natural groundwater flow regime, saline drainage beneath irrigation areas developed close to the rivers, and reductions in dilution flows often occur.

The generally low relief of floodplains means that small changes in the heights and lengths of floods have dramatic effects on the extent, depth and duration of floodplain inundation. Soil salinization caused directly by changes in flooding regime, or indirectly by irrigation, is an often observed consequence of river regulation. The impacts of flow regulation on the hydroecology of floodplain vegetation communities include drowning, water stress, salt stress, reduction in nutrient supply and invasion of exotic species. All arise directly from the changes in hydroperiod which accompany regulation.

The case study provides an example where a number of the above effects have been felt. This study investigated the complex interaction between flow regulation and the resultant river salinization and vegetation dieback due to soil salinization. The results of this study contributed to the development of a conjunctive management plan which jointly addressed both of these major effects.

ACKNOWLEDGEMENTS

The work described in the case study forms part of a large multidisciplinary study headed by Glen Walker and supported by Grant Nos. 88/44 and CWS2 from the Land and Water Resources Research and Development Corporation. Tom Hatton and Kumar Narayan have also made valuable scientific contributions to the project, as have students Peter Thorburn, Lisa Mensforth, Geoff Hodgson, Tania Streeter, Peter Taylor, Ian Overton, Anthony Charlesworth, Craig Simmons and Linda Vader. Worthy of particular thanks are Kerryn McEwan, Andrew Holub and John Dighton who carried out much of the technical and laboratory work. Peter Stace, Bob Newman and Tony Sharley have provided valuable data and discussion throughout the project.

REFERENCES

Allison, G.B., Cook, P.G., Barnett, S.R., Walker, G.R., Jolly, I.D. and Hughes, M.W. (1990) Land clearance and river salinisation in the western Murray Basin, Australia. *J. Hydrol.*, **119**, 1–20.

Belt, C.B. (1975) The 1973 flood and man's constriction of the Mississippi River. *Science*, **189**, 681–684.

Breen, C.M., Rogers, K.H. and Ashton, P.J. (1988) Vegetation processes in swamps and flooded plains. In: Symoens, J.J. (ed.), *Vegetation of Inland Waters*, Kluwer, Dordrecht, 223–247.

Bren, L.J. (1992) Tree invasion of an intermittent wetland in relation to changes in the flooding frequency of the River Murray, Australia. *Aust. J. Ecol.*, **17**, 395–408.

Brinson, M.M. (1990) Riverine forests. In: Lugo, A.E., Brinson, M.M. and Brown, S. (eds), *Forested Wetlands*, Elsevier, Amsterdam, 87–141.

Buchholz, K. (1981) Effects of minor drainages on woody species distributions in a successional floodplain forest. *Can. J. For. Res.*, **11**, 671–676.

Carlson, C.A. and Muth, R.T. (1989) The Colorado River: Lifeline of the American southwest. In: Dodge, D.P (ed.), Proceedings of the International Large Rivers Symposium, Honey Harbour, Ontario, September 1986. *Canadian Special Publication of Fisheries and Aquatic Sciences* No. 106, Department of Fisheries and Oceans, Ottawa, 220–239.

Charlesworth, A.T., Narayan, K.A. and Simmons, C.T. (1994) Modelling of salt accession within the Chowilla anabranch and possible mitigation schemes: a density-dependent model of groundwater flow and solute transport. *CSIRO Division of Water Resources Report* No. 94/7, Canberra, 23 pp.

Chowilla Working Group (1992) *Chowilla Resource Management Plan: Progress Report*. Murray–Darling Basin Commission, Canberra, 33 pp.

Close, A.F. (1985) Assessment of risks for the water resources of the River Murray in South Australia. In: *Proc. Hydrology and Water Resources Symp.*, Sydney, May 1985. Institution of Engineers Australia, Canberra, 131–135.

Close, A.F. (1990a) The impact of man on the natural flow regime. In: Mackay, N. and Eastburn, D. (eds), *The Murray*, Murray–Darling Basin Commission, Canberra, 61–74.

Close, A.F. (1990b) River salinity. In: Mackay, N. and Eastburn, D. (eds), *The Murray*, Murray–Darling Basin Commission, Canberra, 127–144.

Eastburn, D. (1990) The river. In: Mackay, N. and Eastburn, D. (eds), *The Murray*, Murray–Darling Basin Commission, Canberra, 3–15.

Eldridge, S.R. (1991) The role of salinisation in the dieback of black box (*Eucalyptus largiflorens* F. Muell.) on the floodplain of the River Murray, South Australia BSc (Hons) thesis, University of Adelaide.

Fleming, P.M. (1982) The water resources of the Murray–Darling basin. In: *Murray–Darling Basin Project Development Study. Stage 1 – Working Papers*. CSIRO Division of Water and Land Resources, Canberra, 39–54.

Gay, L.W. and Fritschen, L.J. (1979) An energy budget analysis of water use by salt cedar. *Water Resour. Res.*, **15**, 1589–1592.

Graf, W.L. (1985) *The Colorado River – Instability and Basin Management*. Association of American Geographers, Washington, DC, 86 pp.

Griffin, G.F., Stafford Smith, D.M., Morton, S.R., Allan, G.E. and Masters, KA. (1989) Status and implications of the invasion of Tamarisk (*Tamarix aphylla*) on the Finke River, Northern Territory, Australia. *J. Environ. Manage.*, **29**, 297–315.

Grubaugh, J.W. and Anderson, R.V. (1989) Long-term effects of navigation dams on a segment of the upper Mississippi River. *Reg. Rivers, Res. and Manage.*, **4**, 97–104.

Hatton, T.J., Walker, J., Dawes, W.R. and Dunin, F.X. (1992) Simulations of hydroecological responses to elevated CO_2 at the catchment scale. *Aust. J. Bot.*, **40**, 679–696.

Hodgson, G.A. (1993) Predicting the health of *Eucalyptus largiflorens* in the Chowilla anabranch region of the River Murray using GIS and remote sensing. BSc (Hons) thesis, Australian National University, Canberra.

Hollis, G.E. (1990) Environmental impacts of development on wetlands in arid and semi-arid lands. *Hydrol. Sci.*, **35**, 411–428.

Horton, J.S. (1992) Management problems in phreatophyte and riparian zones. *J. Soil and Water Conservation*, **27**(**2**), 57–61.

Jacobs, T. (1990) Regulation and management of the River Murray. In: Mackay, N. and Eastburn, D. (eds), *The Murray*, Murray–Darling Basin Commission, Canberra, 39–58.

Johnson, W.C., Burgess, R.L. and Keammerer, W.R. (1976) Forest overstorey vegetation and environment on the Missouri River floodplain in North Dakota. *Ecol. Monog.*, **46**, 59–84.

Jolly, I.D., McEwan, K.L., Holub, A.N., Walker, G.R., Dighton, J.C. and Thorburn, P.J. (1992a) *Compilation of Groundwater Data from the Chowilla Anabranch Region, South Australia*. CSIRO Division of Water Resources Tech. Mem. No. 92/9, Canberra, 111 pp.

Jolly, I.D., Walker, G.R., Thorburn, P.J. and Hollingsworth, I.D. (1992b) The cause of black box decline in the Chowilla anabranch region, South Australia and possible ameliorative approaches. In: *Catchments of Green. Proc. Nat. Conf. on Vegetation and Water Management*, Adelaide, South Australia, March 1992. Greening Australia, Canberra, 33–41.

Jolly, I.D., WaLker, G.R. and Thorburn, P.J. (1993) Salt accumulation in semi-arid floodplain soils with implications for forest health. *J. Hydrol.*, **150**, 589–614.

Jolly, I.D., Walker, G.R. and Narayan, K.A. (1994a) Floodwater recharge processes in the Chowilla anabranch system. *Aust. J. Soil Res.*, **32**, 417–435.

Jolly, I.D., Walker, G.R. and Thorburn, P.J. (1994b) Salt balances in semi-arid floodplain soils and consequences for riparian vegetation health. In: *Water Down Under 1994. Proc. (Volume 2B) Int. Groundwater and Hydrology Conf.*, Adelaide, South Australia, November 1994. Institution of Engineers, Australia Conf. Publ. No. 94/14, 687–690.

Junk, W.J., Bayley, P.B. and Sparks, R.E. (1989) The flood pulse concept in river–floodplain systems. In: Dodge, D.P (ed.), Proceedings of the International Large Rivers Symposium, Honey Harbour, Ontario, September 1986. *Canadian Special Publication of Fisheries and Aquatic Sciences* No. 106, Department of Fisheries and Oceans, Ouawa, 110–127.

Kondolf, G.T., Webb, J.W., Sale, M.J. and Felando, T. (1987) Basic hydrologic studies for assessing impacts of flow diversions on riparian vegetation: examples from streams of the eastern Sierra Nevada, California, USA. *Environ. Manage.*, **11**, 757–769.

Konikow, L.F. and Person, M. (1985) Assessment of long-term salinity changes in an irrigated stream-aquifer system. *Water Resour. Res.*, **21**, 1611–1624.

Lugo, A.E., Brown, S. and Brinson, M.M. (1990) Concepts in wetland ecology. In: Lugo, A.E., Brinson, M.M. and Brown, S. (eds), *Forested Wetlands*, Elsevier, Amsterdam, 53–85.

Mackay, N.J., Hillman, T.J. and Rolls, J. (1988) *Water Quality of the River Murray. Review of Monitoring, 1978 to 1986*. Water Quality Report No. 3, Murray–Darling Basin Commission, Canberra, 62 pp. Margules and Partners, P. and J. Smith, and Victorian Dept. of Conservation, Forests and Lands (1990) *River Murray Riparian Vegetation Study*. Murray–Darling Basin Commission, Canberra, 187 pp.

Mensforth, L.J. and Walker, G.R. (1994) Water use of plants in periodically saline and waterlogged environments—three case studies. In: *Proc. Third Nat. Conf. on Productive Use of Saline Land*, Echuca, Victoria, March 1994. Murray–Darling Basin Commission, Canberra, 79–86.

Miles, D.L. (1977) *Salinity in the Arkansas Valley of Colorado*. United States Environmental Protection Agency Report No. EPA-1A6-D4-0544, Fort Collins, 62 pp.

Mitsch, W.J., Dorge, C.L. and Wiemhoff, J.R. (1979) Ecosystem dynamics and a phosphorus budget of an alluvial cypress swamp in southern Illinois. *Ecology*, **60**, 1116–1124.

Morton, R. and Cunningham, R.B. (1985) Longitudinal profile of trends in salinity in the River Murray. *Aust. J. Soil. Res.*, **23**, 1–13.

Murray–Darling Basin Ministerial Council (1987) *Murray–Darling Basin Environmental Resources Study*. Murray–Darling Basin Commission, Canberra, 426 pp.

Narayan, K.A., Jolly, I.D. and Walker, G.R. (1993) *Predicting Flood-driven Water Table Fluctuations in a Semi-arid Floodplain of the River Murray Using a Simple Analytical Model*. CSIRO Division of Water Resources Report No. 93/2, Canberra, 16 pp.

National Environmental Consultancy (1988) *Chowilla Salinity Mitigation Scheme—Draft Environmental Impact Statement*. Report prepared by National Environmental Consultancy for the Engineering and Water Supply Dept. of South Australia, Adelaide, 166 pp.

Noyce, T. and Nicolson, K. (1992) *Chowilla Flood and Groundwater Monitoring*. South Australian Dept. Environment and Planning Report, Adelaide, 42 pp.

O'Malley, C. and Sheldon, F. (eds) (1990) *Chowilla Floodplain Biological Study*. Nature Conservation Society of South Australia, Adelaide, 224 pp.

Ohlmeyer, R.G. (1991) *Investigation of the Feasibility of Manipulating Water Levels in the River Murray*. South Australian Engineering and Water Supply Dept. Rep. 91/11, Adelaide, 47 pp.

Orlob, G.T. and Ghorbanzadeh, A. (1981) Impact of water resource development on salinisation of semi-arid lands. *Agric. Water Manage.*, **4**, 275–293.

Overton, I.C., Jolly, I.D., Taylor, P.J., Hatton, T.J. and Walker, G.R. (1994) Developing a GIS for determining flow management to conserve vegetation on the Chowilla floodplain, South Australia. In: *Resource Technology 1994*. Proc. Nat. Conf., Melbourne, September 1994. University of Melbourne, 78–91.

Poff, N.L. and Ward, J.V. (1990) Physical habitat template of lotic systems: recovery in the context of historical pattern of spatiotemporal heterogeneity. *Environ. Manage.*, **14**, 629–645.

Pillsbury, A.F. (1981) The salinity of rivers. *Sci. American*, **245**, 32–45.

Rhoads, B.L. (1990) The impact of stream channelisation on the geomorphic stability of an arid-region river. *Nat. Geog. Res.*, **6**, 157–177.

River Murray Commission (1970) *Murray Valley Salinity Investigations Volume 1*. River Murray Commission, Canberra, 476 pp.

Rogers, B.R. (1981) Fools rush in, Part 3: selected dryland areas of the world. *Arid Lands Newsletter*, **14**, 24–25.

Sharley, T. and Huggan, C. (1995) Chowilla Resource Management Plan, Murray–Darling Basin Commission, Canberra, pp. 140.

Sokolovskiy, S.P. (1967) Water–salt regime of floodplain soils of Ciscaucasia based on the Kuma River Valley. *Soviet Soil Sci.*, **7**, 962–972.

Sparks, R.E., Bayley, P.B., Kohler, S.L. and Osborne, L.L. (1990) Disturbance and recovery of large floodplain rivers. *Environ. Manage.*, **14**, 699–709.

Taylor, P.J., Walker, G.R., Hodgson, G., Hatton, T.J. and Correll, R.L. (1996) Testing of a G.I.S. model of *Eucalyptus largiflorens* health on a semi-arid, saline floodplain. *Environmental Management* (in press).

Thoms, M.C. and Walker, K.F. (1993) Channel changes associated with two adjacent weirs on a regulated lowland alluvial river. *Reg. Riv., Res. Manage.*, **8**, 271–284.

Thorburn, P.J., Walker, G.R. and Woods, P.H. (1992) Comparison of diffuse discharge from shallow water tables in soils and salt flats. *J. Hydrol.*, **136**, 253–274.

Thorburn, P.J., Hatton, T.J. and Walker, G R. (1993) Combining measurements of transpiration and stable isotopes of water to determine groundwater discharge from forests. *J. Hydrol.*, **150**, 563–587.

Thorburn, P.J., Walker, G.R. and Jolly, I.D. (1995) Uptake of saline groundwater by plants: An analytical model for semi-arid and arid areas. *Plant and Soil*, **175**, 1–11.

Vader, L., Jolly, I.D. and Walker, G.R. (1994) *Hydrochemistry and Stable Isotopes as Indicators of Groundwater Mixing and Discharge to Streams in the Chowilla Region, South-eastern Australia*. CSIRO Division of Water Resources Report 94/6, Canberra.

van Hylckarna, T.E.A. (1970) Water use by salt cedar. *Water Resour. Res.*, **6**, 728–735.

Voss, C.I. (1984) SUTRA: *A Finite-Element Simulation Model for Saturated–Unsaturated Fluid-Density-Dependent Groundwater Flow with Energy Transport or Chemically-Reactive Species Solute Transport*. US Geol. Surv. Water Resour. Invest. Rep. 84-4369, Reston, 409 pp.

Walker, K.F. (1992) A semi-arid lowland river: the River Murray, South Australia In: Calow, P.A. and Petts, G.E. (eds), *The Rivers Handbook*, Vol. 1, Blackwell Scientific, Oxford, 472–492.

Walker, K.F. and Thoms, M.C. (1993) Environmental effects of flow regulation on the lower River Murray, Australia. *Reg. Riv., Res. Manage.*, **8**, 103–119.

Walker, K.F., Thorns, M.C. and Sheldon, F. (1992) Effects of weirs on the littoral environment of the River Murray, South Australia In: Boon, P.J., Calow, P. and Petts, G.E. (eds), *River Conservation and Management*, Wiley, Chichester, 271–292.

Warrick, A.W. (1988) Additional solutions for steady-state evaporation from a shallow water table. *Soil Science*, **146**, 63–66.
Waterhouse, J. (1989) *The Hydrogeology of the Chowilla Floodplain – Status Report*. Report to the Murray–Darling Commission, Canberra, 31 pp.

19 Integrated Field, Laboratory and Numerical Investigations of Hydrological Influences on the Establishment of Riparian Tree Species

K.S. RICHARDS*, F.M.R. HUGHES*, A.S. EL-HAMES*, T. HARRIS*, G. PAUTOU†, J.-L. PEIRY‡ and J. GIREL†

*Department of Geography, University of Cambridge, UK
† Institut de Biologie Alpine, Universite Joseph Fourier, Grenoble, France
‡ Instut de Geographie, Universite Joseph Fourier, Grenoble, France

19.1 INTRODUCTION

Floodplains have attracted human occupation since the development of agriculture because the fluxes of water and nutrients between them and their adjacent channels during floods make them highly fertile. There are early records from Roman times in Europe of attempts to manipulate river flows, and therefore these fluxes, in order to suit a range of human purposes including the maintenance of floodplain fertility and agriculture (Petts, 1989, 1990). For example, the practice of "warping" floodplain soils by diverting river flow was common throughout Europe (Girel, 1994). However, extensive channelization has occurred in Europe in the nineteenth and twentieth centuries (Pautou and Bravard, 1982; Decamps *et al.*, 1989; Micha and Borlee, 1989), and in North America, channelization and other forms of river management have occurred since the advent of the first European settlers during the eighteenth century (Turner *et al.*, 1981; Mitsch and Gosselink, 1993). Most of these more recent river management schemes have been concerned primarily with flood control and protection rather than with the health of floodplain vegetation (whether natural or cultivated). Their effect has been to isolate the floodplain from its active channel, both in the case of channelization as a deliberate attempt to move high flows more rapidly through the river system, and in the case of barrages whose purpose is to provide water storage for water supply and hydroelectric power production. This in turn alters hydrograph shapes, in terms of both peak flows and recession curves.

In recent decades there has been a growing interest in understanding the interactions between biotic and abiotic elements of floodplains and their active channels. This has

Floodplain Processes. Edited by Malcolm G. Anderson, Des E. Walling and Paul D. Bates.

been prompted by an increasing appreciation of the roles that floodplain areas play, as buffer zones between agricultural land and adjacent channels (Peterjohn and Correll, 1984; Lowrance *et al.*, 1985; Haycock and Burt, 1993; Haycock *et al.*, 1993), as natural temporary water storage areas during flood periods (Sutcliffe and Parks, 1989; Large and Petts, 1994), and as wildlife habitats and corridors (Anderson and Ohmart, 1985; Harper *et al.*, 1994). Poor river water quality, the destructive impacts of floods on human lives and property, competition for water resources, especially in semi-arid and arid areas, and loss of floodplain habitats from the conservation perspective, have all created a need to improve scientific understanding of these roles of floodplains, to establish ways in which their functions can be restored, and to develop the institutional mechanisms needed to facilitate this.

An understanding of the interactions between river hydrology and geomorphology and the regeneration of floodplain vegetation is critical to the success of any river restoration initiative. Many different approaches to river restoration have been suggested, ranging from complete restoration which involves structural and functional return to a pre-disturbance state, through partial rehabilitation to abandonment of sacrificial cases for which recovery is deemed improbable (Boon, 1992; ECON, 1993). Various methods for implementing these degrees of restoration have been suggested (Clewell and Lea, 1990; NRC, 1992; Petersen *et al.*, 1992; Peterken and Hughes, 1995), and in a number of locations in Europe and North America restoration projects have been initiated. Notable among these for the scale of restoration are the Kissimmee floodplain system in Florida (Boon, 1992) and the Des Plaines River Wetlands Demonstration Project in Illinois, USA (Hey *et al.*, 1989; Sanville and Mitsch, 1994). In Europe, several smaller-scale restoration projects have been or are being implemented, many of them in Germany and Denmark and mostly on small streams. In the UK, the River Restoration Project will eventually have stretches of both a rural river (the River Cole, Oxfordshire) and an urban river (the River Skeme, County Durham) restored to pre-disturbance states. Rehabilitation of sections of large European rivers such as the Rhine is also being considered (Kern, 1992). In all these projects hydrological considerations are the most important design elements, because flooding and water-table recharge patterns, along with the associated nutrient additions to the floodplain, will fundamentally affect the success of the restoration initiative.

Where floodplain restoration is not possible or indeed not necessary, there is nevertheless considerable interest in the ways in which flows are allocated to different uses. In different river basins the potential users of river water vary greatly, but until recently most consideration has been given either to the need to abstract water for industrial, domestic and agricultural use, or for the needs of instream species of fish and aquatic plants. It is only since the interest in floodplain habitats and their restoration has developed that the water needs of riparian species have been seriously considered within the context of river flow allocation. For example, in the Kruger National Park in South Africa, a methodology is being developed which is intended to accommodate the water rights and demands of riparian ecology (Moon *et al.*, 1996). However, the problem is more complex than an identification of the total water needs of riparian vegetation. In order to maintain existing and restored floodplain vegetation communities and to ensure their long-term regeneration, it is also important to determine optimal magnitudes, frequencies, durations and timings of flow favoured by individual species. In the case of most riparian tree species with life cycles of at least 100 years, while minimum flows are necessary for their maintenance once established (Stromberg and Patten, 1990), varied intra-annual, inter-annual and inter-decadal flows are necessary to meet particular life-cycle needs, particularly at the regeneration and establishment stages. These long-term flow sequences are also key determinants of the process of community succession (Pautou and Decamps, 1985; Bradley and Smith, 1986; Baker, 1989; Virginillo *et al.*, 1991; Hughes, 1994).

For instream species, the Instream Flow Incremental Methodology (IFIM) developed by the US Fish and Wildlife Service (Bovee, 1982, 1986) is currently the most sophisticated methodology for allowing the ecological demands of target species to be expressed in the same terms as other water resource demands. It has been applied in a number of situations (Bullock and Guslard, 1992; Gore *et al.*, 1992), but is not without criticism for both its potential to be manipulated to suit user needs (Gan and McMahon, 1990), and its overdetailed and expensive data demands (Swales and Harris, 1994). Simpler and less expensive methods have been developed for instream species, including the Expert Panel Assessment Method (Swales and Harris, 1994). In many countries, the concept of minimum flows has dominated decision-making about water allocation, although there is an increasing appreciation of multiple-use ethics and water-budgeting through planned reservoir storage and release as effective means of meeting multiple user demands (Scudder, 1991; Stalnaker, 1994).

Despite these developments, the means of allocating flows for riparian species as opposed to instream species remains little studied. This is in large part because riparian vegetation patterns have to be defined and explained in terms of the changing and complex interactions between river flows, water-table recharge and available soil moisture, as well as in terms of both competition for above and below-ground resources between different species, and the geomorphological changes that occur over the lifetime of individual plants because of erosion, deposition, channel migration and associated substrate changes. Interpretation of vegetation patterns and the habitats occupied by different species over the life cycle of most riparian trees is further complicated by the many human-induced changes to rivers over the same time period (Hughes, 1994; Pautou and Arens, 1994), and to changes in historical floodplain land use (Girel, 1994). There is a significant need for detailed definition of the optimal habitats of many riparian vegetation species, and for the incorporation of this information into models of floodplain response to changing inputs of water, sediment and nutrients. Several researchers are currently devising models (often reach-based) which will allow prediction of changes in different river reaches in response to changing water and sediment availability. Some concentrate on minimum flows for maintaining channel form, and predict changes in the absence of those minimum flows (e.g. King and Tharme, 1994; Moon *et al.*, 1996). However, in order to include vegetation changes within such models, and to define both minimum and optimal flows for riparian vegetation communities, further detailed research on target species is essential.

The purpose of this chapter is to illustrate how a combination of three approaches (field monitoring, laboratory experiments and numerical modelling) can be used to obtain the baseline information necessary to identify these needs of floodplain vegetation, and to develop ideas for the management of floodplain ecology. The field investigations reported here were carried out in two tributaries of the heavily managed

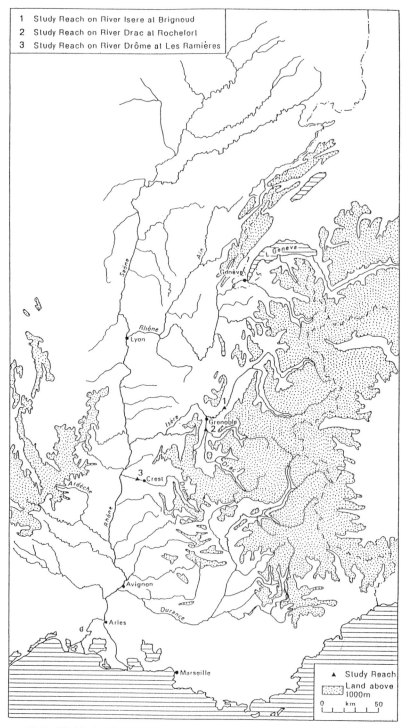

Figure 19.1   Field sites in the Rhone basin on the Isère and Drac rivers near Grenoble

and channelized Rhône River basin (the Isère and the Drac) in the French Alps (sites 1 and 2 in Figure 19.1), and consisted of detailed field measurement of water-table response to different channel flows, and associated vegetation studies of seedling growth and spatial patterns of vegetation in relation to substrate and soil moisture conditions. This research was supplemented by experimental "laboratory" research carried out in greenhouses, which involved subjecting selected riparian tree species from the field area to different water-table conditions in a controlled situation. Finally, a numerical groundwater flow model capable of dealing with saturated–unsaturated flow conditions was used to simulate the available soil moisture in the unsaturated zone in both the experimental studies and the field floodplain sites. Success in simulating the critical soil moisture conditions in the laboratory and field cases then justifies use of such a model to extrapolate to field conditions with different morphologies and sedimentologies, although there remain difficulties associated with appropriate measurement of soil physical properties (common to most physically based hydrological modelling). However, such groundwater–soil moisture models may be applicable in the identification of management decisions related to riparian vegetation, in a similar manner to application of IFIM methods for aquatic species. The integrated tripartite methodology illustrated in this chapter may provide a rigorous basis through which to provide information for river managers on both the water needs of riparian vegetation, and the means of manipulating the flow and recharge processes in order to satisfy those needs.

19.2   FIELD MONITORING

The field installations designed for this research project involved transects of tensiometers consisting of porous pots attached to closed water-filled perspex pipes connected by flexible tubes to piezo-electric pressure transducers; the instrumental details are similar to those described by Dowd and Williams (1989). The transducers were connected by cable laid in underground pipes to a multiplexer and data logger installed in trees above flood level. Figure 19.2a illustrates in section the arrangement at the site on the River Isère; this is similar to that installed on the Drac. The Isère site consists of a patch of incipient floodplain about 15 m wide and 100 m long in the form of a linear, detached bar developing within a channelized reach 15 km north of Grenoble, with a chute channel between it and a narrow (5 m) strip of floodplain at the foot of the embankment (Figure 19.2b). In addition to the three sets of three tensiometers, water-level recorders were installed in the main and chute channels, and other parameters were monitored including temperature and specific electrical conductivity.

The sediments at the Isère site are sandy silts extending to a depth of 1–1.5 m, where channel gravels are encountered. Between the basal gravels and sandier surface deposits, the material is uniformly silty but with a general upwards fining. The surface sands were deposited by over-bar flows during the study period. Resurveys in the Decembers of 1993 and 1994 revealed significant average aggradation of 0.40 m. The topographical data for the site have been combined with water level and discharge data to create a model of the spatial variation in depth and the extent of flooding across the

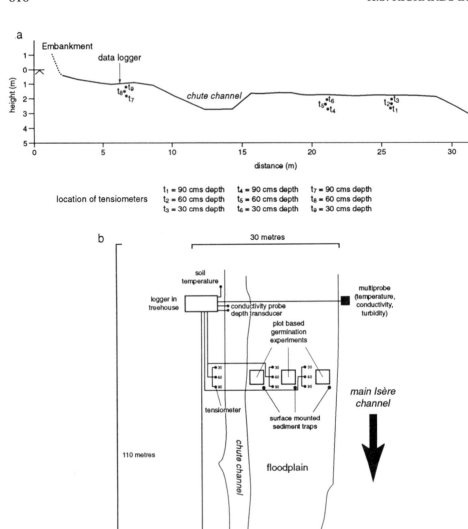

Figure 19.2 (a) Instrumentation at the Isère site, showing the tensiometer transect. (b) A plan view of the Isère site, showing the bar and chute channel

site between January 1994 and August 1995; during June 1994 the bar was completely submerged for a period of 9 hours during a major flood, and partially submerged for over 48 hours. Data collected on suspended sediment transport during this high flow period demonstrate very high concentrations (6.1–11.2 g/l), giving a deposit of 0.20–0.30 m in less than 10 hours at the site. The configuration of the site permits an examination of the relationship between the main and chute channels. Figure 19.3 shows that, when the water elevation in the main channel exceeds ca. 1.50 m (relative to an arbitrary datum at the foot of the main Isère bank), and the whole bar is just submerged, there is a linear relationship with the depth in the chute channel. However, this is not the case at lower stages. The depth in the chute channel varies as the bed is scoured and filled, particularly at its downstream exit. As the downstream extension of the bar blocks the exit from the chute channel, so the water level remains higher than that in the main channel on the falling limb of a hydrograph. However, after scour of the exit region under high flow conditions, this effect is reduced and the linearity of the relationship extends to lower stages. Furthermore, hysteretic behaviour is induced by a lag in the relationship with the main channel; on the rising stage, the chute channel has a lower water level than the main channel until the entrance is flooded, while on the falling stage the level remains higher in the chute unless the exit has been scoured.

Figure 19.4 shows the relationship between the main channel water level and the water pressure and tension recorded during the latter half of June 1994 (including the flood on 26–27 June) by the three tensiometers in the centre of the island. As expected, during the low flow period prior to the flood, the tensiometers at depths of 0.30 and 0.60 m record suctions, which vary in relation to fluctuations in water level (which are here frequent and rapid because of releases from hydro-power reservoirs, and of the order of 0.50 m). The deepest mid-bar tensiometer, however, displays positive water pressures indicating a water table at elevations of about 0.20–0.30 m above the

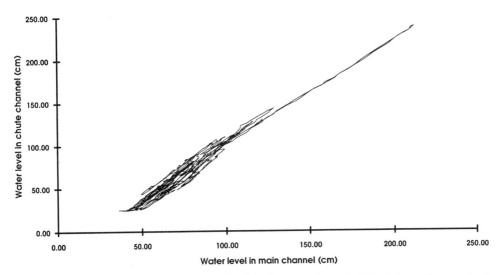

Figure 19.3 Relationship between water level in the main channel of the Isère and water depth in the chute channel

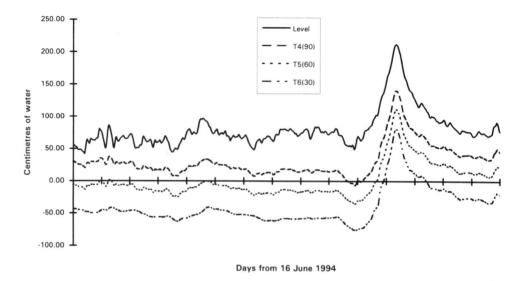

Days from 16 June 1994

Figure 19.4 Water-level variations in the Isère in the second half of June 1994, and water pressure and suction variations monitored by the mid-bar tensiometers at 90 cm, 60 cm and 30 cm beneath the surface

tensiometer (elevations of about 0.80-1.00 m, consistent with the water levels in the channel). Although the conditions are dry in most of the upper soil column, this indicates a close relationship between the groundwater levels and those in the main channel. This is reinforced by the evidence of close correlation in the fluctuations of water level in the channel and in the soil moisture status, the latter being more damped towards the surface of the sediment column. During the flood, the water pressures recorded at each piezometer (tensiometer) are consistent with their different elevations, and soil saturation is maintained during the recession period below 0.30 m although the period of saturation above this is only about 36 hours. Figure 19.5 illustrates an alternative means of depicting the data, by plotting pore pressures/suctions for the three instruments at depths of 0.60 m. T5, in the centre of the island and at an elevation of 1.01 m remains at lower suctions than T2, nearest the main channel and at 1.22 m, and T8 beneath the flood plain against the embankment and at 1.65 m. Both of these appear to maintain suctions of about 70–90 cm, and T8 is only wet for about 16 hours during the flood.

The water-table variations identified in the data from the Isère site suggest rates of water-table lowering during periods of low river water level of between 1.5 cm/day (see Figure 19.4 in the period prior to the flood) and 7.5 cm/day, presumably largely associated with evapotranspiration losses. Rates of drawdown of up to 25 cm/day occur during the early phase of hydrograph recession as the river water level drops. Rates comparable to the former (up to 3.0 cm/day) were employed in experimental studies of the effects of moisture availability on rates of seedling growth.

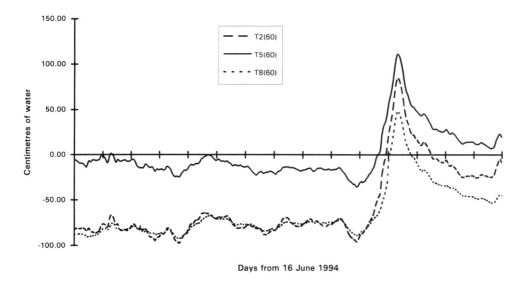

Figure 19.5 Pore water pressures and suctions at three tensiometers in the transect at the Isère site, all at 60 cm beneath the ground surface at their respective positions

19.3 GREENHOUSE EXPERIMENTS

The greenhouse experiments (Hughes *et al.*, in press) were designed to study the response of early successional woody riparian species to different soil moisture conditions, and were initially carried out on *Alnus incana* (L.) Moench, the European grey alder, and subsequently on *Populus nigra* (black poplar). The apparatus used in the experiments is based on a design by Mahoney and Rood (1991), and is referred to as a "rhizopod" (Figure 19.6). Each rhizopod consists of a central water well connected to 16 soil-filled tubes 1.2 m in height. By varying the water levels in the central water wells, it is possible to provide different water table conditions for plants growing in the side tubes of a set of rhizopods. In the experiments conducted on *Alnus incana*, each of five rhizopods had eight growth tubes filled with fine, silty sand sediments from the Isère River and eight filled with coarse, gravelly sand sediments from the Drac River. The stratigraphy of the field sites was recreated in the rhizopod tubes. The drawdown rates applied to the five rhizopods were 3 cm/day, 3 cm/day plus a weekly application of 1 cm rainfall, 1 cm/day, 0.5 cm/day and 0 cm/day. Alder seeds were collected from the field sites and a total of 32 seedlings was used per experimental treatment. The seeds were germinated initially in seed trays, and seedlings of an approximately uniform 2 cm in height were transplanted to the soil-filled tubes in the rhizopods. During the 155 days of the experiment, measurements were made to monitor the growth and health of the alder seedlings, including shoot height, leaf number, leaf length and leaf health. Leaf area, shoot weight, root weight and the weight of nitrogen-fixing nodules were measured at the end of the experiment; root length was measured at the start and end of the experiment. In two of the rhizopods (those with water decline rates

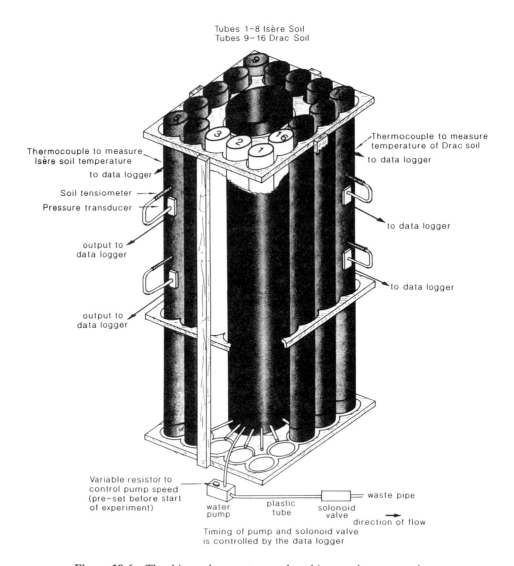

Figure 19.6   The rhizopod apparatus employed in greenhouse experiments

of 1 cm/day and 3 cm/day) soil suction readings were logged automatically using tensiometers inserted in selected growth tubes.

Average shoot growth rates of the *Alnus incana* seedlings are plotted in Figure 19.7. These results indicate that better performance occurs in the finer-grained and generally more moist Isère soils, although this only becomes noticeable after about 50–60 days. In the coarse Drac soils, growth rates are very low throughout the experiment in the rhizopod experiencing the 3 cm/day drawdown regime, and for the first 70 days in soils experiencing no drawdown except through evapotranspiration. Seedlings also show high mortality rates in the soils experiencing rapid water-table drawdown. All rhizopods

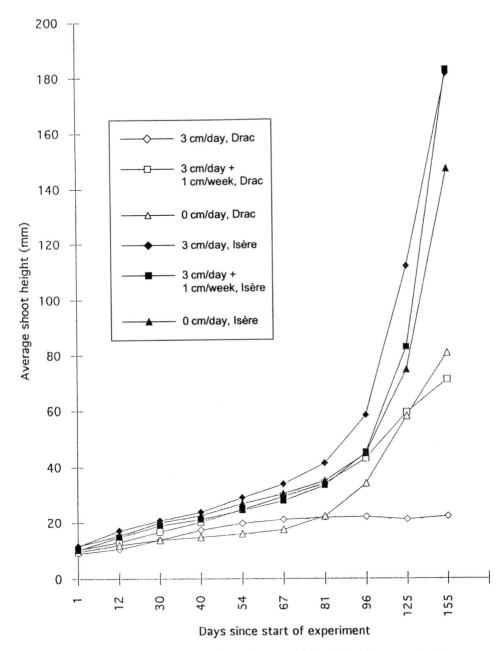

Figure 19.7 Experimental results showing shoot growth by *Alnus incana* under three water-table regimes in two different sediment types (from the Isère and Drac field sites)

have high initial water tables, and it is noticeable that seedling growth rates are low at the start of the experiment, but accelerate once the upper parts of the soil profiles have drained. It therefore appears that both drought and excessive wetness inhibit growth. In the finer Isère soils, average shoot growth rates were generally higher for all water-level treatments than in the Drac soils. *Alnus incana* seedlings in the slowest draining soils (0 cm/day) performed least well while the fastest draining water regimes promoted the best growth in the experiment (both 3 cm/day and 3 cm/day plus 1 cm/week of rain).

In general, *Alnus incana* seedlings grown in the coarser Drac soils produced very long roots relative to their shoot weight compared to those grown in Isère soils, showing that enormous energy resources were spent on tracking water rather than in adding to shoot biomass. Average leaf areas at the end of each experiment were universally higher for plants grown in all Isère soils. Since *Alnus incana* is a nitrogen-fixing species, the relationship between development of nodules and plant performance was also studied. A positive correlation existed between shoot and root weight and weight of nodules. A greater weight of nodules was therefore present on plants grown in Isère soils than Drac soils, but actual weights varied with water-table treatment.

The experimental work has allowed us to identify for *Alnus incana*, and more recently for *Populus nigra*, critical soil suctions in the two different alluvial sediment types. In the coarser Drac soils, once the suction becomes greater than around 60–70 cm of water (equivalent in this soil to a moisture content of <0.10 cm^3/cm^3) in the rooting zone, seedlings become drought stressed and growth tails off. Critical suctions were rapidly reached when water-table drawdown rates were ≥ 21 cm/day. In the field these drawdown rates are frequently experienced on the Drac because of the operation of hydroelectric reservoirs, and relatively localized fine-grained sites are the only ones able to support regeneration of this species. By contrast, in the Isère soils, drawdown rates of 1 cm/day and 3 cm/day gave ideal soil moisture conditions. Soil suctions in these treatments were of the order of 80–110 cm of water in the rooting zone but were associated with moisture contents of between 0.25 and 0.40 cm^3 in the different layers of this fine-grained sediment stratigraphy. These suctions and soil moistures occur in the field for quite long periods from August until March; furthermore, Figures 19.4 and 19.5 illustrate that even in the growing season in June, when river levels are higher, such suctions occur in the upper 0.50 m of the soils in the Isère bar. However, during the critical seedling establishment period (May to July), conditions at this site can be wet during floods which provide saturated soil conditions at the surface. The rhizopod experiments demonstrate an intolerance by *Alnus incana* to such conditions, with lower growth measurements in all seedlings grown in the moister regimes. In fact, regeneration of *Alnus incana* has been recorded only on the highest parts of the Isère field site and at low densities (12–150 individuals per 100 m^2), despite an abundant supply of seed from adjacent trees. It seems likely that unfavourable soil moisture conditions during the first weeks following germination are a contributory factor.

19.4 THE NUMERICAL SIMULATION OF SOIL MOISTURE VARIATION IN FLOODPLAIN SEDIMENTS

An infiltration model was developed to simulate water-table and soil moisture changes in both the rhizopods and the floodplain cross-sections. The purpose of developing such a model was twofold. It was anticipated, firstly, that it would provide soil suction data that are more relevant to the plant physiology than are water-table levels alone, and that it would provide more detailed information on the distribution of suctions above the water table than could be gained using tensiometry. Secondly, the model would provide a physical basis for extrapolation, through which information about water tables and soil suctions could be gained at other locations of interest for which the topography and soil properties could be measured, but at which the installation of soil moisture monitoring equipment was impossible on cost and access grounds.

The model needed to be capable of dealing with an external water-level control, such as an adjacent and variable river water level or a rhizopod reservoir water level, as well as with the ponded head of water that occurs during flood period recharge. It also needed to accommodate the typically stratified soils found in floodplain environments. The general flow equation for the unsaturated–saturated zone (the Richards' equation; Richards, 1931) is physically based, and although computationally demanding, it has been widely shown to be capable of providing realistic predictions of moisture distribution in complex soil profiles under the variety of boundary conditions likely to be encountered in these applications. This equation, in the unsaturated–saturated form and in two dimensions is:

$$C(\psi)\frac{\partial \psi}{\partial t} = \frac{\partial}{\partial x}\left[K(\psi)\frac{\partial \psi}{\partial x}\right] + \frac{\partial}{\partial z}\left[K(\psi)\left(\frac{\partial \psi}{\partial z}+1\right)\right] \qquad (19.1)$$

where C is soil capillary capacity, K is hydraulic conductivity, ψ is hydraulic suction, t is time, x is the horizontal direction, and z the vertical direction. The equation can be defined for a single (vertical) dimension by ignoring the first term on the right hand side. Details of the development of the numerical form of this equation and the solution methods employed may be found in El-hames and Richards (1995). Briefly, the numerical scheme developed to solve Equation 19.1 is a Crank–Nicholson, implicit finite difference scheme in both one (vertical) and two (cross-sectional) dimensions. This involves a system of simultaneously linear algebraic equations which can be solved by the Gauss elimination technique. In the two-dimensional solution an Iterative Alternative Direction Implicit (IADI) method was used to minimize stability problems, and to reduce the complexity of matrix calculation. A full solution of Equation 19,1 can simulate a variety of problems including steady-state flow, vertical leakage, confined and unconfined aquifers, specified head problems, mixed saturated–unsaturated flow and stratified soil columns. Boundary conditions may include a specified flux, a specified head, or a mixed condition; in the floodplain case, the head needs to vary through the simulation.

In this application, an initial moisture content distribution in the soil is assumed and pressure head values (ψ) are estimated from moisture–tension curves ($\theta-\psi$) measured for each soil layer. The Millington–Quirk method (M–Q) is then used to estimate unsaturated hydraulic conductivity $K(\psi)$ values (Millington and Quirk, 1959). This avoids the need to measure the ($K-\theta$) relationship, a largely intractable operation. The input data for the model can be classed into four groups. The first is of data related to the numerical scheme requirements; the time and space grid intervals, and the tolerance value used in simulations (λ). The model can handle variably sized grid cells, to represent different soil horizons. The second group includes the soil physical parameters, including soil layer thicknesses, moisture–tension curves, saturated hydraulic conductivities, and saturated soil moisture contents, and the depth to the lower boundary. In addition, an initial moisture content for each horizon in the soil profile is also required, as is the moisture content of the basal boundary condition. The third is a data group defining the meteorological and hydraulic conditions (rainfall, evaporation and ponded head), and finally, surveyed cross-sectional elevation and distance data are required to represent the floodplain and channel topography.

19.4.1 Initial testing of the models

An initial requirement after development of the one and two-dimensional versions of the numerical model was to test it for efficiency and robustness, and this was undertaken using extreme cases from the literature, ranging from homogeneous soil with a high hydraulic conductivity, through ponded head examples, to simulations for stratified soils with from two to six layers (similar to the Isère and Drac cases). Boundary conditions ranged from very wet to very dry sandy soils, and included simulations comparable to the rhizopod or floodplain problems where an external water-level control exists.

The first test data sets were derived from Hills *et al.* (1989) and Warrick (1991), who used models similar to that developed here to examine 26 different cases (involving varying initial conditions, different numerical parameters and solution techniques, and both pressure head and water content formulations). Three of these (Warrick's cases 1, 9, and 10) were selected to compare with predictions by the model used here; this test is therefore of the validity of the numerical model development (numerical algorithms, solution procedures, and computer implementation). The simulations were for a 100 cm deep soil column of five 20 cm layers, consisting of two alternate soil types; the Berino loamy fine sand and the Glendale clay loam, beginning with the Berino loamy sand layer at the top (the soil depth and soil types are similar to those at the Isère field site; the properties are listed in Table 19.1). In the original simulations, numerical conditions were as listed in Table 19.2, and each simulation applied 0.0014 cm/min of rainfall at the upper boundary, with the lower boundary condition set at the basal soil layer's initial moisture content (Table 19.1). To apply the model developed in this research simulations employed a depth increment, Δz, of 4 cm, and the time increment (Δt) was set at 180 s in one run and 810 s (the maximum time interval the model could handle without instability) in

Table 19.1 Soil properties and initial conditions for numerical tests of the one-dimensional infiltration model

Soil	Saturated hydraulic conductivity K_{sat} (cm min^{-1})	Saturated moisture content θ_{sat} (cm^3 cm^{-3})	Initial pressure (cm)	Initial moisture content (cm^3 cm^{-3})
Berino loamy fine sand	0.376	0.366	−1000	0.034
Glendale clay loam	0.009	0.469	−1000	0.249

Table 19.2 Numerical conditions employed in the test cases used to assess the numerical performance of the developed model

Case	Method	Δz(cm)	Δt(cm)
1	Pressure head	4.0	14.5
9	Pressure head	0.5	1.5
10	Water content	0.5	29.0

another. All of the simulations were run for a total of 5 days. The results of the original simulations published by Warrick (1991) and the outputs of the model developed here, are shown in Figure 19.8. The graphs show close agreement between the published results and the output of the new model, providing confirmation that the one-dimensional numerical solution of the Richards' equation and its computer implementation are reliable.

The robustness of the two-dimensional version of the model was also tested by applying it to the 2D experiment conducted by Vauclin *et al.* (1979). In addition, the model was checked in terms of its ability to conserve mass. Vauclin *et al.* (1979) undertook laboratory experiments on the variation of soil moisture in a 2D

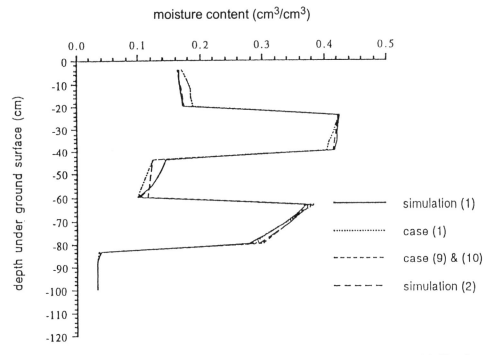

Figure 19.8 Comparison between the model simulations after 5 days of simulated infiltration, and sample one-dimensional problems simulated by the similar model developed by Warrick (1991). Simulation 1 uses a time increment of 180 s, and simulation 2 a time increment of 810 s

homogeneous soil block, 300 cm long, 200 cm high and 5 cm thick, and modelled moisture redistribution in the block. The soil, which had a measured saturated hydraulic conductivity of 0.583 cm min, was packed uniformly between perspex walls, and a water table was imposed at a depth of 135 cm. A constant flux of 0.247 cm min was applied over a width of 100 cm in the centre of the top of the block. The initial condition for the soil moisture content was as shown in Figure 19.9. The experimental and numerical results of Vauclin *et al.* (1979) were compared to predictions from the 2D model. These simulations were continued for 1.5 hours, and the comparison was made at 0.5-hour intervals, at 16, 64 and 80 cm from the centre of flux, as shown in Figure 19.9. The model grid cell size was 5 cm in both the horizontal and vertical directions, and the time increment for the simulation was 1 minute. Comparison between this model's predictions and the published data are again in good agreement; indeed, especially at 1.5 hours the model developed in this research provides a better approximation to the experimental data than the model developed by Vauclin *et al.* (1979) themselves. This confirms the robustness of the model, and justifies its application to 2D problems.

19.4.2 Application to the rhizopod case

Having confirmed the reliability of the model, it was first applied in its one-dimensional form to simulate soil moisture distribution in the rhizopods employed in the experimental greenhouse studies. As described above, these were designed to investigate the effect of moisture content on the growth rate of different plant species,

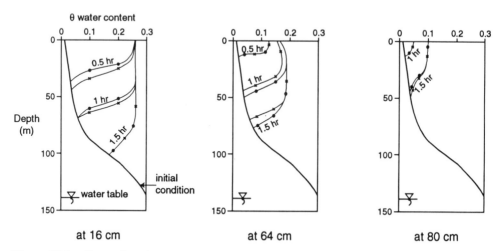

Figure 19.9 Comparison between the model simulations and the two-dimensional experimental results obtained by Vauclin *et al.* (1979). Crosses define the wetting front movement simulated by our model, and dots are the simulations by Vauclin *et al.* Distances (*) are measured from the centre of the experiment, in a horizontal direction.

but their scale allows only limited tensiometry and a numerical model was considered to be a preferable means of describing the suction variation through the rooting zone. The rhizopods contained soil collected from the study sites, arranged in similar layers to those in the field. Some tensiometers were installed at different depths in the soil to measure the suctions developing as the water table was lowered. Water in reservoirs connected to the soil-filled tubes was pumped out at different rates to control lowering of the water table in the soil columns. To simulate the effects of this, it was necessary to alter the model to include this externally-controlled lowering of the water table. Simulations were undertaken for both the Isère and Drac soil columns, using data obtained by laboratory experimentation on saturated hydraulic conductivities and moisture–tension curves for the different soil horizons. Simulations covered a 20 week period, with a time increment of 1 minute and grid increment of 2 cm. Water-table lowering rates were 3 cm/day and 1 cm/day for both the Isère and the Drac cases, and an evapotranspiration rate was assumed of 0.5 cm/day.

Results were printed at 30 and 60 cm depths for the Isère soil and at 15 and 30 cm for the Drac soil, to allow comparison with suction values recorded by tensiometers installed at these depths in the greenhouse rhizopods. Results show good agreement between recorded and measured values in the Isère experiment (Figure 19.10). At 30 cm depth results diverge somewhat, but the general trend is satisfactory. Although this divergence amounts to 15 cm of suction, this is insignificant when converted to

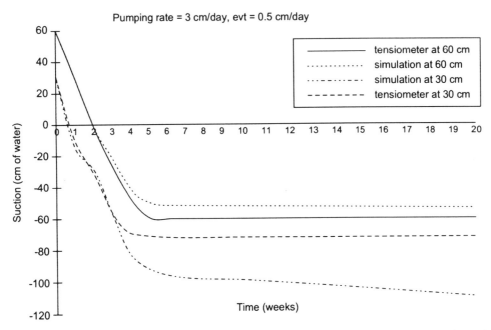

Figure 19.10 Comparison between the tensiometer data collected during the rhizopod experiments for one soil column of Isère sediment, and simulated soil moisture variations (as suctions) for the same soil

moisture content. Results for the Drac experiment at 30 cm were also acceptable, although an instability problem initially caused unrealistic behaviour in the simulations at a depth of 15 cm. This instability was because the depth (15 cm) at which the tensiometer was placed is just at the boundary between two soil layers having different physical characteristics, and the abrupt change in physical parameters at this point caused numerical instability until the change was smoothed.

19.4.3 Application to the Isère cross-section

One requirement of the two-dimensional version of the model was that it had to handle complex channel cross sections with breaks in the soil columns in both the horizontal and vertical directions. These breaks reflect the existence of channels, which may be occupied by either water or air (non-soil media). The model jumps across grid cells defined as containing water or air, but does not ignore their effect on any neighbouring soil cells. This method deals successfully with complex geometries, and results in a rapid solution. To test the model, the geometry of the Isère cross-section was approximated using a 50 cm (horizontal dimension) by 25 cm (vertical dimension) grid (Figure 19.11), resulting in a simulation cross-section 28.5 m wide and 2.75 m deep. Six soil layers, based on those for which field measurements had been made and from which samples had been obtained in order to measure hydraulic conductivities and suction–moisture curves, were defined for the simulation. Table 19.3 outlines their properties.

Starting from a fully saturated initial condition and imposing the late June peak flood level shown in Figure 19.4, the model was used to simulate the drawdown of the water table in the bar sediments in response to the declining water level in the main and chute channels during the flood recession. This resulted in relatively rapid simulated lowering of the water table by about 20 cm in the first 24 hours (largely because of evaporative drying at the surface), but then a much slower change which failed to generate soil water suctions at depths of 30 and 60 cm beneath the surface. A second simulation was undertaken with the initial condition being a water-table level in the centre of the bar at about 90 cm below the surface (Figure 19.11a), and moisture contents and suctions in the sediments above this at values indicated by the field tensiometers immediately prior to the flood event (Figure 19.4). The effect of the late June hydrograph on soil moisture was then simulated. As the water levels rise in the channels, the model simulated the development of saturated zones on the banks (Figure 19.11b). Immediately after the passage of the flood peak, the surface soil is saturated, forming a perched water table, but saturation does not persist for long, and by 48 hours into the simulation (Figure 19.11c), evaporation and drainage from the surface have produced an unsaturated surface layer with a moisture content of above 80%, with a zone below it with a lower moisture content of between 60% and 80%. The simulated water table in this case is lower in the centre of the bar than towards the edges, where recharge has occurred from the banks. Four days into the simulation (Figure 19.11d), the water table has declined, and moisture contents have decreased in the unsaturated zone, although it is still more moist at the surface than beneath. This behaviour is more complex than that which occurs in a sandy gravel with a saturated hydraulic conductivity an order of magnitude higher, such as that at the sites studied on the Drôme and Drac rivers. Accordingly, it is

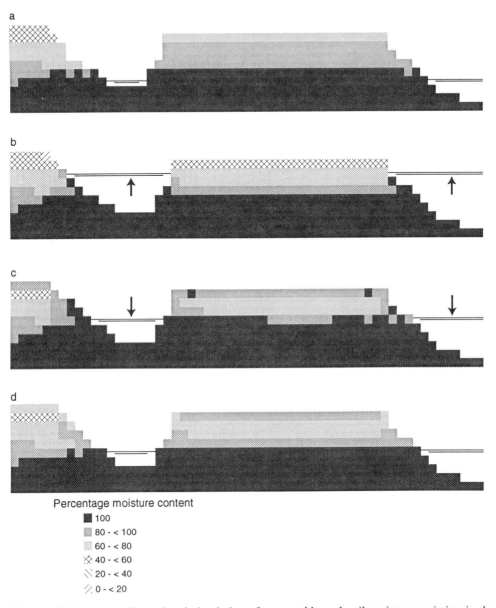

Percentage moisture content
■ 100
▨ 80 - < 100
░ 60 - < 80
✕ 40 - < 60
╲ 20 - < 40
╱ 0 - < 20

Figure 19.11 A two-dimensional simulation of water-table and soil moisture variation in the bar at the Isère site up to and after the flood peak experienced on 27 June 1994. The values mapped are the moisture contents as a percentage of saturated moisture content. (a) is at time $t = 0$ on 26 June, (b) is at $t = 24$ hours on the rising limb of the hydrograph, (c) is at $t = 48$ hours on the recession limb of the hydrograph, (d) is at $t = 96$ hours

Table 19.3 The properties of soil/sediment horizons defined for the simulation of channel–groundwater coupling using the two-dimensional model

Layer	Thickness (cm)	Hydraulic conductivity K_{sat} (cm/min)
1 (surface)	25	0.100
2	50	0.136
3	25	0.150
4	25	0.114
5	25	0.107
6 (Gravel)	75	0.694

evident that decisions about flow management for the riparian vegetation must be based on a proper understanding of the relationships among soil moisture, groundwater recharge, and river water levels in floodplains with different topographic and sedimentological characteristics. While the model provides a basis for achieving this, and of identifying the wide range of floodplain soil moisture behaviour, it is not without its problems. Indeed, although both simulations described above are representative of cases that are likely to be encountered in the field, they do not replicate the Isère data (Figure 19.4) particularly well. Since the tests of the model were satisfactory, and it produces acceptable results in relation to the rhizopod data, it seems likely that the laboratory measurements of the soil moisture characteristics curves do not replicate well the field condition of the sediment, probably because the rapid deposition rate results in low compaction and bulk densities by comparison with those employed in the laboratory experiments to measure the soil properties. Thus successful application of this modelling framework to an environmental management objective necessitates careful planning of the range of model cases, and rigorous sampling and measurement in order to calibrate the model for those cases.

19.5   DISCUSSION AND CONCLUSIONS

One interesting outcome of this research has been, for certain floodplain sediments, identification of the response of groundwater levels and soil moisture status to channel water levels. This makes it possible to match observed or simulated suctions to the limiting soil moisture conditions for seedling growth, and alone could represent a useful input to flow management by defining the timing and maintenance of stage changes for the purposes of encouraging the regeneration of riparian species. However, it is clear that in many cases a cost–benefit analysis of the value of water for various conflicting purposes is likely to prevent the option of reserving water for purely ecological purposes; it would be necessary to maintain particular patterns of water-level change along considerable lengths of river reach and for considerable periods in order to encourage germination and early growth of seedlings. In addition, creation of new sites suitable for regeneration of early successional species would require flood events capable of initiating geomorphological change through bank erosion, bar construction, channel migration, and spatially varied deposition patterns. All of these processes are required in order to create new niches for colonization in order to maintain ecological diversity, but are unlikely to be practically feasible components of river management. The results of this study, nevertheless, suggest practical options for river management that could exploit, firstly, an understanding of groundwater response to river level variation and the ecological impact of that response, and secondly, a recognition of the effects of local geomorphological conditions on the maintenance of conditions conducive to regeneration.

In the first case, the approach adopted in this research could permit simulation of river flow conditions that optimize recharge that is ecologically beneficial. Given knowledge of the critical moisture contents both for seed germination and early growth (limiting wetness), and for the maintenance of moisture contents in the rooting zone that optimize conditions for seedling growth (limiting suctions), different hydrograph forms can be applied in a numerical model in order to identify those which create favourable hydrological conditions. These conditions can then be imposed by appropriate flow manipulation by river managers. The evidence of monitored suctions and seedling surveys in the field, and modelling of groundwater and soil moisture in two dimensions, suggests that in some highly permeable sediments, maintaining flow levels at just below that which would cause inundation could recharge soil moisture without causing waterlogging, and indicates the frequency and duration required for such a flow. Such water levels are commonly created by reservoir releases, and although they may need to be maintained for longer in order to be effective in recharging floodplains, this implies that only relatively small changes to reservoir operating rules might be necessary to produce ecologically beneficial flows. In the case of less permeable sediments, however, it will be difficult to recharge groundwater by this means, and inundation may prove to be necessary in order to produce moist topsoil conditions. In order to apply the modelling strategy outlined above in order to evaluate these alternatives, it will be necessary to gather data on the range of topographic and sedimentological conditions occurring along a reach, and to measure the necessary soil physical properties for a representative set of those conditions in ways that are consistent with the sediment characteristics in the field (itself not an easy task, but one which is commonly encountered in hydrological modelling applications that have predictive purposes).

However, the costs of reach-scale flow management, in terms of the value of the electricity generation or water supply foregone by the sustained release, may militate against its introduction, and the second option based on local exploitation of the geomorphology then becomes attractive. For example, it would appear that certain topographic characteristics of the study sites on the Isère and Drac, together with suitable sedimentological conditions, might form the basis for local projects in which riparian vegetation is conserved through the temporary provision of river water levels that can recharge storage in ways that maintain soil moisture levels. In both the Isère and Drac sites, the areas of riparian woodland studied are on bar forms partially separated from the channel margins by chute channels. In the Isère site in particular, it is possible to identify a complex pattern of interrelationship between water levels in the main and chute channels, particularly because the chute channel exit is intermittently blocked. Local management of such chute channels could encourage their use as reservoirs by building up the levels of their exits with semi-permeable barriers (gravel and boulder dams or artificial log jams), and by encouraging inflow at the upstream end. Natural hydrographs or reservoir releases then provide a means of recharging the storage of water in the chute channel, which then drains slowly through the adjacent floodplain sediments towards the main channel once the water level there has dropped again after the flood peak. It would even be possible to extract water from the main channel upstream during low flow periods in order to recharge the chute-channel reservoir. A pipeline or channel connection with a sluice-gate would permit relatively easy operation of such a scheme, which has much in common with the *aqm*-based irrigation schemes of Arabia (Richards *et al.*, 1987). In addition to the local water level management such a scheme would permit, it would be possible to use the fine deposits of silt that accumulate in the chute channel to "warp" the soils of particularly sandy areas of the floodplain or bar surface in order to encourage moisture retention on otherwise dry surfaces. Furthermore, since waterlogging at critical times may also affect regeneration, the water output from the chute could also be controlled in order to encourage more rapid drying after floods.

Thus, in specific circumstances, either flow management with a reach-scale impact, or exploitation of a suitable geomorphological niche at a more local scale, may represent the preferred means of maintaining the conditions for regeneration of riparian woodland. Field monitoring, laboratory experimentation and numerical modelling of groundwater hydrology have been combined in this chapter to assist in the identification of the river flow management options that are ecologically beneficial. The inverse problem arises when reduction of riparian vegetation is desirable, for example immediately upstream from major urban areas in order to improve the conveyance of channels. Here, informed manipulation of water levels based on the methods outlined in this chapter could equally be applied to the inhibition of germination and establishment of vegetation. Sadly, this is as likely to be the application of the methodology outlined in this chapter as the original objective, that of managing flows to maintain and improve the health of a rapidly declining resource of riparian woodland. However, the methodology applied in this study may have wider applicability; and as several of the chapters in this collection have demonstrated, environmental management is increasingly likely to be informed by evidence provided by the combination of experimental investigations in the field and laboratory with numerical modelling that allows confident extrapolation to the particular field locations for which management is required.

ACKNOWLEDGEMENTS

This research has been funded by the European Commission Environment Programme under Research Grant EV5V-CT93-0261 (DG XII FZPA); the authors are grateful to Dr H. Barth for his interest in the project.

REFERENCES

Anderson, B.W. and Ohmart, R.D. (1985) Riparian vegetation as a mitigating process in stream and river restoration. In: Gore, J.A. (ed.) *The Restoration of Rivers and Streams: Theories and Experience*, Butterworth, Boston, 41–81.

Baker, W.L. (1989) Macro- and micro-scale influences on riparian vegetation in western Colorado. *Annals of the Association of American Geographers*, **79**, 65–78.

Boon, P.J. (1992) Essential elements in the case for river conservation. In: Boon, P.J., Calow, P. and Petts, G.E. (eds), *River Conservation and Management*, John Wiley & Sons, Chichester, 11–34.

Bovee, K.D. (1982) *A Guide to Stream Habitat Analysis Using the Instream Flow Incremental Methodology*. US Fish and Wildlife Service, Instream Flow Information Paper No. 12, FWS/OBS-82/26, 248 pp.

Bovee, K.D. (1986) *Development and Evaluation of Habitat Suitability Criteria for Use in the Instream Flow Incremental Methodology*. US Fish and Wildlife Service, Instream Flow Information Paper No. 21, US Fish and Wildlife Service Biologic Report **86**(7), 235 pp.

Bradley, C.E. and Smith, D.G. (1986) Plains cottonwood recruitment and survival on a prairie meandering river floodplain, Milk River, southern Alberta and northern Montana. *Canadian Journal of Botany*, **64**, 1433–1442.

Bullock, A. and Gustard, A. (1992) Application of the instream flow incremental methodology to assess ecological flow requirements in a British lowland river. In: Carling, P.A. and Petts, G.E. (eds), *Lowland Floodplain Rivers*. John Wiley & Sons, Chichester, 251–277.

Clewell, A.F. and Lea, R. (1990) Creation and restoration of forested wetland vegetation in the southeastern United States. In: Kuslerm, J.A. and Kentula, M.E. (eds), *Wetlands Creation and Restoration: The Status of the Science*. Island Press, Washington, DC, 195–230.

Decamps, H., Fortune, M. and Gazelle, F. (1989) Historical changes of the Garonne River, southern France. In: Pelts, G (ed.), *Historical Change of Large Alluvial Rivers*, John Wiley & Sons, Chichester, 249–268.

Dowd, J.F. and Williams, A.G. (1989) Calibration and use of pressure transducers in soil hydrology. *Hydrological Processes*, **3**, 43–49.

ECON (1993) *River Restoration Project: Phase 1 Feasibility Study*. The River Restoration Project, Huntingdon.

Al-hames, A.S. and Richards, K.S. (1995) Testing the numerical difficulty applying Richards' equation to sandy and clayey soils. *Journal of Hydrology*, **167**, 381–391.

Gan, K. and McMahon, J. (1990) Variability of results from the use of PHABSIM estimating habitat area. *Regulated Rivers: Research and Management*, **5**, 233–239.

Girel, J. (1994) Old distribution procedures of both water and matter fluxes in the floodplains of western Europe. Impact on the present vegetation. *Environmental Management*, **28**, 108–118.

Gore, J.A., Layzer, J.B. and Russell, I.A. (1992) Non-traditional applications of instream flow techniques for conserving habitat of biota in the Sabie River of southern Africa. In: Boon, P.J., Calow, P. and Petts, G.E. (eds), *River Conservation and Management*, John Wiley & Sons, Chichester, 151–160.

Harper, D., Smith, C., Barham, P. and Howell, R. (1994) The ecological basis for the management of the natural river environment. In: Harper, D. and Ferguson, A.J.D. (eds), *The Ecological Basis for River Management*, John Wiley & Sons, Chichester, 219–238.

Haycock, N.E. and Burt, T.P. (1993) Role of floodplain sediments in reducing the nitrate concentration of surface runoff: a case study in the Cotswolds, U.K. *Hydrological Processes*, **7**, 286–295.

Haycock, N.E., Pinay, G. and Walker, C. (1993) Nitrogen retention in river corridors: European perspective. *Ambio*, **22**, 340–346.

Hey, D.L., Cardamone, M.A., Sather, J.H. and Milsch, W.J. (1989) Restoration of riverine wetlands: the Des Plaines river wetlands demonstration project. In: Mitsch, W.J. and Jorgensen, S.E. (eds), *Ecological Engineering: An Introduction to Ecotechnology*, John Wiley & Sons, New York, 159–183.

Hills, R.G., Porro, I., Hudson, D.B. and Wierenga, P.J. (1989) Modelling one-dimensional infiltration into very dry soils, 1. Model development and evaluation. *Water Resources Research*, **25**, 1259–1269.

Hughes, F.M.R. (1994) Environmental change, disturbance and regeneration in semi-arid floodplain forests. In: Millington, A.C. and Pye, K. (eds), *Environmental Change in Drylands: Biogeographical and Geomorphological Perspectives*, John Wiley & Sons, Chichester, 322–345.

Hughes, F.M.R., Harris, T., Richards, K.S., Pautou, G., El-hames, A., Barsoum, N., Girel, J., Piery, J.-L. and Foussadier, R. (in press) Woody riparian species response to different soil moisture conditions: laboratory experiments on *Alnus incana* (L) Moench, *Journal of Global Ecology and Biogeography Letters.*

Kern, K. (1992) Restoration of lowland rivers: the German experience. In: Carling, P.A. and Petts, G.E. (eds), *Lowland Floodplain Rivers: A Geomorphological Perspective*, John Wiley & Sons, Chichester, 279–297.

King, J.M. and Tharme, R.E. (1994) *Assessment of the Instream Flow Incremental Methodology and Initial Development of Alternative Instream Flow Methodologies for South Africa.* Report to the Water Research Commission by the Freshwater Unit, University of Cape Town. WRC Rept. No. 295/1/94.

Large, A.R.G. and Petts, G.E. (1994) Rehabilitation of river margins. In: Calow, P. and Petts, G.E. (eds), *The Rivers Handbook, Hydrological and Ecological Principles*, Blackwell Scientific, Oxford, 401–418.

Lowrance, R.R., Leonard, R.A. and Asmussen, L.E. (1985) Nutrient budgets for agricultural watersheds in the southeastern coastal plain. *Ecology*, **66**, 287–296.

Mahoney, J.M. and Rood, S.B. (1991) A device for studying the influence of declining water table on poplar growth and survival. *Tree Physiology*, **8**, 305–314.

Micha, J.-C. and Borlee, M.C. (1989) Recent historical change on the Belgian Meuse. In: Petts, G. (ed.), *Historical Change of Large Alluvial Rivers*, John Wiley & Sons, Chichester, 269–296.

Millington, R.J. and Quirk, J.P. (1959) Permeability of porous media. *Nature*, **189**, 387–388.

Mitsch, W.J. and Gosselink, J.G. (1993) *Wetlands.* Van Norstrand Reinhold, New York, 722 pp.

Moon, B.P., van Niekerk, A.W., Heritage, G.L., Rogers, K.H. and James, C.S. (1996) A geomorphological approach to the management of rivers in the Kruger National Park: the case of the Sabie River. *Transactions, Institute of British Geographers, New Series* (in press).

NRC (1992) *Restoration of Aquatic Ecosystems: Science, Technology and Public Policy.* National Academy Press, Washington, DC.

Pautou, G. and Arens, M.-F. (1994) Theoretical habitat templates, species traits, and species richness: floodplain vegetation in the Upper Rhone River. *Freshwater Biology*, **31**, 507–522.

Pautou, G. and Bravard, J.P. (1982) L'incidence des activities humaines sur la dynamique de l'eau et l'evolution de la vegetation dans la vallee du Haut-Rhone francais. *Rev. Geogr. Lyon*, **87**, 63–79.

Pautou, G. and Decamps, H. (1985) Ecological interactions between the alluvial forests and hydrology of the Upper Rhone. *Arch. Hydrobiol.*, **104**(1), 13–37.

Peterjohn, W.T. and Correll, D.L. (1984) Nutrient dynamics in an agricultural watershed; observations on the role of a riparian forest. *Ecology*, **65**, 1466–1475.

Peterken, G.F. and Hughes, F.M.R. (1995) Restoration of floodplain forests in Britain. *Forestry*, **68**(3), 187–202.

Petersen, R.C., Petersen, L.B.-M. and Lacoursiere, J. (1992) A building block model for stream restoration. In: Boon, P.J., Calow, P. and Petts, G.E. (eds), *River Conservation and Management*, John Wiley & Sons, Chichester, 293–311.

Petts, G.E. (1989) Historical analysis of fluvial hydrosystems. In: Petts, G. (ed.), *Historical Change of Large Alluvial Rivers*, John Wiley & Sons, Chichester, 1–18.

Petts, G.E. (1990) Forested river corridors: a lost resource. In: Cosgrove, D. and Petts, G.E. (eds), *Water, Engineering and Landscape*, Belhaven Press, London, 23–34.

Richards, K.S., Brunsden, D., Jones, D.K.C. and McCaig, M. (1987) Applied fluvial geomorphology: river engineering project appraisal in a geomorphological context. In: Richards, K.S. (ed.), *River Channels: Environment and Process*, Blackwell, Oxford, 348–382.

Richards, L.A. (1931) Capillary conduction of liquids in porous medium. *Physics*, **1**, 318–333.

Sanville, W. and Mitsch, W.J. (eds) (1994) Creating freshwater marshes in a riparian landscape: research at the Des Plaines River Wetlands Demonstration Project. *Ecological Engineering Special Issue*, **3**(4).

Scudder, T. (1991) The need and justification for maintaining transboundary flood regimes: The Africa case. *Natural Resources Journal*, **31**, 75–107.

Stalnaker, C.B. (1994) Evolution of instream flow habitat modelling. In: Calow, P. and Petts, G.E. (eds), *The Rivers Handbook, Hydrological and Ecological Principles*, Blackwell Scientific, Oxford, 276–286.

Stromberg, J.C. and Patten, D.T. (1990) Riparian vegetation instream flow requirements: a case study from a diverted stream in the eastern Sierra Nevada, California USA. *Environmental Management*, **14**(2), 185–194.

Sutcliffe, J.V. and Parks, Y.P. (1989) Comparative water balances of selected African wetlands. *Hydrological Sciences Journal*, **34**(1), 49–62.

Swales, S. and Harris, J.H. (1994) The Expert Panel Assessment Method (EPAM): a new tool for determining environmental flows in regulated rivers. In: Harper, D. and Ferguson, A.J.D. (eds), *The Ecological Basis for River Management*, John Wiley & Sons, Chichester, 125–134.

Turner, R.E,, Forsythe, S.W. and Craig, N.J. (1981) Bottomland hardwood forest land resources of the southeastern United States. In: Clark, J.R. and Benforado, J. (eds), *Wetlands of Bottomland Hardwood Forests*, Elsevier, Amsterdam, 13–28.

Vauclin, M., Khanji, D. and Vachaud, G. (1979) Experimental and numerical study of a transient, two-dimensional unsaturated–saturated water table recharge problem. *Water Resources Research*, **15**, 1089–1101.

Virginillo, M., Mahoney, J.M. and Rood, S.B. (1991) Establishment and survival of poplar seedlings along the Oldman River, Southern Alberta. In: Rood, S.B. and Mahoney, J.M. (eds), *The Biology and Management of Southern Alberta's Cottonwoods*, University of Lethbridge, Alberta, 55–63.

Warrick, A.W. (1991) Traveling front approximations for infiltration into stratified soils. *Journal of Hydrology*, **128**, 213–222.

Author Index

Subject Index

σ transformation, 207
3D meshes, 206

Avective-diffusion equation, 369
Aggradation, 43
Aggregates, 417
Alder woods, 122
Algal dynamics, 377
Algebraic Stress Model (ASM), 317
Algorithms
  in TELEMAC-2D, 200
Alluvial architecture, 80, 85
  deposits, 113
  fans, 51
  fan deposition, 51
Alnus incana, 619, 620, 621, 622
Ammonia nitrogen distribution, 388
  levels, 389
Ammonium concentrations, 478
Anastomoses, 49
  and avulsive floodplains, 49
Anastomosing channels, 17
Annual flood, 540
  pulse, 546
Anthropogenic activity, 120
Application of numerical models, 71
Aquatic habitats, 537
Artificial compound channels, 282
Astroturf sedimentation traps, 434
Asymmetric compound channel, 317, 320,
  321–22
Atlantic, 118
Atmospheric pollen, 111
Attenuation–discharge curves, 146
Autoregressive model, 341
Avulsion, 49,81

Backswamp environments, 104
Bank erosion, 30, 46
  erodibility, 39, 40

erosion rate law, 31
erosion rates, 47, 48
Bankfull, 171
Bar development, 23
  mode, 23, 24
Bars, 48
Bayes' theorem, 348
Bayesian framework, 335, 342
Bayesian uncertainty estimation, 340
Bed
  friction, 159, 236
  load transport, 374
  shear stress, 312, 316, 317
  shear-stress distribution, 321
  shear velocity, 373
  topography, 31
Biochemical oxygen demand, 377
Biodiversity, 589
Biological indicators
  of water quality, 376
Biological responses, 540
Biota, 541, 544, 557
BOD, 393
Boreal, 118
Bottom friction, 189
Boundary condition relaxation, 245, 247
Boundary shear stress, 151, 152, 153
Boussinesq, 364
  assumption, 299
  relation, 302
  eddy-viscosity concept, 195
Braids, 22
  channels, 29
  kinematics, 29
Braided channels, 24, 52
Braided stream development, 22
Braiding, 25, 27
Bronze Age, 123
Buffer zones, 484, 612
Bulk flow, 223, 241